Hydrology in Practice is an excellent and very successful introductory text for engineering hydrology students who go on to be practitioners in consultancies, the Environment Agency and elsewhere. It has never been superseded by any other text in this respect.

This fourth edition of *Hydrology in Practice* while retaining all that is excellent about its predecessor, by Elizabeth Shaw, replaces the material on the Flood Studies Report with an equivalent section on the methods of the Flood Estimation Handbook and its revisions. Other completely revised sections on instrumentation and modelling reflect the many changes that have occurred over recent years. The updated text has taken advantage of the extensive practical experience of the staff of JBA Consulting who use the methods described on a day-to-day basis. Topical case studies further enhance the text and the way in which students at undergraduate and MSc level can relate to it. The fourth edition will also have a wider appeal outside the UK by including new material on hydrological processes which also relate to courses in geography and environmental science departments. In this respect the book draws on the expertise of Keith J. Beven and Nick A. Chappell who have extensive experience of field hydrological studies in a variety of different environments, and teaching undergraduate hydrology courses for many years.

Second and final year undergraduates (and MSc) students of hydrology in Engineering, Environmental Science and Geography Departments across the globe, as well as professionals in environmental protection agencies and consultancies, will find this book invaluable. It is likely to be the course text for every undergraduate/MSc hydrology course in the UK and in many cases overseas too.

Elizabeth M. Shaw is now retired but has worked as a lecturer at universities in the UK and Australia, as well as having practised as a hydrologist. **Keith J. Beven** and **Nick A. Chappell** are in the Lancaster Environment Centre at the University of Lancaster, UK. **Rob Lamb** is a consultant with JBA Consulting.

Hydrology in Practice

Fourth edition

Elizabeth M. Shaw, Keith J. Beven,
Nick A. Chappell and Rob Lamb

Spon Press
an imprint of Taylor & Francis
LONDON AND NEW YORK

First published in 1983 by Chapman and Hall
Second Edition 1988
Third edition 1994

This edition published 2011 by Spon Press
2 Park Square, Milton Park, Abingdon, Oxon OX14 4RN

Simultaneously published in the USA and Canada by Spon Press
270 Madison Avenue, New York, NY 10016, USA

Spon Press is an imprint of the Taylor & Francis Group, an informa business

© 2011 Elizabeth M. Shaw, Keith J. Beven, Nick A. Chappell and Rob Lamb

Typeset in Sabon by Glyph International Ltd.
Printed and bound in Great Britain by CPI Antony Rowe, Chippenham, Wiltshire

This publication presents material of a broad scope and applicability. Despite
stringent efforts by all concerned in the publishing process, some typographical
or editorial errors may occur, and readers are encouraged to bring these to our
attention where they represent errors of substance. The publisher and author
disclaim any liability, in whole or in part, arising from information contained in
this publication. The reader is urged to consult with an appropriate licensed
professional prior to taking any action or making any interpretation that is within
the realm of a licensed professional practice.

British Library Cataloguing in Publication Data
A catalogue record for this book is available from the British Library

Library of Congress Cataloging in Publication Data
Hydrology in practice / Elizabeth M Shaw ... [et al.]. – 4th ed.
    p. cm.
Includes bibliographical references and index.
1. Hydrology. I. Shaw, Elizabeth M.
GB661.2.S53 2010
551.48024'627–dc22
2009049509

ISBN 13: 978-0-415-37041-7 (hbk)
ISBN 13: 978-0-415-37042-4 (pbk)
ISBN 13: 978-0-203-03023-3 (ebk)

Contents

PART II
HYDROLOGICAL ANALYSIS

PART IV
THE FUTURE OF HYDROLOGY IN PRACTICE

Preface to the fourth edition

When the first edition of Elizabeth M. Shaw's *Hydrology in Practice* appeared in 1983, it was immediately perceived to be a valuable addition to the hydrology texts that were then available. It became a standard text in most, if not all, undergraduate and Master's level hydrology courses in the UK and was widely used elsewhere, despite the strong orientation towards the practical training of UK engineering hydrologists. It was Elizabeth's stated intention to ensure that engineers received the training necessary in hydrological measurements and analysis to be able to manage surface and groundwater water resources, floods and droughts in the UK.

Since that time, the organisation of the water industry has changed dramatically in the UK. No longer are there integrated water authorities dealing with supply, waste water, flood risk and water quality. Now we have regulated utility companies with responsibilities for supply and treatment and separate government agencies with responsibilities for water quality standards, licensing of abstractions and effluents, and flood defence. Much less hydrological analysis is done 'in house' by these bodies; much more is commissioned from consultants. The major insurance companies have also taken much more interest in the hydrology of floods and droughts. European Directives, including the Habitats Directive, Water Framework Directive and Floods Directive, have resulted in environmental and sustainability concerns becoming more important relative to purely engineering design issues in water resource management.

Elizabeth has long retired from teaching at Imperial College, but in talking to her in preparing this edition it is clear that she remains concerned about the training of the next generation of engineers and environmental hydrologists. She particularly feels that the recent developments have resulted in a loss of local expertise dealing with local problems using best practice methods. When we received the invitation of Taylor & Francis to prepare a fourth edition of *Hydrology in Practice*, she generously gave us a free hand to change the text and presentation in any way we wished. We have very much intended, however, to keep the applied nature of the text well to the fore, so that it still adhered to the original aims of the first edition. The need is still there, even if the scope of practical training required for hydrologists is now somewhat wider. Some of the methods in the earlier editions have been superceded but we hope that much of the spirit of the original remains.

Indeed, although we have made extensive changes to the text, there is still a lot that is recognisable from the third edition. Thus we have suggested that Elizabeth should remain as the first author for this fourth edition, especially since everyone has simply referred to the earlier editions as 'Shaw' for as long as we can remember.

We feel that this would be a suitable recognition of her valuable contribution to the training of several generations of hydrology students. We have therefore maintained the structure of the previous editions in grouping the chapters into Parts, concerned with Hydrological Measurements (Chapters 2–8), Hydrological Analysis (Chapters 9–15) and Engineering Applications (Chapters 16–19). We have tried to look to the future in adding an additional chapter on The Future of Hydrology in Practice.

Special thanks are due to Jeremy Benn and the staff of JBA Consulting. This is a consultancy that has grown with the reorganisation of the water industry in the UK and that has been at the forefront in developing and applying new methods of hydro-logical and hydraulic analysis. This edition has greatly benefited from the experience of practical applications within the company and we would particularly like to recognise the contributions of David Archer, Jeremy Benn, Eleanor Charles, Mandy Crossley, John Dudley, Paul Dunning, Paul Ecclestone, Duncan Faulkner, Barry Hankin, Peter Henrys, Neil Hunter, Caroline Keef, John Mawdsley, Steve Rose, Zdenka Rosolova, Vicky Shackle, Judith Stunell, Sebastien Tellier, Simon Waller, Paul Wass and Maxine Zaidman.

We are also grateful to the following individuals for their assistance: Andy Binley and Hao Zhang of Lancaster University, Paul Quinn of Newcastle University, Tracey Haxton, and Andy Young of Wallingford HydroSolutions, Janet Bromley of United Utilities, and Jamie Hannaford, Alison Kay, Gwyn Rees, Nick Reynard of CEH Wallingford.

We are grateful to many organisations and individuals for permission to use their line drawings and photographs. Each image is individually acknowledged within the associated figure caption. While we have illustrated the text on hydrological measure-ments with representative equipment that is currently available in the UK, it should be noted that equipment of similar or alternative design will be available from other companies and no specific endorsement by the authors is implied.

Keith J. Beven
Nick A. Chappell
Rob Lamb

Chapter 1

The hydrological cycle and hydrometeorology

The history of the evolution of hydrology as a multi-disciplinary subject, dealing with the occurrence, circulation and distribution of the waters of the Earth, has been presented by Biswas (1970). Man's need for water to sustain life and grow food crops was well appreciated throughout the world wherever early civilization developed. Detailed knowledge of the water management practices of the Sumarians and Egyptians in the Middle East, of the Chinese along the banks of the Hwang-Ho and of the Aztecs in South America continues to grow as archaeologists uncover and interpret the artefacts of such centres of cultural development. It was the Greek philosophers who were the first serious students of hydrology, and thereafter, scholars continued to advance the understanding of the separate phases of water in the natural environment. However, it was not until the seventeenth century that the work of the Frenchman, Perrault, provided convincing evidence of the form of the hydrological cycle which is currently accepted: measurements of rainfall and river flow in the catchment of the upper Seine published in 1694 (Dooge, 1959) proved that quantities of rainfall were sufficient to sustain river flow.

Hydrology as an academic subject became established within institutions of higher education in the 1940s. Valuable research contributions to the subject had been reported earlier but the expansion in the more widespread applications of hydrology resulted in at least five textbooks being published in that decade in the United States.

Over the last 50 years, advances in sensor technology coupled with the development of numerical models representing hydrological processes have led to a reappraisal of the content and definition of hydrology. Today's scientific hydrologists and engineering hydrologists now appreciate the need to combine accurate field measurement with appropriate numerical models. Equally, there is an awareness of the controlling influence of hydrometeorology on the water pathways that comprise the hydrological cycle at catchment and global scales.

1.1 The hydrological cycle and water pathways

The driving force of the natural circulation of water is derived from the radiant energy received from the Sun. The bulk of the Earth's water is stored on the surface in the oceans (Table 1.1) and hence it is logical to consider the hydrological cycle as beginning with the direct effect of the Sun's radiation on this largest reservoir. Heating of the sea surface causes *evaporation*, the transfer of water from the liquid to the gaseous

Table 1.1 One estimate of global water distribution

Store	Volume (1000 km³)	Per cent of total water	Per cent of fresh water
Oceans, seas and bays	1 338 000	96.5	–
Ice caps, glaciers and permanent snow	24 064	1.74	68.7
Groundwater	23 400	1.7	–
Fresh	(10 530)	(0.76)	30.1
Saline	(12 870)	(0.94)	–
Soil moisture	16.5	0.001	0.05
Ground ice and permafrost	300	0.022	0.86
Lakes	176.4	0.013	–
Fresh	(91.0)	(0.007)	0.26
Saline	(85.4)	(0.006)	–
Atmosphere	12.9	0.001	0.04
Swamp water	11.47	0.0008	0.03
Rivers	2.12	0.0002	0.006
Biological water	1.12	0.0001	0.003

Source: Gleick, P. H. (1996) Water resources. In *Encyclopedia of Climate and Weather*, ed. by S. H. Schneider, Oxford University Press, New York, vol. 2, pp. 817–823.

state, to form part of the atmosphere. It remains mainly unseen in atmospheric storage for an average of 10 days. Through a combination of circumstances, the water vapour changes back to the liquid state again through the process of *condensation* to form clouds and, with favourable atmospheric conditions, *precipitation* (rain, snow, etc.) is produced either to return directly to the ocean storage or to embark on a more devious route to the oceans via the land surface. Snow may accumulate in polar regions or on high mountains and consolidate into ice, in which state water may be stored naturally for very long periods. In more temperate lands, rainfall may be intercepted by vegetation from which some of the intercepted water may return at once to the air by wet-canopy evaporation. A significant proportion of the rainfall that reaches the land surface will return to the atmosphere by transpiration via plants, while the remainder travels over or beneath the land surface towards rivers by mechanisms described as runoff[1] generation pathways.

1.2 Pathways generating river flow

In the later nineteenth and early twentieth century, it began to be recognised that different parts of a catchment area might produce different amounts of river flow. Again this was perhaps expressed first in France in the work of Imbeaux in a study of the Durance basin published in 1892. He tried to take account of the role of river flow generation at different distances from a catchment outlet in controlling the shape of the hydrograph, and of elevation in controlling the patterns of snowmelt during the melt season. The idea of delay in runoff reaching the catchment outlet can be represented in terms of a *time–area histogram*. It was later developed into the first storm event rainfall to river flow model to be widely used around the world that we now know as the *unit hydrograph* (see Section 12.5).

Use of the unit hydrograph for practical applications, however, requires that we try to estimate the proportion of storm rainfall that contributes to the *storm hydrograph* for a particular event (i.e. that river flow which appears soon after rainfall). It has long been known that not all the rainfall falling in an event contributes to the storm hydrograph. Some contributes to a much slower subsurface pathway or is lost back to the atmosphere by evapo-transpiration. Thus application of the unit hydrograph concepts required a rather arbitrary separation of the hydrograph into so-called *stormflow* and *baseflow*. This led to a rather easy assumption that the storm runoff in a river was made up of rainfall from the particular rain-event or snowmelt water.

In fact we now know that the situation is somewhat more complicated than that easy assumption because analyses of environmental tracers since the 1970s have shown that, in many environments, not all the storm hydrograph is made up of rainfall that fell in that storm (see Section 11.3). Some of the hydrograph comes from water that was already stored in the catchment prior to the rainfall event. That water is displaced from storage into the stream channels during the event as a result of rainfall infiltrating into the subsurface. This is really one of the most important conceptual advances in scientific hydrology since it has very important implications for understanding *hillslope hydrology*, water quality variations and ecological impacts of storm events.

The history of river flow generation concepts often starts with the ideas of Robert Elmer Horton, probably the most influential American hydrologist of the twentieth century. In a paper published in 1933 (and reproduced in Beven, 2006) Horton first expressed a concept of hillslope hydrology based on the idea that the storm hydrograph is made up of rainfall in excess of the infiltration capacity of the soil. This is the concept of *infiltration excess (or Hortonian) overland flow*. This idea leads to a nice simple interpretation of catchment response. If we know the volume of rainfall in an event, and we know the volume of *stormflow* (here meaning that river flow proportion above the hydrograph separation line; Fig. 12.5, Section 12.3) recorded at a river gauging site, then the difference must be what was infiltrated (on average) into the soil. This allows the infiltration capacity of the soil to be back-calculated (subject to some simple assumptions of how it might vary over time). This information can then be used in a simple runoff model to predict what might happen under different conditions. Combining this prediction of how much runoff will be generated by a given rainfall, and the unit hydrograph to predict the timing of the runoff is a technique that is still used to the present day (it still underlies some aspects of the UK Flood Estimation Handbook, for example, see Chapter 16).

The problem is, of course, that it is wrong. Some catchments in arid areas under high rainfall intensities, or the extreme case of impermeable surfaces in urban areas, might work like this, though even then it is unlikely that infiltration excess overland flow will occur everywhere, and overland flow generated on one part of a slope might later infiltrate further downslope. It is even unlikely that Horton saw this type of infiltration-excess overland flow in his own experimental catchment in New York State (Beven, 2004). However, as we have already noted, in very many catchments, much of the storm hydrograph is made up of displaced pre-event storage. Thus, other concepts of river flow generation are needed.

The first real reconsideration of the Hortonian concept was by Roger P. Betson in 1964. Betson worked for the Tennessee Valley Authority in the United States and realised that in the forested catchments of the Appalachians, there was no way that infiltration excess overland flow could occur everywhere, except perhaps in the most extreme rainfall events. He therefore suggested that overland flow would be generated on only part of the hillslopes and that since infiltration rates of soils tend to be lower when the soil is wetter (Section 5.1.1) and, as a result of downslope flows between events, soils will tend to be wetter in the lower parts of hillslopes, then the runoff generation would be most likely at the bottom of hillslopes close to the stream channels. He inferred from the analyses of storm runoff volumes that the proportion of the catchment generating overland flow could be quite small (as low as 2–4 per cent) in some catchments.

At about the same time, John Hewlett, working at the Coweeta catchments in North Carolina, suggested that the infiltration capacities of the soils in that area were so large that it was extremely unlikely that any runoff would be seen over the surface of the soil. Yet storm hydrographs were still recorded. He suggested that the storm runoff therefore must be generated by subsurface flows and by rainfall directly on to the stream channel and immediate riverside area. He also suggested that the water contributing to the river was not necessarily the rainfall, invoking a concept of so-called 'translatory flow' to explain the displacement of stored water by the infiltration of the rainwater. In fact, a previous Director of the Coweeta Laboratory, Charles Hursh, had expressed much the same idea in the 1930s, and had coined the term *subsurface stormflow* for this type of river flow generation mechanism. The concept of translatory flow had also already been mentioned in the 1930s but these concepts had been dominated by the Hortonian paradigm in the later engineering literature.

There are some circumstances, however, when overland flow can be generated on soils with high infiltration capacities. This is when the soil becomes saturated by a combination of downslope flow within a hillslope and rain falling on saturated areas. In fact, downslope flows can maintain the lower parts of hillslopes at, or close to, saturation for long periods of time in some circumstances, so that only small amounts of rainwater might be required before overland flow is generated. This will particularly be the case in relatively shallow soils overlying an impermeable base, and where there are convergent flow lines into the hillslope hollows. This was first demonstrated by the work of Tom Dunne in the Sleepers River catchments in Vermont in the late 1960s. He showed how saturated areas in the catchment could persist for long periods of time, how they varied seasonally, being most extensive at the end of the snowmelt season in Vermont, and how they were largely controlled by the patterns of downslope flow on the hillslopes. This type of overland flow became known as *saturation overland flow* that was generated on a *variable contributing area* in the catchment. In some cases the resulting overland flow will also have a component of *return flow*, which subsurface water forced back on to the surface through a seepage face; and one of the earliest studies of environmental tracers in storm runoff, by Mike Sklash and Bob Farvolden (1979), provided evidence to reinforce this concept. In an area of river flow generation in one of their study catchments, they showed that tracer concentrations were sometimes indicative of a rainfall source and at other times indicative of a subsurface, pre-event storage source.

Work elsewhere has revealed further complications. Darrell Weyman (1970) working in the East Twin catchment in the Mendips, UK, showed that saturated contributing areas could arise without the soil being completely saturated but where saturation built up above a soil horizon of lower permeability. He also showed that subsurface contributions to the stream channel could be hugely variable in space and might be associated with zones of higher soil permeability within a very heterogeneous soil. Bunting (1961) had earlier called such preferential pathways *percolines* and he treated them as a subsurface extension of the dendritic (tree-like) channel network. Recent modelling work has shown how, in heterogeneous soils, subsurface flow might be simulated as being channelled into channel network-like structures (e.g. Weiler and McDonnell, 2004), while tracer work has shown that, in some catchments, fast responding subsurface pipes produce storm runoff that may be made up predominantly of pre-event stored water rather than rainwater (Sklash *et al.*, 1996).

It is perhaps worth finishing this brief summary of river flow generation pathways by saying that the different major concepts shown in Fig. 1.1 are not mutually exclusive. They might all occur in different events in the same catchment, or in the same event in different parts of a catchment, depending on the rainfall intensities; prior wetness of the catchment (*antecedent conditions*); topography of the hillslopes; type, structure and heterogeneity of the soil, regolith and rock; existence of percolines; channel

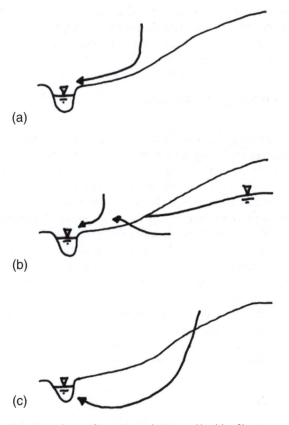

Fig. 1.1 River flow generation pathways for systems dominated by (a) infiltration-excess overland flow, (b) saturation overland flow, and (c) subsurface flow.

density; and other factors. An excellent review of where and when different types of runoff production might occur is given by Tom Dunne (1978) and Wilfred Brutsaert (2005).

There is an underlying research question about how, over long periods of time, these different factors might be linked to the long-term development of the catchment soils, topography and vegetation cover, and how, in recent times, people might have affected the nature of the river flow generation processes through land management practices and urbanisation. Such questions are not yet fully resolved, and it is perhaps unlikely that they will ever be properly resolved given the complexities of short-term and long-term changes to which catchments have been subjected in different environments.

Fortunately, this is not a barrier to hydrological analysis and prediction. In many cases we are only interested in predicting river flow, and do not need to worry too much about the water pathways. This is one reason why unit hydrograph concepts have survived so long: if we can match river flow volumes and timings using these simple concepts, then we may be able to make some useful predictions even if the details of the pathways are incorrect. There are situations, however, particularly in understanding water quality variations, where it may be critical to appreciate the different surface and subsurface pathways. In such cases, an appreciation for the different mechanisms of river flow generation described above will be important. We will return to this in the discussion of predictive rainfall-runoff models in Chapter 12. First, we need to give an overview of how hydrometeorology regulates these river flow generation pathways and the pathways of evapo-transpiration (Chapter 10).

1.3 Hydrometeorological control of hydrological pathways

The science of meteorology has long been recognised as a separate discipline, though students of the subject usually come to it from a rigorous training in physics or mathematics. The study of *hydrometeorology* may be seen as a branch of hydrology linking the fundamental knowledge of the meteorologist with the needs of the hydrologist. In this text, hydrometeorology is taken to be the study of precipitation and evaporation, the two fundamental phases in the hydrological cycle, which involve processes in the atmosphere, and at the Earth's surface/atmosphere interface.

The hydrologist will usually be able to call upon the services of a professional meteorologist for weather forecasts and for special studies, e.g. the magnitude of extreme rainfalls. However, a general understanding of precipitation and evaporation is essential if the hydrologist is to appreciate the complexities of the atmosphere and the difficulties that the meteorologist often has in providing answers to questions of quantities and timing. A description of the properties of the atmosphere and of the main features of solar radiation will provide the bases for considering the physics of evaporation and the formation of precipitation.

1.3.1 The atmosphere

The atmosphere forms a distinctive protective layer about 100 km thick around the Earth. Although both air pressure and density decrease rapidly and continuously with

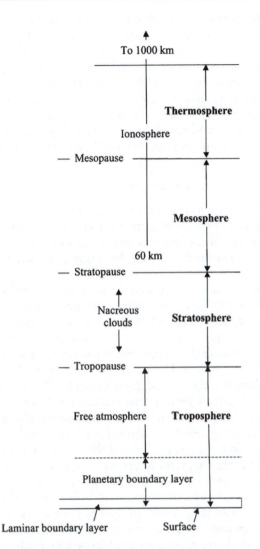

Fig. 1.2 Structure of the atmosphere. (Adapted from Strangeways, I. (2007) *Precipitation: Theory, Measurement and Distribution*, Cambridge University Press, Cambridge.)

increasing altitude, the temperature varies in an irregular but characteristic way. The layers of the atmosphere, 'spheres', are defined by this temperature profile. After a general decrease in temperature through the *troposphere* (Fig. 1.2), the rise in temperature from heights of 20–50 km is caused by a layer of ozone, which absorbs short-wave solar radiation, releasing some of the energy as heat.

To the hydrologist, the troposphere is the most important layer because it contains 75 per cent of the weight of the atmosphere and virtually all its moisture. The meteorologist, however, is becoming increasingly interested in the stratosphere and mesosphere, since it is in these outer regions that some of the disturbances affecting the troposphere and the Earth's surface have their origins.

The height of the *tropopause*, the boundary zone between the troposphere and the stratosphere, is at about 11 km, but this is an average figure, which ranges from about 8 km at the Poles to about 16 km at the Equator. Seasonal variations also are caused by changes in pressure and air temperature in the atmosphere. In general, when surface temperatures are high and there is a high sea-level pressure, then there is a tendency for the tropopause to be at a high level. On average, the temperature from ground level to the tropopause falls steadily with increasing altitude at the rate of $6.5°C \, km^{-1}$. This is known as the *lapse rate*. Some of the more hydrologically pertinent characteristics of the atmosphere as a whole are now defined in more precise terms.

1.3.1.1 Atmospheric pressure and density

The meteorologist's definition of atmospheric pressure is 'the weight of a column of air of unit area of cross-section from the level of measurement to the top of the atmosphere'. More specifically, pressure may be considered to be the downward force on a unit horizontal area resulting from the action of gravity (g) on the mass (m) of air vertically above.

At sea level, the average atmospheric pressure (p) is 100 kPa (1 bar or 100 000 $N \, m^{-2}$). A pressure of 100 kPa is equivalent to 760 mm of mercury; the average reading on a standard mercury barometer. Measurements of atmospheric pressure are usually given in millibars (mb). It is common meteorological practice to refer to heights in the atmosphere by their average pressure in millibars, e.g. the top of the stratosphere (the stratopause) is at the 1 mb level. The air density (ρ) may be obtained from the expression $\rho = p/RT$, where R is the specific gas constant for dry air ($0.29 \, kJ \, kg^{-1} \, K^{-1}$) and $T \, K$ is the air temperature. At sea level, the average $T = 288 \, K$ and thus $\rho = 1.2 \, kg \, m^{-3}$ (or $1.2 \times 10^{-3} \, g \, cm^{-3}$) on average at sea level. Air density falls off rapidly with height. Unenclosed air, a compressible fluid, can expand freely, and as pressure and density decrease with height indefinitely, the limit of the atmosphere becomes indeterminate. Within the troposphere however, the lower pressure limit is about 100 mb. At sea level, pressure variations range from about 940 to 1050 mb; the average sea level pressure around the British Isles is 1013 mb. Pressure records form the basis of the meteorologist's synoptic charts with the patterns formed by the *isobars* (lines of equal pressure) defining areas of high and low pressure (anticyclones and depressions, respectively). Interpretation of the charts plotted from observations made at successive specified times enables the changes in weather systems to be identified and to be forecast ahead. In addition to the sea level measurements, upper air data are plotted and analysed for different levels in the atmosphere.

1.3.1.2 Chemical composition

Dry air has a very consistent chemical composition throughout the atmosphere up to the mesopause at 80 km. The proportions of the major constituents are as shown in Table 1.2. The last category contains small proportions of other inert gases and, of particular importance, the stratospheric layer of ozone which filters the Sun's radiation. Small quantities of hydrocarbons, ammonia and nitrates may also exist temporarily in the atmosphere. Man-made gaseous and particulate pollutants are found particularly in areas of heavy industry, and can have considerable effects on local

Table 1.2 Major constituents of air

	Percentage (by mass)
Nitrogen	75.51
Oxygen	23.15
Argon	1.28
Carbon dioxide etc	0.06

weather conditions. Traces of radioactive isotopes from nuclear fission also contaminate the atmosphere. Although there is no evidence that isotopes have a significant effect on weather, their presence has been found useful in tracing the movement of water through the hydrological cycle.

1.3.1.3 Water vapour

The amount of water vapour in the atmosphere (Table 1.3) is directly related to the temperature and thus, although lighter than air, water vapour is restricted to the lower layers of the troposphere because temperature decreases with altitude. The distribution of water vapour also varies over the Earth's surface according to temperature, and is lowest at the Poles and highest in equatorial regions. The water vapour content or *humidity* of air is usually measured as a vapour pressure, and the units used are millibars (mb).

Several well-recognised physical properties concerned with water in the atmosphere are defined to assist understanding of the complex changes that occur in the meteorological phases of the hydrological cycle.

(a) *Saturation.* Air is said to be saturated when it contains the maximum amount of water vapour it can hold at its prevailing temperature. The relationship between saturation vapour pressure (e) and air temperature is shown in Fig. 1.3. At typical temperatures near the ground, e ranges from 5 to 50 mb. At any temperature $T = T_a$, saturation occurs at corresponding vapour pressure $e = e_a$.

Table 1.3 Average water vapour values for latitudes with temperate climates (volume %)

Height (km)	Water vapour
0	1.3
1	1.0
2	0.69
3	0.49
4	0.37
5	0.27
6	0.15
7	0.09
8	0.05

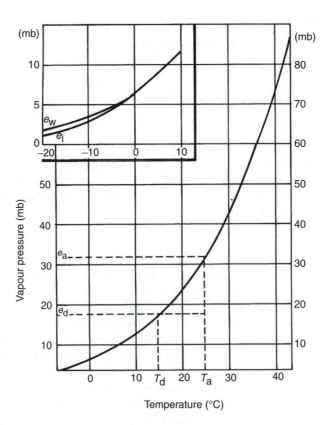

Fig. 1.3 Saturation pressure and air temperature, where $e_a - e_d$ is the saturation deficit and T_d is the dew point temperature.

Meteorologists acknowledge that saturated air may take up even more water vapour and become *supersaturated* if it is in contact with liquid water in a sufficiently finely divided state (e.g. very small water droplets in clouds). At sub-zero temperatures, there are two saturation vapour pressure curves, one with respect to water (e_w) and one with respect to ice (e_i; Fig. 1.3, inset). In the zone between the curves, the air is unsaturated with respect to water but supersaturated with respect to ice. This is a common condition in the atmosphere as will be seen later.

(b) *Dew point* is the temperature, T_d, at which a mass of unsaturated air becomes saturated when cooled, with the pressure remaining constant. In Fig. 1.3, if the air at temperature T_a is cooled to T_d, the corresponding saturation vapour pressure, e_d, represents the amount of water vapour in the air.

(c) *Saturation deficit* is the difference between the saturation vapour pressure at air temperature, T_a, and the actual vapour pressure represented by the saturation vapour pressure at T_d, the dew point. The saturation deficit, $e_a - e_d$, represents the further amount of water vapour that the air can hold at the temperature, T_a, before becoming saturated.

(d) *Relative humidity* is the relative measure of the amount of moisture in the air to the amount needed to saturate the air at the same temperature, i.e. e_d/e_a, represented as a percentage. Thus, if $T_a = 30°C$ and $T_d = 20°C$, relative humidity is

$$\frac{e_d}{e_a} \times 100 = \frac{23\,\text{mb}}{42.5\,\text{mb}} \times 100 = 54 \text{ per cent}$$

(e) *Absolute humidity* (ρ_w) is generally expressed as the mass of water vapour per unit volume of air at a given temperature and is equivalent to the water vapour density. Thus, if a volume $V\,\text{m}^3$ of air contains m_{wv} g of water vapour,

$$\rho_w = \frac{\text{Mass of water vapour (g)}}{\text{Volume of air (m}^3)} = \frac{m_{wv}}{V}(\text{g m}^{-3})$$

(f) *Specific humidity* (*SH*) relates the mass of water vapour $(m_{wv}\text{ g})$ to the mass of moist air (in kg) in a given volume; this is the same as relating the absolute humidity (g m^{-3}) to the density of the same volume of unsaturated air $(\rho\text{ kg m}^{-3})$:

$$SH = \frac{m_{wv}(\text{g})}{(m_{wv} + m_d)(\text{kg})} = \frac{\rho_w}{\rho}(\text{g kg}^{-1})$$

where m_d is the mass (kg) of the dry air.

(g) *Precipitable water* is the total amount of water vapour in a column of air expressed as the depth of liquid water in millimetres over the base area of the column. Assessing this amount is a specialised task for the meteorologist. The precipitable water gives an estimate of maximum possible rainfall, though has the unreal assumption of total condensation and neglects the effect of advection.

In a column of unit cross-sectional area, a small thickness, dz, of moist air contains a mass of water given by:

$$dm_{wv} = \rho_w\,dz$$

Thus, in a column of air from heights $z1$ to $z2$, corresponding to pressures $p1$ and $p2$: the total mass of water m_w is

$$\int_{z1}^{z2} \rho_w\,dz$$

Also, $dp = -\rho g dz$ and, by rearrangement, $dz = -dp/\rho g$. Thus:

$$m_w = -\int_{p1}^{p2} \frac{\rho_w}{\rho g}\,dp$$

$$= \frac{1}{g}\int_{p1}^{p2} q\,dp$$

Allowing for the conversion of the mass of water (m_w) to equivalent depth over a unit cross-sectional area, the precipitable water is given by:

$$W \text{ (mm)} = \frac{0.1}{g} \int_{p_1}^{p_2} SH dp$$

where p is in mb, SH in $g\,kg^{-1}$ and $g = 9.81\,m\,s^{-2}$.

In practice, the integration cannot be performed since q is not known as a function of p. A value of W is obtained by *summing* the contributions for a sequence of layers in the troposphere from a series of measurements of the specific humidity $\bar{q}$ at different heights and using the average specific humidity $\bar{q}$ over each layer with the appropriate pressure difference:

$$W \text{ (mm)} = \frac{0.1}{g} \sum_{p_1}^{p_2} \overline{SH} \Delta p$$

Example. From a radiosonde (balloon) ascent, the pairs of measurements of pressure and specific humidity shown in Table 1.4 were obtained. The precipitable water in a column of air up to the 250 mb level is calculated ($g = 9.81\,m\,s^{-2}$).

1.3.2 Solar radiation

The main source of energy at the Earth's surface is radiant energy from the Sun, termed solar radiation or insolation. It is the solar radiation impinging on the Earth that fuels the heat engine driving the hydrological cycle. The amount of radiant energy received at any point on the Earth's surface (assuming no atmosphere) is governed by the following well-defined factors.

(a) *The solar output.* The Sun, a globe of incandescent matter, has a gaseous outer layer about 320 km thick and transmits light and other radiations towards the Earth from a distance of 145 million km. The rate of emission of energy is shown in Fig. 1.4 but only a small fraction of this is intercepted by the Earth. Half the total energy emitted by the Sun is in the visible light range, with wavelengths from 0.4 to 0.7 μm. The rest arrives as ultraviolet or infrared waves, from 0.25 up to 3.0 μm.

The maximum rate of the Sun's emission (10 500 kW m^{-2}) occurs at 0.5 μm wavelength in the visible light range. Although there are changes in the solar output

Table 1.4

Pressure (mb)	1005	850	750	700	620	600	500	400	250
Specific humidity q ($g\,kg^{-1}$)	14.2	12.4	9.5	7.0	6.3	5.6	3.8	1.7	0.2
$P_n - P_{n+1} = \Delta p$		155	100	50	80	20	100	100	150
Mean SH $= \overline{SH}$		13.30	10.95	8.25	6.65	5.95	4.70	2.75	0.95
$\overline{SH}\Delta p$		2061.5	1095.0	412.5	532.0	119.0	470.0	275.0	142.5

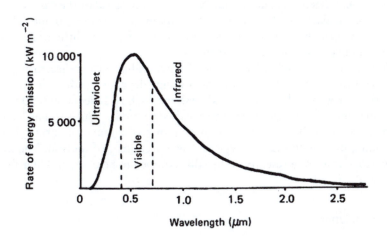

Fig. 1.4 Solar radiation.

associated with the occurrence of sunspots and solar flares, these are disregarded in assessing the amount of energy received by the Earth. The total solar radiation received in unit time on unit area of a surface placed at right angles to the Sun's rays at the Earth's mean distance from the Sun is known as the *solar constant*. The average value of the solar constant is $1.39\,\text{kW}\,\text{m}^{-2}$ ($1.99\,\text{cal}\,\text{cm}^{-2}\,\text{min}^{-2}$).

(b) *Distance from the Sun.* The distance of any point on the Earth's surface from the Sun is changing continuously owing to the Earth's eccentric orbit. The Earth is nearest the Sun in January at perihelion and furthest from it in July at aphelion. The solar constant varies accordingly.

(c) *Altitude of the Sun.* The Sun's altitude above the horizon has a marked influence on the rate of solar radiation received at any point on the Earth. The factors determining the Sun's altitude are latitude, season and time of day.

(d) *Length of day.* The total amount of radiation falling on a point of the Earth's surface is governed by the length of the day, which itself depends on latitude and season.

1.3.2.1 Atmospheric effects on solar radiation

The atmosphere has a marked effect on the energy balance at the surface of the Earth. In one respect it acts as a shield protecting the Earth from extreme external influences, but it also prevents immediate direct loss of heat. Thus it operates as an energy filter in both directions. The interchanges of heat between the incoming solar radiation and the Earth's surface are many and complex. There is a loss of energy from the solar radiation as it passes through the atmosphere known as *attenuation*. Attenuation is brought about in three principal ways as follows.

(a) *Scattering.* About 9 per cent of incoming radiation is scattered back into space through collisions with molecules of air or water vapour. A further 16 per cent are also scattered, but reach the Earth as diffuse radiation, especially in the shorter wavelengths, giving the sky a blue appearance.

(b) *Absorption.* Fifteen per cent of solar radiation is absorbed by the gases of the atmosphere, particularly by the ozone, water vapour and carbon dioxide. These gases absorb wavelengths of less than 0.3 μm only, and so very little of this radiation penetrates below an altitude of 40 km.

(c) *Reflection.* On average, 33 per cent of solar radiation is reflected from clouds and the ground back into space. The amount depends on the *albedo* (α) of the reflecting surfaces. White clouds and fresh white snow reflect about 90 per cent of the radiation ($r = 0.9$), but a dark tropical ocean under a high sun absorbs nearly all of it ($\alpha \to 0$). Between these two extremes is a range of surface conditions depending on roughness, soil type and water content of the soil. The albedo of the water surface of a reservoir is usually assumed to be 0.05, and of a short grass surface, 0.25.

1.3.2.2 Net radiation

As a result of the various atmospheric losses, only about 43 per cent of solar (short-wave) radiation reaches the Earth's surface, where most is absorbed and heats the land and oceans. The Earth itself radiates energy in the long-wave range (Fig. 1.5) and this long-wave radiation is readily absorbed by the atmosphere. The Earth's surface emits more than twice as much energy in the infrared range as it receives in short-wave solar radiation.

The balance between incoming and outgoing radiation varies from the Poles to the Equator. There is a net heat gain in equatorial regions and a net heat loss in polar regions. Hence, heat energy travels through circulation of the atmosphere from lower to higher latitudes. Further variations occur because the distribution of the continents and oceans leads to differential heating of land and water.

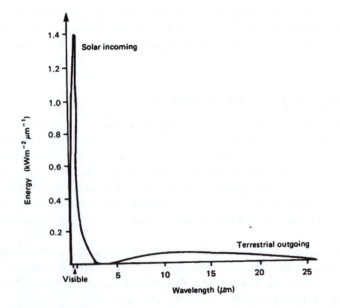

Fig. 1.5 Solar and terrestrial radiation.

Table 1.5 Average radiation values for selected latitudes ($W\,m^{-2}$)

	July season			January season		
	R_a	R_o	R_N	R_a	R_o	R_N
50°N	250	210	40	70	190	−120
Equator	280	240	40	310	240	70
30°S	170	220	−50	320	230	90

The amount of energy available at any particular point on the Earth's surface for heating the ground and lower air layers, and for the evaporation of water, is called the *net radiation*.

The net radiation R_N may be defined by the equation:

$$R_N = R_a - \alpha(R_a) + R_l - R_o$$

where R_a is the incoming short-wave (solar) radiation, α is the albedo, R_l is the incoming long-wave radiation and R_o is the outgoing long-wave radiation.

Incoming long-wave radiation comes from clouds (from absorbed solar radiation), and this has the following effects in the net radiation equations. In clear conditions, $R_l \approx (0.6 \text{ to } 0.8)R_o$, thus $R_l - R_o$ gives a net loss of long-wave radiation. For cloudy conditions, $R_l \approx R_o$ and $R_l - R_o$ becomes 0.

More significant are diurnal variations in net radiation, which is the primary energy source for evaporation. At night, $S = 0$ and R_l is smaller or negligible so that $R_N \approx R_o$. In other words, net radiation is negative, and there is a marked heat loss, which is particularly noticeable when the sky is clear.

Some average values of solar (R_a), terrestrial (R_o) and net (R_N) radiation for points on the earth's surface are given in Table 1.5.

1.4 Evaporation

Evaporation is the primary process of water transfer in the hydrological cycle. The oceans contain 95 per cent of the Earth's water and constitute a vast reservoir that remains comparatively undisturbed. From the surface of the seas and oceans, water is evaporated and transferred to temporary storage in the atmosphere, the first stage in the hydrological cycle.

1.4.1 Factors affecting evaporation

To convert liquid water into gaseous water vapour at the same temperature a supply of energy is required (in addition to that possibly needed to raise the liquid water to that temperature). The *latent heat of vaporization* ($2.6 \times 10^6\,J\,kg^{-1}$) must be added to the liquid molecules to bring about the change of state. The energy available for evaporation is the net radiation obtaining at the water surface and is governed by local conditions of solar and terrestrial radiation.

The rate of evaporation is dependent on the temperature at the evaporating surface and that of the ambient air. It also depends on the vapour pressure of the existing water vapour in the air, since this determines the amount of additional water vapour that the air can absorb. From the saturation vapour pressure and air temperature relationship shown in Fig. 1.3, it is clear that the rate of evaporation is dependent on the saturation deficit. If the water surface temperature, T_s, is equal to the air temperature, T_a, then the saturated vapour pressure at the surface, e_s, is equal to e_a. The saturation deficit of the air is given by $(e_s - e_d)$, where e_d is the measure of the actual vapour pressure of the air at T_a.

As evaporation proceeds, the air above the water gradually becomes saturated and, when it is unable to take up any more moisture, evaporation ceases. The replacement of saturated air by drier air would enable evaporation to continue. Thus, wind speed is an important factor in controlling the rate of evaporation. The roughness of the evaporating surface is a subsidiary factor in controlling the evaporation rate because it affects the turbulence of the air flow.

In summary, evaporation from an open water surface is a function of available energy, the net radiation, the temperatures of surface and air, the saturation deficit and the wind speed. The evaporation from a vegetated surface is a function of the same meteorological variables, but it is also dependent on the presence of *negative pressure potential* (Section 6.2) within the soil or regolith. From a land surface, it is a combination of the evaporation of liquid water from precipitation collected on the land surface, from wetted vegetation surfaces and the transpiration of water by plants. Methods for the measurement of evaporation quantities are presented in detail in Chapter 4, and methods of analysis in Chapter 10.

1.5 Precipitation

The moisture in the atmosphere, although forming one of the smallest storages of the Earth's water, is the most vital source of fresh water for mankind. Water is present in the air in its gaseous, liquid and solid states as water vapour, cloud droplets and ice crystals, respectively.

The formation of precipitation from the water as it exists in the air is a complex and delicately balanced process. If the air was pure, condensation of the water vapour to form liquid water droplets would occur only when the air became greatly super-saturated. However, the presence of small airborne particles called *aerosols* provides nuclei around which water vapour in normal saturated air can condense. Many experiments, both in the laboratory and in the open air, have been carried out to investigate the requisite conditions for the change of state. Aitken (Mason, 1975) distinguished two main types of condensation nuclei: *hygroscopic particles* having an affinity for water vapour, on which condensation begins before the air becomes saturated (mainly salt particles from the oceans); and *non-hygroscopic particles* needing some degree of supersaturation, depending on their size, before attracting condensation. This latter group derives from natural dust and grit from land surfaces and from man-made smoke, soot and ash particles.

Condensation nuclei range in size from a radius 10^{-3} μm for small ions to 10 μm for large salt particles. The concentration of aerosols in time and space varies considerably. A typical number for the smallest particles is 40 000 per cm^3, whereas for giant nuclei of more than 1 μm radius there might be only 1 per cm^3. Large hygroscopic salt nuclei are normally confined to maritime regions, but the tiny particles called *Aitken nuclei* can travel across continents and even circumnavigate the Earth. Although condensation nuclei are essential for widespread condensation of water vapour, only a small fraction of the nuclei present in the air take part in cloud droplet formation at any one time.

Other conditions must be fulfilled before precipitation occurs. First, moist air must be cooled to near its dew point. This can be brought about in several ways as follows.

(a) By an adiabatic expansion of rising air. A volume of air may be forced to rise by an impeding mountain range. The reduction in pressure causes a lowering of temperature without any transference of heat.
(b) By a meeting of two very different air masses. For example, when a warm, moist mass of air converges with a cold mass of air, the warm air is forced to rise and may cool to the dew point. Any mixing of the contrasting masses of air would also lower the overall temperature.
(c) By contact between a moist air mass and a cold object such as the ground.

Once cloud droplets are formed, their growth depends on hygroscopic and surface tension forces, the humidity of the air, rates of transfer of vapour to the water droplets and the latent heat of condensation released. A large population of droplets competes for the available water vapour and so their growth rate depends on their origins and on the cooling rate of air providing the supply of moisture (Fig. 1.6).

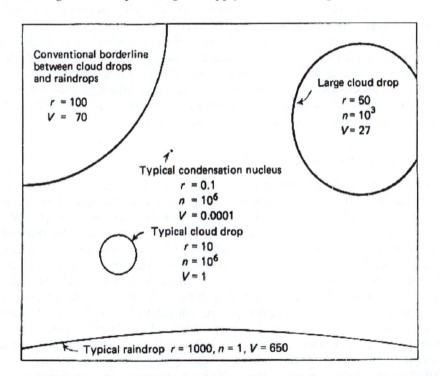

Fig. 1.6 Comparative sizes, concentrations and terminal falling velocities of some particles involved in condensation and precipitation processes, where r = radius (μm); n = number per dm^3 (10^3 cm^3); V = terminal velocity (cm s^{-1}). (Reproduced from B. J. Mason (1975) *Clouds, Rain and Rainmaking*, 2nd edn, by permission of Cambridge University Press.)

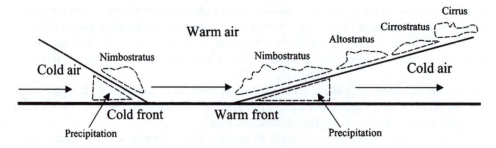

Fig. 1.7 Frontal weather conditions, showing cloud and precipitation around cold and warm fronts. (Reproduced with permission from I. Strangeways (2007) *Precipitation: Theory, Measurement and Distribution*, Cambridge University Press, Cambridge.)

The mechanism becomes complicated when the temperature reaches freezing point. Pure water can be supercooled to about $-40°$C (233 K) before freezing spontaneously. Cloud droplets are unlikely to freeze in normal air conditions until cooled below $-10°$C (263 K) and commonly exist down to $-20°$C (253 K). They freeze only in the presence of small particles called *ice nuclei*, retaining their spherical shape and becoming solid ice crystals. Water vapour may then be deposited directly on to the ice surfaces. The crystals grow into various shapes depending on temperature and the degree of supersaturation of the air with respect to the ice.

Condensed water vapour appears in the atmosphere as clouds in various characteristic forms; a standard classification of clouds is shown in Fig. 1.7. The high clouds are composed of ice crystals, the middle clouds of either water droplets or ice crystals, and the low clouds mainly of water droplets, many of them supercooled. Clouds with vigorous upwards vertical development, such as cumulonimbus, consist of cloud droplets in their lower layers and ice crystals at the top.

1.5.1 Theories of raindrop growth

Considerable research has been carried out by cloud physicists on the various stages involved in the transference of atmospheric water vapour into precipitable raindrops or snowflakes. A cloud droplet is not able to grow to raindrop size by the simple addition of water vapour condensing from the air. It is worth bearing in mind that one million droplets of radius 10 μm are equivalent to a single small raindrop of radius 1 mm. Fig. 1.6 shows the principal characteristics of nuclei, cloud droplets and raindrops.

Cloud droplets can grow naturally to about 100 μm in radius, and although tiny drops from 100 to 500 μm may, under very calm conditions, reach the ground, other factors are at work in forming raindrops large enough to fall to the ground in appreciable quantities. There are several theories of how cloud droplets grow to become raindrops, and investigations into the details of several proposed methods continue to claim the attention of research workers.

The Bergeron process, named after the famous Norwegian meteorologist, requires the coexistence in a cloud of supercooled droplets and ice particles and a temperature less than $0°$C (273 K). The air is saturated with respect to water but supersaturated with respect to ice. Hence water vapour is deposited on the ice particles to form ice crystals.

The air then becomes unsaturated with respect to water so droplets evaporate. This process continues until either all the droplets have evaporated or the ice crystals have become large enough to drop out of the cloud to melt and fall as rain as they reach lower levels. Thus the crystals grow at the expense of the droplets. This mechanism operates best in clouds with temperatures in the range -10 to $-30°C$ (263–243 K) with a small liquid water content.

Growth by collision: in clouds where the temperature is above $0°C$ (273 K), there are no ice particles present and cloud droplets collide with each other and grow by coalescence. The sizes of these droplets vary enormously and depend on the size of the initial condensation nuclei. Larger droplets fall with greater speeds through the smaller droplets with which they collide and coalesce. As larger droplets are more often formed from large sea-salt nuclei, growth by coalescence operates more frequently in maritime than in continental clouds. In addition, as a result of the dual requirements of a relatively high temperature and generous liquid water content, the growth of raindrops by coalescence operates largely in summer months in low-level clouds.

When cloud temperatures are below $0°C$ (273 K) and the cloud is composed of ice particles, their collision causes growth by *aggregation* to form snowflakes. The most favourable clouds are those in the 0 to $-4°C$ (269 K) range and the size of snowflakes decreases with the cloud temperature and water content.

Growth by accretion occurs in clouds containing a mixture of droplets and ice particles. Snow grains, ice pellets or hail are formed as cloud droplets fuse on to ice particles. Accretion takes place most readily in the same type of cloud that favours the Bergeron process, except that a large content of liquid water is necessary for the water droplets to collide with the ice particles.

Even when raindrops and snowflakes have grown large enough for their gravity weight to overcome up-draughts of air and fall steadily towards the ground, their progress is impeded by changing air conditions below the clouds. The temperature may rise considerably near the Earth's surface and the air may become unsaturated. As a result snowflakes usually melt to raindrops and the raindrops may evaporate in the drier air. On a summer's day it is not uncommon to see cumulus clouds trailing streams of rain which disappear before they reach the ground. With dry air below a high cloud base of about 3 km, all precipitation will evaporate. Hence it is rare to see rainfall from altocumulus, altostratus and higher clouds (see Fig. 1.7). Snowflakes rarely reach the ground if the surface air temperature is above $4°C$, but showers of fine snow can occur with the temperature as high as $7°C$, if the air is very dry.

Further explanation of the processes involved in raindrop formation is given in Sumner (1988) and Strangeways (2007).

1.6 Weather patterns producing precipitation

The main concern of the meteorologist is an understanding of the general circulation of the atmosphere with the aim of forecasting the movements of pressure patterns and their associated winds and weather. It is sufficient for the hydrologist to be able to identify the situations that provide the precipitation, and for the practising civil engineer to keep a 'weather eye' open for adverse conditions that may affect his site work.

The average distribution and seasonal changes of areas of high atmospheric pressure (*anticyclones*) and of low-pressure areas (*depressions*) can be found in most good atlases. Associated with the location of anticyclones is the development of homogeneous air masses. A *homogeneous air mass* is a large volume of air, generally covering an area greater than 1000 km in diameter, which shows little horizontal variation in temperature or humidity. It develops in the stagnant conditions of a high-pressure area and takes on the properties of its location (known as a *source region*). In general, homogeneous air masses are either cold and stable, taking on the characteristics of the polar regions from where they originate, or they are warm and unstable, revealing their tropical source of origin. Their humidity depends on whether they are centred over a large continent or over the ocean. The principal air masses are summarized in Table 1.6. Differences in atmospheric pressure cause air masses to move from high- to low-pressure regions and they become modified by the environments over which they pass. Although they remain homogeneous, they may travel so far and become so modified that they warrant reclassification. For example, when polar maritime air reaches the British Isles from a south-westerly direction, having circled well to the south over warm subtropical seas, its character will have changed dramatically.

Precipitation can come directly from a maritime air mass that cools when obliged to rise over mountains in its path. Such precipitation is known as *orographic rainfall* (or snowfall, if the temperature is sufficiently low), and is an important feature of the western mountains of the British Isles, which lie across the track of the prevailing winds bringing moisture from the Atlantic Ocean. Orographic rain falls similarly on most hills and mountains in the world, with similar locational characteristics, though it may occur only in particular seasons.

When air is cooled as a result of the converging of two contrasting air masses, it can produce more widespread rainfall independent of surface land features. The boundary between two air masses is called a *frontal zone*. It intersects the ground at the *front*, a band of about 200 km across. The character of the front depends on the difference between the air masses. A steep temperature gradient results in a strong or *active front* and much rain, but a small temperature difference produces only a *weak front* with less or even no rain. The juxtaposition of air masses across a frontal zone gives rise to two principal types of front according to the direction of movement.

Fig. 1.8 illustrates cloud patterns and the weather associated firstly with a *warm front*, in which warm air is replacing cold air, and secondly with a *cold front*, in which cold air is pushing under a warm air mass. In both cases, the warm air is made to rise and hence cool, and the condensation of water vapour forms characteristic clouds

Table 1.6 Classification of air masses

Air mass	Source region	Properties of source
Polar maritime (Pm)	Oceans; 50° latitude	Cool, rather moist, unstable
Polar continental (Pc)	Continents in vicinity of Arctic Circle; Antarctica	Cool, dry, stable
Arctic or Antarctic (A)	Arctic Basin and Central Antarctica in winter	Very cold, dry, stable
Tropical maritime (Tm)	Sub-tropical oceans	Warm and moist; unstable inversion common feature
Tropical continental (Tc)	Deserts in low latitude; primarily the Sahara and Australian deserts	Hot and dry

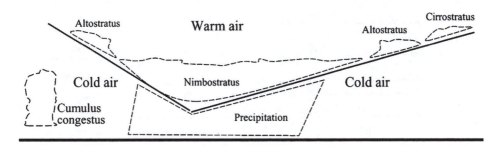

Fig. 1.8 Frontal weather conditions, showing cloud and precipitation development around an occluded front. (Reproduced from Strangeways, I. (2007) *Precipitation: Theory, Measurement and Distribution*, Cambridge University Press, Cambridge.)

and rainfall. The precipitation at a warm front is usually prolonged with gradually increasing intensity. At a cold front, however, it is heavy and short-lived. Naturally, these are average conditions; sometimes no rain is produced at all.

Over the world as a whole there are distinctive regions between areas of high pressure where differing air masses confront each other. These are principally in the mid-latitudes between 30° and 60° in both hemispheres, where the main boundary, the *polar front*, separates air masses having their origins in polar regions from the tropical air masses.

In addition, there is a varying boundary between air masses originating in the northern and southern hemispheres known as the *intertropical convergence zone* (ITCZ). The seasonal migration of the ITCZ plays a large part in the formation of the monsoon rains in south-east Asia and in the islands of Indonesia.

Four major weather patterns producing precipitation have been selected for more detailed explanation.

1.6.1 Mid-latitude cyclones or depressions

Depressions are the major weather pattern for producing precipitation in the temperate regions. More than 60 per cent of the annual rainfall in the British Isles comes from such disturbances and their associated features. They develop along the zone of the polar front between the polar and tropical air masses. Knowledge of the growth of depressions, the recognition of air masses and the definition of fronts all owe much to the work of the Norwegian meteorologists Wilhelm and Jacob Bjerknes in the 1920s.

The main features in the development and life of a mid-latitude cyclone are shown in Fig. 1.9. The first diagram illustrates in plan view the isobars of a steady-state condition at the polar front between contrasting air masses. The succeeding diagrams show the sequential stages in the average life of a depression. A slight perturbation caused by irregular surface conditions, or perhaps a disturbance in the lower stratosphere, results in a shallow wave developing in the frontal zone. The initial wave, moving along the line of the front at $15-20 \, \mathrm{m \, s^{-1}}$ (30–40 knots), may travel up to 1000 km without further development. If the wavelength is more than 500 km, the wave usually increases in amplitude, warm air pushes into the cold air mass and active fronts are formed.

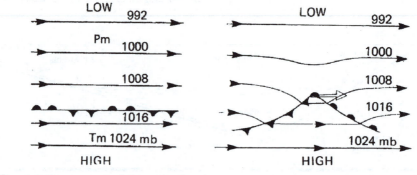

(a) Quasi-stationary part of the polar front

(b) Frontal wave

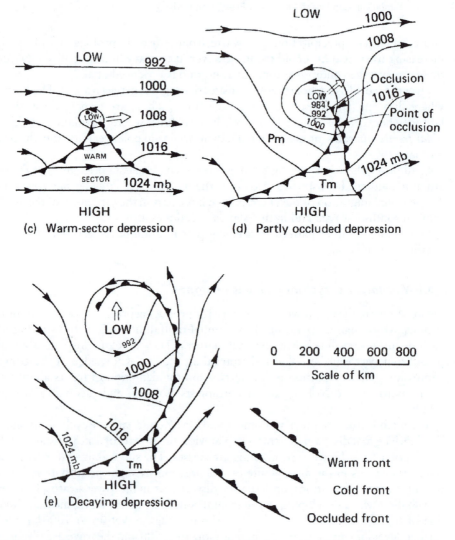

(c) Warm-sector depression

(d) Partly occluded depression

(e) Decaying depression

0   200   400   600   800
Scale of km

Warm front

Cold front

Occluded front

Fig. 1.9 Life cycle of a model occluding depression. (Adapted from Met Office (1962) *A Course in Elementary Meteorology*, Her Majesty's Stationary Office.)

As a result, the air pressure falls and a 'cell' of low pressure becomes trapped within the cold air mass. Gradually the cold front overtakes the warm front, the warm air is forced aloft, and the depression becomes *occluded*. The low-pressure centre then begins to fill and the depression dies as the pressure rises. On average, the sequence of growth from the first perturbation of the frontal zone to the occlusion takes 3–4 days. Precipitation usually occurs along the fronts and, in a very active depression, large amounts can be produced by the occlusion, especially if its speed of passage is retarded by increased friction at the Earth's surface. At all stages, orographic influences can increase the rainfall as the depression crosses land areas. A range of mountains can delay the passage of a front and cause longer periods of rainfall. In addition, if mountains delay the passage of a warm front, the occlusion of the depression may be speeded up.

1.6.2 Waves in the easterlies and tropical cyclones

Small disturbances are generated in the trade wind belts in latitudes 5–25° both north and south of the Equator. Irregular wind patterns showing as isobaric waves on a weather map develop in the tropical maritime air masses on the equatorial side of the subtropical high-pressure areas. They have been studied most in the Atlantic Ocean to the north of the South American continent. A typical easterly wave is shown in Fig. 1.10. A trough of low pressure is shown moving westwards on the southern flanks of the Azores anticyclone. The length of the wave extends over 15–20° longitude (1500 km) and, moving with an average speed of 6.7 m s^{-1} (13 knots), takes 3–4 days to pass. The weather sequence associated with the wave is indicated beneath

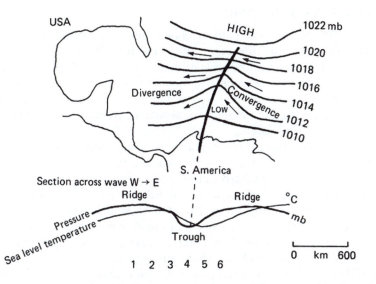

Fig. 1.10 A wave in the easterlies. Weather sequence: 1 – small Cu, no pp.; 2 – Cu, a few build-ups, haze, no pp.; 3 – larger Cu, Ci and Ac, better visibility, pr ... pr ... pr (showers); 4 – very large Cu, overcast Ci Ac, prpr or rr (continuous rain); 5 – Cu and Cb, Sc, As, Ac, Ci, pRpR (heavy showers), (thunderstorm); 6 – large Cu, occasional Cb, some Sc, Ac, Ci, pRpr – prpr; Cu = Cumulus; Cb = Cumulonimbus; Ci = Cirrus; Ac = Altocumulus; As = Altostratus; Sc = Stratucumulus.

the diagram. In the tropics, the cloud-forming activity from such disturbances is vigorous and subsequent rainfall can be very heavy: up to 300 mm may fall in 24 h.

As in mid-latitudes, the wave may simply pass by and gradually die away, but the low pressure may *deepen* with the formation of a closed circulation with encircling winds. The cyclonic circulation may simply continue as a shallow depression giving increased precipitation but nothing much else. However, rapidly deepening pressure below 1000 mb usually generates hurricane-force winds blowing round a small centre of 30–50 km radius, known as the *eye*. At its mature stage, a *hurricane* centre may have a pressure of less than 950 mb. Eventually the circulation spreads to a radius of about 300 km and the winds decline. Copious rainfall can occur with the passage of a hurricane; record amounts have been measured in the region of Southeast Asia, where the effects of the storms have been accentuated by orography. However, the rainfall is difficult to measure in such high winds. In fact, slower moving storms usually give the higher records. Hurricanes in the region of Central America often turn northwards over the United States and die out over land as they lose their moisture. On rare occasions disturbances moving along the eastern coastal areas of the United States are carried into westerly air-streams and become vigorous mid-latitude depressions.

Hurricanes tend to be seasonal events occurring in late summer when the sea temperatures in the areas where they form are at a maximum. They are called *typhoons* in the China Seas and *cyclones* in the Indian Ocean and off the coasts of Australasia. These tropical disturbances develop in well-defined areas and usually follow regular tracks; an important fact when assessing extreme rainfalls in tropical regions (McGregor and Nieuwolt, 1998).

1.6.3 Convectional precipitation

A great deal of the precipitation in the tropics is caused by local conditions that cannot be plotted on the world's weather maps. When a tropical maritime air mass moves over land at a higher temperature, the air is heated and forced to rise by convection. Very deep cumulus clouds form, becoming cumulonimbus extending up to the tropopause. Fig. 1.11 shows the stages in the life cycle of a typical cumulonimbus. Sometimes these occur in isolation, but more usually several such convective cells grow together and the sky is completely overcast.

The development of convective cells is a regular daily feature of the weather throughout the year in many parts of the tropics, although they do not always provide rain. Cumulus clouds may be produced but evaporate again when the air ceases to rise. With greater vertical air velocities, a large supply of moisture is carried upwards. As it cools to condensation temperatures, rainfall of great intensity occurs. In extreme conditions, hail is formed by the sequential movement of particles up and down in the cloud, freezing in the upper layers and increasing in size by gathering up further moisture. As the rain and hail fall, they cause vigorous down draughts, and when these exceed the vertical movements, the supply of moisture is reduced, condensation diminishes and precipitation gradually dies away. Thunder and lightning are common features of convectional storms with the interaction of opposing electrical charges in the clouds. The atmospheric pressure typically is irregular during the course of a storm.

Convectional activity is not confined to the tropics; it is a common local rain-forming phenomenon in higher latitudes, particularly in the summer. Recent studies have shown

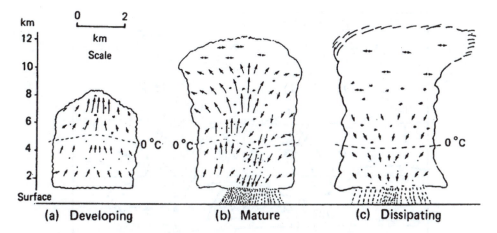

Fig. 1.11 Convective cells – stages in the life cycle. Time scales: (a) approximately 20 min; (b) approximately 20 min, heavy rain and hail, thunder may develop; (c) 30 min to 2 h, rainfall intensity decreasing. Total life cycle 1–2 h.↔, ice; *, snow; ⋯⋯, rain and hail; ↑↓, winds.

that convection takes place along frontal zones thus adding to rainfall intensities. Wherever strong convectional forces act on warm moist air, rain is likely to form and it is usually of high intensity over a limited area.

1.6.4 Monsoons

Monsoons are weather patterns of a seasonal nature caused by widespread changes in atmospheric pressure. The most familiar example is the monsoon of Southeast Asia where the dry, cool or cold winter winds blowing outwards from the Eurasian anticyclone are replaced in summer by warm or hot winds carrying moist air from the surrounding oceans being drawn into a low-pressure area over northern India. The seasonal movements of the ITCZ play a large part in the development and characteristics of the weather conditions in the monsoon areas. The circulation of the whole atmosphere has a direct bearing on the migration of the ITCZ, but in general the regularity of the onset of the rainy seasons is a marked feature of the monsoon. Precipitation, governed by the changing seasonal winds, can be caused by confrontation of differing air masses, low-pressure disturbances, convection and orographic effects. A map of the monsoon areas is shown in Fig. 1.12. Actual quantities of rain vary, but as in most tropical and semi-tropical countries, intensities are high (McGregor and Nieuwolt, 1998).

Further explanation of the processes producing the weather systems described is given in Holton (2004).

1.7 Climate

Following the appreciation of the meteorological mechanisms that affect evaporation and produce precipitation, it is pertinent to consider these hydrological processes

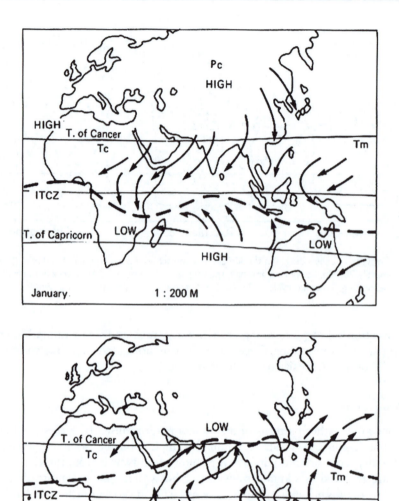

Fig. 1.12 Monsoon lands – pressure systems and winds.

on a longer time scale. Evaporation was presented as an instantaneous process. The precipitation-forming mechanisms extended into weather patterns that may last up to about a week. The study of climate is based on *average weather conditions*, specified usually by measures of temperature and precipitation over one or more months though other phenomena may also be aggregated. Statistics gathered for each month over a period of years and averaged give a representation of the climate of the location.

Table 1.7 Köppen climate classification

		Estimated percentages	
		Land surfaces	Total surface
A	Tropical rain climates – forests	20	36
B	Arid climates	26	11
C	Warm temperate rain climates – trees	15	27
D	Boreal forest and snow climates	21	7
E	Treeless cold snow climates	17	19

The most renowned classification of climates is that of Köppen who categorized climates according to their effect on vegetation.

The major groupings are given in Table 1.7. Subdivisions of these main groups are defined by thresholds of temperature and rainfall values; the details are given in most books of climatology. Their geographical distribution is shown in Fig. 1.13, though Peel *et al.* (2007) provide a revised (colour) version of this map.

These broad definitions of climatic regions are built up from the instrumental records of observing stations which are thus providing sample statistics representing conditions over varying areas. Such meteorological records have only been made with any reliability since the advancing development of instruments in the seventeenth century (Manley, 1970), and world coverage was limited until the late nineteenth century.

Before instrumental records, knowledge of the climate of different regions has been built up by the study of what is now called proxy data. For example, in the UK the proportion of certain tree pollens found in layers of lake sediments or upland peats give indications of the existence of tree cover in earlier times. Similarly, varying layers of clays and silts in surface deposits, as in Sweden, help to differentiate between warm and cold periods. In the western United States, the study of growth rings in the trunks of very old trees, allow climatologists to extend climatic information to periods before instrumental observations. On the global scene, the analyses of deep-sea sediments and ice cores are of increasing importance in the assemblage of climate knowledge.

In addition, archaeological and historical records of transient events, such as the extent of sea ice round the Poles, the fluctuation of mountain glaciers and even the variation of man's activities in the extent of vine growing and the abundance of the wheat harvests, all contribute clues to the climate of former times.

The assimilation and interpretation of such variable information gathered worldwide has occupied climatologists for many years and a broadly agreed sequence of climatic events has been established, aided by the findings of the geologists. However, the worldwide coverage of climatic information before this century was far from representative of all land regions and even less was known of the much larger oceanic areas. The recent concern over man-made changes in the composition of the atmosphere and the increasing ability to model changes in climate has led scientists to study the dynamic components contributing to climate as distinct from current weather, and the resultant impacts on hydrology (Bates *et al.*, 2008). Detailed discussion of climate change impacts on hydrology are presented within Chapter 19.

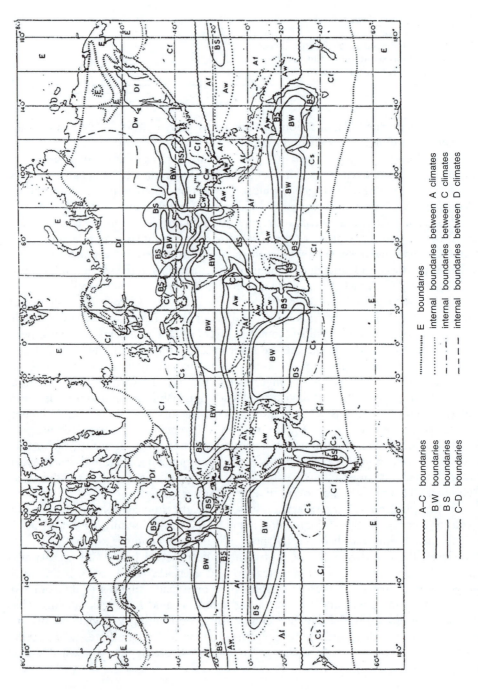

Fig. 1.13 Köppen's world classification of climates.

A–C boundaries
BW boundaries
B S boundaries
C–D boundaries

E boundaries
internal boundaries between A climates
internal boundaries between C climates
internal boundaries between D climates

Note

1 Runoff is the river discharge ($m^3 s^{-1}$) per unit catchment area, hence has the units of $m s^{-1}$ or $mm h^{-1}$.

References

Bates, B. C., Kundzewicz, Z. W., Wu, S. and Palutikof, J. P. (eds) (2008) *Climate Change and Water*. Technical Paper of the Intergovernmental Panel on Climate Change, IPCC Secretariat, Geneva, 210 pp.

Beven, K. J. (2004) Robert Horton's perceptual model of infiltration. *Hydrological Processes* 18, 3447–3460.

Beven, K. J. (2006) *Benchmark Papers in Storm Runoff Generation*. IAHS Press, Wallingford.

Biswas, A. K. (1970) *History of Hydrology*. North-Holland Pub. Co., Amsterdam, 336 pp.

Brustseart, W. (2005) *Hydrology – An Introduction*. Cambridge University Press, Cambridge.

Bunting, B. T. (1961) The role of seepage moisture in soil formation, slope development and stream initiation. *American Journal of Science* 259, 503–518.

Dooge, J. C. I. (1959) *Un bilan hydrologique au XVIIe siecle*. Houille Blanche, 14e annee No. 6, 799–807.

Dunne, T. (1978) Field studies of hillslope flow processes. In: Kirkby, M. J. (ed.) *Hillslope Hydrology*, John Wiley & Sons, pp. 227–294.

Gleick, P. H. (1996) Water resources. In: Schneider, S. H. (ed.) *Encyclopedia of Climate and Weather*, Vol. 2. Oxford University Press, New York, pp. 817–823.

Holton, J. R. (2004) *An Introduction to Dynamical Meteorology*, 4th edn. Academic Press, San Diego, 535 pp.

Manley, G. (1970) The climate of the British Isles. In: Wallen, C. C. (ed.) *Climates of Northern and Western Europe*, Elsevier, Amsterdam, 81–133. (*World Survey of Climatology*, Vol. 5)

Mason, B. J. (1975) *Clouds, Rain and Rainmaking*, 2nd edn. Cambridge University Press, Cambridge, 189 pp.

McGregor, G. R. and Nieuwolt, S. (1998) *Tropical Climatology: An Introduction to the Climates of the Low Latitudes*. Wiley, Chichester, 339 pp.

Met Office (1978) *A Course in Elementary Meteorology*, 2nd edn. HMSO, London, 208 pp.

Peel, M. C., Finlayson, B. L. and McMahon, T. A. (2007). Updated world map of the Köppen–Geiger climate classification. *Hydrology and Earth System Science* 1, 1633–1644.

Sklash, M. G. and Farvolden, R. N. (1979) The role of groundwater in storm runoff. *Journal of Hydrology* 43, 45–65.

Sklash, M. G., Beven, K. J., Gilman, K. and Darling, W. G. (1996) Isotope studies of pipeflow in Plynlimon, Wales, UK. *Hydrological Processes* 10, 921–944.

Strangeways, I. (2007) *Precipitation: Theory, Measurement and Distribution*. Cambridge University Press, Cambridge.

Sumner, G. (1988) *Precipitation: Process and Analysis*. John Wiley & Sons, Chichester, 455 pp.

Weiler, M. and McDonnell, J. (2004) Virtual experiments: a new approach for improving process conceptualization in hillslope hydrology. *Journal of Hydrology* 285, 3–18.

Weyman, D. R. (1970) Throughflow on hillslopes and its relation to the stream hydrograph. *International Association of Scientific Hydrology Bulletin* 15, 25–33.

Chapter 2

Hydrometric networks

The concept of the hydrological cycle forms the basis for the hydrologist's understanding of the sources of water at or under the Earth's surface and its consequent movement by various pathways back to the principal storage in the oceans. Two of the greatest problems for the hydrologist are quantifying the amount of water in the different phases in the cycle and evaluating the rate of transfer of water from one phase to another within the cycle. Thus measurement of the components of the cycle is a major objective of the engineering hydrologist and scientific hydrologist.

Nationwide schemes to measure hydrological variables are now considered essential for the development and management of the water resources of a country. As a result, responsibility for measurement stations is focused on central or regional government agencies and detailed considerations are afforded to the planning of hydrological measurements. Cost–benefit assessments are also being made on the effectiveness of data gathering, and hence scientific planning is being recommended to ensure optimum networks to provide the required information.

2.1 Gauging networks

One of the main activities stimulated by the International Hydrological Decade (IHD, 1965–1974) was the consideration of hydrological network design, a subject that, it was felt, had been previously neglected (Rodda, 1969). It was recognized that most networks, even in developed countries, were inadequate to provide the data required for the increasing need of hydrologists charged with the task of evaluating water resources for expanding populations. While the situation has improved considerably within developed countries, the poor coverage of hydrometric networks within many tropical countries remains a concern.

Before approaching the problem, it is pertinent to define a network. Langbein (1965) gave a broad definition: 'A network is an organized system for the collection of information of a specific kind: that is, each station, point or region of observation must fill one or more definite niches in either space or time'.

The design of the optimum hydrometric network must be based on quantified objectives wherever possible, with costs and benefits included in the design procedure. One approach is the evaluation of the worth of the data collected, which sometimes means realizing the benefits lost through lack of data. Closely connected with network design and data collection is an appreciation of the errors within hydrometric data (Herschy, 1999).

2.2 Design considerations

There are several well-defined stages in the design of a network of gauging stations for the measurement of hydrological variables. The first comprises initial background research on the location and known characteristics of the area to be studied. The size of the area and whether it is a political entity or a natural drainage basin are of prime importance. When assessing the design problem, it is advisable to think in terms of natural catchment areas even if the total area is defined by political boundaries. The physical features of the area should be studied. These include the drainage pattern, the surface relief (altitudinal differences), the geological structure and the vegetation. The general features of the climate should be noted; seasonal differences in temperature and precipitation can be identified from good atlases or standard climatological texts. The characteristics of the precipitation also affect network design and the principal meteorological causes of the rainfall or snowfall should be investigated.

The second stage in network design involves the practical planning. Existing measuring stations should be identified, visited for site inspection and to determine observational practice, and all available data assembled. The station sites should be plotted on a topographical map of the area or, if the area is too large, one overall locational map should be made and separate topographical maps compiled for individual catchment areas. The distribution of the measuring stations should be studied with regard to physical features and data requirements, and new sites chosen to fill in any gaps or provide more detailed information for special purposes. The number of new gauging sites required depends on the density of stations considered to be an optimum for the area. (Indications of desirable station densities are given in the following sections.) Any new sites in the network are chosen on the map, but then they must be identified on the ground. Visits to proposed locations are essential for detailed planning and selected sites may have to be adjusted to accord with ground conditions.

The third stage involves the detailed planning and design of required installations on the new sites. These vary in complexity according to the hydrological variable to be measured, ranging from the simple siting of a single storage rain gauge to the detailed designing of a compound weir for stream measurements or the drilling of boreholes for monitoring groundwater levels. The costing of the hydrometric scheme is usually done at this stage and when this is approved and the finance is available, steps can be taken to execute the designed scheme.

A procedure that may be carried out at any stage is the testing of the validity of the data produced by the network with or without any new stations, provided that there are enough measurements available from the existing measurement stations to allow significant statistical analyses. These may take various forms depending on the variability of the measurements being tested. The worth of data produced is now an important factor in network design, but such cost–benefit evaluation is complicated by the many uses made of the data and by the unknown applications that may arise in the future.

The ideal hydrometric scheme includes plans for the measurement of all the many different hydrological variables, including water quality. Designed networks of water quality monitoring stations are now being established in conjunction with arrangements for measurements of quantity. In the following sections, further particulars of network design for the more usual variables, precipitation, evaporation, overland flow, subsurface flow, river flow and groundwater will be given.

2.3 Precipitation networks

The design of a network of precipitation gauging stations is of major importance to the hydrologist since it is intended to provide a measure of the water input to the river catchment system. The rainfall input is irregularly distributed both over the catchment area and in time. Another consideration in precipitation network design must be the rainfall type as demonstrated in the areal rainfall errors obtained over a catchment of 500 km^2 having ten gauges (Table 2.1). This also shows that a higher density of gauges is necessary to give acceptable areal values on a daily basis.

As a general guide to the density of precipitation stations required, Table 2.2 gives the absolute minimum density for different parts of the world. The more variable the areal distribution of precipitation, as in mountainous areas, the more gauges are needed to give an adequate sample. In regions of low rainfall totals, the occurrence is variable but the infrequent rainfall events tend to be of higher intensity and thus network designers should ensure adequate sampling over areas that would be prone to serious flooding (see Chapter 16).

In the UK, recommended minimum numbers of rain gauges for reservoired moorland areas were laid down by water engineers many years ago (Institute of Water Engineers, 1937; Table 2.3). For real-time operation of an upland impounding reservoir at least one recording gauge would now be recommended to record heavy falls over short periods.

From measurement theory for any random variable, the recommended number of rain gauges should be based on the standard error in rainfall for the particular location

Table 2.1 Areal rainfall errors (%). Reproduced from J.C. Rodda (1969) *Hydrological Network Design – Needs, Problems and Approaches*. WMO/IHD Report No. 12.

Type of rainfall	Day	10 days	Month	Season
Frontal	19	8	4	2
Convective	46	17	10	4

Table 2.2 Minimum density of precipitation stations. (Reproduced from World Meterological Organization (1965) *Guide to Hydrometeorological Practices*)

Region	Minimum density range (km^2/gauge)
Temperate, Mediterranean and tropical zones	
Flat areas	600–900
Mountainous areas	100–250
Small mountainous islands (<20 000 km^2)	25
Arid and polar zones	1500–10 000

Table 2.3 Rain gauge networks for the UK: minimum numbers of rain gauges required in reservoired moorland areas. (Reproduced from Institute of Water Engineers (1937) *Transactions of the Institute of Water Engineers* XLII, 231–259)

Area (km²)	Rain gauges		
	Daily	Monthly	Total
2	1	2	3
4	2	4	6
20	3	7	10
41	4	11	15
81	5	15	20
122	6	19	25
162	8	22	30

and sampling period. The standard error of the mean, σ_{err}, may be used to estimate the closeness of the sample mean to the true mean,

$$\sigma_{err} = \frac{\sigma_s}{\sqrt{n}}$$

where σ_s is the standard deviation, and n is the number of independent observations. Thus, if the rainfall is expected to have a standard deviation of 50 mm, and we wish to approximate the population mean to a standard error of 10 mm, then the number of rain gauges required is 25. This is, however, an underestimate of the minimum gauge density if the spatial pattern of rainfall is sought (rather than just a catchment average). For this, an estimate of the geostatistical structure of the rainfall field could be used to derive the required rain gauge density (Moore *et al.*, 2000; Nour *et al.*, 2006; Villarini *et al.*, 2008; see also Section 9.8.1).

Ground-based radars are used within many nations to measure rainfall patterns. Each rainfall radar measures an area of approximately 15 000 km². The British Isles are currently covered by 16 rainfall radars, with a further four stations proposed (Holehead, Munduff Hill, Old Buckingham, High Moorsley; Section 3.6). The real-time combination of radar and rain gauge data can produce reasonable detail over very large areas (Alpuim and Barbosa, 1999) and is discussed further in Section 3.7.

2.4 Evaporation networks

The assessment of evaporation loss over a catchment by means of local measurements is the next to be considered. The various recommended methods of measurement are outlined in Chapter 4. Since evaporation and transpiration over an area are relatively conservative quantities in the hydrological cycle, fewer gauging stations are required to give areal evaporation estimates than for areal rainfalls. Evaporation and transpiration are dependent on altitude, and thus a network of measuring stations should sample different altitudinal zones within a catchment area. To give some idea of numbers, for experimental catchments covering 18 km² in Wales with a range in altitude of 460 m, where 25 rain gauges are needed for evaluations of areal rainfall with a 2 per cent error, only three or four evaporation stations would be necessary for areal evaporation estimates (McCulloch, 1965). A reliable single station would provide adequate information over a flat plain or plateau. Variations could still occur, however, from differing types of vegetation.

Over 400 towers supporting sensors for measuring evaporation (Chapter 4) have been installed across the globe, with data shared across the FLUXNET network (Baldocchi *et al.*, 2001; Section 6.5). Further, the representativeness of these stations at describing evaporation across continental regions has been assessed recently (Yang *et al.*, 2008). Additionally, the data from this network have been combined with satellite data (e.g. MODIS temperature and humidity data) to derive regional estimates of evaporation (Cleugh *et al.*, 2007).

2.5 Overland flow networks

Networks of overland flow plots are important in the derivation of regional and national estimates of erosion rate. Estimates of the minimum number of overland flow troughs (Section 6.5) needed to characterise the variability in overland flow across landscapes can be obtained from the erosion literature (e.g. Nearing *et al.*, 1999).

2.6 Subsurface water networks

The main purpose of subsurface water investigation for water companies to identify the extent of productive *aquifers* (i.e. groundwater bodies with high porosity and high saturated hydraulic conductivity), to determine their hydraulic properties, and to make arrangements for monitoring the water levels within the aquifers. The sites of existing wells should be noted, as these may have water level records which give the long-term fluctuations of the water table. Furthermore, siting of boreholes for observations must take into account differences in hydraulic properties within an aquifer in addition to variations between aquifers. For example, in the UK, fairly homogeneous Triassic sandstone may require a basic network of one borehole per $260 \, km^2$. However, chalk aquifers can be very variable in saturated hydraulic conductivity and consequently a denser network of boreholes, say 1 in $5 \, km^2$, may be needed to record water level fluctuations (Ineson, 1965). A more comprehensive overview of the design criteria for subsurface monitoring is given in Sara (2006), including the use of geostatistical methods (Carrera *et al.*, 1984).

The monitored borehole network within the UK is detailed within Marsh and Hannaford (2008), and the importance of this network has been enhanced by the development of groundwater resources and the recharge of aquifers depleted by over-pumping.

2.7 River gauging networks

The establishment of river gauging stations is often the most costly item in a hydrometric scheme and, as such, river gauging is usually the responsibility of a national or regional authority, e.g. the Environment Agency of England and Wales. The density of gauging stations depends on the nature of the terrain and for water resources on the population creating a water demand. In England and Wales, it was proposed that there should be 400 primary gauging stations, equivalent to a density of 1 in 375 km^2 (Boulton, 1965). When the Water Resources Act (1963) came into force, the number of gauging stations producing records for publication was approaching this figure, and coordination of further planning of the nationwide network was undertaken by the central authority, the Water Resources Board. The then individual River Authorities were advised on the status of the gauging stations required:

- primary or principal stations defined as permanent stations to measure all ranges of discharges and observations and records to be accurate and complete;
- secondary or subsidiary stations to operate for as long as necessary to obtain a satisfactory correlation with the record of a primary station; their function is to provide hydrological knowledge of streams likely to be used for water supply abstractions; the range of a secondary station should be as comprehensive as possible and the observations and records should be of primary station standard;
- special stations are those serving particular needs, such as reservoir levels and dry weather flow stations for controlling abstractions; these may be permanent or temporary stations according to requirements and they can be related to primary and secondary stations.

Currently within the UK, a combination of the Environment Agency, the Scottish Environmental Protection Agency, the Rivers Agency of Northern Ireland and many water companies maintain the network of (primary, secondary and special) river gauging stations and associated data. The UK Hydrometric Register (Marsh and Hannaford, 2008) currently reports data for around 1500 river gauging stations in the UK; this is equivalent to a density of 1 in 163 km^2.

For water resources evaluation, 20 years of records from a secondary station would suffice to give an acceptable correlation coefficient between the monthly discharges of the secondary station and a primary station. Then the secondary station could be discontinued. Extension of discharge information for a short-term secondary station can also be made by relating the *flow duration curves* (Section 11.4).

The ultimate design and establishment of a river gauging network depends on the data requirements, the hydrological characteristics of the area and the achievement of an acceptable cost–benefit relationship for the scheme.

There are increasing efforts to make data from river gauging stations available internationally. The Global Runoff Data Centre (GRDC),[1] maintained by the German Federal Institute of Hydrology (BfG), currently holds daily or monthly river-flow data for 7332 river gauging stations in 156 countries across the globe.

The following chapters of Part I dealing with hydrological measurements describe the methods of measurement of the different hydrological variables and the instruments in most common use in the UK.

Note

1 See http://www.bafg.de/GRDC/EN/Home/homepage__node.html

References

Alpuim, T. and Barbosa, S. (1999) The Kalman filter in the estimation of area precipitation. *Environmetrics*, 10, 377–394.

Baldocchi, D., Falge, E., Gu, L., Olson, R., Hollinger, D., Running, S., Anthoni, P., Bernhofer, Ch., Davis, K., Fuentes, J., Goldstein, A., Katul, G., Law, B., Lee, X., Malhi, Y., Meyers, T., Munger, J. W., Oechel, W., Pilegaard, K., Schmid, H. P., Valentini, R., Verma, S., Vesala, T., Wilson, K. and Wofsy, S. (2001) FLUXNET: a new tool to study the temporal and spatial variability of ecosystem-scale carbon dioxide, water vapour and energy flux densities. *Bulletin of the American Meteorological Society* 82, 2415–2435.

Boulton, A. G. (1965) *Surface Water, Basic Principles Related to Network Design.* IASH Pub. No. 67, 234–244.

Carrera, J., Usunoff, E. and Szidarovszky, F. (1984) A method for optimal observation network design for groundwater management. *Journal of Hydrology* 73, 147–163.

Cleugh, H. A., Leuning, R., Mu, Q. and Running, S. W. (2007) Regional evaporation estimates from flux tower and MODIS satellite data. *Remote Sensing of Environment* 106, 285–304 (doi: 10.1016/j.rse.2006.07.007).

Herschy, R. W. (1999) *Hydrometry: Principles and Practices,* 2nd edn. John Wiley, New York, 376 pp.

Ineson, J. (1965) *Ground Water Principles of Network Design.* IASH Pub. No. 68, 476–483.

Institute of Water Engineers (1937) Report of Joint Committee to Consider Methods of Determining General Rainfall Over Any Area, Transactions of the Institute of Water Engineers, XLII, 231–259.

Langbein, W. B. (1965) *National Networks of Hydrological Data.* IASH Pub. No. 67, 5–11.

Marsh, T. and Hannaford, J. (eds) (2008) UK Hydrographic Register. Hydrological data UK series. Centre for Ecology & Hydrology, Wallingford, 200pp.

McCulloch, J. S. G. (1965) *Hydrological Networks for Measurement of Evaporation and Soil Moisture.* IASH Pub. No. 68, 579–584.

Moore, R. J., Jones, D. A., Cox, D. R. and Isham, V. S. (2000) Design of the HYREX raingauge network. *Hydrology and Earth System Sciences* 4, 523–530.

Nearing, M. A., Govers, G. and Norton, L. D. (1999) Variability in soil erosion data from replicated plots. *Soil Science Society of America Journal* 63, 1829–1835.

Nour, M. H., Smit, D. W. and Gamal El-Din, M. (2006) Geostatistical mapping of precipitation: implications for rain gauge network design. *Water Science and Technology* 53, 101–110.

Rodda, J. C. (1969) *Hydrological Network Design – Needs, Problems and Approaches.* WMO/IHD Report No. 12, 57 pp.

Sara, M. N. (2006) Groundwater monitoring system design. In: Nielsen, D. M. (ed.) *Practical Handbook of Environmental Site Characterisation and Ground-Water Monitoring,* 2nd edn. Taylor & Francis, Boca Raton, 517–572.

Villarini, G., Mandapaka, P. V., Krajewski, W. F. and Moore, R. J. (2008) Rainfall and sampling uncertainties: a rain gauge perspective. *Journal of Geophysical Research - Atmospheres* 113, D11102. 12, pp. 10.1029/2007JD009214.

World Meteorological Organization (WMO; 1965) *Guide to Hydrometeorological Practices.* WMO, Geneva.

Yang, F., Zhu, A., Ichii, K., White, M. A., Hashimoto, H. and Nemani, R. R. (2008) Assessing the representativeness of the AmeriFlux network using MODIS and GOES data. *Journal of Geophysical Research* 113(g4), G04036.

Chapter 3

Precipitation

Of all the components of the hydrological cycle, the elements of precipitation, particularly rain and snow, are the most commonly measured. Sevruk and Klemm (1989) have estimated that there are 150 000 storage rain gauges in use worldwide. It would appear to be a straightforward procedure to catch rain as it falls and the depth of snow lying can be determined easily by readings on a graduated rod. People have been making these simple measurements for more than 2000 years; indeed, the first recorded mention of rainfall measurement came from India as early as 400 BC. The first rain gauges were used in Korea in the 1400s AD (as a means to plan farming and set taxes), and 200 years later, in ca. 1680 in England, Sir Christopher Wren and Robert Hooke described designs for the self-recording rain gauge.

Climatologists and water engineers appreciate that making an acceptable precipitation measurement is not as easy as it may first appear. It is not physically possible to catch all the rainfall or snowfall over a catchment; the precipitation over the area can only be *sampled* by rain gauges. The measurements are made at several selected points representative of the area and values of the total volume (Ml) or equivalent areal depth (mm) over the catchment are calculated later. Such are the problems in obtaining representative samples of the precipitation reaching the ground that, over the years, a comprehensive set of rules has evolved. The principal aim of these rules is to ensure that all measurements are comparable and consistent. All observers are recommended to use standard instruments installed uniformly in representative locations and to adopt regular observational procedures (as set within the particular country).

Many investigations carried out in England into the problems of rainfall measurement owe their origin to the enthusiasm of one man, G. J. Symons. Symons, a civil servant in the Meteorological Department of the Board of Trade in the 1850s, instigated and encouraged formal scientific experiments by such volunteers as retired army officers or clergymen whose spare time interests included observations of the weather and measurements of meteorological variables (Mill, 1901). The results of this work were incorporated by Symons into his *Rules for Rainfall Observers*. Symons' rules continue to form the basis of the practice of precipitation measurement in the UK today (Met Office, 2006).

The Met Office, which in 1919 inherited the advisory functions of Symons and his successors in the British Rainfall Organization, has approved instruments of several designs having the salient features recommended as a result of the early experiments. These include various types of storage rain gauge and automatic rain gauges. For the assessment of water resources, monthly totals may suffice; for evaluating flood peaks in urban areas (Chapters 16 and 18), rainfall intensities over an hour or even minutes could be required, so automatic rain gauges are used.

3.1 Non-recording (storage) rain gauges

Rain gauges vary in capacity depending on whether they are to be read daily or monthly. The period most generally sampled is the day, and most precipitation measurements are the accumulated depths of water caught in simple storage gauges over 24 h.

For many years, the UK's recognized standard daily rain gauge has been the Met Office Mark II instrument (Fig. 3.1a; Met Office, 1980; British Standard 7843, 1996). The gauge has a sampling orifice of diameter 127 mm. The 12.7 mm rim is made of brass, the traditional material for precision instruments, and the sharply tooled knife edge defines a permanent accurate orifice. The Snowdon funnel forming the top part of the gauge has a special design. A straight-sided drop of 102 mm above the funnel prevents losses from out-splash in heavy rain. Sleet and light snowfall also collect readily in the deep funnel and, except in very low temperatures, the melted water runs down to join the rain in the collector. The Snowdon funnel, the main outer casing of the gauge and an inner can are all made of copper, a material that has a smooth surface, wets easily and whose surface, once oxidized, does not change. The inside of the collecting orifice funnel should never be painted, since the paint soon cracks, water adheres to the resulting rough surface and there are subsequent losses by evaporation. The main collector of the rain water is a glass bottle with a narrow neck to limit evaporation losses. The gauge is set into the ground with its rim level at 300 mm above the ground surface, which should ideally be covered with short grass, chippings or gravel to prevent in-splash in heavy rain.

During very wet weather, the rain collected in the bottle may overflow into the inner can. Bottle and can together hold the equivalent of 150 mm rainfall depth.

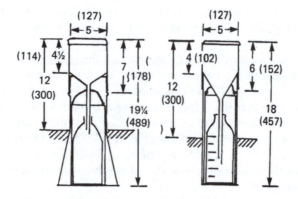

Fig. 3.1 Two daily storage rain gauges in use in the UK: (a) a Met Office Mark II type, and (b) a Snowdon type. Units outside of the parentheses are the original design in inches, while those in parentheses are the equivalent millimetres.

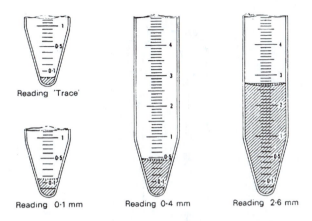

Fig. 3.2 Reading the rain measure. Millimetre graduations. (Adapted from Met Office (1980) *Handbook of Meteorological Instruments*, Vol. 5. Her Majesty's Stationery Office.)

The inner can is easily removed from the outer casing and its contents can be emptied and measured without disturbing the installation.

The gauge is inspected each day at 0900 h GMT, even if it is thought that no precipitation has occurred. Any water in the bottle and inner can is poured into a glass measure (Fig. 3.2) and the reading taken at the lowest point of the meniscus. The glass measure is graduated in relation to the orifice area of the rain gauge and so gives a direct reading of the depth of rain that fell on the area contained by the brass rim. The glass measure has a capacity of 10 mm; if more than 10 mm of rain has fallen, the water in the gauge must be measured in two or more operations. The glass measure is tapered at the bottom so that small quantities can be measured accurately. If no water is found in the gauge and precipitation is known to have fallen, this should be noted as a 'trace' in the records. The glass bottle and inner can should be quite empty before they are returned to the outer case. It is advisable to check the instrument regularly for any signs of external damage, or general wear and tear. Severe frosts can sometimes loosen the joints of the copper casing and, if this is suspected, testing for leaks should be carried out.

The *Snowdon* gauge (Fig. 3.1b), a Met Office Mark I instrument, remains in favour among private observers in the UK, since without the splayed base it is easily maintained in a garden lawn. It is, however, more difficult to keep rigid with the rim level. Globally, the daily storage gauge in most common use is the German *Hellmann* gauge, with over 30 000 gauges of this type in use (Sevruk and Klemm, 1989). This gauge is similar in design to the Snowdon gauge, but with a larger funnel diameter of 159.6 mm.

Monthly rain gauges hold larger quantities of precipitation than daily gauges. The catch is measured using an appropriately graduated glass measure holding 50 mm. Monthly gauges are designed for remote mountain areas and are invaluable on the higher parts of reservoired catchments. Measurements are made on the first day of each month to give the previous month's total and corrections may need to be made

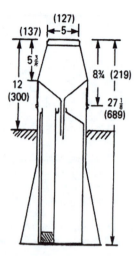

Fig. 3.3 An Octapent monthly storage gauge in use in the UK. Units outside of the parentheses are the original design in inches, while those in parentheses are the equivalent millimetres.

to readings obtained from remote gauges recorded late in the day in wet weather. The *Octapent* monthly rain gauge (Fig. 3.3), a hybrid of the 5-inch and old 8-inch diameter gauges, is made in two sizes with capacities of 685 mm and 1270 mm.

3.2 Recording rain gauges

The need for the continuous recording of precipitation arose from the need to know not just how much rain has fallen, but when it fell and over what period. Numerous instruments have been invented with two main types being widely used: the tilting-siphon rain recorder developed by Dines, and the tipping-bucket gauge, which had its origins with Sir Christopher Wren and Robert Hooke.

The *Dines tilting-syphon* rain recorder (Fig. 3.4) is installed with its rim 500 mm above ground level. The rain falling into the 287 mm diameter funnel is led down to a collecting chamber containing a float. A pen attached to the top of the plastic float marks a chart on a revolving drum driven by clockwork. The collecting chamber is balanced on a knife edge. When there is no rain falling, the pen draws a continuous horizontal line on the chart; during rainfall, the float rises and the pen trace on the chart slopes upwards according to the intensity of the rainfall. When the chamber is full, the pen arm lifts off the top of the chart and the rising float releases a trigger disturbing the balance of the chamber, which tips over and activates the syphon. A counter-weight brings the empty chamber back into the upright position and the pen returns to the bottom of the chart. With double syphon tubes, syphoning should be completed within 8 s, but the rain trap reduces the loss during heavy rainfall. It is recommended, however, that a standard daily storage gauge is installed nearby and that quantities recorded are amended to match the daily total. Each filling of the float chamber is equivalent to 5 mm of precipitation.

Fig. 3.4 The internal mechanism of a tilting-syphon rain gauge. The design is described in Met Office (1982) *Observer's Handbook*, 4th edition. Her Majesty's Stationery Office.

Charts are normally record by the day, but modifications to the instrument can allow a strip chart to be used which gives continuous measurements for as long as a month and which has an extended timescale for intense falls over very short periods. In cold weather, the contents of the float chamber may freeze and special insulation with thermostatically controlled heating equipment, the simplest being a low-wattage bulb, can be installed. The provision of heating assists in the melting of snow, but in very cold weather or during heavy snow, existing low-powered heating devices will not be adequate and there will be a time lag in the melted water being recorded on the chart. Care must be taken to avoid too much heating since evaporation of the melted snow would result in low measurements. Adequate drainage below the gauge should be provided during installation, especially in heavy clay soils and in areas liable to heavy storms, for the syphon system will fail if the delivery pipe enters flood water in the soak away. A model for use in the tropics has a 128-mm diameter receiving aperture and the filling of the float chamber represents 25 mm on the chart. Despite the increased use of tipping-bucket rain gauges with data loggers, tilting-syphon gauges with charts remain in widespread use throughout the tropics.

The principle of the *tipping-bucket* rain gauge is shown in Fig. 3.5. Rain is led down a funnel into a wedge-shaped bucket of fixed capacity. When full, the bucket tips to empty and a twin adjoining bucket begins to fill. At each tip, a magnet attached to the connecting pivot closes a circuit and the ensuing pulse is recorded on a data logger. The mechanism can be used in a variety of gauges. The 15 g of water in one bucketful represents 1 mm of rain caught in a 150 cm^2 gauge, and 0.2 mm in a 750 cm^2 gauge. A small adjustment allows the tipping buckets to be calibrated precisely. It is advisable to install an adjacent storage gauge (sometimes called a 'check gauge') so that a day's or month's total can be measured if the recording mechanism fails. Ceramic resistors connected to a high-capacity battery can be used to reduce the likelihood of the mechanism freezing during cold weather. Other errors specific to tipping-bucket rain gauges are detailed within Hodgkinson *et al.* (2005).

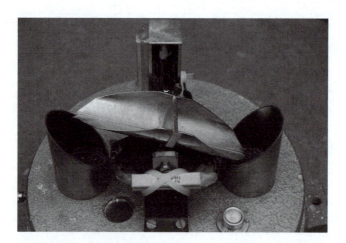

Fig. 3.5 The internal mechanism of a tipping bucket rain gauge. The design is described in BS7843. 1996. Guide to the acquisition and management of precipitation data. British Standards Institution.

Many other types of automatic rain gauge are either in development or use within the UK, including *electronic weighing* rain gauges, *capacitance* rain gauges, *drop-counting* rain gauges and *optical present weather detectors*. *Disdrometers* are particularly notable, given that they give raindrop size distribution as well as rainfall intensity, which is useful for rainfall radar calibration and erosion studies (Strangeways, 2003, 2007).

3.3 Siting the rain gauge

Choosing a suitable site for a rain gauge is not easy. The amount measured by the gauge should be representative of the rainfall on the surrounding area. What is actually caught as a sample is the amount that falls over the orifice area of a standard gauge, that is, $150\,cm^2$. Compared with the area of even a small river catchment of $15\,km^2$, for example, this 'point' measurement represents only a 1 in 10^9 fraction of the total catchment area. Thus even a small error in the gauged measurement due to poor siting represents a very substantial volume of water over a catchment.

It is best to find some level ground if possible, definitely avoiding steep hillsides, especially those sloping down towards the prevailing wind. In the UK, the wind comes mainly from westerly directions. A sheltered, but not over-sheltered, site is the ideal (Fig. 3.6). It is advisable to measure the height of sheltering objects in determining the best site, taking into account anticipated growth of surrounding vegetation.

In over-exposed locations on moorlands, plateaus and extensive plains, where natural shelter may be scarce, a turf wall of the kind designed by Hudleston is recommended (Fig. 3.7; Hudleston, 1934). The surrounding small embankment prevents wind eddies, which can inhibit rain drops from falling into an unprotected gauge. The disadvantages of this enclosure are that drifting snow may engulf the gauge and very heavy rain may flood it if there is no drainage channel beneath the wall.

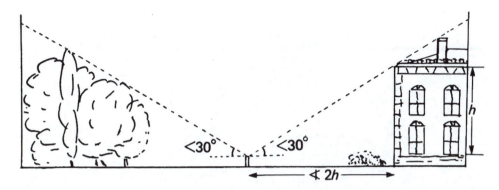

Fig. 3.6 Minimum shelter allowed at a rain gauge site.

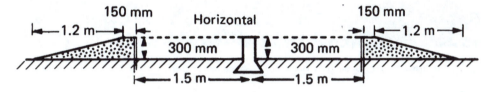

Fig. 3.7 A Hudleston turf wall installation.

It has always been appreciated in the UK that the compromise setting of the gauge rim 300 mm above the ground surface is not altogether satisfactory. Hydrologists have led the move to require gauges to be set with the rim at ground level and various methods to prevent in-splash have been developed. The most acceptable installation is one in which the gauge is set in the centre of a pit about 1 m square, which is then covered with a metal grid with a hole in the middle to accommodate the funnel. Ideally, the square grid slats should be less than 1 mm thick at the top edge and be 50 mm deep with 50 mm spacing. This installation is known as an Institute of Hydrology Ground-Level Gauge (Fig. 3.8; Institute of Hydrology, 1977).

It has been shown by several researchers that a standard daily gauge in its conventional setting typically catches 6–8 per cent less rain than a properly installed ground-level gauge, though undercatch can be considerably more on windy mountain slopes (Rodda, 1967; Sevruk and Hamon, 1984). If turf-wall or ground-level installations are not used, then wind speed data should be used to correct rainfall totals dynamically (Sevruk, 1996).

Rain-gauge sites should be examined occasionally to note any possible changes in the exposure of the instrument. Removal of neighbouring trees or the growths of adjacent plants are modifications of the natural surroundings that could affect the rain gauge record. Observers should be encouraged to report any major structural changes to buildings near the gauge because they could result in changing wind patterns in the vicinity of the instrument which could also affect the homogeneity of the catch record. When inconsistencies in a record caused by such changes in the exposure of a gauge are reported or discovered, the data processors make suitable amendments to the measurements (World Meteorological Organization; WMO, 2008).

Fig. 3.8 An Institute of Hydrology Ground-Level Gauge installation.

3.4 Horizontal rain and occult precipitation gauges

On steep mountain slopes, particularly close to ridge tops, a significant proportion of the precipitation can have a horizontal component that is intercepted by trees but poorly measured by conventional rain gauges. Where these mountains slopes are in excess of 1500 to 1200 m in height, the total precipitation comprises a small but measurable quantity of occult precipitation or fog interception. The *modified Juvik fog gauge* can be used to measure both wind-driven horizontal rainfall and fog interception (Frumau *et al.*, 2006); however, there are no permanent installations of these gauges within the UK.

3.5 Snowfall gauges

There are various solid forms of precipitation, and all except hail require the surface air temperature to be lower than about 4°C if they are to reach the ground.

Small quantities of snow, sleet or ice particles fall into a rain gauge and eventually melt to yield their water equivalent. If the snow remains in the collecting funnel, it must be melted to combine the catch with any liquid in the gauge. If practicable, the gauge may be taken indoors to aid melting but any loss by evaporation should be avoided. Alternatively, a quantity of warm water measured in the graduated rain measure for the rain gauge type can be added to the snow in the funnel and this amount subtracted from the measured total.

Large wind errors arise with the use of rain gauges to measure snowfall. The WMO designed the *octagonal, vertical, double-fence shield* (or *double-fence inter-comparison reference*) to reduce this effect and act as the international reference (WMO, 1998).

When snow has accumulated on the ground, its depth can be measured. A representative smooth cover of the ground, not subject to drifting is selected and sample depths are taken with a metre stick held vertically. For a rough estimate of water equivalent, the average snow depth is converted taking 300 mm of *fresh* snow equal to 25 mm of rain. However, the density of fresh snow may range between 50 and $200 \, \mathrm{g \, L^{-1}}$ according to the character of the snow flakes. When compacted snow lies for several days and there are subsequent accumulations, the observer is advised to take density measurements at selected points over the higher parts of important catchment areas. The density of snow increases with compaction to around $300 \, \mathrm{g \, L^{-1}}$. Sample volumes of the snow are taken at different depths (WMO ___ e weighed and the density calculated. Thence the wate___ snow can be obtained. If the snow is ___ to give an overall density or melted t___

A continu___ snow is essential to promote warni___ *ow pillows* can be used continuously___ her and Stewart, 1995). These device___ snow weight and hence accumulated___

3.6 Ground

Unlike satellite___ ___surement. There are several types o___ ___e UK uses a conical radar beam in t___ ___lected and scattered back from the p___ ___een rainfall rate, *R*, and radar reflect___

$$Z = aR^b$$

where *b* varies between 1.4 and 1.7, and *a* varies from 140 for drizzle to 500 for heavy showers. There are, however, many environmental factors, such as *bright band* (from melting atmospheric snow) and permanent ground obstructions that need to be taken into account to provide the correct calibrations. These calibrations incorporate adjustments to observations from tipping-bucket rain gauges and Disdrometers; see Collier (1989) for further discussion. The present operational network of radar stations measuring rainfall in the UK is shown in Fig. 3.9; and the calibrated coloured displays are seen regularly on TV weather programmes.

The detailed rainfall intensity displays are received by the regional offices of the Environment Agency and Scottish Environmental Protection Agency and provide front-line information for flood warning. The central advantage of the rainfall radar method to the hydrologist is that it produces a measure of the rainfall over the whole of a catchment area as it is falling.

3.7 Combined radar and satellite observations

Thermal-infrared and visible wavelength spectrometers mounted on satellite platforms have the ability to measure cloud-top brightness, temperature and texture. These characteristics help distinguish cloud type, and this knowledge can improve the real-time calibration of ground-based radar. Within the Met Office Nimrod system, these spectra from the Meteosat satellite are combined with the ground-based radar

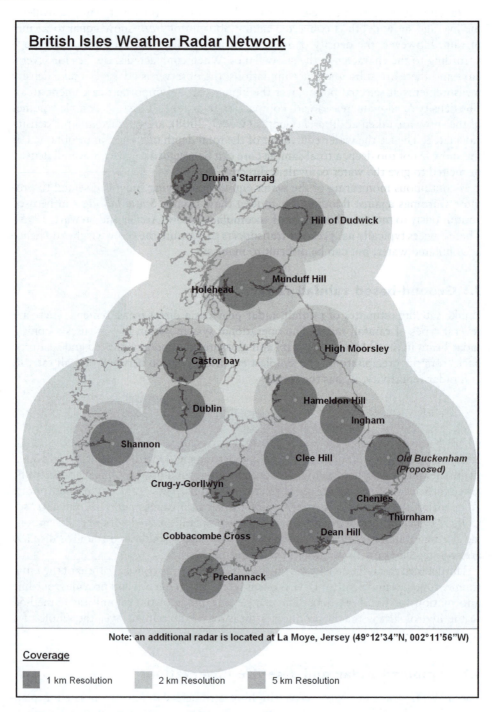

Fig. 3.9 Rainfall radar coverage in the British Isles. (Image provided by and reproduced with permission of the Met Office. © Crown copyright 2009.)

and telemetered tipping-bucket rainfall data to produce 1 km resolution rainfall for the whole of the UK. These are available to hydrologists via the British Atmospheric Data Centre, BADC (http://badc.nerc.ac.uk/data/nimrod/). Additionally, the more specialised Met Office GANDOLF system provides 2 km resolution data for periods when convective thunderstorms are present. These data are also integrated with a nowcast (12-h forecast) modelling system called STEPS (Pierce *et al.*, 2004).

Outside of the UK, Meteosat data have been used to observe rainfall without the use of ground-based radar (Symeonakis *et al.*, 2008). The first radar system designed to measure reflectivity of rainfall from space has been operational since 1997. Mounted on the TRMM (Tropical Rainfall Measuring Mission) satellite, this radar provides rainfall intensity estimates for tropical regions, with a pass of each location every 3 h.

3.8 Net precipitation gauges

The proportion of precipitation that reaches the ground beneath vegetation (i.e. that which is not lost by wet-canopy evaporation; Section 10.3) is called *net precipitation*. This precipitation reaches the ground either by running down vegetation stems, and is called *stemflow*, or drips through or off vegetation canopies as *throughfall*. Stemflow can be measured using a stemflow collar that directs the flow into a tipping-bucket rain gauge (Institute of Hydrology, 1977) or into a larger (1-L) tipping-bucket mechanism. Throughfall can be measured using a network of storage or tipping-bucket rain gauges placed beneath the canopy. The greater spatial variability of stemflow and throughfall in comparison to rainfall means that 50 such gauges are needed for comparison with rainfalls measured in the open or above the canopy. Alternatively, a smaller number of the larger throughfall troughs (4 m in length, 0.1 m width and 0.3 m depth) or plastic-sheet net rainfall gauges (Institute of Hydrology, 1977) can be used.

Note

1 http://nora.nerc.ac.uk/5770/

References

Archer, D. R. and Stewart, D. (1995) The installation and use of a snow pillow to monitor snow water equivalent. *Journal of the CIWEM* 9, 221–230.

British Standard 7843 (1996) *Guide to the Acquisition and Management of Precipitation Data*. British Standards Institution, London.

Collier, C. G. (1989) Applications of Weather Radar Systems. *A guide to uses of radar data in meteorology and hydrology*. Horwood, Chichester, 249pp.

Frumau, A., Bruijnzeel, S. and Tobon, C. (2006) *Hydrological Measurement Protocol for Montane Cloud Forest*. Annex 2. Final Technical Report DFID-FRP Project No. R7991. Free University, Amsterdam.

Hodgkinson, R. A., Pepper, T. J. and Wilson, D. W. (2005) *Evaluation of Tipping Bucket Rain Gauge Performance and Data Quality*. Environment Agency Science Report W6-048/SR, 54pp.

Hudleston, F. (1934) A summary of seven years' experiments with rain gauge shields in exposed positions, 1926–1932 at Hutton John, Penrith. *British Rainfall* 1933, 274–293.

Institute of Hydrology (1977)[1] *Selected Measurement Techniques in Use at the Plynlimon Experimental Catchments*. Institute of Hydrology Report 43. Natural Environment Research Council, Wallingford.

Met Office (1982) *Observer's Handbook*, 4th edn. HMSO, London.

Met Office (1980) *Handbook of Meteorological Instruments*, Vol. 5, 2nd edn, HMSO, London, 34 pp.

Met Office (2006) *SPOT-ON Observers Guide: Precipitation v3 9/06*. Met Office, Exeter, 12 pp.

Mill, H. R. (1901) The development of rainfall measurement in the last 40 years. *British Rainfall* 1900, 23–41.

Pierce, C., Bowler, N., Seed, A., Jones, A., Jones, D. and Moore, R. (2004) Use of a stochastic precipitation nowcast scheme for fluvial flood warning and prediction. Success stories in radar hydrology. *6th International Conference on Hydrological Applications of Weather Radar, Melbourne, Australia, 2–4 February 2004*. Bureau of Meteorology, Melbourne.

Rodda, J. C. (1967) The systematic error in rainfall measurement. *Journal of the Institute of Water Engineers* 21, 173–177.

Sevruk, B. (1996) Adjustment of tipping bucket precipitation gauge measurements. *Atmospheric Research* 42, 237–246.

Sevruk, B. and Hamon, W. R. (1984) *International Comparisons of National Precipitation Gauges with a Reference Pit Gauge*. WMO Instruments and Observing Methods, Report No. 17, WMO/TD, No. 38.

Sevruk, B. and Klemm, S. (1989) Types of standard precipitation gauges. In: *WMO/IAHS/ETH International Workshop on Precipitation Measurement*, St Moritz.

Symeonakis, E., Bonifacio, R. and Drake, N. (2008) A comparison of rainfall estimation techniques for sub-Saharan Africa. *International Journal of Applied Earth Observation and Geoinformation* 11, 15—26.

Strangeways, I. (2003) *Measuring the Natural Environment*, 2nd edn. Cambridge University Press, Cambridge.

Strangeways, I. (2007) *Precipitation: Theory, Measurement and Distribution*, 2nd edn. Cambridge University Press, Cambridge.

World Meteorological Organisation (WMO; 1998) *WMO Solid Precipitation Intercomparison Measurement*. WMO/TD, No. 872. WMO, Geneva.

World Meteorological Organisation (2008) *Guide to Meteorological Instruments and Methods of Observation, 1997*. WMO-No. 8, 7th edn. WMO, Geneva, 681.

Chapter 4

Evaporation

Of the several phases in the hydrological cycle, that of evaporation is one of the most difficult to quantify. Certainly, it is difficult to define the unseen amounts of water stored or moving underground, but above the ground surface, the great complexities of evaporation make it an even more elusive quantity to define; yet evaporation can account for the large differences that occur between incoming precipitation and water available in the rivers. In the UK, if annual totals are considered, evaporation would appear to deprive parts of south-east England of all of its rainfall; in actual practice, evaporation amounts vary seasonally and surplus surface water feeds the rivers in winter. In hotter climates with seasonal rainfall, evaporation losses cause rivers to dry up and river flows are dependent on excessive, heavy rainfall in the wet season.

For the hydrologist, the loss of water by evaporation must be considered from two main aspects. The first, *open water evaporation* from an open water surface, E_o, is the direct transfer of water from lakes, reservoirs and rivers to the atmosphere. This can be relatively easily assessed if the water body has known capacity and does not leak. The second form of evaporation loss is *evapotranspiration*, E_t, and comprises a loss of intercepted precipitation (wet canopy evaporation) and transpired water from plant surfaces. The value of E_t varies according to the type of vegetation, its ability to transpire and to the availability of water in the soil. It is much more difficult to quantify E_t than E_o, since wet-canopy evaporation and transpiration rates can vary considerably over an area and the source of water from the ground for the plants requires careful definition.

Both forms of evaporation, E_o and E_t, are influenced by the local climatic conditions. Although the instrumental measurements are not simple and straightforward as for rainfall, it is a compensating factor that evaporation quantities are less variable in time and therefore more easily predicted than rainfall amounts. With unlimited supplies of water, evaporation is one of the more consistent elements in the hydrological cycle.

4.1 Factors affecting evaporation

The physical process in the change of state from liquid to vapour operates in both E_o and E_t, and thus the general physical conditions influencing evaporation rates are common to both.

(a) Latent heat is required to change a liquid into its gaseous form and, in nature, this is provided primarily by energy from the Sun. The latent heat of vaporisation comes from solar (short-wave) and terrestrial (long-wave) radiation. The incoming *solar radiation* is the dominant source of heat and affects evaporation amounts over the surface of the Earth according to the latitude and season.

(b) The *temperature* of both the *air* and the *evaporating surface* is important and is also dependent on the major energy source, the Sun. The higher the air temperature, the more water vapour it can hold, and similarly, if the temperature of the evaporating water is high, it can more readily vaporise. Thus evaporation amounts are high in tropical climates and tend to be low in the polar regions. Similar contrasts are found between summer and winter evaporation quantities in mid-latitudes.

(c) Directly related to temperature is the water vapour capacity of the air. A measure of the amount of water vapour in the air is given by the vapour pressure, and a unique relationship exists between the saturated vapour pressure and the air temperature. Evaporation is dependent on the *saturation deficit* of the air, which is the amount of water vapour that can be taken up by the air before it becomes saturated. The saturation deficit is given by the difference between the saturation vapour pressure at the air temperature and the actual vapour pressure of the air. Hence more evaporation occurs in inland areas where the air tends to be drier than in coastal regions with damp air from the sea.

(d) As water evaporates, the air above the evaporating surface gradually becomes more humid until finally it is saturated and can hold no more vapour. If the air is moving, however, the amount of evaporation is increased as drier air replaces the humid air. Thus *wind speed* at the surface is an important factor. Evaporation is greater in exposed areas that enjoy plenty of air movement than in sheltered localities where air tends to stagnate. It will be noted that the temperature and wind speed factors may be in conflict in affecting evaporation, since windy areas tend to be cooler and sheltered areas are often warmer. Over a large catchment area, it is the general characteristics of the prevailing air mass that will have the major affect on evaporation (apart from the direct solar radiation). The principal influences on the physical process of evaporation enumerated above are in their turn affected by wider considerations. The following factors outline more generally larger-scale influences.

(e) The prevailing weather pattern indicated by the *atmospheric pressure* affects evaporation. The edge of an anticyclone provides ideal conditions for evaporation as long as some air movement is operating in conjunction with the high air pressure. Low atmospheric pressure usually has associated with it damp unsettled weather in which the air is already well charged with water vapour and conditions are not conducive to aid evaporation.

(f) The nature of the *evaporating surface* affects evaporation by modifying the wind pattern. Over a rough irregular surface, friction reduces wind speed but has a tendency to cause turbulence so, with an induced vertical component in the wind, evaporation is enhanced. Over an open water surface, strong winds cause waves, which provide an increased surface for evaporation in addition to causing turbulence. As wind passes over smooth even surfaces there is little friction and turbulence, and the evaporation is affected predominantly by the horizontal velocity.

Variations in some of the dominant factors operating over different surfaces can result in noticeable changes in evaporation rates over small adjacent areas in short time periods. Diurnal fluctuations are considerable since during the night there is no solar radiation. However, evaporation totals over neighbouring areas show relatively smaller differences over periods of a week or a month.

Evaporation is necessarily dependent on a supply of water and thus the availability of moisture is a crucial factor. With all the other factors acting favourably, once the body of water disappears, then open water evaporation E_o ceases. For E_t, the availability of water is not so easily observed. Plants draw their supply from the soil where the moisture is held under negative pressure potential, and their rate of transpiration is governed by the stomata in the leaves, which act like valves to regulate the passage of water through the pores according to the incidence of light. The pores are closed in darkness and hence transpiration ceases at night. When there is a shortage of water in the soil, the stomata regulate the pores and reduce transpiration. Thus E_t is controlled by soil moisture content and potential (Chapter 5) and the capacity of the plants to transpire, which is conditioned by the meteorological factors.

If there is a continuous supply and the rate of evaporation is unaffected by lack of water, then both E_o and E_t are regulated by the meteorological variables: radiation, saturation deficit and wind speed. The wet canopy evaporation plus transpiration from a vegetated surface with unlimited water supply is known as *potential evaporation* (PE) and it constitutes the maximum possible loss rate due to the prevailing meteorological conditions.

4.2 Measurement of open water evaporation

An indirect measurement of evaporation from open water can be made by taking the difference in storage of a body of water measured at two known times, which gives a measure of the evaporated water over the time interval. If rain has fallen during the time period, then the rainfall quantity must be taken into account. In practice, this *water budget* method is used on two widely differing spatial scales, by measurements at reservoirs and by measurements with specially designed instruments maintained at meteorological stations.

4.2.1 Water budget of reservoirs

The evaporation from a reservoir over a time period is given by:

$$E_o = I - O \pm \Delta S$$

where I = riverflow into the reservoir plus precipitation on to the reservoir surface, O = outflow from the reservoir (i.e. drawoff to supply and overflow) plus subsurface seepage and ΔS = change in reservoir storage. Although water engineers are anxious to assess evaporation losses from their surface reservoir sources of supply, few impounding reservoirs are instrumented to give the measurements required for the water budget equation. In the UK, it is rare to find river flow gauging stations on the streams flowing into reservoirs. There are usually several feeder tributaries, which add to the complexity and cost of total inflow measurements.

The evaluation of outflow from a reservoir is, however, usually made regularly. Measurements of drawoff to supply compensation water releases to the river and overflow are made regularly and the water levels in the reservoir give changes in storage. Difficulties sometimes arise in the assessment of flood flows over spillways while the amount of leakage beneath the dam and through the sides and bottom of a reservoir can only be roughly estimated. The measurement of evaporation from an operating reservoir using the water budget method can only give a broad approximation to water loss unless a thorough knowledge of the different components is available.

A valuable, if old, study of reservoir evaporation was made by Lapworth for the Kempton Park Reservoir from 1956 to 1962 (Lapworth, 1965). During these years, there was no inflow and no outflow from this storage reservoir on the Thames flood plain. Hence, it was expected that E_o would be equal to ΔS, the change in storage with rainfall deducted. In addition to the necessary rainfall measurements, meteorological stations were set up to make the observations required for calculating evaporation (to be described later) so that comparisons of the different methods of evaporation evaluation could be made. The results showed that there were marked seasonal differences between the measured changes in storage E_{Res} and the calculated E_o. The average annual evaporation total over the 7 years was 663 mm E_{Res} and the monthly means are given in Table 4.1. In explanation, the values for E_{Res}, the observed evaporation from the reservoir, have two components, E_o, calculated evaporation due to surface water conditions, plus E_{St} (i.e. $E_{Res} - E_o$), calculated evaporation taking into account by the changing heat storage of the water in the reservoir. As seen in Table 4.1, during the autumn months, evaporation from the reservoir due to heat storage effects is enhanced. This results from heat diffusing and convecting to the surface from the lower water layers, which had absorbed the energy of the summer sun. In the spring, the temperature of the water body is low following the colder winter months and the evaporation is reduced as some of the available incoming energy is absorbed by the cold lower layers. The overall effects on reservoir loss by these seasonal fluctuations of stored energy are dependent on the dimensions of the reservoir. Wide shallow bodies of water are more readily affected by marked seasonal temperature changes, whereas in deep narrow reservoirs, the smaller seasonal fluctuations of stored energy will have less effect on water loss. For an operational impounding reservoir, heat storage is also affected by water temperatures of inflows and outflows. To obtain monthly estimates of E_{Res} in practice, calculated values of E_o can be obtained by other methods, to which are added estimated values of E_{St}, either positive or negative according to season.

Table 4.1 Kempton Park Reservoir evaporation 1956–62 (mm)

	J	F	M	A	M	J	J	A	S	O	N	D
E_{Res}	15	18	28	48	76	94	107	94	74	58	33	18
E_{St}	+3	−3	−13	−20	−18	−23	−5	+5	+15	+23	+20	+10
E_o	12	21	41	68	94	117	112	89	59	35	13	8
$E_{o\ tank}$	5	13	30	56	89	109	103	84	58	33	15	8

4.2.2 Evaporation tanks and pans

Although there may be difficulties in relating the measurements of evaporation from small bodies of water to the real losses from a large reservoir, the advantages in using tanks and pans are numerous. These relatively small instruments, with either circular or square plan sections, are easily managed and can be transported to any required location for simple installation. Originally designed to be kept at meteorological stations where readings are made regularly at a fixed time each day, their operation has been improved by the attachment of self-filling devices and by the continuous measurement of the water level (Chow, 1994). However, the general opinion in the UK is that this method of evaporation measurement is unreliable and the data collected are incapable of being adequately quality controlled. The current tendency is to use calculated estimates of evaporation (Chapter 10).

Of the many evaporimeters used experimentally in the 1860s, the *tank* ascribed to *Symons* became the British standard instrument (Fig. 4.1a). It is a galvanized iron tank, 1.83 m square and 0.61 m deep and set in the ground with the rim 100 mm above ground level. The tank holds about 1.8 m^3, the water level being kept at near ground level and never allowed to fall more than 100 mm below the rim. Measurements of the water level are made daily using a hook gauge attached to a vernier scale and any rainfall measured in the previous 24 h must be added. The depth of evaporation is evaluated as shown in the example in Table 4.2. Records compiled in this way from a British standard tank kept at Kempton Park are given in the last row of Table 4.1. It should be noted that rainfall observations made at 0900 h are normally allocated to the previous day.

The most widely used instrument nowadays is the *American or US Class A pan* (Fig. 4.1b). This is circular with a diameter of 1.21 m and is 255 mm deep. It is set with the base 150 mm above the ground surface on an open wooden frame so that the air circulates freely round and under the pan. The water level in the pan is kept to about 50 mm below the rim. The level is measured daily with a hook gauge and the difference between two readings gives a daily value of evaporation. Alternatively, evaporation can be obtained by bringing the water level in the pan back to a fixed level with a measured amount of water. Again any rainfall must be allowed for. Since the sides of the pan are exposed to the sun, the contained water tends to attain a higher temperature than in pans set in the ground and thus the measured evaporation is higher than otherwise. For example, in the Kempton Park study, a US Class A pan was installed in 1959, and for the 4 years of records, 1959–62, the average annual evaporation measured by the pan was 963 mm, compared with 673 mm from the reservoir and 625 mm from a Symons tank. On an annual basis, the reservoir evaporation was 0.7 times the pan measurement. This factor, 0.7, is known as the *pan coefficient* and its value varies slightly over different climatic regions. If seasonal evaporation values are required, the heat storage effects cause greater differences between US Class A pan and reservoir. Monthly pan coefficients must be obtained and used to give monthly estimates of reservoir evaporation from pan measurements.

Many experiments with modified installations of the US Class A pan have been made in attempts to inhibit the exaggerated evaporation due to the overheating of the water. In India, the outside has been painted white to increase radiation reflection and some studies have recommended setting the pan in the ground. In arid regions, the pan

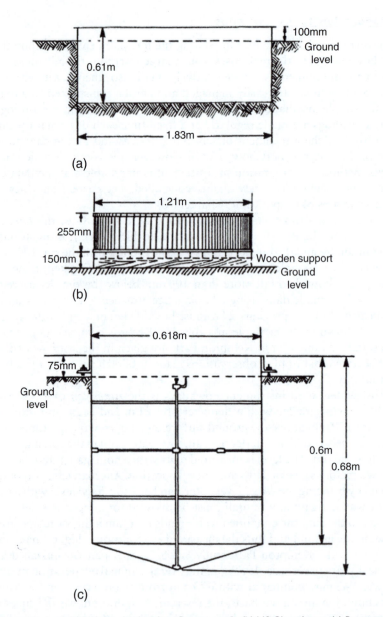

Fig. 4.1 Evaporimeters: (a) UK standard tank or Symons tank; (b) US Class A pan; (c) Russian GGI-3000 tank.

is an attraction to birds and animals and is usually covered with a wire mesh. Such a screen of 25 mm chicken wire mesh, while preventing the bulk removal of water, gave an average reduction of 14 per cent in measurements of monthly mean evaporation over 2 years of measurements in Kenya, owing to reduction in radiation. In using evaporation measurements from a US Class A pan, careful note must be made of its siting and installation. Another instrument that has been accepted by many countries

Table 4.2 Example calculation of the depth of evaporation

Date		Observations made at 0900 h GMT (mm)		Daily records (mm) (from 0900 h)	
		Rain	Hook gauge	Evaporation	Rainfall
June	1	0.3	21.1	1.0	–
	2	–	20.1	1.6	8.9
	3	8.9	27.4	0.7	12.7
	4	12.7	39.4	1.8	–
	5	–	37.6	2.3	–
	6	–	35.3	2.0	–
	7	–	33.3	–	–

is the Russian tank (Fig. 4.1c). The Russian *GGI-3000 tank* has a smaller surface area (0.3 m², 0.618 m diameter) than the other instruments, but has the depth of the British tank (0.60–0.685 m). It is cylindrical with a conical base and is made of galvanized iron. The tank is installed in the ground with the rim about 75 mm above the surface. A comparison between the GGI-3000 tank and the Class A pan was made at Valday (USSR) over 11 summer seasons and an average ratio (tank/pan) for the seasonal evaporation totals was 0.78. Thus like the Symons tank, the GGI-3000 tank gives a measure of E_o of the correct order of magnitude, but a measure of heat storage effects is required before the reservoir loss can be evaluated. Because of the length of records available from networks of evaporation pans, their use remains valuable for identifying long-term trends in evaporation, as within the recent studies published within *Science* and the *Journal of Geophysical Research* (Roderick and Farquhar, 2002; Liu *et al.*, 2004).

4.2.3 Atmometers

These are devices that can give direct measurement of evaporation. A water supply is connected to a porous surface and the amount of evaporation over a designated time period is given by a measure of the change in water stored. Thus $E_o = \Delta S$. This evaporation mechanism has been likened to transpiration from leaves, but as the biological control is not simulated, atmometer data are considered as measures of E_o. It is essential to have a constant instrument exposure to ensure consistent observations and it has been found satisfactory to have atmometers set in a well-ventilated screen as is used for exposing thermometers to register air temperature. Atmometers are simple, inexpensive and easy to operate, but care must be taken to see that the porous surfaces from which the evaporation takes place are kept clean. Two types are described here.

The *Piche* evaporimeter consists of a glass tube 14 mm in diameter and 225 mm long with one end closed. A circular disc of 32-mm diameter absorbent blotting paper is held against the open end by a small circular metal disc with a spring collar. The evaporating surface area is 1300 mm² and this is fed constantly by the water in the tube hung up by its closed end. The tube is graduated to give a direct reading of evaporation (E_o) over a chosen time period, usually a day. The measurement in millimetres is related to the evaporating surface of both sides of the paper. The tube holds an equivalent of 20 mm of evaporation; the water is replenished when necessary. When the Piche evaporimeter is exposed in a standard temperature screen, the annual values have been found to be approximately equivalent to the open water evaporation from a US Class A pan. This type of instrument is used widely within Africa and the Near East.

In the *Bellani* atmometer, the porous surface is provided by a thin ceramic disc, 85 mm in diameter. This is attached to a graduated burette holding the water supply. As with the Piche evaporimeter, the difference in burette readings over a specified time gives the measure of evaporation. Bellani atmometers remain in use by research scientists (Gavilán and Castillo-Llanque, 2009) and are particularly attractive when fitted with electronic sensors (Giambelluca *et al.*, 1992)

4.3 Measurement of evapotranspiration

The measurement of evaporation loss from a vegetated land surface is even more complex than the measurement of loss from an open water body. The extra mechanism of plant transpiration must be added to considerations of water availability and the ability of the atmosphere to absorb and carry away the water vapour. However, similar approaches for the measurement of E_t may be adopted.

4.3.1 Water budget method

To establish the E_t loss from a catchment area draining to a gauging station on a river, the water balance over a selected time period can be evaluated:

$$E_t = P - Q - G \pm \Delta S$$

where P is precipitation, Q river discharge, G discharge of groundwater across basin divides and ΔS change in storage. For a natural catchment, measurements of the precipitation and river discharge may be made satisfactorily with some degree of accuracy, but the measurement of groundwater movement into or out of the drainage area cannot be made easily. In water balance studies, it is usually assumed that the catchment is watertight and that no subsurface movement of water across the defined watershed is occurring. However, if there are aquifers noted in the area, groundwater movements across divides should be investigated (Genereux and Jordan, 2006). The evaluation of change in storage depends on the time period over which the water balance is being made. On an annual basis, the time at which the balance is least affected is chosen so that the water stored in the ground and in surface storage is approximately the same each year and thus in the equation, $\Delta S = 0$. Nevertheless, significant differences in the amount stored may occur from one year to another within certain basins. In the UK, the end of September marks the end of the Water Year when most of the transpiration of the summer season is over and the groundwater replenishment of the winter months is about to begin. If monthly losses are required, then values of ΔS must be obtained. Measurements of soil moisture content can be made regularly each month (Section 6.1) or can be budgeted from potential evaporation calculations and rainfall measurements.

Measurements of the components of the water balance equation for a catchment are even less reliable than for a reservoir and hence its use for evaluating E_t is recommended only for annual values. For shorter time periods, changes in storage should be measured and, in all instances, a thorough knowledge of the catchment area is essential.

4.3.2 Percolation gauges

These are instruments specially designed for measuring evaporation and transpiration from a vegetated surface, E_t, and are comparable with the tanks and pans used for measuring E_o. Similarly, there are very many different designs and, in general, these are regarded as research tools rather than standard instruments to be installed at every meteorological station. A cylindrical or rectangular tank about 1 m deep is filled with a representative soil sample supporting a vegetated surface and is then set in the ground. A pipe from the bottom of the tank leads surplus percolating water to a collecting container. The surface of the gauge should be indistinguishable from the surrounding grass or crop covered ground. A rain gauge is sited nearby and the evapotranspiration is given by the following equation:

$$E_t = \text{Rainfall} - \text{vertical percolation}$$

Percolation gauges do not take into account changes in the soil moisture storage and thus measurements should be made over a time period defined by instances when the gauge is saturated so that any difference in the soil moisture storage is small. Records are generally compiled on a monthly basis in climates with rainfall all the year round.

4.3.3 Weighing lysimeters

By taking into account change in water storage in the ground, lysimeters improve on the E_t measurements of percolation gauges. Compared with the latter, lysimeters are much more complex, more expensive to construct and maintain and therefore have a greater association with research installations.

A large block of undisturbed soil covered by representative vegetation is surrounded by a watertight container driven into the ground. A sealing base with a drain pipe is secured to the bottom of the block and a weighing device established underneath. Then:

$$E_t = \text{Rainfall} - \text{vertical percolation} \pm \text{weight change}$$

All units of measurements are referred to the area of the lysimeter orifice at ground level. The accuracy in the measurement of actual evaporation by lysimetry is dependent on the sensitivity of the weighing mechanism and, to detect small changes in soil moisture storage, large block samples are required. However, once the complications of the elaborate installations have been overcome with a suitable balance, the lysimeters can be easy to run and produce accurate E_t estimates (Vaughan *et al.*, 2007).

4.4 Measurement of meteorological variables for evaporation estimation

The factors affecting evaporation have already been described. The principal source of energy, the Sun, transmits its radiation through the atmosphere. This is measured by solarimeters maintained by the UK Met Office at observatories and major meteorological stations, and these data are available from the British Atmospheric Data Centre. However, the measurement of net radiation, the difference between incoming radiation (short and long wave) and outgoing radiation (reflected short wave and ground-emitted long wave) is of particular concern in the calculation of evaporation. In particular, several modelling techniques described in Chapter 10 require closure of the energy balance equation and hence accurate measurements of these components.

The setting up of a net radiometer for hydrological studies is fairly straightforward and will be mentioned further in connection with automatic weather stations. The more conventional measurements of air temperature, humidity, wind speed and direction, and sunshine are made at most reference meteorological stations. The hydrologist can use all these observations for his own purpose, but increasingly it is found that there are limitations of coverage for particular areas and it may be necessary to set up special stations to provide the data required for evaporation estimations. Wherever possible, it is recommended that the rules and guidelines developed over the years by the expert organizations should be followed by the hydrologist establishing a new meteorological station or automatic weather station.

4.4.1 Siting a reference meteorological station

For a reference meteorological station, level ground about 10 m by 7 m in extent covered with short grass is selected and enclosed by open fencing or railings. The site should not have any steep slopes in the immediate vicinity and should not be located near trees or buildings; a very open site is advisable for sunshine recorders and anemometers.

A recommended plan for the instrument enclosure is shown in Fig. 4.2. The geographical coordinates of the station latitude and longitude should be determined from the relevant topographical map, if available, and the height above sea level established from the nearest Ordnance Survey bench mark. Particular attention should be paid to noting the exact orientation of the enclosure since the setting up of the anemometer and sunshine recorder is dependent on direction.

4.4.2 Evaporation-related instruments at a reference meteorological station

The most prominent feature of a meteorological station is the Stevenson screen, which houses the *air thermometers*. The ordinary screen provides the standard exposure for the air thermometers with their bulbs 1.25 m above the ground surface. The double-louvered screen is painted white and set firmly in the ground so that it opens away from the direction of the midday sun (i.e. to the north in the northern hemisphere). The two vertically hung thermometers are for the direct reading of the air temperature (dry bulb) and the reading of the wet bulb, covered with muslin kept moist by a wick leading from a small reservoir of distilled water. With these two temperature readings, the dew point, vapour pressure and relative humidity of the air are obtained from hygrometric tables, a humidity slide rule or equations. Supported horizontally are maximum and minimum thermometers. The four thermometers are

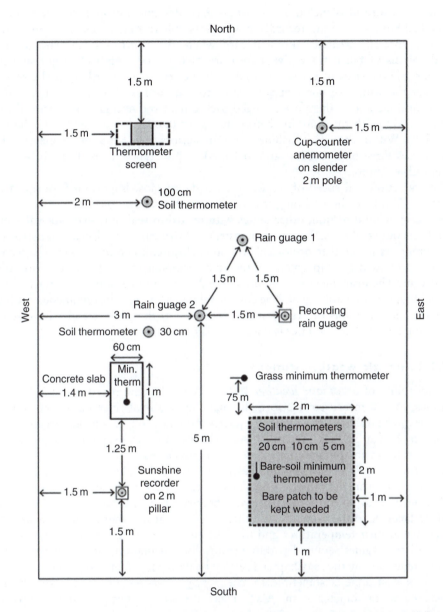

North

1.5 m   1.5 m

←— 1.5 m —→   ⊙ ←— 1.5 m —→
    Thermometer   Cup-counter
      screen   anemometer
                          on slender
                           2 m pole

←——— 2 m ———→ ⊙ 100 cm
                 Soil thermometer

                        ⊙ Rain guage 1

                   1.5 m   1.5 m

        Rain guage 2
←——— 3 m ———→ ⊙ ←— 1.5 m —→ ◎ Recording
                                rain guage
Soil thermometer ⊙ 30 cm

      60 cm

Concrete slab ┌─ Min. ─┐
←— 1.4 m —→│ therm │ 1 m   ● Grass minimum thermometer
           │   ●   │   75 m
           └────────┘   ↕ ←———— 2 m ————→

      1.25 m   ┌──────────────────┐
                             │ Soil thermometers │ ↑
           5 m   │ 20 cm 10 cm 5 cm │ │
                             │● Bare-soil minimum│ 2 m
        Sunshine   │   thermometer   │ │
←— 1.5 m —→ ◎ recorder   │ Bare patch to be │←1 m→
           on 2 m   │ kept weeded   │ ↓
           pillar   └──────────────────┘

      1.5 m   1 m

South

West / East

Fig. 4.2 Plan of a meteorological station for the northern hemisphere (dimensions in metres). (Reproduced with permission from *WMO 2008. Guide to Meteorological Instruments and Methods of Observation.* WMO-No. 8, 7th edn. World Meteorological Organisation, Geneva, 681.)

read at 0900 h GMT each day in the UK and, at this time, the maximum and minimum thermometers are reset. At some meteorological stations, the dry and wet bulb thermometers are read again at 1500 h, but where more detailed observations are required, manual measurements can be supplemented with an automatic weather station (Section 4.4.3).

The enclosure also includes a *sunshine recorder* and *anemometer*. A standard Campbell-Stokes sunshine recorder has a glass sphere that focuses the Sun's rays on to a specially treated calibrated card where they burn a trace. The accumulated lengths of burnt trace give a measure of the total length of bright sunshine in hours. Three sizes of cards are used with the recorder according to the season, i.e. over the winter or summer solstice or the equinoxes. A single card records a day's sunshine and is therefore changed each day at the normal observational time 0900 h GMT with the sunshine before and after 0900 h being credited to the correct days. Within the UK, sunshine duration sensors (e.g. Kipp and Zonen CSD-3, which uses three photodiodes) are now used in preference to the Campbell-Stokes (Met Office, 2006).

The direction and speed of the wind are some of most important features of the weather. Although the hydrologist concerned with evaporation may not be unduly worried by wind direction, other duties with regard to real-time hydrological events should encourage the installation of an instrument to measure both characteristics. For measurements in the meteorological station enclosure, a *cup anemometer* is recommended and are often cup generator anemometers incorporating a remote-indicating wind vane. The instrument is fixed on a pole 2 m from the ground and attached to a simple counter or a data logger. The cup anemometer can give instantaneous readings of wind velocity (knots or $m\,s^{-1}$) or provide a run-of-the-wind, a collective distance in kilometres when the counter is read each day.

4.4.3 Automatic weather stations

The reliability of *automatic weather station* (AWS) has improved over the last two decades, so that measurements formally made manually at reference meteorological stations can be obtained at a higher frequency with data logged sensors. An annotated photograph of an AWS is given in Fig. 4.3. The instruments, excluding the rain gauge, are mounted on an aluminium mast with two cross arms.

A *net radiometer* or *net pyrradiometer* (Fig. 4.4) is installed at the end of one of the arms to measure net radiation. Specifically, they measure the short- and long-wave radiation from around 0.3 to 70 μm via the heating of a black surface. One black surface faces down horizontally, another faces upward; a thermopile measures the difference in their temperatures and thus the energy exchange (Strangeways, 2003). Net radiation values required within evaporation calculations of the energy budget and combination methods (Chapter 10) may be derived from measurements of short-wave radiation alone, so sometimes the cheaper *pyranometer* (also called a *solarimeter*) replaces a net radiometer on the AWS. There are two types of pyranometer used for measuring short-wave radiation: the thermal solarimeter and the photodiodes sensor. The thermal solarimeter, like the net radiometer, senses radiation by measuring the heating effect of a black surface. The cheaper photodiode sensors use a light-sensitive diode, photodiode or photovoltaic cell to measure solar radiation, where the response is to individual incoming photons directly, rather than a heating effect of the thermal sensors (Strangeways, 2003).

Automatic weather stations typically use cup anemometers (Fig. 4.4) to measure wind velocity required by the mass transfer and combination methods of calculating evapotranspiration (Chapter 10).

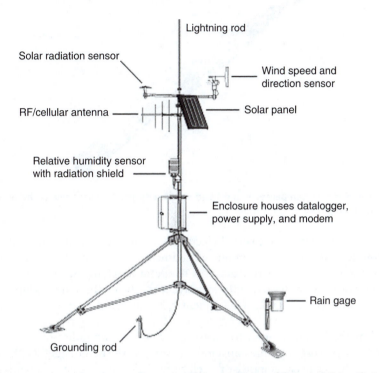

Lightning rod

Solar radiation sensor

Wind speed and
direction sensor

RF/cellular antenna

Solar panel

Relative humidity sensor
with radiation shield

Enclosure houses datalogger,
power supply, and modem

Rain gage

Grounding rod

Fig. 4.3 An automatic weather station. (Reproduced by permission of Campbell Scientific Ltd.)

Kipp & Zonen

Fig. 4.4 A net radiometer or net pyrradiometer. (Reproduced by permission of Kipp and Zonen BV.)

Fig. 4.5 A combined air temperature and relative humidity probe. (Reproduced by permission of Vaisala Ltd.)

Air temperature can be measured electronically using resistance thermometers or thermocouples. Resistance thermometers (either platinum resistance thermometers or thermistors) change their resistance as temperature changes. In contrast, thermocouples measure the potential difference resulting from the temperature difference between two different metals. Consequently, the latter device is less suited to use on an AWS.

Several methods can be used to measure the relative humidity of the air, though most AWS systems use thin-film capacitive sensors. As the name suggests, these devices measure the change in capacitance of a cell comprising an organic polymer layer sandwiched between a metal electrode and an upper metal film (Strangeways, 2003). The air temperature and relative humidity sensors need to be housed in a small specially designed thermal radiation screen (Fig. 4.5) instead of a conventional-type Stevenson screen.

All the instruments are wired to a data logger. The Met Office maintains a network of over 150 such automatic weather stations across the UK, with an average density of more than 1 every 1500 km^2.

4.5 Direct measurement of evapotranspiration by eddy flux instruments

In contrast to the energy balance, mass transfer and combination methods (Chapter 10) that utilise data from automatic weather stations or manually read meteorological instruments, the *eddy flux* or *eddy covariance* method (Chapter 10) requires dedicated sensors. Eddy flux calculations require data on vertical wind velocity and humidity sampled at sub-second (i.e. Hertz) frequencies. These data can be combined to give a direct measurement of the evapotranspiration. A three-dimensional sonic anemometer can be used to sample the wind velocities (Fig. 4.6), while a closed-path gas analyser or open-path krypton hygrometer (Fig. 4.7) can be used to measure humidity. A network of over 400 towers supporting these sensors has been deployed across the globe (Baldocchi *et al.*, 2001), including 16 towers within the UK.

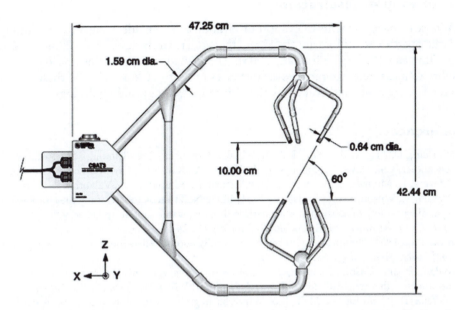

Fig. 4.6 A three-dimensional sonic anemometer. (Reproduced by permission of Campbell Scientific Ltd.)

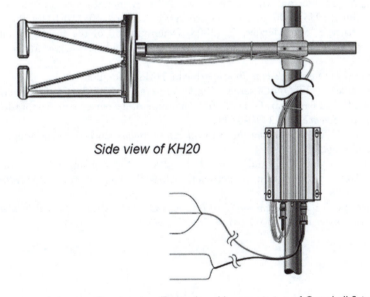

Side view of KH20

Fig. 4.7 An open-path krypton hygrometer. (Reproduced by permission of Campbell Scientific Ltd.)

4.6 Scintillometer measurements of evapotranspiration

Large-scale measurements of evapotranspiration can be made using a combination of a large-aperture scintillometer (LAS) and the more recently developed millimetre-wave scintillometer (MWS; Ludi *et al.*, 2005). The key advantage of this approach is the ability to measure evaporation over distances of up to 10 km. Trials of these systems in the UK are being undertaken by the Centre for Ecology and Hydrology.

References

Baldocchi, D., Falge, E., Gu, L., Olson, R., Hollinger, D., Running, S., Anthoni, P., Bernhofer, Ch., Davis, K., Fuentes, J., Goldstein, A., Katul, G., Law, B., Lee, X., Malhi, Y., Meyers, T., Munger, J. W., Oechel, W., Pilegaard, K., Schmid, H. P., Valentini, R., Verma, S., Vesala, T., Wilson, K. and Wofsy, S. (2001) FLUXNET: a new tool to study the temporal and spatial variability of ecosystem-scale carbon dioxide, water vapour and energy flux densities. *Bulletin of the American Meteorological Society* 82, 2415–2435.

Chow, T.L. (1994) Design and performance of a fully automated evaporation pan. *Agricultural and Forest Meteorology* 68, 187–200.

Gavilán, P. and Castillo-Llanque, F. (2009) Estimating reference evapotranspiration with atmometers in a semiarid environment. *Agricultural Water Management* 96, 465–472.

Genereux, D. P. and Jordan, M. T. (2006) Interbasin groundwater flow and groundwater interaction with surface water in a lowland rainforest, Costa Rica. *Journal of Hydrology* 320, 385–399.

Giambelluca, T. W., McKenna, D. and Ekern, P. C. (1992) An automated recording atmometer: 1. Calibration and testing. *Agricultural and Forest Meteorology* 62, 109–125.

Lapworth, C. F. (1965) Evaporation from a reservoir near London. *Journal of the Institute of Water Engineers* 19, 163–181.

Liu, B., Xu, M., Henderson, M. and Gong, W. (2004) A spatial analysis of pan evaporation trends in China, 1955-2000. *Journal of Geophysical Research* 109, 10.1029/2004JD004511.

Ludi, A., Beyrich, F. and Matzler, C. (2005) Determination of the turbulent temperature-humidity correlation from scintillometric measurement. *Boundary-Layer Meteorology* 117, 525–550.

Met Office (1982) *Observer's Handbook,* 4th edn. HMSO, London.

Met Office (2006) *SPOT-ON Observers Guide: Sunshine* v3 7/06. Met Office, Exeter, 4 pp.

Roderick, M. L. and Farquhar, G. D. (2002) The cause of decreased pan evaporation over the past 50 years. *Science* 298, 1410–1411.

Strangeways, I. (2003) *Measuring the Natural Environment,* 2nd edn. Cambridge University Press, Cambridge.

Vaughan, P. J., Trout, T. J. and Ayars, J. E. (2007) A processing method for weighing lysimeter data and comparison to micrometeorological ETo predictions. *Agricultural Water Management* 88, 141–146.

World Meteorological Organisation (WMO) (2008) *Guide to Meteorological Instruments and Methods of Observation.* WMO-No. 8, 7th edn. WMO, Geneva, 681.

Chapter 5

Hillslope and aquifer hydraulic parameters

In the general progression of the hydrological cycle beginning with atmospheric water vapour and ensuing precipitation (Chapters 1 and 3), it is then subsurface flow that transports most of the *residual rainfall* (rainfall less evapotranspiration) to rivers. Overland flow on hillslopes (Section 6.5) transports a much smaller proportion of the flow to rivers. The study of subsurface flow within soil, regolith and rock strata is of vital importance to the assessment of available groundwater resources (Chapter 17), the migration of nutrients and contaminants (Section 8.1), and in the regulation of both evapotranspiration (Chapter 10) and runoff generation, as simulated by catchment models (Chapter 12). While the proportion of precipitation travelling to channels by overland flow is now considered to be much less than that travelling to channels by subsurface pathways (Section 1.2), overland flow observations are of fundamental importance to the quantification of floodplain inundation (Bates *et al.*, 2006), hillslope erosion and sediment delivery (Owens and Collins, 2006), and the migration of phosphorus across agricultural landscapes (Withers and Bailey, 2003).

Most assessments of subsurface flow do not measure flows directly, but instead measure the parameters of *saturated hydraulic conductivity* (Section 5.1) and the *unsaturated hydraulic conductivity curve* (Section 5.2) and combine these with the variables or states of subsurface moisture content and pressure potential (Sections 6.1 and 6.2, respectively). These parameters and states are combined within the Darcy–Richards equation to give a steady-state estimate of subsurface flow (Section 5.1) or simulate the dynamics of subsurface flow (Chapter 15). Early subsurface flow studies did attempt to measure flow directly within soil strata using devices called *throughflow troughs* (e.g. Whipkey, 1965). The excavation of a soil pit to allow insertion of these troughs was however found to alter the direction and magnitude of the flow, making interpretation of such data difficult (Knapp, 1970). By contrast, the volume of overland flow generated by a bounded plot can be measured directly using an *overland flow trough* (Section 6.5.2). Developments in methods of geo-electrical measurement (Kirsch, 2006; Section 6.1.7), water quality sampling (Section 8.3) and water quality analyses (Section 8.4) have however led to the increased use of *water tracers* for the *direct measurement* of subsurface flow (McGuire and McDonnell, 2007; Chapter 6). Relating the migration of a water tracer to the water velocity estimated by the Darcy–Richards Equation (so called *Darcy velocity*) does however require measurement of the additional hydraulic parameters of *porosity* (Section 5.4) and *dispersion coefficient* (Section 5.5). These parameters are contained within the advection–dispersion equation (Nielsen and Biggar, 1961; Chapter 15).

This Chapter describes the measurement of the key parameters that regulate the magnitude and direction of subsurface flow, namely the saturated hydraulic conductivity (Section 5.1) and the unsaturated hydraulic conductivity curve (Section 5.2). The measurement of *infiltration capacity* that regulates the amount of overland flow and its relationship with the saturated hydraulic conductivity is described in Section 5.1.1. Saturated hydraulic conductivity is affected by the variations in the viscosity and density of the water within the soil, regolith or rock. The *intrinsic permeability* is unaffected by these variations, and its derivation from the saturated hydraulic conductivity is described in Section 5.3. Lastly, the measurement of porosity and dispersion coefficient for use in hillslope tracer studies (e.g. Section 6.4 and Chapter 15) is described within Sections 5.4 and 5.5, respectively.

5.1 Saturated hydraulic conductivity

Darcy's law (Darcy, 1856) shows how subsurface flow in saturated ground is directly proportional to the gradient in the *hydraulic head, dh/L* (or gradient in the total head; see Section 6.2), which is called the *hydraulic gradient*,

$$Q = K_s A \frac{dh}{L}$$

where Q is the volumetric discharge of subsurface water, K_s is the *saturated hydraulic conductivity* and A is the cross-sectional area through which the water is flowing. The same equation also defines the term *saturated hydraulic conductivity* of the soil, regolith or rock, namely the subsurface water velocity per unit cross-sectional area and unit hydraulic gradient; however, this is conventionally written as,

$$K_s = \frac{Q}{A} \frac{L}{dh}$$

The hydrological characteristic of the saturated hydraulic conductivity is sometimes called the *coefficient of permeability*, and models of subsurface flow are typically most sensitive to spatial variations in this property. Where only saturated rock is under study, the term *saturated hydraulic conductivity* is sometimes simply described as *hydraulic conductivity* or K. Some typical values of saturated hydraulic conductivity are given in Table 5.1.

Table 5.1 Some typical values of saturated hydraulic conductivity, K_s

Strata	K_s $(cm\,h^{-1})$	Reference
Well-drained topsoil	1–100	Chappell and Ternan (1992)
Well-drained subsoil	1–100	Chappell and Ternan (1992)
Poorly drained subsoil	0.001	Chappell and Ternan (1992)
Rock (highly weathered)	>10 000	Bear (1972)
Rock (unweathered granite)	<0.000 001	Bear (1972)

There are many ways of measuring the saturated hydraulic conductivity, though they can be divided into techniques for testing undisturbed cores, and those undertaken on cased or uncased holes. Tests on repacked soil or repacked regolith give K_s values not normally considered representative of the field situation.

5.1.1 Tests on undisturbed cores

Undisturbed cores of soil, regolith or rock can returned to the laboratory and the saturated hydraulic conductivity determined using a laboratory permeameter. With a *constant-head laboratory permeameter* (Fig. 5.1, left), terms within this last equation are measured and the equation solved directly. Alternatively, a *falling-head laboratory permeameter* (Fig. 5.1, right) can be used and the saturated hydraulic conductivity derived using,

$$K_s = \frac{(r_t^2 L / r_c^2 t)\ln h_1}{h_2}$$

where r_t is the radius of the tube used to apply water to the core, r_c is the radius of the soil core, L is the length of the soil core, t is the time taken for the *pressure head*

Fig. 5.1 Laboratory permeameters.

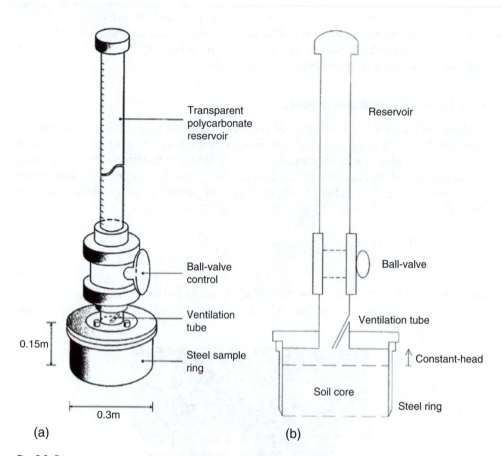

Fig. 5.2 Ring permeameter shown in (a) three-dimensional oblique view and (b) schematic cross-section. (Reproduced from Chappell and Ternan (1997) with permission of Wiley-Blackwell.)

(observed within the water supply tube; see Section 6.2) to fall from level 1 (h_1) to level 2 (h_2) (BS 1377-5, 1990).

Other permeameters are available to allow undisturbed soil cores to be tested in the field. One such method is *ring permeametry*, which applies a constant-head to relatively large, undisturbed cores (7000 cm^3) to derive the saturated hydraulic conductivity directly (Fig. 5.2). The large size of the cores has the advantage that error due to leakage between the soil core and metal ring can be minimised (Chappell and Lancaster, 2007). With *ring permeametry*, the soil core is first lifted from the ground and tested on a stand with atmospheric pressure being maintained at the base of the core. If the metal ring is inserted into the ground (using a press or hammer) and the test undertaken with the core still in the ground, the strata beneath the core, if less permeable that that within the core, will affect (reduce) the value obtained.

Inserting a ring into the ground surface (to a depth of 10 cm), and then inserting a larger outer ring (e.g. 53-cm diameter) to encompass the inner test ring

Fig. 5.3 Double-ring infiltrometer. (Reproduced with permission of Eijkelkamp Agrisearch Equipment BV, Giesbeek, The Netherlands.)

(e.g. 28-cm diameter) gives a *double-ring infiltrometer* (Fig. 5.3). By flooding the outer ring (as well as the inner ring) with water, the flow out of the base of the inner ring is primarily vertical. Note: with small-diameter infiltrometer rings, predominantly vertical flows cannot be maintained and so should not be used. Once the test has saturated the soil within and beneath the rings, the rate of infiltration will reduce to a constant value; this constant rate is called the *infiltration capacity*. The infiltration capacity determined by the standard method (using large diameter rings) is equivalent to the saturated hydraulic conductivity of the ground surface or topsoil (where permeameter cores have been tested with the core in the ground).

5.1.2 Tests on piezometers

Tests can be undertaken using individual *piezometers* (see Section 6.2.1) to derive a saturated hydraulic conductivity. One such test was developed by Hvorslev (1951) and remains a standard test within the UK (BS 6316, 1992). Within this test the original water level within a piezometer is raised or lowered artificially, and the return to the original level, which occurs at an exponential rate, is dependent on the saturated hydraulic conductivity. Consequently, if the height to which the water level is raised at the start of the test is h_o and the height of the water level above the original water level is h after time t, then a semi-logarithmic plot of the ratio h/h_o versus time should yield a straight line.

If the length of the piezometer screen, L, is more than eight times the radius of the filter pack (borehole), R, then the saturated hydraulic conductivity can be found from,

$$K_s = \frac{r^2 \ln(L/R)}{2LT_o}$$

where r is the piezometer radius and T_o is the time taken for the water level to rise to 37 per cent of the initial change. As the water level within the piezometer can be raised or lowered by introducing a solid object (or slug), such tests are often called *slug tests*. Further details of such tests can be found within Butler (1997).

5.1.3 Tests on observation wells

Tests on *observation wells* (see Section 6.2.2) and wells used for procuring ground-water is an essential prerequisite to any exploitation of well field for public water supply. These tests form part of a hydrological sub-discipline known as *well hydraulics*. Different approaches apply depending upon whether the aquifer is confined (by an overlying impeding layer) or unconfined and whether the conditions are unsteady or in steady state.

5.1.3.1 Steady flow in a confined aquifer

When the well fully penetrates a horizontal confined aquifer (Fig. 5.4), flow to the well is also horizontal from all directions (i.e. radial two-dimensional flow). To ensure a steady flow, there must be continuous recharge to the aquifer from sources distant to the well. Assuming also that the aquifer is homogeneous and isotropic and is not affected by compression in dewatering, the flow to the well at any radius r can be expressed by Darcy's law,

$$Q = 2\pi K_s b \frac{dh}{dr}$$

where Q is the pumping rate. Integrating over the radius distance r_w to r_1 and h_w to h_1 gives,

$$Q = \frac{2\pi K b (h_1 - h_2)}{\ln(r_1/r_w)}$$

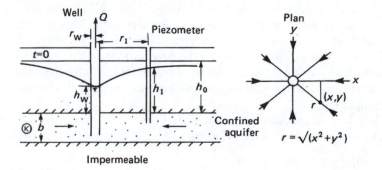

Fig. 5.4 Steady flow in a confined aquifer.

or:

$$Q = \frac{2\pi T(h_1 - h_2)}{\ln(r_1/r_w)}$$

where T is the *transmissivity* of the aquifer, or the product of the *saturated hydraulic conductivity* and *saturated zone thickness, b*. This is called the *Thiem* equation.

By pumping the well at a steady rate and waiting until the well level h_w is constant, observation of the drawdown level h_1 at an observation well at a known distance r_1, from the pumped well, allows estimation of the transmissivity of the aquifer. In practice, the observations from two or more observation wells at different radii are more useful since head losses in the well, caused by friction in the well casing, can then be allowed for.

Example. A well in a confined aquifer was pumped at a steady rate of $0.0311 \ \mathrm{m^3 \ s^{-1}}$. When the well level remained constant at 85.48 m, the observation well level at a distance of 10.4 m was 86.52 m. Calculate the transmissivity,

$$Q = \frac{2\pi Kb(h_1 - h_2)}{\ln(r_1/r_w)}$$

$$T = \frac{Q\ln(r_1/r_w)}{2\pi(h_1 - h_w)}$$

$$= \frac{0.0311\ln(10.4/0.3)}{2\pi(86.52 - 85.48)}$$

$$T = 0.0169 \ \mathrm{m^2 \ s^{-1}}$$

5.1.3.2 Steady flow in an unconfined aquifer

The groundwater flow to a well in an unconfined aquifer may be complicated by the downward movement of recharge water from ground surface infiltration, but here only distant sources in the aquifer are assumed to maintain the steady-state flow. In Fig. 5.5, the water table intersects the well at hr with the water level in the well at hw. Between the two levels is a seepage zone. However, in estimating the flow to the well, the Dupuit–Forchheimer assumption of horizontal flow may be used, thus,

$$Q = 2\pi r K h \frac{dh}{dr}$$

with h the height of the water table replacing b the thickness of the confined aquifer. Integrating over the radius distance $r1$ to $r2$ with corresponding values h_1 and h_2 then:

$$Q = \frac{2\pi K(h_1^2 - h_2^2)}{\ln(r_1/r_2)}$$

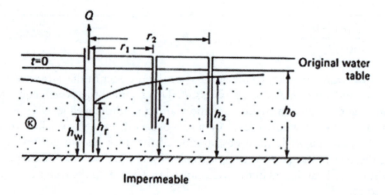

Fig. 5.5 Steady flow to a well in an unconfined aquifer showing drawdown and cone of depression in water table from original position.

from which K can also be evaluated. In applying this equation to an unconfined aquifer, the effects of the seepage from the water table into the well (h_w and r_w) are negligible if the distance of the nearest observation well r_1 is greater than 1.5 times the original water table height h_o.

5.1.3.3 Non-steady flow to wells

The steady-state groundwater flow situations analysed above, and consequent steady-state positions for the piezometric surface or water table, are only achieved after what may be very long periods of time after the start of pumping. During such periods, even though the pumping rate may be steady, the piezometric surface or water table will be falling with time until the steady-state position is reached. Theoretically, if the aquifer were infinite in extent (with no vertical recharge), a steady state would never be completely achieved. Consequently, an ability to analyse such non-steady groundwater flow situations for wells being pumped at a constant rate, is required.

In confined aquifers, the relief of pressure as a piezometric surface falls introduces two compressibility effects: the pore water expands owing to the smaller water pressure, and simultaneously the pore space contracts owing to a greater mechanical stress from the overburden as the reduced water pressure takes less of the load. Thus water is released *from storage* over the aquifer to make up the pumped abstraction; no dewatering of the pore spaces occurs.

In unconfined aquifers, on the other hand, compressibility effects are usually negligible, but dewatering of the pore spaces does occur as the water table falls. The water being released from the whole aquifer integrates to equal the constant pumping abstraction.

In both types of aquifer, the head, h, varies both with distance and time after the start of pumping. Even with a constant pumping rate, the situation is described as being 'non-steady' flow!

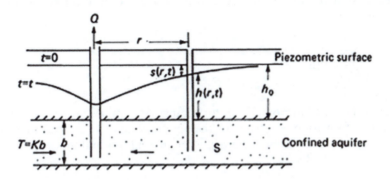

Fig. 5.6 Definition diagram for non-steady flow to a well in a confined aquifer.

5.1.3.4 Non-steady flow in a confined aquifer

Assuming negligible recharge and no head gradient in the vertical, the horizontal, two-dimensional, non-steady flow can be described by,

$$K\left(\frac{\partial^2 h}{\partial x^2} + \frac{\partial^2 h}{\partial y^2}\right) = S_s \frac{\partial h}{\partial t}$$

For a homogeneous, isotropic, confined aquifer of thickness b, this can be written,

$$\frac{\partial^2 h}{\partial x^2} + \frac{\partial^2 h}{\partial y^2} = \frac{S}{T} \frac{\partial h}{\partial t}$$

with T and $S(S = S_s b, T = Kb)$, the transmissivity and storativity, respectively. For non-steady flow to a well, this equation may be transformed to radial coordinates. Thus with $r = \sqrt{(x^2 + y^2)}$:

$$\frac{\partial^2 h}{\partial r^2} + \frac{1}{r} \frac{\partial h}{\partial r} = \frac{S}{T} \frac{\partial h}{\partial t}$$

The solution of this equation yields $h(r, t)$ (Fig. 5.6), the hydraulic head at distance r from a well at time t after the commencement of steady pumping at a rate Q.

For practical purposes, what is usually required is $\{h_o - h(r, t)\}$, which is the drawdown, $s(r, t)$, from the initial rest level head, h_o. The solution of the equation is,

$$s(r, t) = \frac{Q}{4\pi T} \int_u^\infty \exp(u)/u\, du$$

(known as the *Theis equation*),

where Q is the steady pumping rate and $u = r^2 S/4Tt$. Expansion of the integral gives,

$$s(r, t) = \frac{Q}{4\pi T}\left[-0.577216 - \ln u + u - \frac{u^2}{2.2!} + \frac{u^3}{3.3!} - \cdots\right]$$

The expression within the brackets is usually denoted by $W(u)$ known as the *well function* so that:

$$s(r, t) = \frac{Q}{4\pi T} W(u)$$

Values of the well function, $W(u)$, for a range of u values are given in Table 5.2. Knowing the 'formation constants' of the aquifer, S and T, and for a given Q, the drawdown $s(r, t)$ can be estimated directly for any radius r and time t.

Example. To calculate the drawdown in a confined aquifer at $r = 25\,\text{m}$ after 6 hours of pumping. Water with a constant discharge of $0.0311\,\text{m}^3\,\text{s}^{-1}$. The aquifer constants are $S = 0.005$ and $T = 0.0092\,\text{m}^2\,\text{s}^{-1}$.

$$u = \frac{r^s S}{4Tt}$$

$$= \frac{25^2 \times 0.005}{4 \times 0.0092(6 \times 3600)}$$

$$= \frac{3.125}{794.88}$$

$$= 3.93 \times 10^{-3}$$

So that,

$$s(r, t) = \frac{Q}{4\pi T} W(u)$$

$$= \frac{0.0311}{4 \times \pi \times 0.0092} \times 4.97$$

$$= 1.337\,\text{m}$$

The well function can also be used to derive values of S and T by matching measurements of drawdown against time at an observation well against the well function (see e.g. Fetter, 2001).

The Theis solution of the non-steady radial flow equation has, for certain conditions, a simplified form due to *Jacob*. If u is small (< 0.01) only the first two terms in the series of the equation steady flow in an unconfined aquifer need be used (to within 1 per cent of the full series) so that:

$$s(r, t) = \frac{Q}{4\pi T}\left[-0.577216 - \ln\left(\frac{r^2 S}{4Tt}\right)\right]$$

Table 5.2 The well function, $W(u)$

u	1.0	2.0	3.0	4.0	5.0	6.0	7.0	8.0	9.0
X1	0.219	0.049	0.013	0.004	0.001				
X10^{-1}	1.82	1.22	0.91	0.70	0.56	0.45	0.37	0.31	0.26
X10^{-2}	4.04	3.35	2.96	2.68	2.47	2.30	2.15	2.03	1.92
X10^{-3}	6.33	5.64	5.23	4.95	4.73	4.54	4.39	4.26	4.14
X10^{-4}	8.63	7.94	7.53	7.25	7.02	6.84	6.69	6.55	6.44
X10^{-5}	10.94	10.24	9.84	9.55	9.33	9.14	8.99	8.86	8.74
X10^{-6}	13.24	12.55	12.14	11.85	11.63	11.45	11.29	11.16	11.04
X10^{-7}	15.54	14.85	14.44	14.15	13.93	13.75	13.60	13.46	13.34
X10^{-8}	17.84	17.15	16.74	16.46	16.23	16.05	15.90	15.76	15.65
X10^{-9}	20.15	19.45	19.05	18.76	18.54	18.35	18.20	18.07	17.95
X10^{-10}	22.45	21.76	21.35	21.06	20.84	20.66	20.50	20.37	20.25
X10^{-11}	24.75	24.06	23.65	23.36	23.14	22.96	22.81	22.67	22.55
X10^{-12}	27.05	26.36	25.96	25.67	25.44	25.26	25.11	24.97	24.86
X10^{-13}	29.36	28.66	28.26	27.97	27.75	27.56	27.41	27.28	27.16
X10^{-14}	31.66	30.97	30.56	30.27	30.05	29.87	29.71	29.58	29.46
X10^{-15}	33.96	33.27	32.86	32.58	32.35	32.17	32.02	31.88	31.76

from which

$$s(r,t) = \frac{Q}{4\pi T} \ln\left(\frac{2.25Tt}{r^2 S}\right)$$

$$= \underbrace{\frac{Q}{4\pi T} \ln\left(\frac{2.25T}{r^2 S}\right)}_{\text{Intercept}} + \underbrace{\frac{Q}{4\pi T} \ln t}_{\text{Slope}}$$

Intercept   Slope

A semi-log plot of $s(r,t)$ versus $\ln(t/r^2)$ allows T and S to be estimated from the slope and intercept of a best-fit straight line, drawn given greater regard to the points at longer times (Fig. 5.7).

Then, given the slope of the fitted straight line, m

$$T = \frac{Q}{4\pi m}$$

And given T and the intercept, c

$$\frac{Q}{4\pi T} \ln\left(\frac{2.25Tc}{r^2 S}\right) = 0$$

$$\frac{2.25Tc}{r^2 S} = 1$$

$$S = \frac{2.25Tc}{r^2}$$

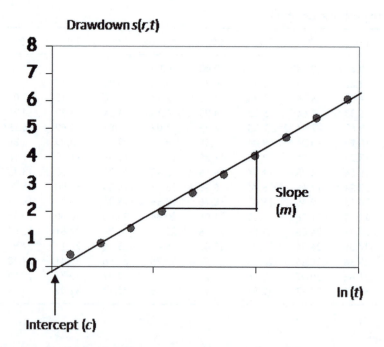

Fig. 5.7 Fitted straight line to drawdown versus log time data for the case of non-steady flow in a confined aquifer. (Adapted from a figure prepared by Andrew Binley).

5.1.3.5 Non-steady flow in an unconfined aquifer

In an unconfined aquifer, it has been seen that steady flow, based on the Dupuit–Forchheimer assumptions, becomes somewhat unrealistic close to the well when the slope of the water table becomes steep. In assessing non-steady flow, dewatering of the medium results in changes in transmissivity and, when the water table is lowered, the storativity is decreased. Thus the application of the non-steady flow equations becomes difficult. However, for small drawdowns over short time periods, the Theis equation may be used for rough estimates of $s(r, t)$.

Various methods have been derived to deal with the problems of analysis of observations in more complex flow situations, including non-steady flow in unconfined aquifers, and for these the reader is referred to the specialist texts on groundwater (e.g. Fetter, 2001; Todd and Mays, 2005), which also deal with the added difficulties encountered with non-homogeneous and anisotropic aquifers, more complex boundary conditions in aquifers of limited extent, and the rather common situation where leaky aquifers receive recharge or lose water through a low permeability aquitard.

5.2 Unsaturated hydraulic conductivity curve

Darcy's law describing subsurface flow within the saturated zone can be extended to describe flow with unsaturated soil, regolith and rock by replacing the saturated

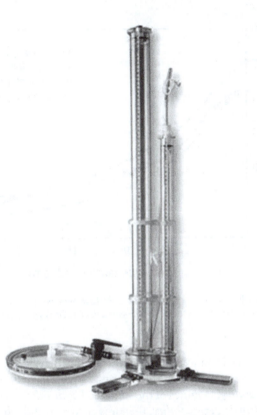

Fig. 5.8 A tension infiltrometer. Reproduced with permission of Soil Measurement Systems.

hydraulic conductivity with the *unsaturated hydraulic conductivity curve (K(h))*. For small pressure heads of up to about −30 cm H$_2$O below atmospheric pressure (Fig. 5.8), this curve can be measured directly using a *tension infiltrometer* (Fig. 5.9).

For larger negative pressure heads, the *unsaturated hydraulic conductivity curve* is normally derived by modelling the data within the *moisture release curve* (Section 6.3). The models of Brooks and Corey (1964) and van Genuchten (1980) are most commonly used.

5.3 Intrinsic permeability

According to Hubbert (1940), the *saturated hydraulic conductivity (K$_s$)* is actually a compound property dependent upon the *intrinsic permeability (k)* of the soil, regolith or rock, and the density (ρ_f) and dynamic viscosity (μ) of the fluid (i.e. water):

$$K_s = \frac{k\rho_f}{\mu}$$

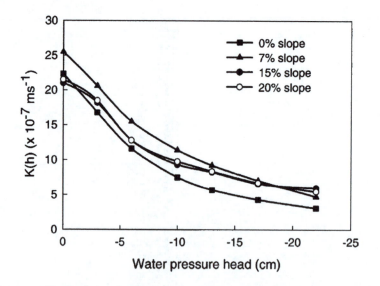

Fig. 5.9 Unsaturated hydraulic conductivity curve over low negative pressure heads for four example slope locations. Reproduced from Bodhinayake *et al.* (2004) with permission of the Soil Science Society of America.

Consequently, changes in salt content of water (thus density) and the temperature of the water (thus viscosity) will affect the *saturated hydraulic conductivity* but not affect the *intrinsic permeability*. If values of the *saturated hydraulic conductivity* are not to be converted to an *intrinsic permeability*, then it is clear that they should be corrected to a standard temperature, and hence a known water viscosity. This standard temperature is typically 20°C (Chappell and Ternan, 1997).

5.4 Porosity

The porosity (η) of a soil, regolith or rock is the fraction of an undisturbed volume of porous media that is occupied by pores rather than solids. Porosity can be determined using the gas pycnometer method, where the effect of adding dry solids into a gas-filled chamber is measured (Danielson and Sutherland, 1986). More commonly, porosity is determined directly from the measurement of the volumetric moisture content (Section 6.1) when the ground (or a sample) is saturated. Measurements of porosity and a dispersion coefficient are needed to equate observations of tracer migration to the subsurface velocity calculated by the Darcy–Richards equation (see Section 6.4).

5.5 Dispersion coefficient

The migration of a substance dissolved within water is described by the advection–dispersion equation (Nielsen and Biggar, 1961; Section 15.6.1). The key terms within this equation are the *longitudinal dispersion coefficient* (D_L) and the mean pore water velocity ($\overline{v_{pore}}$). The mean pore-water velocity is the velocity calculated by the Darcy equation (i.e. Q/A; Section 5.1) divided by the porosity (Section 5.4). Values of the dispersion coefficient for saturated soil, regolith or rock under consideration can be determined by application of a pulse of a tracer to a saturated core or below the water level within a piezometer, followed by an inverse simulation of the advection–dispersion equation (Parker and van Genuchten, 1984; van Genuchten, 1986). The derived values of the dispersion coefficient can then be used in the interpretation of subsurface tracing experiments (Section 6.4).

References

Bates, P. D., Wilson, M. D., Horritt, M. S., Mason, D., Holden, N. and Currie, C. (2006) Reach scale floodplain inundation dynamics observed using airborne Synthetic Aperture Radar imagery: data analysis and modelling. *Journal of Hydrology* 328, 306–318.

Bear, J. (1972) *Dynamics of Fluids in Porous Media*. Dover Publications, New York.

Bodhinayake, W., Si, B. C. and Noborio, K. (2004) Determination of hydraulic properties in sloping landscapes from tension and double-ring infiltrometers. *Vadose Zone Journal* 3, 964–970.

Brooks, R. H. and Corey, A. T. (1964) Hydraulic properties of porous media. Hydrology Paper No. 3, Colorado State University, Fort Collins.

BS 1377-5 (1990) *Methods of Test for Soils for Civil Engineering Purposes. Compressibility, Permeability and Durability Tests*. British Standards Institution, London.

BS 6316 (1992) *Code of Practice for Test Pumping of Water Wells*. British Standards Institution, London.

Butler, J. J. (1997) *The Design, Performance and Analysis of Slug Tests*. Lewis Publishers, New York.

Chappell, N. A. and Lancaster, J. W. (2007) Comparison of methodological uncertainty within permeability measurements. *Hydrological Processes* 21(18), 2504–2514.

Chappell, N. A. and Sherlock, M. D. (2005) Contrasting flow pathways within tropical forest slopes of Ultisol soil. *Earth Surface Processes and Landforms* 30, 735–753.

Chappell, N. A. and Ternan, J. L. (1992) Flow-path dimensionality and hydrological modelling. *Hydrological Processes* 6, 327–345.

Chappell, N. A. and Ternan, J. L. (1997) Ring permeametry: design, operation and error analysis. *Earth Surface Processes and Landforms* 22, 1197–1205.

Danielson, R. E. and Sutherland, P. L. (1986) Porosity. In: Klute, A. (ed.) *Methods of Soil Analysis, Part 1 Physical and Mineralogical Methods*, 2nd edn. American Society of Agronomy, Madison, pp. 443–461.

Darcy, H. (1856) *Les Fonataines publiques de la Ville de Dijon*. V Dalmont, Paris.

Fetter, C. W. (2001) *Applied Hydrogeology*, 4th edn. Prentice-Hall, Upper Saddle River, 598 pp.

Hubbert, M. K. (1940) The theory of ground-water motion. *Journal of Geology* 48, 785–944.

Hvorslev, M. J. (1951) *Time Lag and Soil Permeability in Ground-water Observations*, Bulletin No. 36. Waterways Experiment Station, US Corporation of Engineers: Vicksburg, MS.

Kirsch, R. (2006) *Groundwater Geophysics*. Springer, Berlin, 493 pp.

Knapp, B. J. (1970) *Patterns of Water Movement on a Steep Upland Hillside, Plynlimon, Central Wales*. Ph.D. Thesis, University of Reading, unpublished.

McGuire, K. and McDonnell, J. J. (2007) Stable isotope tracers in watershed hydrology. In: Lajtha, K. and Michener, W. (eds) *Stable Isotopes in Ecology and Environmental Sciences*, 2nd edn. Blackwell Publishing, Oxford.

Nielsen, D. R. and Biggar, J. W. (1961) Miscible displacement in soils. I. Experimental information. *Soil Science Society of America Proceedings* 25, 1–5.

Owens, P. N. and Collins, A. J. (2006) *Soil Erosion and Sediment Redistribution in River Catchments: Measurement, Modelling and Management*. CABI Publishing, Wallingford.

Parker, J. C. and van Genuchten, M. Th. (1984) *Determining Transport Parameters from Laboratory and Field Tracer Experiments*. Virginia Agricultural Experiment Station Bulletin 84–3, Blacksburg.

Todd, D. K. and Mays, L. W. (2005) *Groundwater Hydrology*, 3rd edn. Wiley, New York.

van Genuchten, M. Th. (1980) A closed-form equation for predicting the hydraulic conductivity of unsaturated soils. *Soil Science Society of America Journal* 44, 892–898.

van Genuchten, M. Th. (1986) Solute dispersion coefficients and retardation factors. In: Klute, A. (ed.) *Methods of Soil Analysis, Part 1, Physical and Mineralogical Methods*, 2nd edn. American Society of Agronomy, Madison, pp. 1025–1054.

Whipkey, R. Z. (1965) Subsurface stormflow from forested slopes. *International Association of Scientific Hydrology*. Bulletin 10, 74–85.

Withers, P. J. A. and Bailey, G. A. (2003) Sediment and phosphorus transfer in overland flow from a maize field receiving manure. *Soil Use and Management* 19, 28–35.

Chapter 6

Hillslope moisture states and flows

As noted within the introduction to Chapter 5, most assessments of subsurface flow do not measure flows directly, but instead measure the variables or states of moisture content (Section 6.1) and pressure potential (Section 6.2) and estimate flow using the Darcy–Richards equation (see Chapter 14). This contrasts with the assessment of overland flow, which is normally measured directly using *overland flow troughs* (Section 6.5.2). Developments in methods of geo-electrical measurement (Kirsch, 2006; Section 6.1.7), water-quality sampling (Section 8.3) and water-quality analyses (Section 8.4) have, however, led to the increased use of *water tracers* for the *direct measurement* of subsurface flow (McGuire and McDonnell, 2007; Section 6.4).

This chapter will explain the methods used to measure: (1) the moisture states within the Darcy–Richards flow equation, namely moisture content (Section 6.1), pressure potential (Section 6.2) and their interrelation (Section 6.3); (2) subsurface tracer flow (Section 6.4), and (3) the incidence and volume of overland flow (Section 6.5).

6.1 Subsurface moisture content

The moisture content can be expressed as the mass of water (m_w) within a mass of dry soil, regolith or rock (m_s),

$$\theta_m = \frac{m_w}{m_s}$$

This is called the *mass wetness* (θ_m, g g^{-1}, also called *gravimetric wetness*; the term avoided here to prevent confusion with the *gravimetric method* discussed later). This measure is rarely used by hydrologists because of the influence of unmeasured variations in the mass of dry soil, regolith or rock (m_s) within undisturbed volumes of soil, regolith or rock (V_s), i.e. the dry bulk density of the soil (g cm^{-3}). Consequently, the most commonly expressed measure of moisture content is the *volumetric wetness* (θ_w, cm cm^{-3}) where the mass wetness is multiplied by the dry bulk density (ρ_d) of the same sample. Given that 1 cm^3 of water has a mass of 1 g, this means that the volumetric wetness is equivalent to the volume of water (V_w) within an undisturbed volume of soil (V_s),

$$\theta_v = \frac{m_w}{m_s}\frac{m_s}{V_s} = \frac{V_w}{V_s}$$

The porosity of a soil, regolith or rock (η, $cm\,cm^{-3}$) is the total pore space within the porous media (Section 5.4), and is equivalent to the maximum volumetric wetness, i.e. all pore space is filled with water. The *saturation wetness* (θ_s, limits $0 \rightarrow 1$) is the proportion of the pore space that is filled with water,

$$\theta_s = \left(\frac{\theta_v}{\eta}\right)$$

Direct measurement of moisture content can be achieved by only a few methods, notably the *gravimetric method* and *carbide gas method*.

6.1.1 Carbide gas method

The *carbide gas method* is not commonly used within hydrology, but is commonly used within agriculture. With this method, a known wet mass of soil can be added to a pressure vessel containing a quantity of calcium carbide (CaC_2). Once sealed and the two components brought together, water reacts with the calcium carbide to produce ethyne (C_2H_2, acetylene) gas,

$$CaC_2 + 2H_2O \rightarrow C_2H_2 + Ca(OH)_2$$

The increased pressure within the vessel (called a 'Speedy Soil Moisture Meter') caused by the production of ethyne is then measured with a pressure gauge, calibrated to values of mass wetness.

6.1.2 Gravimetric method

The *gravimetric method* can be used to measure the mass wetness, volumetric wetness and saturation wetness. If volumetric wetness is required, an undisturbed soil or regolith core of known volume (V_s) must be first removed from the ground. This sample is then returned to the laboratory where is it first weighed before being transferred to an oven. Following the British Standard (BS1377-3, 1990), the sample is dried in the oven for 24 h at 105°C, cooled in a desiccator and then reweighed to determine by the volume of water lost (V_w) and dry soil or regolith mass (m_s). For sands, further drying would not give further moisture loss. However, with clays moisture would continue to be lost, primarily from the structure of the clays. Thus, for clay soils the British Standard is designed to give a reference moisture content rather than an absolute value. Particular consideration needs to be given to the determination of moisture content within organic soils (e.g. peat), as drying above 50°C also results in loss of organic materials by volatilisation. Particular care also needs to be given to the sampling of well-structured soils, as disturbances to dry bulk density (i.e. amount of soil collected in the sample tin) affect the apparent volumetric wetness. Once dried at 105°C, the structure of soil sample will not return to the natural state, so that a sample cannot be returned to the field for re-sampling. The gravimetric method is therefore, described as a destructive sampling method. Many hydrological applications do, however, require repeated sampling of the same soil volume to derive a time series of moisture content dynamics. This need has led to the development of several indirect or analogue methods of determining, typically, volumetric wetness. The most commonly used methods in the UK are *neutron moderation* (Fig. 6.1), *time-domain reflectometry, satellite and airborne radiometry* and *electrical resistivity*. Where undisturbed samples can be taken for gravimetric analysis, it is often recommended that the use of these methods is accompanied by local calibration against the gravimetric method.

6.1.3 Neutron moderation method

Neutron moderation (also called *neutron scattering* or *attenuation*) is a technique developed in the 1950s whereby fast neutrons from a sealed americium-241:beryllium source bombard a similarly sized mass, such as hydrogen nuclei in the form of water, to give a cloud of thermalised slow neutrons. The density of this cloud can then be measured with a detector (e.g. boron trifluoride or helium-3). One such device is the Institute of Hydrology neutron probe (Institute of Hydrology, 1979). With this device, the source and detector are lowered into the ground to the required measurement depth using sealed aluminium 'access tubes'. After a day's measurements, a long count is taken within a water reference to correct for the effects of radioactive decay of the source. This device was shown to give reliable measurements of volumetric wetness, except within the topsoil, where neutron losses from the soil surface make the technique very sensitive to the exact depth of sampling. Installation of access tubes and measurements down to 3 m has been possible. From the 1970s to early 1990s, the neutron probe was the preferred method of measuring a time series of soil moisture content by UK hydrologists. Increasing regulation of radioactive sources and the inability to automate large field arrays has meant that most UK hydrologists now prefer to measure volumetric wetness using techniques based upon *time-domain reflectometry*. The exception to this is the civil engineering community who continue to use neutron moderation to measure near-surface (0–0.3 m) wetness using surface moisture-density gauges (Fig. 6.1).

6.1.4 Time-domain reflectometry method

Time-domain reflectometry (TDR) is based upon the transmission and reflection of electromagnetic (EM) signals along parallel wave-guides. The rate of propagation of a reflection (caused by a large impedance change) from the bottom of the wave-guide to the top depends upon the dielectric properties of the material surrounding the wave-guide (i.e. soil and water). For signals of between 50 MHz and 1 GHz, the propagation velocity (v_p; m s^{-1}) is simply,

$$v_p = \frac{c}{\sqrt{\varepsilon}}$$

where c is the speed of light in a vacuum (3×10^8 m s^{-1}) and ε (or Ka) is the apparent dielectric constant or relative permittivity of the material surrounding the wave-guide (dimensionless). If the top and bottom of the wave-guide can be identified in a trace on a *time-domain reflectometer* oscilloscope, then the velocity of an EM wave returning

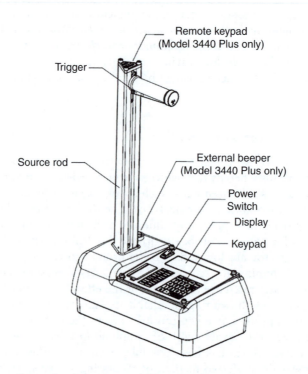

Remote keypad
(Model 3440 Plus only)

Trigger

Source rod

External beeper
(Model 3440 Plus only)

Power
Switch

Display

Keypad

Fig. 6.1 A surface moisture-density gauge used for measuring volumetric wetness within surface (0–0.3 m) soil layers. (Reproduced with permission of Troxler Electronic Laboratories.)

from the bottom to the top of the wave-guide is,

$$v = \frac{L}{t}$$

where L is the length of the wave-guide (m; for standard cable tester, e.g. Tektronix 1502C) and t is the travel time (seconds). Combining the last two equations gives the dielectric constant as,

$$\varepsilon = \left(\frac{ct}{L}\right)^2$$

The distance ct (metres) can be derived directly from the trace on the oscilloscope, manually or by data-logging and processing the data. Where such high propagation velocities are used, the apparent dielectric constant of soil varies largely with changes in moisture content and Topp *et al.* (1980) give the following relationship for mineral soils,

$$\theta_v = -5.3 \times 10^{-2} + 2.92 \times 10^{-2}\varepsilon - 5.5 \times 10^{-4}\varepsilon^2 + 4.3 \times 10^{-6}\varepsilon^3$$

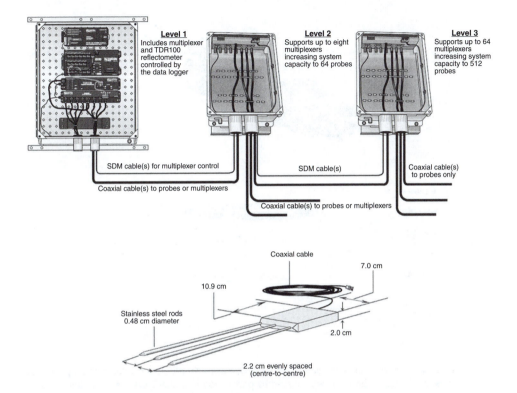

Fig. 6.2 A time-domain reflectometry system for measuring volumetric wetness at several locations, showing a time-domain reflectometer, datalogger, RF-MUX and a wave-guide. (Reproduced by permission of Campbell Scientific Ltd.)

where θ_v is the volumetric wetness, and ε is the relative permittivity or apparent dielectric constant (dimensionless). The wave-guides are metal rods, pushed into the ground no more than 0.05 m apart. For sandy soil, wave-guides can be up to 0.7 m in length, before the signal is so attenuated that the distance ct cannot be identified on the oscilloscope trace (or by an analysis program). For clay soils, the maximum wave-guide length may be only 0.3 m. It should be noted that the inflection in the trace produced by the lower end of the wave-guide is not always sharp or simple. Where such conditions arise, significant errors can be introduced into the ct measurement and hence volumetric wetness measurement.

By connecting several wave-guides to radio frequency multiplexors (RF-MUX), the volumetric wetness at many locations in the field or laboratory can be monitored automatically with a data logger (Fig. 6.2).

6.1.5 Simplified time-domain reflectometry method

High cost is the main disadvantage of time-domain reflectometers operating at gigahertz frequencies, though there is also some risk of vermin damage (then water damage) to the coaxial cables connecting the wave-guides to the reflectometer and

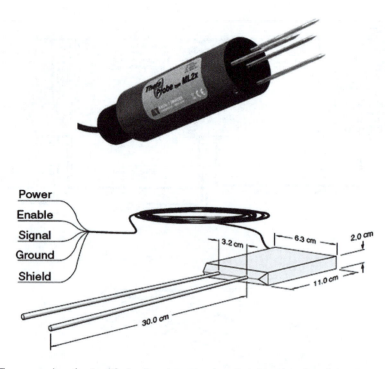

Fig. 6.3 Two examples of a simplified soil moisture probe using time-domain reflectometry principles, namely a Theta probe (top: reproduced by permission of Delta-T Devices Ltd), and a CS616 probe (bottom: reproduced by permission of Campbell Scientific Ltd).

radio frequency multiplexer (RF-MUX). As a consequence, many hydrologists opt to use simplified sensors using time-domain reflectometry principles that are one-tenth of the cost and lack coaxial cables.

With *simplified TDR moisture probes*, a continuous 100 MHz outgoing wave is sent down a wave-guide and a reflection generated at the lower end. These outgoing and returning signals interfere, producing a composite standing wave. The ratio of the outgoing wave to the composite standing wave is dependent on the dielectric constant of the soil around the wave-guide and can be calibrated to the volumetric wetness. Two such *simplified TDR moisture probes* are shown in Fig. 6.3. The main disadvantage of these simplified probes is the greater uncertainty in the volumetric wetness readings partly due to a greater sensitivity to local soil characteristics.

6.1.6 Satellite and airborne radiometry method

As with TDR (and simplified TDR), airborne and satellite radiometry measure changes in the dielectric properties of the ground surface (and the ground surface roughness). With active microwave radiometry, an EM signal produced by a remote power source is propagated to the target ground surface and a proportion is reflected and returns to a remote sensor. The amplitude of the received/transmitted power ratio, also called the *backscatter coefficient* (dB), is proportional to the dielectric constant of the ground

(Ulaby *et al.*, 1986). An example of a study using active microwave radiometery involves the 5.3 GHz (C-band) AMI-SAR instrument on the ERS-2 satellite platform (Walker *et al.*, 2004). This study was able to map volumetric soil moisture content over the 0.1 km² Nerrigundah catchment in Australia. Similar radiometers have been used on airborne platforms. For example, the polarimetric scanning radiometer (PSR/CX), was flown aboard a NASA P-3 aircraft to measure moisture content over a large basin in Mexico (Vivoni *et al.*, 2008).

6.1.7 Electrical resistivity method

The problem with all moisture probes using the principles of TDR is the maximum volume of ground that can be sampled with a single wave-guide. The well-established geophysical technique of *electrical resistivity* can overcome the volume constraint on moisture measurement. The bulk electrical resistivity of the soil, regolith or rock (ρ_b, Ωm) is related to moisture content via Archie's law,

$$\rho_b = \frac{\rho_w}{\eta^m \theta_s^n}$$

where ρ_b is the electrical resistivity of the pore water, m and n are fitting parameters, sometimes known as the cementation exponent and saturation exponent, respectively. Alternatively, the relation can be expressed as the inverse of resistivity, namely conductivity,

$$\sigma_b = \sigma_w \eta^m \theta_s^n$$

where σ_b is the bulk electrical conductivity (S m^{-1}; note $1\,\mathrm{S\,m^{-1}} = 1 \times 10^{-5}\,\mu\mathrm{S\,cm^{-1}}$), and σ_w is the electrical conductivity of the pore water (Sm^{-1}; Archie, 1942). Given that the volumetric wetness is the product of the porosity and saturation wetness, this relation has been simplified to give the expression,

$$\sigma_b = A\sigma_w \theta_v^B$$

or

$$\theta v = \left(\frac{\sigma_b}{A\sigma_w}\right)^{1/B}$$

where A and B are alternative fitting parameters (Shah and Singh, 2005; Schwartz *et al.*, 2008).

The bulk electrical conductivity (or bulk resistivity) of the soil, regolith or rock can be determined using various electrode configurations installed on the ground surface or in boreholes. A Wenner configuration uses two current and two potential electrodes; these are equally spaced in a line, with the current being applied to the outer electrodes (Fig. 6.4). The resistance measured between the two potential electrodes usually being most influenced by the ground at a depth equivalent to approximately half

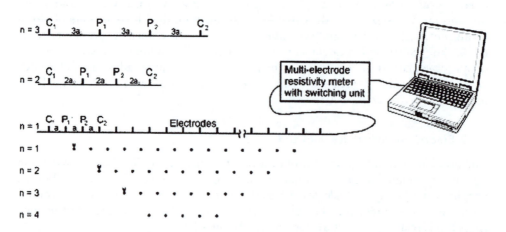

Fig. 6.4 A Wenner electrode configuration used for measuring subsurface moisture content by electrical resistivity. (Reproduced with permission from Kirsch, R. (2006) *Groundwater Geophysics*. Springer-Verlag, Berlin.)

the electrode spacing. With this configuration, the bulk resistivity or bulk conductivity is derived from the measured resistance using,

$$\rho_v - \frac{1}{\sigma_b} - 2\pi R a$$

where R is the measured resistance (Ω) and a is the electrode spacing (m). Further details of this and other resistivity or geo-electrical methods can be found in Binley *et al.* (2002), Kirsch (2006), Vereecken *et al.* (2006) and Robinson *et al.* (2008).

6.2 Pressure potential, pressure head and hydraulic head

There are two principal forms of energy: kinetic and potential. Because subsurface water flow is relatively slow (with the exception of that within *natural soil pipes*), its kinetic energy, which is proportional to velocity squared, is considered negligible. In contrast, the potential energy, which is due to position or internal condition, is of critical importance in determining the state and movement of water in the soil. Water, like other forms of matter, flows from areas of high potential to areas where it is lower.

The unit quantity of water on which potential is based can be volume, mass or weight. A potential per unit weight of h metres is equivalent to gh J kg^{-1} on a mass basis or $\rho_w gh$ Pa on a volume basis (note acceleration due to gravity, g, is 9.80665 m s^{-1} and density of pure water, ρ_w, is approximately 10^3 kg m^{-3}). The *pressure potential* is the amount of useful work that must be done per unit quantity of pure water to transfer reversibly and isothermally an infinitesimal quantity of water from a pool at a standard atmospheric pressure that contains a solution identical in composition to the subsurface water and is at the elevation of the point under consideration. Thus, the pressure potential per unit volume,

$$\phi_{p(volume)} = p = \rho_w g h$$

where p is the pressure, ρ_w is the density of the water, g is the acceleration of gravity and h is the *pressure head*. Thus the pressure potential per unit mass,

$$\phi_{p(mass)} = \frac{p}{\rho} = g h$$

and per unit weight,

$$\phi_{p(weight)} = \frac{p}{\rho g} = h$$

When the hydrostatic pressure of water within the subsurface is greater than the atmospheric pressure, the *pressure potential* is positive. Where it is less than atmospheric pressure, the subsurface water system has a *negative pressure potential*, which is also termed the *matric potential, capillary potential, suction* or *tension*.

A further potential term is involved in subsurface water movement, namely the *gravitational potential*. To raise water against gravity requires work to be expended, and this energy is stored within water body as gravitational potential energy. The *gravitational potential* of subsurface water at each point is normally determined at a point relative to an arbitrary reference level. It is customary to set this level below the subsurface system being studied so that values are always positive. As with pressure potential, the *gravitational potential* can be expressed in three ways,

$$\phi_{g(volume)} = \rho_w g z$$

$$\phi_{\varepsilon(mass)} = g z$$

$$\phi_{g(weight)} = z$$

where z is the *elevation head*. The combination of the pressure head and gravitational head gives the *hydraulic head* (H),

$$H = p + z$$

As the other terms of *osmotic head, pneumatic head* and *envelope head* are considered negligible for many applications, the hydraulic head often approximates the *total head* (H_t), which is equivalent to the *total potential* per unit weight (Φ).

6.2.1 Piezometer design

Within soil, regolith or rock that is fully saturated, the pressure potential can be measured with a *piezometer*. A *piezometer* is a narrow tube installed within the ground, where only the lower section is in contact with subsurface water system via a screened section and gravel pack (Fig. 6.5).

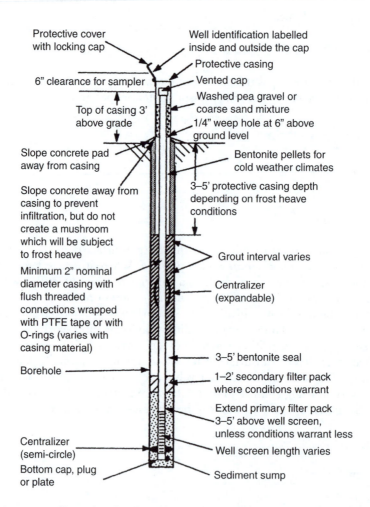

Fig. 6.5 A piezometer. (Reproduced with permission from Nielsen, D. M. (2006) *Practical Handbook of Environmental Site Characterization and Ground-water Monitoring*, 2nd edn. Taylor & Francis, London.)

The top of the piezometer can be capped to prevent ingress of rainfall, but must not be sealed as the upper water surface in the piezometer must be at atmospheric pressure. For piezometers to be installed within soil horizons (the solum) of say 0–3 m, the hole for the piezometer can be hand-augered. Where the piezometer is to be installed to greater depths, mechanical drilling is required. This drilling can be rotary or percussive drilling; alternatively jetting methods or drive-in piezometers can be used. Todd and Mays (2005) provide comprehensive details of drilling methods. To prevent a leakage of water between the piezometer tube and ground, bentonite clay seals are added above the screened section, and sometimes also at the surface (Fig. 6.5). The height of the free-water surface above the screened piezometer base is called the *piezometric head*, and is equivalent to the *pressure head, h*. This pressure head can be measured using a manual dip meter, or continuously using a pressure transducer connected to a data logger, as used for river stage measurement (Section 7.3.3). Where the profile contains impeding strata, the pressure head within each stratum may be measured by installing several piezometers in close proximity, each with a piezometer screen in contact with a different stratum. This configuration is described as a *piezometer nest*. Comprehensive details of piezometer design and installation are given within Nielsen (2006).

6.2.2 Observation well design

Measurement of the pressure head at depth within rock aquifers (i.e. groundwater bodies with high porosity and high saturated hydraulic conductivity) is more commonly measured within an *observation well* rather than a piezometer. In contrast to the situation with a piezometer, the water column within an *observation well* is normally in direct contact with the whole depth of the rock aquifer via an extended screen and gravel pack. These devices often use larger diameter cases, allowing float-operated water-level recorders to be used; these float systems can be monitored electronically using shaft encoders (Section 7.3.2). Pressure transducers (Section 7.3.3) are, however, more commonly used for level measurements within observation wells.

If there are no confining strata along this screened length, then these devices give values of pressure head similar to those of the piezometer. If confining strata are present, then perched water tables may develop, which leak into the observation well; under these circumstances piezometers should be used.

6.2.3 Tensiometer design

Where saturated conditions are absent, pressure potential can be measured with a device called a *tensiometer* (Richards, 1928). A *tensiometer* can be used to measure both positive and negative pressure potential (Pa; this can be converted to a pressure head for comparison with piezometer data). A tensiometer consists of a porous cup, generally of ceramic material connected to a sealed tube (Fig. 6.6). Once installed within the soil and filled with (de-aired) water, the water within the tube equilibrates with the pressure potential in the soil surrounding the porous cup. The pressure potential within the tensiometer can be measured manually using an attached mercury manometer or vacuum/pressure gauge, and such devices are still manufactured. More commonly, tensiometers are fitted with a pressure transducer and the pressure potential

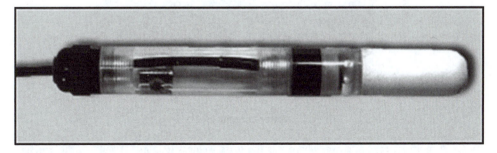

Fig. 6.6 A tensiometer with integral transducer. (Reproduced with permission of UMS GmbH)

data logged. The range of negative pressure potential that can be measured is normally 0–80 kPa (equivalent to 800 millibars) due to the air-entry point of the porous cup. When pressure transducers are used, the maximum value of positive pressure potential recorded is limited by the type and range of transducer used.

6.3 Moisture release curve

The relationship between the subsurface moisture content (volumetric wetness) and the negative pressure potential (Pa) is called the *moisture release curve* (Fig. 6.7); it is also known as the *moisture retention curve* and *moisture characteristic curve*.

The simplest expression of this empirical relationship is,

$$\phi_{p(volume)}{}^{-1} = a\theta_w{}^b$$

where $\phi_{p(volume)}{}^{-1}$ is the negative pressure potential, and a and b are the parameters that can be determined for a particular soil from the intercept and slope of a straight line fitted to the plot of log $\phi_{p(volume)}{}^{-1}$ against log θ_w (Visser, 1969). This curve is used to estimate one moisture variable from the other and in the estimation of the unsaturated hydraulic conductivity curve (Section 5.2).

Over the range of pressure potential 0 to -20 kPa, a *sand tension table* (Fig. 6.8a) can be used to derive the curve. With this device, undisturbed soil cores are placed on the sand surface and allowed to equilibrate with the negative pressure potential created in the device. These cores are then weighed as part of a gravimetric measurement of soil moisture content (Section 6.1.2).

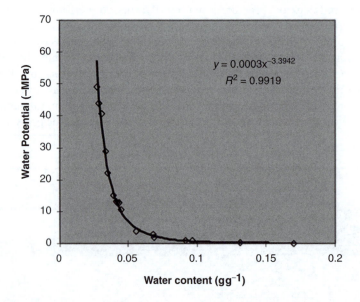

Fig. 6.7 A moisture release curve, where x is the mass wetness (θ_m, gg^{-1}) and y is the pressure potential ($\phi_{p(volume)}$ in $-$MPa). (Reproduced with permission of Decagon Devices Inc.)

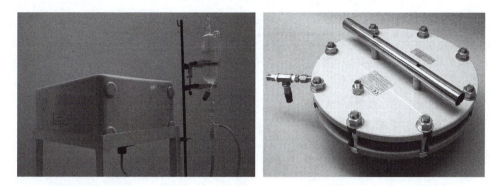

Fig. 6.8 Devices for measuring the moisture release curve: (a) a sand tension table; and (b) a pressure plate apparatus (reproduced with permission of Soil Moisture Equipment Corporation, Santa Barbara, USA).

Over the range of pressure potential -20 to $-150\,000\,\text{kPa}$, a *pressure plate device* (Fig 6.8b) can be used to derive the curve. The undisturbed samples are placed on a porous plate within a pressure vessel. A pump is used to set the pressure potential, and once in equilibrium with the pressure plate device, samples are weighed and finally dried as part of gravimetric analysis.

6.4 Subsurface flow tracing

A *conservative tracer* of water pathways has the physical, chemical or biological characteristics (Section 8.1) that allow it to move at the same velocity as the water, but not allow it to bind (e.g. sorb) on to any surrounding media (i.e. organics, soil, regolith, rock or river bed). Tracers include sodium chloride, bromide, fluorescent dyes (e.g. Rhodamine WT) and the hydrogen- and oxygen-isotopes of water. Tagging subsurface flow with the isotopes of water has the advantage that they are the water itself, but the disadvantage that the analysis using a *mass spectrometer* (Section 8.4) is expensive and fractionation can occur during sampling (Kendall and McDonnell, 1998; McGuire and McDonnell, 2007). Common salt is often used as a tracer for *dilution gauging* of small rivers and streams (Section 7.5) and has the advantage that it can be traced *in situ* (or non-invasively) using *electrical resistivity methods* (Section 6.1.7; Osiensky and Donaldson, 1995; Vanderborght *et al.*, 2005) or by using *electrical conductivity probes* within piezometers (Section 8.5). Additionally, tracers can be sampled from within piezometers or extracted using *suction lysimeters* (also called *vacuum samplers*; Section 8.3.2: Nielsen and Nielsen, 2006). At natural exposures of the soil–bedrock interface, tracers may be collected from the same location as the rate of water exfiltration is measured (Tromp-van Meerveld *et al.*, 2007).

Tracers can be artificially added to the subsurface system using a surface irrigation system, line-source injection (drip or spray), via a trench or via piezometers and observation wells.

An example injection and sampling array for tracer tests on a hillslope section is shown in Fig. 6.9. Within this study Chappell and Sherlock (2005) applied a NaCl tracer as a line-source spray at the start of individual rainstorms. Tracer was then

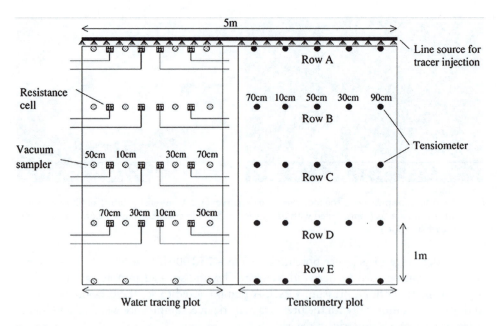

Fig. 6.9 An experimental design for subsurface water tracing within a hillslope soil. (Reproduced from Chappell and Sherlock, 2005, with permission of Wiley-Blackwell.)

extracted from the subsurface using *vacuum samplers* at four depths (Section 8.3.2) and the electrical conductivity measured using an *electrical conductivity probe*. The NaCl concentration was then determined using a site-specific calibration. Gold-coated resistance cells (Coleman, 1946) were used to monitor the electrical conductivity of the subsurface water *in situ*. *Tensiometers* (Section 6.2.3) were also installed within this array and the data combined with *unsaturated hydraulic conductivity* data (Section 5.2) to calculate Darcian velocities (Chapter 14).

Subsurface flow tracing can be used to determine a mean pore-water velocity ($\overline{v_{\mathrm{pore}}}$) and thence the Darcy velocity (q_{Darcy}) as derived by the Darcy–Richards equation,

$$q_{Darcy} = \overline{v_{pore}}\,\eta$$

where η is the porosity (Section 5.4). Derivation of the mean pore-water velocity from a tracer plume does, however, require the measurement of the local *longitudinal dispersion coefficient* and inversion of the advection–dispersion equation (Section 5.5; Chapter 14).

6.5 Overland flow measurement

Water moving over slopes towards stream channels is called *overland flow*, and is generated either by rainfall failing to infiltrate because of a low *infiltration capacity* (Section 5.1.1) or by precipitation falling on to already saturated ground or the exfiltration of subsurface flow prior to reaching a channel (see Section 1.2) or by fluvial flooding, where the river overtops its banks. Strictly, overland flow is any lateral flow above mineral soil horizons, and thus includes flow within the litter layer. While the proportion of precipitation travelling to channels by overland flow is now considered to be much less than that travelling to channels by subsurface pathways (Section 1.2), overland flow observations are of fundamental importance to the quantification of floodplain inundation (Bates *et al.*, 2006), hillslope erosion and sediment delivery (Owens and Collins, 2006), and the migration of phosphorus across agricultural landscapes (Withers and Bailey, 2003). This brief section will describe methods of measuring the presence and rates of overland flow, and is included to support later modelling (Sections 12.8.4 and 12.8.5) and to assist those hydrologists working on hydro-geomorphological or water-quality problems.

6.5.1 Measurement of incidence and spatial extent

The spatial distribution of areas likely to generate overland flow can be measured using a network of *crest stage recorders* (Fig. 6.10; Beven and Kirkby, 1979; Burt *et al.*, 1983; Holden and Burt, 2003; Bracken and Kirkby, 2005). These are small cups that fill with water when overland flow is present on slopes. Nests of cups can be used to identify different depths of overland flow.

In areas where overland flow is produced in areas of saturated topsoil rather than areas of low *infiltration capacity* (see Section 5.1.1), then measurements of *moisture content* of the organic surface horizons or topsoil (also called A soil horizon) are useful for mapping the likelihood of overland flow (Western and Grayson, 2000; Lin, 2006). Explanation of techniques for measuring soil moisture content is given in Section 6.1.

For erosion studies, knowledge of the micro-scale patterns of overland flow can be important. On soil slopes, overland flow is more likely to be present within micro-rills rather than as thin sheets of water. As erosivity of overland flow is dependent on the depth of flow, concentrating flow into micro-rills makes the water more erosive. Quantification of the three-dimensional structure of these micro-rills can be obtained by surveying with a total station, differential global positioning system (GPS) or laser scanning system (Chihua *et al.*, 2002).

At a much larger scale, remote sensing can be used to identify areas that are covered by overland flow. Airborne synthetic aperture radar (SAR) imagery has been used to map the extent of overland flow caused by fluvial flooding (Bates *et al.*, 2006).

6.5.2 Direct measurement of overland flow

The volume of overland flow generated by a bounded plot can be measured by directing the water from an *overland flow trough* into either a large tipping-bucket device or a flume. Where a tipping-bucket device is used (Fig. 6.11), the bucket tips when it has filled to a known (calibrated) volume. As it tips, it moves a magnet past a reed switch, which makes an electrical circuit.

The time that this circuit is made can be recorded using a data logger. For details of the alternative flume-based approach, including ways of measuring stage within the flume and calibrating stage to spot discharges, see Chapter 7. The plot is best bounded or isolated with the use of steel or plastic sheets. By bounding the plot,

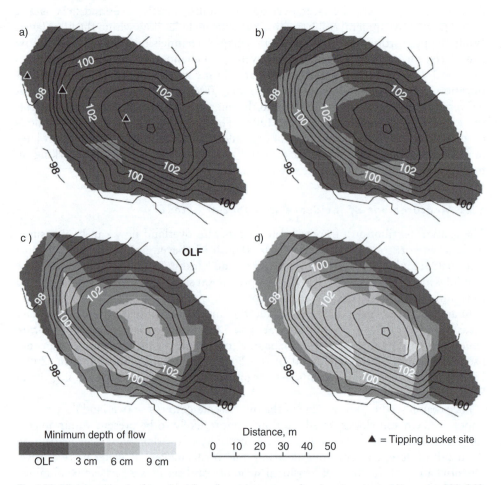

Fig. 6.10 Minimum depth of overland flow from the peat surface in micro-basin H1 on day 239–240, 1999 as monitored by crest-stage tubes: (a) 0300 day 239; (b) 0900 day 239; (c) 2100 day 239; and (d) 0900 day 240. (Reproduced from Holden and Burt, 2003, with permission of the American Geophysical Union.)

the volume of overland flow per time interval per unit area (hence a depth per time interval, e.g. mm h^{-1}) can be compared with the rainfall depth per time interval or river runoff per time interval. Fig. 6.12 shows a time series of overland flow per unit area recorded by the tipping-bucket shown in Fig. 6.11 (site 'E8') during an example storm-event, together with flows from a road drain ('E2'), a channel-head gully ('E5') and the third-order stream ('P1').

The contact between the soil and the trough that directs the overland flow into the tipping-bucket or flume is important. A large drop between the soil and the trough would produce local erosion of soil immediately upslope, while a step-up can produce artificial ponding and sedimentation (Hudson, 1957). Consequently, the presence of substantial lowering of the soil surface within the bounded plot by erosion would necessitate adjustment of the height of the sill of the trough.

Fig. 6.11 A 3-L tipping-bucket device used for measuring overland flow.

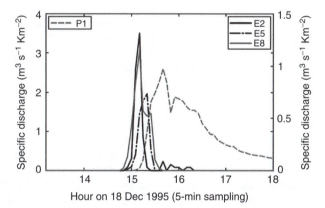

Fig. 6.12 Overland flow per unit area recorded by the tipping-bucket shown in Fig. 6.11 (site 'E8') during an example storm-event, together with flows from a road drain ('E2'), a channel-head gully ('E5') and the third-order stream ('P1'). (Reproduced from Chappell *et al.*, 2004 with permission of Wiley-Blackwell.)

Overland flow, even within similar topographic locations, has a high spatial variability. Consequently, many bounded plots are needed to determine an accurate catchment estimate of overland flow; care is also needed to avoid biasing the sampling to areas with a high overland flow (Hudson, 1993).

It can be noted that similar troughs inserted into the upslope wall of soil pits (called *throughflow troughs*) have been used to examine subsurface flow. Knapp (1970) has however shown that the excavation of a pit substantially alters the incidence, magnitude and patterns of subsurface flow, making data from throughflow troughs very difficult to interpret. The exception to this is where throughflow troughs can be inserted in to natural soil faces, such as those studied by Woods and Rowe (1996).

References

Archie, G. (1942) Electrical resistivity log as an aid in determining some reservoir characteristics. *American Institute of Mining and Metallurgical Engineers* 146, 55–62.

Bates, P. D., Wilson, M. D., Horritt, M. S., Mason, D., Holden, N. and Currie, C. (2006) Reach scale floodplain inundation dynamics observed using airborne Synthetic Aperture Radar imagery: data analysis and modelling. *Journal of Hydrology* 328, 306–318.

Beven, K. J. and Kirkby, M. J. (1979) A physically-based variable contributing area model of basin hydrology. *Hydrological Sciences Bulletin* 1, 43–69.

Binley, A., Cassiani, G., Middleton, R. and Winship, P. (2002) Vadose zone model parameterisation using cross-borehole radar and resistivity imaging. *Journal of Hydrology* 267(3–4), 147–159.

Bracken, L. J. and Kirkby, M. J. (2005) Differences in hillslope runoff and sediment transport rates within two semi-arid catchments in southeast Spain. Geomorphology 68, 183–200.

BS 1377–3 (1990) *Methods of Test for Soils for Civil Engineering Purposes. Chemical and Electro-Chemical Tests*. British Standards Institution, London.

Burt, T. P., Butcher, T. P., Coles, N. and Thomas, A. D. (1983) The natural history of Slapton Ley Nature Reserve XV: hydrological processes in the Slapton Wood catchment. *Field Studies* 5, 731–752.

Chappell, N. A. and Sherlock, M. D. (2005) Contrasting flow pathways within tropical forest slopes of Ultisol soil. *Earth Surface Processes and Landforms* 30, 735–753.

Chappell, N. A., Douglas, I., Hanapi, J. M. and Tych, W. (2004) Source of suspended-sediment within a tropical catchment recovering from selective logging. *Hydrological Processes* 18, 685–701.

Chihua, H., Fenli, Z. and Darboux, F. (2002) How surface conditions affect sediment and chemical transport. *Twelfth ISCO Conference*, Beijing.

Childs, E. C. (1969) *An Introduction to the Physical Basis of Soil Water Phenomena*. John Wiley, London, 493 pp.

Coleman, E. A. (1946) The place of electrical soil moisture meters in hydrological research. *Transactions of the American Geophysical Union* 27, 847–853.

Hillel, D. (1980) *Fundamentals of Soil Physics*. Academic Press, New York, 413 pp.

Holden, J. and Burt, T. P. (2003) Runoff production in blanket peat covered catchments. *Water Resources Research* 39, 1191, doi:10.1029/2002WR001956.

Hudson, N. W. (1957) The design of field experiments on soil erosion. *Journal of Agricultural Engineering Research* 2, 56–65.

Hudson, N. W. (1993) *Field Measurement of Soil Erosion and Runoff*. Food and Agriculture Organization of the United Nations, Rome.

Institute of Hydrology (1979) *Neutron Probe System IH II. Instruction Manual*. Natural Environment Research Council, Swindon, 32 pp.

Kendall, C. and McDonnell, J. J. (1998) *Isotope Tracers in Catchment Hydrology*. Elsevier Science B.V., Amsterdam.

Kirsch, R. (2006) *Groundwater Geophysics*. Springer, Berlin. 493 pp.

Knapp, B. J. (1970) *Patterns of Water Movement on a Steep Upland Hillside, Plynlimon, Central Wales*. Ph.D. thesis, University of Reading, unpublished.

Lin, H. (2006) Temporal stability of soil moisture spatial pattern and subsurface preferential flow pathways in the Shale Hills Catchment. *Vadose Zone Journal* 5, 317–340.

McGuire, K. and McDonnell, J. J. (2007) Stable isotope tracers in watershed hydrology. In: Lajtha, K. and Michener, W. (eds) *Stable Isotopes in Ecology and Environmental Sciences*, 2nd edn. Blackwell Publishing, Oxford.

Nielsen, D. M. (2006) *Practical Handbook of Environmental Site Characterisation and Ground-Water Monitoring*. Taylor & Francis, Boca Raton.

Nielsen, D. M. and Nielsen, G. L. (2006) Ground-water sampling. In: Nielsen, M. (ed.) *Practical Handbook of Environmental Site Characterisation and Ground-Water Monitoring*, 2nd edn. Taylor and Francis, Boca Raton, pp. 959–1112.

Osiensky, J. L. and Donaldson, P. R. (1995) Electrical flow through an aquifer for contaminant source leak detection and delineation of plume evolution. *Journal of Hydrology* 169, 243–263.

Owens, P. N. and Collins, A. J. (2006) *Soil Erosion and Sediment Redistribution in River Catchments: Measurement, Modelling and Management*. CABI Publishing, Wallingford.

Richards, L. A. (1928) The usefulness of capillary potential to soil moisture and plant investigators. *Journal of Agricultural Research* 37, 719–742.

Robinson, D. A., Binley, A., Crook, N., Day-Lewis, F., Ferré, P. T., Grauch, V. J. S., Knight, R., Knoll, M., Lakshmi, V., Miller, R., Nyquist, J., Pellerin, L., Singha, K. and Slater, L. (2008) Advancing process-based watershed hydrological research using near-surface geophysics: a vision for, and review of, electrical and magnetic geophysical methods. *Hydrological Processes* 22, 3604–3635.

Shah, P. H. and Singh, D. N. (2005) Generalized Archie's Law for estimation of soil electrical conductivity. *Journal of ASTM International* 2 (5), 1–20.

Schwartz, B. F., Schreiber, M. E. and Yan, T. (2008) Quantifying field-scale soil moisture using electrical resistivity imaging. *Journal of Hydrology* 362, 234–246.

Todd, D. K. and Mays, L. W. (2005) *Groundwater Hydrology*, 3rd edn. Wiley, New York.

Topp, G. C., Davis, J. L. and Annan, A. P. (1980) Electromagnetic determination of soil water content: measurements in coaxial transmission lines. *Water Resources Research* 16, 574–582.

Tromp-van Meerveld, H. J., Peters, N. E. and McDonnell, J. J. (2007) Effect of bedrock permeability on subsurface stormflow and the water balance of a trenched hillslope at the Panola Mountain Research Watershed, Georgia, USA. *Hydrological Processes* 21, 750–769.

Ulaby, F. T., Moore, R. K. and Fung, A. K. (1986) *Microwave Remote Sensing: Active and Passive, Vol. III: From Theory to Applications*. Artech House, Boston, London, p. 1098.

Vanderborght, J., Kemna, A., Hardelauf, H. and Vereecken, H. (2005) Potential of electrical resistivity tomography to infer aquifer transport characteristics from tracer studies: a synthetic case study. *Water Resources Research* 41, W06013, doi:10.1029/2004WR003774.

van Genuchten, M. T. (1980) A closed-form equation for predicting the hydraulic conductivity of unsaturated soils. *Soil Science Society of America Journal* 44, 892–898.

Vereecken, H., Binley, A., Cassiani, G., Revil, A. and Titov, K. (2006) *Applied Hydrogeophysics*. Springer, Berlin, 383 pp.

Visser, W. C. (1969) An empirical expression for the desorption curve. In: Rijtema, P. E. and Wassink, H. (eds) *Water in the Unsaturated Zone*, Unesco, Paris, 329–335.

Vivoni, E. R., Gebremichael, M., Watts, C. J., Bindlish, R. and Jackson, T. J. (2008) Comparison of ground-based and remotely-sensed surface soil moisture estimates over complex terrain using SMEX04. *Remote Sensing of Environment* 112, 314–325.

Walker, J. P., Houser, P. R. and Willgoose, G. R. (2004) Active microwave remote sensing for soil moisture measurement: a field evaluation using ERS-2. Hydrological Processes 1811, 1975–1997.

Western, A. W. and Grayson, R. B. (2000) Soil moisture and runoff processes at Tarrawarra. In: Grayson, R. B. and Blöschl, G. (eds) *Spatial Patterns in Catchment Hydrology – Observations and Modelling*. Cambridge University Press, Cambridge, pp. 209–246.

Whipkey, R. Z. (1965) Subsurface stormflow of forested slopes. *International Association of Scientific Hydrology Bulletin* 10, 74–85.

Withers, P. J. A. and Bailey, G. A. (2003) Sediment and phosphorus transfer in overland flow from a maize field receiving manure. *Soil Use and Management* 19, 28–35.

Woods, R. and Rowe, L. K. (1996) The changing spatial variability of subsurface flow across a hillside. *Journal of Hydrology (New Zealand)* 5, 51–86.

Chapter 7

River flow

To introduce the measurement of river discharge, a summary of the salient features of open channel flow is given here. The main topic of this chapter is *river hydrometry*, which in its restricted sense means river level, velocity and discharge measurement. A proper understanding of river flow measurement requires a basic knowledge of the mechanics of *open channel flow*. A more detailed analysis of unsteady open channel flow is contained in Chapter 14.

7.1 Open channel flow

Water in an open channel is effectively an incompressible fluid that is contained but can change its form according to the shape of the container. In nature, the bulk of fresh surface water either occupies hollows in the ground, as lakes, or flows in well-defined channels. Open channel flow also occurs in more regular man-made sewers and pipes as long as there is a free water surface and gravity flow.

The hydrologist is interested primarily in discharge of a river in terms of cubic metres per second ($m^3 s^{-1}$), but in the study of open channel flow, although the complexity of the cross-sectional area of the channel may be readily determined, the velocity of the water in metres per second ($m s^{-1}$) is also a characteristic of prime importance. The variations of velocity both in space and in time provide bases for the standard classifications of flow.

7.1.1 Uniform flow

In practice, uniform flow usually means that the velocity pattern within a constant cross-section does not change in the direction of the flow. Thus in Fig. 7.1, the flow shown is uniform from A to B in which the depth of flow, yo, called the *normal depth*, is constant. The values of velocity, v, remain the same at equivalent depths. Between B and C, the flow shown is non-uniform; both the depth of flow and the velocity pattern have changed. In Fig. 7.1, the depth is shown as decreasing in the direction of flow (y1 < yo). A flow with depth increasing (y1 > yo) with distance would also be non-uniform.

7.1.2 Velocity distributions

Over the cross-section of an open channel, the velocity distribution depends on the character of the river banks and of the bed and on the shape of the channel.

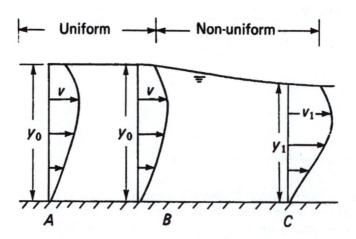

Fig. 7.1 Uniform and non-uniform flow.

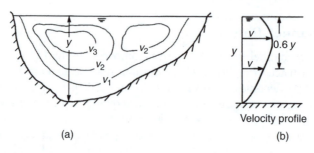

(a)   (b)

Fig. 7.2 Velocity distributions.

The maximum velocities tend to be found just below the water surface and away from the retarding friction of the banks. In Fig. 7.2a, lines of equal velocity show the velocity pattern across a stream with the deepest part and the maximum velocities typical of conditions on the outside bend of a river. A plot of the velocities in the vertical section at depth y is shown in Fig. 7.2b. The average velocity of such a profile is often assumed to occur at or near 0.6 depth.

7.1.3 Laminar and turbulent flow

When fluid particles move in smooth paths without lateral mixing, the flow is said to be *laminar*. Viscous forces dominate other forces in laminar flow and it occurs only at very small depths and low velocities. It is seen in thin films over smooth paved surfaces. Laminar flow is identified by the Reynolds number $Re = \rho_w \upsilon y / \mu$, where ρ_w is the water density and μ the dynamic viscosity. (For laminar flow in open channels, Re is less than about 500). As the velocity and depth increase, Re increases and the flow becomes *turbulent*, with considerable mixing laterally and vertically in the channel. Nearly all open channel flows are turbulent.

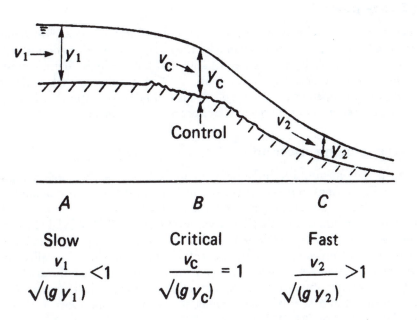

Fig. 7.3 The occurrence of critical flow as defined by the Froude number, $Fr = v/\sqrt{(gy)}$, where v is the velocity, g is gravitational acceleration and y is the depth of flow. Slow flow is also called subcritical flow, and fast flow is also called supercritical flow.

7.1.4 Critical, slow and fast flow

Flow in an open channel is also classified according to an energy criterion. For a given discharge, the energy of flow is a function of its depth and velocity, and this energy is a minimum at one particular depth, the critical depth, y_c (Fig. 7.3). It can be shown (Akan, 2006) that the flow is characterized by the dimensionless Froude number:

$$Fr = \frac{v}{\sqrt{(gy)}}$$

where v is the velocity, g is gravitational acceleration and y is the depth of flow. For $Fr < 1$, flow is said to be subcritical (slow, gentle or tranquil). For $Fr = 1$, flow is critical, with depth equal to y_c the critical depth. For $Fr > 1$, flow is supercritical (fast or shooting) (Fig. 7.3). Larger flows have larger values of v_c and y_c. The occurrence of *critical flow* is very important in the measurement of river discharge because, at the point of *critical flow* for a given discharge, there is a unique relationship between the velocity and the discharge as $v = \sqrt{(gy)}$. Thus only depth has to be measured to calculate velocity. Elsewhere, the flow might be either a subcritical or supercritical state, and both velocity and depth would have to be measured to derive discharge. As discussed later in Section 7.6, if a cross-section where critical flow occurs naturally cannot be found, the installation of a weir or flume can force the flow to become critical.

7.1.5 Steady flow

This occurs when the velocity at any point does not change with time. Flow is unsteady in surges and flood waves in open channels. The analytical equations of unsteady flow are complex and difficult to solve (see Chapter 14) but the hydrologist is most often concerned with these unsteady flow conditions. With the more simple conditions of steady flow, some open channel flow problems can be solved using the principles of continuity, conservation of energy and conservation of momentum.

7.2 River gauging methods

As in the measurement of precipitation, measurement of river discharge is a sampling procedure. For springs and very small streams, accurate volumetric quantities over timed intervals can be measured, and is called *volumetric gauging*. For a large stream, a continuous measure of one variable, river level (Section 7.3), is related to the spot measurements of discharge collected by *dilution gauging methods* (Section 7.5) or calculated from sampled values of the variables, velocity and area (so-called *velocity-area methods*; Section 7.4). Where *velocity–area methods* are used, the discharge of a river, Q, is normally obtained from the summation of the product of mean velocities in the vertical, $\bar{v}$, and area of related segments, a, of the total cross-sectional area, A (Fig. 7.4; see Section 7.4.3). Thus,

$$Q = \sum (\bar{v}, b, y) = \sum (\bar{v}, a)$$

The fixed cross-sectional area is determined with relative ease, but it is much more difficult to ensure consistent measurements of the flow velocities to obtain values of $\bar{v}$.

A single estimate of river discharge can be made readily on occasions when access to the whole width of the river is feasible and the necessary velocities and depths can be measured. However, such 'one-off' values are of limited use to the hydrologist. Continuous monitoring of the river flow is essential for assessing water availability. The continuous recording of velocities across a river is normally not a practical proposition for large rivers, *ultrasonic (doppler) flow meters* are sometimes used to measure a continuous record of velocity in small streams. It is, however, relatively simple to arrange for the continuous measurement of the river level. A fixed and constant relationship is required between the river level (called *stage*) and the discharge at the gauging site. This occurs along stretches of a regular channel where the flow is slow and uniform and the stage-discharge relationship is under 'channel control'. In reaches where the flow is usually non-uniform, it is important to arrange a unique relationship between water level and discharge. It is therefore necessary either to find a natural 'bed control' as in Fig. 7.3, where critical flow occurs over some rapids with a tranquil pool upstream, or to build a control structure across the bed of the river making the flow pass through critical conditions (Fig. 7.4b). Where a control structure has been built, discharge is sometimes described as being measured by *structural methods* (Section 7.6). In both cases, the discharge, Q, is a unique function of y_c and hence of the water level just upstream of the control. In establishing a permanent critical section gauging station, care has to be taken to verify that the bed or structural control regulates the upstream flow for all discharges. At very high flows, the section of critical flow may be 'drowned

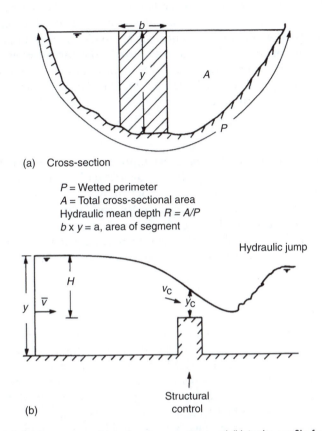

(a)   Cross-section

P = Wetted perimeter
A = Total cross-sectional area
Hydraulic mean depth $R = A/P$
$b \times y = a$, area of segment

(b)

Fig. 7.4 Channel definitions, where (a) is the cross-section, and (b) is the profile for $y \gg y_c$ and H is head.

out' as higher levels downstream of the control eliminate the critical depth. Then the flow depths will be greater than y_c throughout the control and the relationship between the upstream water level and discharge reverts to 'channel control'.

At a gauging site, when the flow is contained within the known cross-section and is controlled by a bed structure, then the discharge Q is a function of H (head), the difference in height between the water level upstream and the crest level of the bed control (Fig. 7.4b). The functional stage-discharge relationship or *rating curve* (see Section 7.7) is established by either measuring discharge directly (*dilution methods*) or by estimating Q from sampled measurements of either velocity across the channel (*velocity–area methods*), when it is convenient, for different values of H. Regularly observed or continuously recorded stages or river levels can then be converted to corresponding discharge time series. For a structural control, e.g. a weir built to standard specifications, the stage-discharge or $Q \sim H$ relationship is known, and velocity–area or dilution measurements are used only as a check on the weir construction and calibration.

After flood flows, cross-sectional dimensions at a gauging station should be checked for erosion–deposition-related changes and, if necessary, the river level-discharge relationships amended by a further series of velocity-area measurements.

The type of river gauging station depends very much on the site and character of the river. To a lesser extent, its design is influenced by the data requirements, since most stations established on a permanent basis are made to serve all purposes. Great care must therefore be afforded to initial surveys of the chosen river reach and the behaviour of the flow in both extreme conditions of floods and low flows should be observed if possible. Details of methods used for stage and discharge measurements will be given in the following sections.

7.3 Stage

The water level at a gauging station, the most important measurement in river hydrometry, is generally known as the *stage*. It is measured with respect to a datum, either a local bench mark or the crest level of the control, which in turn should be levelled into the geodetic survey datum of the country (Ordnance Survey datum in the UK). All continuous estimates of the discharge derived from a continuous stage record depend on the accuracy of the stage values. The instruments and installations range from the most primitive to the highly sophisticated, but can be grouped into a few important categories.

7.3.1 The staff gauge

This is a permanent graduated staff generally fixed vertically to the river bank at a stable point in the river unaffected by turbulence or wave action. It could be conveniently attached to the upstream side of a bridge buttress but is more likely to be fixed firmly to piles set in concrete at a point upstream of the river flow control. The metre graduations, resembling a survey staff, are shown in Fig. 7.5 and they should extend from the datum or lowest stage to the highest stage expected. The stage is read to an accuracy of ±3 mm. Where there is a large range in the stage with a shelving river bank, a series of vertical staff gauges can be stepped up the bank side with appropriate overlaps to give continuity. For regular river banks or smooth man-made channel sides, specially made staff gauges can be attached to the bank slope with their graduations, extended according to the angle of slope, to conform to the vertical scale of heights. All staff gauges should be made of durable material insensitive to temperature changes and they should be kept clean especially in the range of average water levels.

Depending on the regime of the river and the availability of reliable observers, single readings of the stage at fixed times of the day could provide a useful regular record. Such measurements may be adequate on large mature rivers, but for flashy streams and rivers in times of flood critical peak levels may be missed. Additionally, in these days of increasing modification of river flow by man, the sudden surges due to releases from reservoirs or to effluent discharges could cause unexpected irregular discharges at a gauging station downstream, which could be misleading if coincident with a fixed-time staff reading. To monitor irregular flows, either natural storm flows or man-made interferences, continuous level recording is essential.

7.3.2 Float-operated recorders

A reliable means of recording water level is provided by a float-operated recorder. To ensure accurate sensing of small changes in water level, the float must be installed

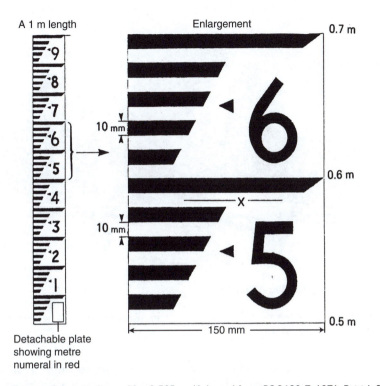

A 1 m length

Enlargement

0.7 m

10 mm

10 mm

0.6 m

X

0.5 m

150 mm

Detachable plate
showing metre
numeral in red

Fig. 7.5 A staff gauge. Stage reading at $X = 0.585$ m. (Adapted from BS 3680-7: 1971, British Standards Institution.)

in a *stilling well* to exclude waves and turbulence from the main river flow. Two different mechanisms are used to record the travel of the float:

7.3.2.1 Chart recorder

The float with its geared pulley and counterweight turns the charted drum set horizontally and the pen arm is moved across the chart by clockwork or an electrical mechanism (Fig. 7.6). The timescale of the chart is usually designed to serve a week, but the trace continues round the drum until the chart is changed or the clock stops. With this instrument all levels are recorded, but the timescale is limited. On visiting a gauging station, a hydrologist can see at once whether or not a current storm event has peaked. However, chart records require careful analysis and time must be spent in abstracting data on a digitiser tablet.

7.3.2.2 Shaft encoder

The float with its geared pulley and counterweight turns a *shaft encoder* (also called *rotary encoder*). The encoder comprises of a disc with concentric, metal rings with breaks, plus a series of fixed contact wipers. As the disc rotates on the shaft, those

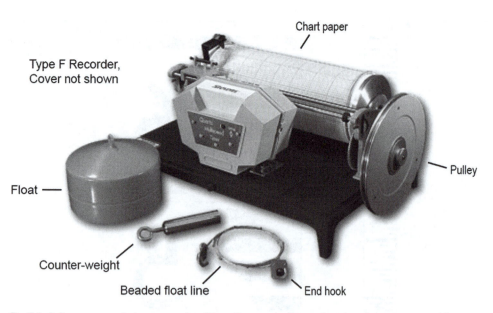

Fig. 7.6 A float-operated chart recorder (Type F system). (Reproduced with permission of Stevens Water Monitoring Systems Inc.)

wipers that are in electrical contact with the metal ring sections create a unique binary code, which can be recorded on an eight-bit data logger. Some chart recorders (e.g. Stevens units) can be retro fitted with shaft recorders. An example of a shaft encoder is shown in Fig. 7.7, and similar devices are used within the Environment Agency's network of river gauges in England and Wales.

7.3.3 Electronic pressure sensor

The measurement of stage by pressure sensors, an indirect method converting the hydrostatic pressure at a submerged datum to the water level above, are widely used for gauging small rivers and streams. Those pressure sensors used within these applications typically use piezo-resistive, silicon strain gauges, and are called *pressure transducers* (Fig. 7.8). Versions of these sensors with on-board signal amplification and current output are called *pressure transmitters*. Differential pressure transducers (or transmitters), where the differential pressure between water-level pressure and atmospheric pressure (observed by means of an air pipe running inside the cable connecting the sensor to data logger) are normally used. The calibration of the pressure sensor may change over time and this is normally checked annually.

7.3.4 Gas purge (bubbler) gauge

With gas purge devices (Fig. 7.9), nitrogen from a cylinder or air compressed from a pump, is allowed to bubble slowly out of the end of a tube located close to the river bed.

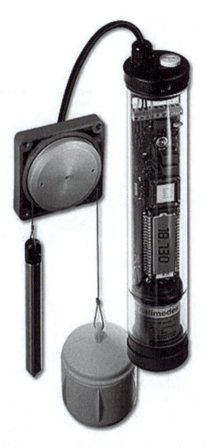

Fig. 7.7 Shaft encoder-based float-operated recorder. (Reproduced with permission of OTT Hydrometry Ltd.)

Fig. 7.8 A submersible pressure transducer used for river-level monitoring in the UK. (Reproduced with permission of Campbell Scientific Ltd.)

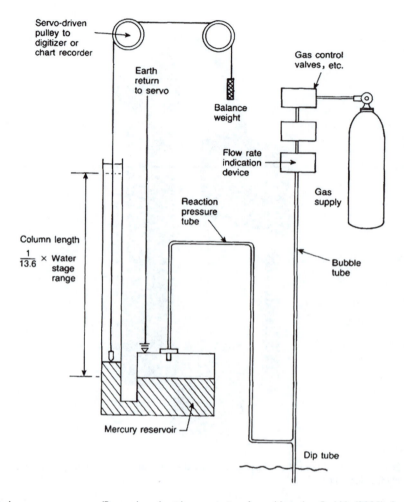

Fig. 7.9 A gas purge gauge. (Reproduced with permission from Herschy, R. W. (2009) *Streamflow Measurement*, 3rd edn. Taylor & Francis, Abingdon.)

When the rate of bubble production is sufficiently small, the pressure in the line is static so that the pressure at the orifice is the same as the pressure at the other end of the tube in the instrument itself. This allows the pressure to be measured in the instrument rather than in the river, and is usually measured with an electromechanical balance (using bellows or mercury-float device) or pressure transducer. The bellows-based electromechanical balance comprises an arm connected to pressure-activated bellows and a counterweight, and a servo-mechanism. The servo-mechanism is used to balance an arm, and its movement is transferred via gears to a pen on a chart. With the mercury-float device, the servo is used to balance a float and counter-weight system where the float is contained within a mercury reservoir. The principal advantage of the gas purge gauge is that no sensors need to be installed within or near the river, only a plastic pipe; the sensors can be housed within a building at some distance from

Fig. 7.10 A radar-based river level sensor. (Reproduced with permission from Vega UK.)

the river. This means that the expensive sensors and recording devices can be more easily protected from damage during flood flows. Gas purge gauges using a cylinder gas supply are widely used on large rivers within the tropics, while those using a pump (e.g. Seba PS-Light-2) are increasingly used on European rivers.

7.3.5 Ultrasonic and radar gauges

Ultrasonic-level gauges are inexpensive devices mounted above the surface of the river. This means that they can be used in contaminated rivers (or sewer systems) where there would be a high risk of fouling or corrosion of floats and pressure transducers. River-level gauges using radar have the advantage over ultrasonic gauges in that they are not significantly affected by air turbulence, temperature, surface angle or dust. Relatively inexpensive radar gauges have been developed recently (e.g. Vegapuls 68) that are in use within estuarine reaches of rivers around the UK (Fig. 7.10).

Details of recommended methods for measuring river level in the UK are given in BS EN ISO 4373 (2008).

7.4 Discharge by velocity–area methods

The most direct method of obtaining a value of discharge to correspond with a stage measurement is by the *velocity–area* method in which the river velocity is measured at selected verticals of known depth across a measured section of the river. Around 90 per cent of the world's rivers gauging sites depend on this method (Shaw, 1994).

At a river gauging station, the cross-section of the channel is surveyed and considered constant unless major modifications during flood flows are suspected, after which it must be resurveyed. The more difficult component of the discharge computation is the series of velocity measurements across the section. The variability in velocity both across the channel and in the vertical must be considered. To ensure adequate sampling of velocity across the river, the ideal measuring section should have a symmetrical flow distribution about the mid-vertical, and this requires a straight and uniform approach-channel upstream, in length at least twice the maximum river width. Then measurements are made over verticals spaced at intervals no greater than 1/15th of the width across the flow. With any irregularities in the banks or bed, the spacings should be no greater than 1/20th of the width (BS EN ISO 748, 2007). Guidance in the number and location of sampling points is obtained from the form of the cross-section with verticals being sited at peaks or troughs.

7.4.1 Measurement of velocity

The simplest method for determining a velocity of flow is by timing the movement of a *float* over a known distance (sometimes called *float gauging*). Surface floats comprising any available floating object are often used in rough preliminary surveys; these measurements give only the surface velocity and a correction factor must be applied to give the average velocity over a depth. A factor of 0.7 is recommended for a river of 1 m depth with a factor of 0.8 for 6 m or greater (BS EN ISO 748, 2007). Specially designed floats can be made to travel at the mean velocity of the stream (Fig. 7.11). The individual timing of a series of floats placed across a stream to determine the cross-sectional mean velocity pattern could become a complex procedure with no control of the float movements. Therefore, this method is recommended only for reconnaissance discharge estimates.

The determination of discharge at a permanent river gauging station is best made by measuring the flow velocities with a *current meter*. This is a reasonably accurate

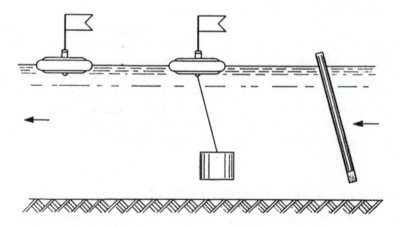

Fig. 7.11 Floats: (a) surface float; (b) canister float for mean velocity; (c) rod float by mean velocity. (Reproduced with permission from R. W. Herschy (ed.) (2009) *Streamflow Measurement*, 3rd edn, © 2009, by permission of Taylor & Francis, Oxford.)

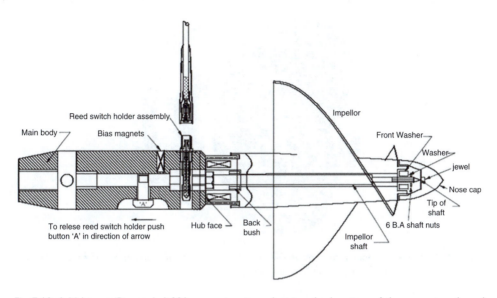

Fig. 7.12 A Valeport 'Braystoke' 001 current meter, showing the location of the magnet and reed switch. (Reproduced with permission of Valeport Ltd.)

instrument that can give a nearly instantaneous and consistent response to velocity changes. There are two main types of meter in current use: the impeller type, which has a single impeller rotating on a horizontal axis, and the electromagnetic type.

The *impeller current meter* (BS ISO 2537, 2007; Fig. 7.12) records the true normal velocity component with actual velocities up to 15° from the normal direction. Following use of a calibration of impellor revolutions to river velocity, this method gives an attainable accuracy of ± 1.5 per cent that can be obtained with the impeller-type current meter in the range of velocities between 0.3 and 10 m s^{-1} (125-mm diameter impellor).

The operation of an *electromagnetic current meter* (Fig. 7.13) utilises the Faraday principle, where water flow cuts lines of magnetic flux, inducing an electromagnetic force (emf) that is sensed by two electrodes. These current meters can be used to

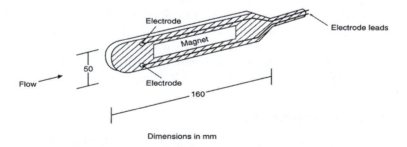

Fig. 7.13 A Valeport electromagnetic (EM) current meter showing the location of two electrodes and magnet. (Reproduced with permission of Valeport Ltd.)

Fig. 7.14 An acoustic Doppler current profiler (ADCP) showing the transducers. (Reproduced with permission of Teledyne RD Instruments.)

measure river velocities as slow as $0.03\,\mathrm{m\,s^{-1}}$ (and up to $4\,\mathrm{m\,s^{-1}}$). They also have the advantage of not having moving parts that can be caught in weeds or damaged against rocks.

In addition to the use of impellor-based and electromagnetic current meters, *ultrasonic (Doppler) flow meters* (BS EN ISO 6416, 2005) are used increasingly within the UK. These devices use transducers ('loud speakers'), and measure the sound returned (echo) by scatters (e.g. fine sediment) within the water. The Unidata Starflow unit can measure a river velocity of range of 0.02–$4.5\,\mathrm{m\,s^{-1}}$ with an accuracy of $\pm\,2$ per cent.

The *acoustic Doppler current profiler* (ADCP; Fig. 7.14) uses the same principle, but can give a very detailed distribution of river velocity at many locations over the river cross-section when mounted on a boat or float (Fig. 7.15).

To ensure that the mean velocity in a cross-section is estimated with good accuracy, it is recommended that the velocities are averaged over at least five transects across the river. This technique has recently become very popular within agencies such as the United States Geological Survey and the Environment Agency of England and Wales for discharge measurement.

Ultrasonic gauges are permanently installed at over 150 gauging stations within the network of the Environment Agency of England and Wales. These installations comprise pairs of ultrasonic transducers mounted at different depths on both river banks and determine the average velocity at each depth across the channel width (Fig. 7.16). These devices are particularly suitable for rivers subject to tidal flows and where gauging structures are impractical (see Herschy, 2009).

7.4.2 Operational considerations

The method of velocity sampling across a gauging section depends on the size of the river and its accessibility. Methods include the use of wading rods, bridges, boats, floats and cableways. With *wading* the current meter is carried on special rods and held in

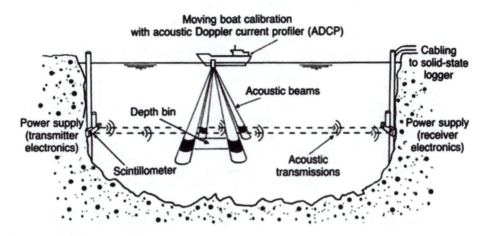

Fig. 7.15 A diagrammatic sketch of an acoustic Doppler current profiler (ADCP). (Reproduced with permission from Herschy, R. W. (2009) *Streamflow Measurement*, 3rd edn. Taylor & Francis, Abingdon.)

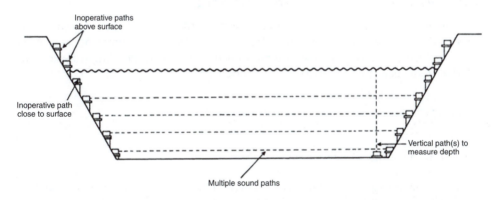

Fig. 7.16 A diagrammatic representation of an ultrasonic system. (Reproduced with permission from Herschy, R. W. (2009) *Streamflow Measurement*, 3rd edn. Taylor & Francis, Abingdon.)

position by the gauger standing on the stream bed a little to the side and downstream of the instrument. This is the ideal method, since the gauger is in full control of the operation, but it is only practicable in shallow streams with low or moderate velocities. Where there is a clear span *bridge* aligned straight across the river near the gauging station, the current meter can be lowered on a line from a gauging reel carried on a trolley. Care must be taken to sample a section with even flow. For very wide rivers, gaugings may have to be made from a *boat* either held in position along a fixed wire or under power across the section. Velocity surveys with an ADCP can be undertaken with the ADCP mounted on a *float* drawn across the channel; though boats and cableways are also used. With a *cableway*, the gauger remains on the river bank, but winds the current meter across the section on a cableway, lowering it to the desired depth by

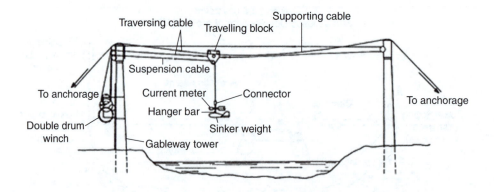

Fig. 7.17 Cableway system. (Reproduced from BS EN ISO 4375: 2004, by permission of BSI.)

remote mechanical control (Fig. 7.17). Normally, the cableway controls are installed in the recorder house alongside the stilling well installed for stage recording.

7.4.3 Gauging procedure for current metering

At the gauging station or selected river cross-section, the mean velocities for small sub-areas of the cross-section $(\bar{v}_i)$ obtained from point velocity measurements at selected sampling verticals across the river are multiplied by the corresponding sub-areas (a_i) and the products summed to give the total discharge,

$$Q = \sum_{i=1}^{n} \bar{v}_i a_i$$

where n = the number of sub-areas.

(a) The estimate Q is the discharge related to the stage at the time of gauging; therefore, before beginning a series of current-meter measurements the *stage* must be *read* and *recorded*.

(b) The width of the river is divided into about 20 sub-sections so that no sub-section has more than 10 per cent of the flow.

(c) At each of the selected sub-division points, the water depth is measured by sounding and the current meter operated at selected points in the vertical to find the mean velocity in the vertical, e.g. at 0.6 depth (one-point method) or at 0.2 and 0.8 depths (two-point method). At a new gauging station where the vertical velocity distribution is at first unknown, more readings should be taken to establish that the best sampling points to give the mean are those of the usual one or two point methods.

(d) For each velocity measurement, the number of complete revolutions of the meter over a measured time period (about 60 s) is recorded using a stopwatch. If pulsations are noticed, then a mean of three such counts should be taken.

(e) When velocities at all the sub-division points across the river have been measured, the *stage* is read again.

Should there have been a difference in stage readings over the period of the gaugings, a mean of the two stages is taken to relate to the calculated discharge. Once gaugers have gained experience of a river section at various river stages, the procedure can be speeded up and one velocity reading only at 0.6 depth taken quickly at each point across the stream, with the depths relating to the stage already known. Such an expedited procedure is absolutely essential when gauging flood flows with rapid changes in stage.

7.4.4 Calculating the discharge from current metering data

The calculation of the discharge from the velocity and depth measurements can be made in several ways. Two of these are illustrated in Fig. 7.18. In the *mean section* method, averages of the mean velocities in the verticals and of the depths at the boundaries of a section sub-division are taken and multiplied by the width of the sub-division, or segment,

$$Q = \sum q_i = \sum \bar{v}a = \sum_{i=1}^{n} \frac{(\bar{v}_{i-1} - \bar{v}_i)}{2} \frac{(d_{i-1} - d_i)}{2} (b_i - b_{i-1})$$

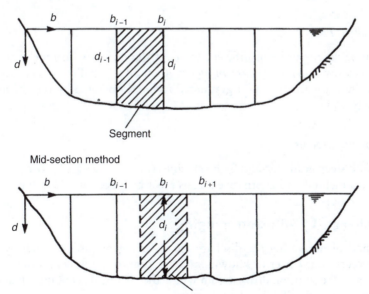

Fig. 7.18 Calculating discharge.

where b_i is the distance of the measuring point (i) from a bank datum and there are n sub-areas. In the *mid-section* method, the mean velocity and depth measured at a sub-division point are multiplied by the segment width measured between the mid-points of neighbouring segments:

$$Q = \sum q_i = \sum \bar{v}a = \sum_{i=1}^{n} \bar{v}_i d_i \frac{(d_{i-1} - d_i)}{2}$$

with n being the number of measured verticals and sub-areas. In the *mid-section* calculation, some flow is omitted at the edges of the cross-section, and therefore the first and last verticals should be sited as near to the banks as possible.

7.4.5 Problems with velocity–area methods

7.4.5.1 Large rivers

Across wide rivers, there is always difficulty in locating the instruments accurately at the sampling points and inaccuracies invariably occur. Problems in locating the bed of the river may also arise in deep and fast flows, and a satisfactory gauging across such a river may take many hours to complete. Check readings of the stage *during* such an operation are advisable. In deep swift-flowing rivers, heavy weights according to the velocity are attached, but the force of the current usually causes a drag downstream from the vertical. Measurements of depth have to be corrected using the measured angle of inclination of the meter cable. For detailed instructions on the gauging methods used in large rivers using the *moving boat method*, see Herschy (2009).

7.4.5.2 Shallow rivers

The depth of flow may be insufficient to cover the ordinary current meter. Smaller instruments known as *pygmy current meters* are used for shallow rivers and low-flow gaugings. They are attached to a graduated rod and operated by the gauger wading across the section.

7.4.5.3 Upland streams

Streams with steep gradients and high velocities cannot be gauged satisfactorily by the velocity-area method and alternative means must be used, e.g. *dilution gauging*.

7.5 Discharge by dilution gauging

This method of measuring the discharge in a stream or pipe is made by adding a chemical solution or tracer of known concentration to the flow and then measuring the dilution of the solution downstream where the chemical is completely mixed with the stream water (BS 3680-2A, 1995).

In Fig. 7.19, c_o, c_1 and c_2 are chemical concentrations (e.g. $g\,L^{-1}$). c_o is the 'background' concentration already present in the water (and may be negligible); $c1$ is the

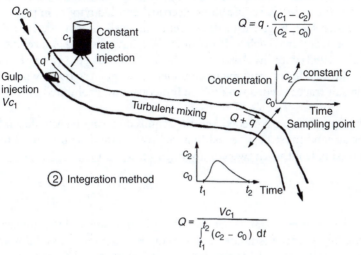

Fig. 7.19 Dilution gauging: two basic methods.

known concentration of tracer added to the stream at a *constant rate q*; and c_2 is a sustained final concentration of the chemical in the well-mixed flow. Thus,

$$Qc_o + qc_1 = (Q+q)c_2$$

whence,

$$Q = \frac{(c_1 - c_2)}{(c_2 - c_o)} q$$

An alternative to this constant rate injection method is the *'gulp'* injection, or *integration*, method. A known volume of the tracer V of concentration c_1 is added in bulk to the stream and, at the sampling point, the varying concentration, c_2, is measured regularly during the passage of the tracer cloud. Then:

$$Vc_1 = Q \int_{t_1}^{t_2} (c_2 - c_o)dt$$

So that,

$$Q = \frac{Vc_1}{\int_{t_1}^{t_2}(c_2 - c_o)dt}$$

The chemical used should have a high solubility, be stable in water and be capable of accurate quantitative analysis in dilute concentrations. It should also be non-toxic to fish and other forms of river life, and be unaffected itself by sediment and other natural chemicals in the water. A favoured chemical is common salt (NaCl), which has a solubility of 3.6 kg in 10 L water at 15°C. This has the added advantage that a solution containing common salt at its water solubility has a high electrical conductivity. It allows electrical conductivity monitoring to be used instead of water sampling, with salt concentrations then being derived from a calibration. Fluorescent dyes such as Rhodamine WT have also been developed as tracers (BS 3680-2D, 1993) with the advantage of being easily detected at very low concentrations. This allows dilution gauging to be undertaken on large rivers where the volume of common salt tracer needed would be impractical and prohibited on environmental grounds.

Careful preparations are needed and the required mixing length, dependent on the state of the stream, must be assessed first. Several methods can be used to assess the minimum mixing length (L); two commonly used equations,

$$L = \beta Q^{1/3}$$

where L is the required distance between the injection site and the downstream sampling station (m), Q is the estimate of stream discharge (m^3 s^{-1}), and β is an empirical coefficient, typically set to 14 for a mid-stream injection and 60 for a injection from one bank, and,

$$L = \frac{0.13b^2C(0.7C + 2\sqrt{g})}{gd}$$

where b is the average width of the channel (m), d is the average depth of the channel (m), and C is the estimate of the Chezy roughness coefficient (Dingman, 2002).

Further details of the dilution gauging method are given in Herschy (2009).

7.6 Structural methods: flumes and weirs

The reliability of the stage-discharge relationship (Section 7.7) can be greatly improved if the river flow can be controlled by a rigid, indestructible cross-channel structure of standardized shape and characteristics. Of course, this adds to the cost of a river gauging station, but where continuous accurate values of discharge are required, particularly for compensation water and other low flows, a special measuring structure may be justified. The type of structure depends on the size of the stream or river and the range of flows it is expected to measure (BS ISO 8368, 1999). The sediment load of the stream also has to be considered.

The basic hydraulic mechanism applied in all measuring flumes and weirs is the setting up of critical flow conditions for which there is a unique and stable relationship between depth of flow and discharge. Flow in the channel upstream is sub-critical, passes through critical conditions in a constricted region of the flume or weir and enters the downstream channel as supercritical flow. It is better to measure the water level (stage) a short distance upstream of the critical-flow section, this stage having a unique relationship to the discharge.

7.6.1 Flumes

Flumes are particularly suitable for small streams carrying a considerable fine sediment load. The upstream sub-critical flow is constricted by narrowing the channel, thereby causing increased velocity and a decrease in the depth. With a sufficient contraction of the channel width, the flow becomes critical in the throat of the flume and a standing wave is formed further downstream. The water level upstream of the flume can then be related directly to the discharge. A typical design is shown in Fig. 7.20. Such *critical depth* flumes can have a variety of cross-sectional shapes.

The illustration shows a plain rectangular section with a horizontal invert (bed profile), but trapezoidal sections are used to contain a wider range of discharges and U-shaped sections are favoured in urban areas for more confined flows and sewage effluents. Where there are only small quantities of sediment, the length of the flume can be shortened by introducing a hump in the invert to reduce the depth of flow and thus

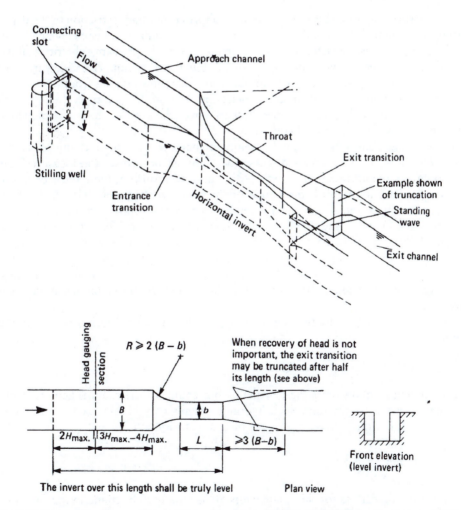

Fig. 7.20 Rectangular throated flume. (Reproduced from BS ISO 4359:1983, by permission of BSI.)

induce critical flow more quickly in the contraction, but the flume must be inspected regularly and any sediment deposits cleared. Relating the discharge for a rectangular cross-section to the measured head, H, the general form of the equation is:

$$Q = KbH^{3/2}$$

where b is the throat width and K is a coefficient based on analysis and experiment. For the derivation of K, the reader is referred to BS ISO 4359 (1983). With flumes built to that standard, Q may be assessed to within 2 per cent without the need for any field calibration. There are many different flume designs that have been built to serve various purposes in measuring a range of flow conditions. These are described fully in Herschy (2009).

7.6.2 Weirs

Weirs constitute a more versatile group of structures providing restriction to the depth rather than the width of the flow in a river or stream channel. A distinct sharp break in the bed profile is constructed and this creates a raised upstream sub-critical flow, a critical flow over the weir and super-critical flow downstream. The wide variety of weir types can provide for the measurement of discharges ranging from a few litres per second to many hundreds of cubic metres per second. In each type, the upstream head is again uniquely related to the discharge over the crest of the structure where the flow passes through critical conditions.

For gauging clear water in small streams or narrow man-made channels, *sharp-crested* or *thin-plate* weirs are used. These give highly accurate discharge measurements but to ensure the accuracy of the stage–discharge relationship, there must be atmospheric pressure underneath the nappe of the flow over the weir (Fig. 7.21). Thin plate weirs can be *full-width* weirs extending across the total width of a rectangular approach channel (Fig. 7.21a) or contracted weirs as in Fig. 7.21b and c. The shape of the weir may be *rectangular* or *trapezoidal* or have a triangular cross-section, a *V-notch*. The angle of the V-notch, θ, may have various values, the most common being 90° and 45°, though narrower angles are used for drainage discharge recorders and 120° angle weirs are more common on flashy tropical streams.

The basic discharge equation for a rectangular sharp crested weir again takes the form,

$$Q = KbH^{3/2}$$

but in finding K, allowances must be made to account for the channel geometry and the nature of the contraction. Such hydraulic details may be obtained from Ackers *et al.* (1978). For the V-notch weirs, the discharge formula becomes,

$$Q = K tan\left(\frac{\theta}{2}\right) H^{3/2}$$

Tables of coefficients for thin-plate weirs are normally to be found in the specialist references (e.g. BS ISO 1438, 2008; Herschy, 2009).

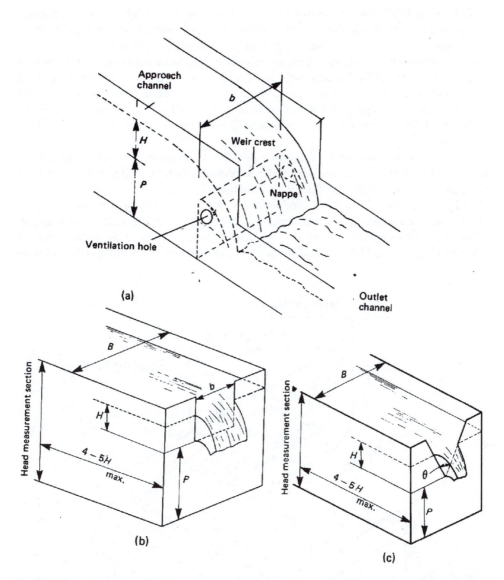

Fig. 7.21 Thin plate weirs. (Reproduced (a) from P. Ackers, *et al.* (1978) *Weirs and Flumes for Flow Measurement*, by permission of John Wiley & Sons, Inc.; (b) and (c) from BS ISO 1438: 2008, by permission of BSI.)

For larger channels and natural rivers, there are several designs recommended for gauging stations and these are usually constructed in concrete. One of the simplest to build is the *broad-crested* (square-edged) or the *rectangular-profile* weir (Fig. 7.22). The discharge in terms of gauged head H is given by,

$$Q = KbH^{3/2}$$

The length L of the weir, related to H and to P, the weir height, is very important since critical flow should be well established over the weir. However, separation of flow may occur at the upstream edge, and with increase in H, the pattern of flow and the coefficient, K, change. Considerable research has been done on the calibration of these weirs. The broad-crested weir with a curved upstream edge, also called the *round-nosed horizontal-crested* weir, gives an improved flow pattern over the weir with no flow separation at the upstream edge, and it is also less vulnerable to damage. The discharge formula is similarly dependent on the establishment of weir coefficients.

A special form of weir with a *triangular profile* was designed by E. S. Crump in 1952 (BS ISO 4360:2008). This ensured that the pattern of flow remained similar throughout the range of discharges and thus weir coefficients remained constant. In addition, by making additional head measurements just below the crest, as well as upstream, the *Crump* weir allows flow measurements to be estimated above the modular limit when the weir has drowned out at high flows. The geometry of the Crump weir is shown in Fig. 7.22a. The upstream slope of 1:2 and downstream slope of 1:5 produce

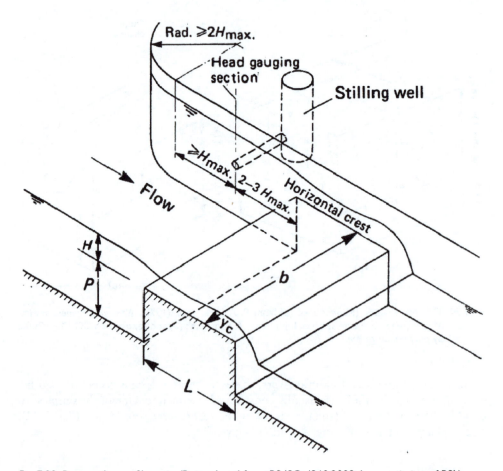

Fig. 7.22 Rectangular profile weir. (Reproduced from BS ISO 4360:2008, by permission of BSI.)

a well-controlled hydraulic jump on the downstream slope in the modular range. Improvement in the accuracy of very low flow measurement has been brought about by the compounding of the Crump weir across the width of the channel. Two or more separate crest sections at different levels may be built with sub-dividing piers to separate the flow. Such structures are designed individually to match the channel and flow conditions. A great deal of research effort has been put into the development of the Crump weir and many have been built in the UK.

The flat-V weir (Fig. 7.23b) is developed from an improvement on the Crump weir (BS ISO 4377, 2002: Fig. 7.23a). By making the shape of the crest across the channel into a shallow V-shape, low flows are measured more accurately in the confined central portion without the need for compounding. The triangular profile may be the same as the Crump weir, 1:2 upstream face and 1:5 downstream, but a profile with both slopes 1:2 is also used. This weir can also operate in the high non-modular flow range and several crest cross-sectional slopes have been calibrated. An extra advantage of the flat V-weir is that it passes sediment more readily than the Crump. Flat V-weirs, including those needing velocity–area measurements at higher stages, are also popular in the UK.

All these structures have a clear upper limit in their ability to measure the stream flow. Usually as the flow rate increases, downstream channel control causes such an increased downstream water level that a flume or weir is drowned out; the unique relationship hitherto existing between the stage or upstream level and the discharge in the so-called 'modular' range is thereafter lost. It is not always practicable to set crest levels in flumes and weirs sufficiently high to avoid the drowning out process at high flows since upstream riparian interests would object to raised water levels and out-of-bank flows at discharges previously within banks. In some cases where non-modular flow occurs regularly, discharges can be calculated by installing level gauges both upstream and downstream of the structure and using the conservation of momentum equation to calculate flow velocity (Ackers *et al.*, 1978).

These structures are generally used for measuring low and medium flows; flood flows are not usually measurable with flumes and weirs. However, in a world-wide context, they are well suited to the smaller rivers of the UK. However, the potential effects of weir construction on fish migration can restrict the establishment of new structures, and the high cost of maintenance can lead to the closure of some stations where data are under-utilised. Furthermore, such structures are considered impracticable and/or prohibitively costly for single-purpose river gauging in rivers of continental proportions.

7.7 Stage–discharge relationship

The establishment of a reliable relationship between the monitored variable stage and the corresponding discharge is essential at all river gauging stations when continuous-flow data are required from the continuous stage record. This calibration of the gauging station is dependent on the nature of the channel section and of the length of channel between the site of the staff gauge and discharge measuring cross-section. Conditions in a natural river are rarely stable for any length of time and thus the stage–discharge relationship must be checked regularly and, certainly after flood flows, new discharge measurements should be made throughout the range of stages. In most organizations

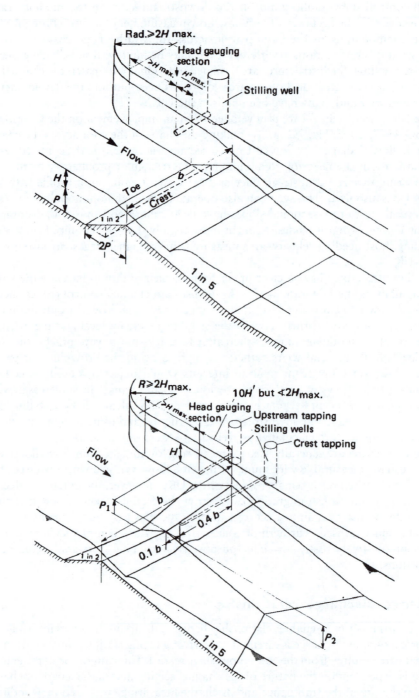

Fig. 7.23 (a) Triangular-profile weir (Crump weir) and (b) triangular-profile flat V-weir. (Reproduced from (a) BS ISO 4360:2008, (b) BS ISO 4377:2002 by permission of BSI.)

responsible for hydrometry, maintaining an up-to-date relationship is a continuous function of the hydrologist.

The stage–discharge relationship can be represented in three ways: as a graphical plot of stage versus discharge (the *rating curve*); in a tabular form (*rating table*); and as a mathematical equation, discharge, Q, in terms of stage, H *(rating equation)*.

7.7.1 The rating curve

All the discharge measurements, Q, are plotted against the corresponding mean stages, H, on suitable arithmetic scales. The array of points usually lies on a curve that is approximately parabolic and a best-fit curve should be drawn through the points by eye. At most gauging stations, the zero stage does not correspond to zero flow. If the points do not describe a single smooth curve, then the channel control governing the Q versus H relationship has some variation in its nature. For example, the stage height at which a small waterfall acting as a natural control is drowned out at higher flows is usually indicated by a distinct change in slope of the rating curve. Another break in the curve at high stages can often be related to the normal bank-full level above which the Q versus H relationship could be markedly different from the within-banks curve owing to the very different hydraulics of flood plain flow.

7.7.2 The rating table

When a satisfactory rating curve has been established, values of H and Q may be read off the curve at convenient intervals and a rating table is constructed by interpolation for required intervals of stage. This is the simplest and most convenient form of the stage–discharge relationship for manual processing of sequential stage records.

7.7.3 The rating equation

The rating curve can often be represented approximately by an equation of the form,

$$Q = aH^b$$

If Q is not zero when $H = 0$, then a stage correction, a realistic value of H_0, for $Q = 0$ must be included,

$$Q = u(H - H_0)^b$$

The values of the constants a, b and H_0 can be found by a least-squares fit using the measured data and trial values of H_0 and with b expected to be within limits depending on the shape of the cross-section. The equation can then be used for converting data-logged stage values into discharge.

When the rating curve does not plot as a simple curve on arithmetic scales, plotting the values of Q and $(H - H_0)$ on logarithmic scales helps in identifying the effects of different channel controls. The logarithmic form of the rating equation,

$$log Q = log a + b log(H - H_0)$$

may then plot as a series of straight lines, and changes in slope can be seen more clearly. Values for *a* and *b* obtained by least-squares fit of the data have been rounded off; the fitted H_0 values are in metres (BS ISO 1100-2, 1998).

When distinctive parts of the stage–discharge relationship can be related to observed different permanent physical controls in the river channel, then separate straight-line logarithmic equations are justified and can be evaluated by least-squares fitting to apply to corresponding ranges in the stage heights.

7.7.4 Irregularities and corrections

In the middle and lower reaches of rivers where the beds consist of sands and gravels and there is no stable channel control, the stage–discharge relationship may be unreliable owing to the alternate scouring and depositing of the loose bed material. If the general long-term rating curve remains reasonably constant, Stout's technique allows individual shift corrections to be made to evaluations of *Q* in accordance with the most recent gaugings by adjusting the stage readings (Fig. 7.24).

Discharges depend both on the stage and on the slope of the water surface; the latter is not the same for rising and falling stages as a non-steady flow passes a gauging station. From gaugings at a particular stage value there are thus two different values of discharge. From gaugings made on a rising stage, the corresponding discharges will be greater than those measured at the same stage levels on the falling stage. Thus, there can be produced a looped rating curve. When the river sustains a steady flow at a particular stage, an average of the two discharges may be taken, but otherwise the values of the relevant rising or falling side of the looped rating curve should be used.

Other irregularities in the stage–discharge relationship may be caused by non-uniform flow generated by interference in the channel downstream of the gauging section, thus overriding the flow control. Such interference can result from the backing up of flow in a main channel owing to flow coming in from a downstream tributary, or from the operation of sluice gates on the main channel. Vegetation growth in the gauging reach (particularly in lowland rivers in southern England) will also interfere with the *Q* versus *H* relation. In certain rivers it may be advisable to have rating curves for different seasons of the year. More permanent interference may be provided by changes in the cross-section owing to the scouring or the deposition of an exceptional flood and, if discharges have been obtained for stages which include inundation of the flood plain, any further developments on the flood plain would affect these values.

7.7.5 Extension of rating curves

It is always extremely difficult to obtain velocity measurements and hence estimated discharges at high stages. The range of the stage–discharge relationship derived from measurements is nearly always exceeded by flood flows. Hydrologists responsible for river gauging should make determined attempts to measure flood peaks, particularly at stations where the rating at high flows is in doubt. However, there are several techniques that can be adopted to assess the discharge at stages beyond the measured limit of the rating curve, but all extensions are strictly only valid for the same shape of cross-section and same boundary roughness.

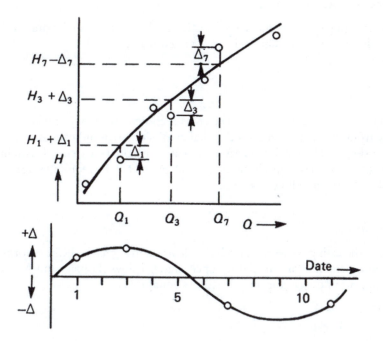

Fig. 7.24 Stout's method.

Shift correction

Date	Recorded stage	Measured discharge	Stage correction	Corrected stage	Estimated discharge
May 1	H_1	Q_1	Δ_1	$H_1 + \Delta_1$	Q_1
2	H_2		Δ'_2	$H_2 + \Delta'_2$	Q'_2
3	H_3	Q_3	Δ_3	$H_3 + \Delta_3$	Q_3
4	H_4		Δ'_4	$H_4 + \Delta'_4$	Q'_4
5	H_5		Δ'_5	$H_5 + \Delta'_5$	Q'_5
6	H_6		$-\Delta'_6$	$H_6 - \Delta'_6$	Q'_6
7	H_7	Q_7	$-\Delta_7$	$H_7 - \Delta_7$	Q_7

7.7.5.1 Logarithmic extrapolation

If the rating curve plots satisfactorily as a straight line on log-log paper, it may be extrapolated easily to the higher stages. However, in using this method, especially if the extrapolation exceeds 20 per cent of the largest gauged discharge, other methods should be applied to check the result. Alternatively, the straight-line equation fitted to the logarithmic rating curve could be used to calculate higher discharges, with similar reservations to check by another method.

7.7.5.2 Velocity-area method

In addition to the rating curve, plots of A versus H and $\overline{V}$ versus H can be drawn. The overall mean velocity across the channel $\overline{V}$ calculated from,

$$Q = A\overline{V}$$

within the range of the measured stages. The cross-sectional area curve can be extrapolated reliably from the survey data (and a change of slope will indicate any change in cross-section shape). The mean velocity curve can normally be extended with little error. Then the higher values of discharge can be calculated for the required stage from the product of the corresponding $\overline{V}$ and A.

7.7.5.3 Stevens method

Extrapolation of the rating curve can also be made using the empirical Chezy formula (or other friction formula) for calculating open channel flow (Francis and Minton, 1984). Under uniform flow conditions, the Chezy equation is,

$$Q = A\overline{V} = AC\sqrt{(RS_c)}$$

where $\overline{V}$ is the overall mean velocity across the channel, R is the hydraulic mean depth (see Fig. 7.4), S_o is the bed slope and C is a coefficient. R is obtained for all required stages from $R = A/P$ with A and P measured in the cross-sectional survey. If $C\sqrt{S_o}$ is taken to be constant (k), then Q versus $kA\sqrt{R}$ plots as a straight line. For gauged stage values of H, corresponding values of $A\sqrt{R}$ and Q are obtained and plotted. The extended straight line can then be used to give discharges for higher stages (Fig. 7.25). The method relies on the doubtful assumption of C remaining constant for all stage values.

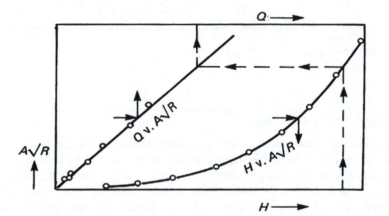

Fig. 7.25 The Stevens method.

7.7.5.4 The Manning formula

This can be used instead of the Chezy formula for extending rating curves, but it is also applied more widely in engineering practice for calculating flows. The formula, where quantities are in SI units, is,

$$Q = A\overline{V} = \frac{AR^{2/3}S_o^{1/2}}{n}$$

$$S_o^{1/2}/n$$

It is applied in a similar way, with $S_o^{1/2}/n$ assumed constant, and Q being plotted against $AR^{2/3}$.

In both the Stevens and Manning formula methods, when a flood discharge exceeds bank-full stage, the roughness factor, C or n, can be changed to model the different flow conditions and the separate parts of the extended flow over the flood plain are calculated with the modified formula. It is generally accepted that the Manning equation is superior to the Chezy equation, since n changes less than C as R varies (Fortune et al., 2004).

Estimates of flood discharges at strategic locations along a river are usually made by this Manning-based method. After notable flood events, surveyors can measure the required cross-sectional area, the wetted perimeter and the bed slope of the affected reach; the peak surface water level is assessed from debris or wrack marks. Selecting an appropriate value of n (Table 7.1 or, for greater detail, see BS 3680-5:1992) for the channel roughness, an estimated peak discharge is calculated. Considerable experience is needed in using this method since the validity of the formula depends on the nature of the flow and the appropriate corresponding line of slope.

7.7.5.5 The conveyance and afflux estimation system (CES/AES)

The CES/AES software toolbox[1] has been developed recently for application in the UK (Abril and Knight, 2004; Knight et al., 2010). It contains a 'conveyance generator' that can be used to extend rating curves by utilising channel roughness information from 700 UK sources (including the UK River Habitat Survey) together with descriptions of the channel cross-sections and the effects of vegetation growth within the channel. It also makes use of experimental data from UK Flood Channel Facility at the University of Birmingham in allowing for additional momentum losses at boundary between channel and overbank flow.

Within the UK, the Hiflows-UK database holds the rating curve information for approximately 1000 gauging stations.

Table 7.1 Sample values of Manning's n

Concrete-lined channel	0.013
Unlined earth channel	0.020
Straight, stable deep natural channel	0.030
Winding natural streams	0.035
Variable rivers, vegetated banks	0.050
Mountainous streams, rocky beds	0.050

Note

1 http://www.river-conveyance.net/

References

Abril, J. B. and Knight, D. W. (2004) Stage–discharge prediction for rivers in flood applying a depth-averaged model. *Journal of Hydraulic Research* 42, 616–629.

Ackers, P., White, W. R., Perkins, J. A. and Harrison, A. J. M. (1978) *Weirs and Flumes for Flow Measurement*. John Wiley & Sons, Chichester, 327 pp.

Akan, A. O. (2006) *Open Channel Hydraulic*. Butterworth Heinemann, Oxford, 384 pp.

BS 3680-2A (1995) *Measurement of Liquid Flow in Open Channels. Dilution Methods. General*. British Standards Institution, London.

BS 3680-2D (1993) *Measurement of Liquid Flow in Open Channels. Dilution Methods. Methods of Measurement Using Fluorescent Tracers*. British Standards Institution, London.

BS 3680-5 (1992) *Measurement of Liquid Flow in Open Channels. Slope Area Method of Estimation*. British Standards Institution, London.

BS EN ISO 4373 (2008) *Hydrometry. Water Level Measuring Devices*. British Standards Institution, London.

BS EN ISO 4375 (2004) *Hydrometric Determinations. Cableway Systems for Stream Gauging*. British Standards Institution, London.

BS EN ISO 6416 (2005) *Hydrometry. Measurement of Discharge by the Ultrasonic (Acoustic) Method*. British Standards Institution, London.

BS EN ISO 748 (2007) *Hydrometry. Measurement of Liquid Flow in Open Channels Using Current-Meters or Floats*. British Standards Institution, London.

BS ISO 1100-2 (1998) *Measurement of Liquid Flow in Open Channels. Determination of the Stage-Discharge Relation*. British Standards Institution, London.

BS ISO 1438 (2008) *Hydrometry. Open Channel Flow Measurement Using Thin-Plate Weirs*. British Standards Institution, London.

BS ISO 2537 (2007) *Hydrometry. Rotating-Element Current-Meters*. British Standards Institution, London.

BS ISO 4359 (1983) *Liquid Flow Measurement in Open Channels. Rectangular, Trapezoidal and U-shaped Flumes*. British Standards Institution, London.

BS ISO 4360 (2008) *Hydrometry. Open Channel Flow Measurement Using Triangular Profile Weirs*. British Standards Institution, London.

BS ISO 4377 (2002) *Hydrometric Determinations. Flow Measurement in Open Channels Using Structures. Flat-V Weirs*. British Standards Institution, London.

BS ISO 8368 (1999) *Hydrometric Determinations. Flow Measurements in Open Channels Using Structures. Guidelines for the Selection of Structure*. British Standards Institution, London.

Dingman, S. L. (2002) *Physical Hydrology*, 2nd edn. Waveland Press Inc., Long Grove.

Fortune, D., McGahey, C. and Nex, A. (2004) Improved conveyance assessment. In: Liong, S.-Y., Phoon, K. K. and Babovic, V. (eds) *Sixth International Conference on Hydroinformatics*. World Scientific Publishing Company, Singapore.

Francis, J. R. D. and Minton, P. (1984) *Civil Engineering Hydraulics*, 5th Edition. Hodder Arnold, London, 400pp.

Herschy, R. W. (2009) *Streamflow Measurement*, 3rd edn. Taylor & Francis, Abingdon.

Knight, D. W., McGahey, C., Lamb, R. and Samuels, P. (2010) *Practical Channel Hydraulics: Roughness, Conveyance and Afflux*. CRC Press, Boca Raton, 470 pp.

Shaw, E. M. (1994) *Hydrology in Practice*, 3rd edn. Chapman and Hall, London.

Chapter 8

Water-quality measurement

Water for consumption in a city may be taken directly from a river and waste water from the city returned to the river. If this is repeated at each large centre from the upper reaches of a river to its mouth, as is the case in the River Thames, for example, then the same water will be treated and reused several times. Water engineers are charged with the duty of ensuring that the water supplied is non-toxic and of a sufficiently high standard for human consumption and that the waste waters are treated to remove pollutants to bring the water to an acceptable quality for return to the natural river.

In countries within tropical regions, the problem of water quality is usually even more acute, since the collection and treatment of waste water is often less advanced (Gleick, 2008). In small settlements, sources of drinking water and disposal of sewage are sometimes scarcely separated, and in congested cities lacking adequate drainage, the dangers from water-borne diseases are compounded.

Increasing industrialization of the world's communities has also led to greater pollution of natural sources of water. Some of the large industrial consumers, such as the electricity-generating stations, may only modify the river temperature, but many manufacturing industries, may strongly pollute the natural rivers. Agricultural intensification has also led to a deterioration of water quality within rivers through the widespread use of fertilisers and pesticides.

Within wealthy countries experiencing a decline of heavy industry, the quality of rivers has improved to such a degree that further improvements aim to enhance the ecological quality of the natural environment. The Europe-wide Water Framework Directive is aimed more at restoring a good ecological status of designated water bodies than ensuring rivers are capable of cost-effective treatment for human consumption (United Kingdom Technical Advisory Group of the Water Framework Directive, 2008; Section 8.2.2).

Within hydrological research, there is an increasing awareness of the merits of using the physical, chemical or biological characteristics of water to help quantify the sources, pathways and ages of water within catchments (McGuire and McDonnell, 2007; see Section 6.4). Equally, a quantification of river discharge (Chapter 7) is fundamental to many studies of river water quality, given that the *mass flux* or *load* of particular physical or chemical characteristics within a river (e.g. milligrams per second) at a specific time is the product of the river discharge (e.g. litres per second) and concentration of the particular characteristic (e.g. milligrams per litre) at a specific time (Section 8.7). Consequently, the hydrologist is often concerned with the quality of rivers and subsurface water, and the water-quality scientist with the measurement of water flow.

8.1 Water-quality characteristics

The chemical composition of water, H_2O, is one of the first formulae learnt in chemistry, and its existence in the gaseous, liquid and solid states according to temperature is readily understood. However, it is in its liquid form that the quality of water is of most importance both for the nature of the pollutants it may carry and for the greater use that it affords. The principal features of water quality in the streams, rivers and lakes with which the water engineer is most concerned may be considered in three main groups: physical, chemical and biological (Tebbutt, 1998).

8.1.1 Physical characteristics

Suspended solids form the most obvious extraneous matter to be carried along by a flowing river. The quantity and type of solids depend on the volumetric discharge and velocity. They range from tree trunks, boulders and other trash dislodged and carried away by floods to minute particles suspended in a tranquilly flowing stream. The solid pollutants of a river derive from organic and inorganic sources. When evaluating quality for the potential use of water, suspended solids (SS) are measured in milligrams per litre.

Turbidity is the term for the cloudiness of water due to fine suspended particles of clay or silt, waste effluents or micro-organisms, and is measured by nephelometric turbidity units (NTU) or formazin turbidity units (FTU) based on the comparison of the scattering of light by a water sample with that of a standard suspension of formazin. An estimate of the suspended solids can be derived from a calibration with the turbidity.

Electrical conductivity is a physical property of water (EC or σ_w; Section 6.1.7) that is dependent on the dissolved salts. Thus its measurement in microsiemens per centimetre ($\mu S\,cm^{-1}$) gives a good estimate of the total dissolved solids (TDS, $mg\,L^{-1}$) content of a river, where

$$TDS = k(EC)$$

and k is a calibration factor for the specific water, and typically varies from 0.55 to 0.75, but values of 1 have been reported by Grove (1972) for some West African rivers. Additionally, migration of applied electrical tracers (e.g. NaCl) through subsurface systems are often tracked with measurements of the EC within the water or when bulked with the ground, σ_b (see Section 6.1.7 and 6.4).

Colour, taste and odour are aesthetic properties of water caused by dissolved impurities either from natural sources, like the peaty waters from upland moors, or from the discharge of noxious substances into the water course by industry or agriculture, e.g. phenols and chlorophenols.

Temperature is a standard physical characteristic that is important in the consideration of the chemical properties of water. Its measurement, in $^\circ C$, in natural rivers is also necessary for assessing the effects of temperature changes on living organisms.

Water comprises two *stable isotopes* of hydrogen (^1H and ^2H) and three of oxygen (^{16}O, ^{17}O and ^{18}O). Heavy isotopes occur preferentially in the liquid phase and the light isotopes in the gaseous phase, resulting in precipitation with different isotopic signatures. These characteristics are increasing used as natural tracers of the residence time of water within catchment and atmospheric systems (see Section 6.8).

Radioactivity in water bodies has received increasing attention as its harmful effects on life in all forms becomes better recognized, but its measurement remains a specialized procedure adopted when dangerous doses are suspected.

8.1.2 Chemical characteristics

Water chemistry is a very extensive subject, since water is the most common solvent and many chemical compounds can be found in solution at the temperatures of naturally occurring water bodies. Only a selection of the more significant chemical features will be mentioned here; more detailed discussion is given in Stumm and Morgan (1996).

The *pH* is a measure of the concentration (or activity) of hydrogen ions (H^+) and indicates the degree of acidity or alkalinity of the water, where

$$pH = -log_{10}[H^+] = log_{10}\frac{1}{[H^+]}$$

and, on the scale from 0 to 14, a pH of 7 is indicative of a neutral solution. If the pH value is less than 7, then the water is acidic, and if the pH value is greater than 7, the water is alkaline.

Alkalinity is caused by the presence of bicarbonate (HCO_3^-), carbonate (CO_3^-) or hydroxide (OH^-), and within natural waters is mostly from the dissolution of chalk or limestone, where

$$CaCO_3 + H_2O + CO_2 \rightarrow Ca(HCO_3)_2$$

The *hardness* of water is due largely to the presence of the cations Ca^{2+} and Mg^{2+}, although Fe^{2+} and Sr^{2+} are also responsible. The metals are normally associated with HCO_3^-, SO_4^{2-}, Cl^- or NO_3^-. Hardness is expressed in terms of $CaCO_3$ and has two forms: (a) carbonate hardness, due to metals associated with HCO_3^-, and (b) non-carbonate hardness, due to metals associated with SO_4^{2-}, Cl^- or NO_3^-. The non-carbonate hardness is estimated by subtracting the alkalinity from the total hardness.

The potential required to transfer electrons from an oxidant to a reductant within water is called the *oxidation–reduction potential* (ORP), where

$$ORP = E^0 - \frac{0.509}{z}log_{10}\frac{[products]}{[reactants]}$$

where E^0 is the cell oxidation potential and z the number of electrons in the reaction.

Dissolved oxygen (DO) plays a large part in the assessment of water quality, since it is an essential ingredient for the sustenance of fish and all other forms of aquatic life. It also affects the taste of water, and a high concentration of dissolved oxygen in domestic supplies is encouraged by aeration. Values of dissolved oxygen are given in milligrams per litre (O_2). *Biochemical oxygen demand* (BOD) is a measure of the consumption of oxygen by micro-organisms (Section 8.1.3) in the oxidation of organic matter. Thus a high BOD (mg L^{-1}; O_2), indicates a high concentration of organic matter usually from waste water discharges.

Nitrogen may be present in water in several forms: in organic compounds (usually from domestic wastes), as *ammonia* (NH_3) nitrogen, in *nitrite* (NO_2) nitrogen or fully oxidized *nitrates* (NO_3). Measures of nitrogen, (mg L^{-1}; N), give indications of the state of pollution by organic wastes with larger quantities in the nitrate form being an indication of oxidation (purification). Nitrification is the oxidation of nitrogen compounds and can be described thus:

$$\text{Organic nitrogen} + O_2 \rightarrow \text{ammonia nitrogen} + O_2 \rightarrow NO_2 \text{nitrogen}$$
$$+ O_2 \rightarrow NO_3 \text{nitrogen}$$

Reduction of nitrogen, or denitrification, is the reversal of this process, thus

$$NO_3^- \rightarrow NO_2^- \rightarrow NH_3 + N_2$$

A spike in the level of ammonia within rivers often indicates a point source input of livestock slurry from overland flow or drainage systems.

There is an increasing awareness that low concentrations of *phosphorus* within river water as a result of overland flow from agricultural land or sewage discharges can have a large impact on the ecological status of rivers.

Chlorides, most often occurring in the NaCl common salt form, are found in brackish water bodies contaminated by sea water or in groundwater aquifers with high salt content. The presence of chlorides (mg L^{-1}; Cl) in a river is also indicative of sewage pollution from other chloride compounds.

Many organic compounds related to industrial discharges or other human activities can be found in trace concentrations in water. These *trace organics* include pesticides, polynuclear aromatic hydrocarbons (PAHs), trihalomethanes (THMs), chlorophenols, oestrogen and benzene.

8.1.3 Biological characteristics

The existence of plant and animal life in rivers and other water bodies is a prime indicator of water quality, and it has a different significance for the ecologist and the water supply engineer. The former is interested in those physico-chemical characteristics that affect the presence of macro-invertebrates, diatoms and fish populations. The presence of micro-organisms in the water to be abstracted for public water supply is of particular concern. These micro-organisms can be viruses, bacteria, fungi, actinomycetea, algae and protozoa.

Viruses are basic organisms (consisting primarily of nucleic acid and protein) that range in size from 0.004 to 0.3 μm. They are all parasitic and cannot grow outside of living organisms. Viruses found within water include poliovirus, which causes paralysis, meningitis and fever, and human torovirus, which causes gastroenteritis (Bosch, 1998).

Bacteria, by contrast are single-cell organisms ranging in size from 0.5 to 5 μm. Most bacteria are necessary; however, some species are responsible for human infection. These bacteria include *Escherichia coli* (*E. coli*), coliforms, aerobic bacteria, *Aeromonas* spp. and *Salmonella* spp.

Fungi are aerobic, multi-cellular organisms that can be responsible for tastes and odours in water supplies, and some are human pathogens, e.g. *Aspergillus* spp.

Actinomycetea are similar in size to bacteria but have a filamentous structure. As with fungi, *Actinomycetea* spp. can affect the taste and odour of water (Zaitlin and Watson, 2006).

Algae are all photosynthetic plants and are mostly multi-cellular. Many algae species can cause taste and odour problems in water. Blue–green algae release toxins that can produce gastrointestinal illness if consumed by humans or domestic animals.

Protozoa are unicellular organisms that are 10–100 μm in length. Some protozoans are pathogenic, for example, *Giardia* and *Cryptosporidium*, and can be resistant to disinfection due to the formation of spores or cysts.

On a routine basis, the common organism *E. coli* found in all human excreta is taken as an indicator of sewage pollution. The measure of concentration in a water sample is the most probable number (MPN) per 100 mL, which is derived statistically from a number of samples. All supplies of water destined for human consumption must have regular bacteriological examination.

8.2 Water-quality standards

The quality of river water and groundwater varies considerably in space and time. Typical values of physical, chemical and biological characteristics of rivers and groundwater used as a source of public water supply (PWS) in the UK are given in Table 8.1.

As a consequence of the differences in the characteristics of upland rivers from lowland rivers or groundwater, these potential sources of PWS require different types and degrees of water treatment; however, all three supplies need to be treated to the same drinking water standards. Rivers, particularly lowland rivers, tend to be more polluted than groundwater, and so greater efforts have been directed towards measures to improve the status of rivers. A key aspect of this attempt to restore river water quality has been the refinement and enforcement of environmental standards for river water.

8.2.1 UK drinking water standards

Water procured for drinking water must comply with specific water-quality standards. Some of the key standards applicable in the UK in comparison with equivalent standards from elsewhere are given in Table 8.2.

8.2.2 UK environmental standards for river water quality

The Water Framework Directive, or WFD (European Commission, 2000: 2000/60/EC) came into force across the European Union on 22 December 2000. This Directive establishes new environmental objectives for the water environment. Two key objectives

Table 8.1 Typical physical, chemical and biological characteristics of water sources in the UK. (Adapted from Tebbutt, 1998)

Characteristic	Source		
	Upland basin	Lowland basin	Chalk aquifer
Physical			
Turbidity (NTU)	5	50	<5
Electrical conductivity (μS cm^{-1})	45	700	600
Total solids (mg L^{-1})	50	400	300
Colour (°H)	70	40	<5
Chemical			
pH	6.0	7.5	7.2
Alkalinity, total (m L^{-1} HCO$_3$)	20	175	110
Hardness, total (m L^{-1} Ca)	10	200	200
Dissolved oxygen (% saturation)	100	75	2
Biochemical oxygen demand (mg L^{-1})	2	4	2
Ammonia nitrogen (mg L^{-1})	0.05	0.5	0.05
Nitrate nitrogen (mg L^{-1})	0.1	2.0	0.5
Chloride (mg L^{-1})	10	50	25
Biological			
Colonies/ml at 22°C	100	30 000	10
Colonies/ml at 37°C	10	5 000	5
Coliform organisms/100 ml	20	20 000	5

NTU, nephelometric turbidity units.

are: (1) to prevent deterioration of the status of all surface and groundwater bodies, and (2) to protect, enhance and restore all bodies of surface water and groundwater with the aim of achieving *good status* for surface and groundwater by 2015. The term 'good status' refers to the ecological status of water bodies, and five classes are defined for rivers, namely *high, good, moderate, poor and bad*.

Within this new system, the first stage of classifying a water quality of a river reach into one of the five classes, is to define its *type*. Seven river types are defined according to altitude (namely height above or below 80 m) and alkalinity (Table 8.3). These two catchment descriptors are used because they have been shown to determine the distribution of biota within UK rivers. Further, this typology is undertaken to allow different environmental standards of water quality to be set for different reach types. However, a slightly different typology is adopted for phosphorus standards (Table 8.4). Rivers of type 2, 4 and 6 naturally support populations of salmonid fish, while type 3, 5 and 7 rivers are classed as cyprinid water within the Freshwater Fish Directive.

The ecological status of each river reach is then established for six water-quality characteristics: (1) dissolved oxygen; (2) biochemical oxygen demand; (3) ammonia; (4) acid conditions; (5) phosphorus; and (6) temperature. These characteristics are defined largely because of confidence in their biological impact. Aquatic macro-invertebrate populations are sensitive to levels of dissolved oxygen, BOD and ammonia, while fish are sensitive to pH, and diatoms are particularly sensitive to

Table 8.2 UK drinking water standards (UK Water Supply [Water Quality] Regulations 1989) for selected characteristics together with WHO (1993) guideline levels, values set for monitoring purposes in EC Directive 98/83/EC and Secondary Maximum Contaminant Levels of the USEPA 1996 Safe Drinking Water Act. (Adapted from Twort *et al.*, 2000)

Characteristic	UK[a]	WHO[b]	EC Directive[c]	USEPA[d]
Physical				
Turbidity	4 FTU	5 NTU	<1 NTU (TS)	5 NTU max
Electrical conductivity ($\mu S\,cm^{-1}$ at 20°C)	1500	–	2500	–
Colour	20 $mg\,L^{-1}$ Pt/Co	15 true colour	Acceptable to consumers	3
Chemical				
pH	5.5–9.5	<8.0 (TS)	≤ 6.5 and ≥ 9.5	6.5–8.5
Alkalinity, total ($mg\,L^{-1}$ HCO_3)	>30	–	–	–
Hardness, total ($mg\,L^{-1}$ Ca)	>60	–	–	–
Ammonia nitrogen ($mg\,L^{-1}$ NH_4)	0.5	1.5	0.5	–
Nitrate nitrogen ($mg\,L^{-1}$ NO_3)	50	50	50	10
Phosphorus ($\mu g\,L^{-1}$)	2200	–	–	–
Aluminium	200 $\mu g\,L^{-1}$	0.2 $mg\,L^{-1}$	200 $\mu g\,L^{-1}$	0.05–0.2 $mg\,L^{-1}$
Iron	200 $\mu g\,L^{-1}$	0.3 $mg\,L^{-1}$	200 $\mu g\,L^{-1}$	0.3 $mg\,L^{-1}$
Manganese	50 $\mu g\,L^{-1}$	0.10 $mg\,L^{-1}$	50 $\mu g\,L^{-1}$	0.05 $mg\,L^{-1}$

[a] UK Water Supply [Water Quality] Regulations 1989.
[b] World Health Organisation Guidelines 1993: Guidelines levels above which consumer complaints may arise.
[c] European Commission Directive 98/83/EC November 1998: Indicator parameters with values set for monitoring purposes, where any exceedances must be investigated.
[d] United States Environmental Protection Agency (USEPA) Regulations under 1996 Safe Drinking Water Act amendments: Secondary maximum contaminant levels
 (SMCL), which are not mandatory.

Table 8.3 Criteria for identifying the types of UK river to which the dissolved oxygen, biochemical oxygen demand, and ammonia standards for rivers apply under the Water Framework Directive. (Adapted from UTAG, 2008)

Site altitude	Alkalinity (as $mg\,L^{-1}$ $CaCO_3$)				
	<10	10–50	50–100	100–200	>200
<80 m	Type 1	Type 2	Type 3	Type 5	Type 7
>80 m	Type 1	Type 2	Type 4	Type 6	Type 7

phosphorus levels (UTAG, 2008). The ecological status of a river reach with reference to each of the six characteristics and the typology is given in Table 8.5. Placing a reach within a class depends on either the mean of all samples being within the class, or a specified proportion of the water samples (i.e. 10th, 90th, 95th or 98th percentile) being within the class. For example, the 95th percentile could be estimated from 1000

Table 8.4 Criteria for identifying the types of UK river to which the soluble reactive phosphorus standards for rivers apply under the Water Framework Directive. (Adapted from UTAG, 2008)

Site altitude	Annual mean alkalinity (as $mg L^{-1}$ $CaCO_3$)	
	<50	>50
<80 m	Type 1n	Type 3n
>80 m	Type 2n	Type 4n

Table 8.5 Environmental standards for river water for levels of (1) dissolved oxygen; (2) biochemical oxygen demand (BOD); (3) ammonia; (4) acid conditions; (5) phosphorus; and (6) temperature under the Water Framework Directive. (Adapted from UTAG, 2008)

Status	High	Good	Moderate	Poor
Dissolved oxygen (% saturation; 10th percentile)				
1, 2, 4, and 6 salmonid	80	75	64	50
3, 5 and 7	70	60	54	45
Biochemical oxygen demand ($mg L^{-1}$; 90th percentile)				
1, 2, 4, 6 and salmonid	3	4	6	7.5
3, 5 and 7	4	5	6.5	9
Total ammonia ($m L^{-1}$; 90th percentile)				
1, 2, 4, 6 and salmonid	0.2	0.3	0.75	1.1
3, 5 and 7	0.3	0.6	1.1	2.5
Acid conditions (pH; [a]5th, [b]10th and [c]95th percentile)				
1–7	6[a]; 9[c]	5.2[b]	–	–
Soluble reactive phosphorus (μLl^{-1} as annual means)				
1n	30	50	150	500
2n	20	40	150	500
3n and 4n	50	120	250	1000
Temperature (°C)				
Annual 98th percentile				
Non-cyprinid	20	23	28	30
Cyprinid	25	28	30	32
Increase or decrease as annual 98th percentile				
Non-cyprinid	2	3	–	–
Cyprinid	2	3	–	–

values, by ranking the values and selecting the 950th largest. Utilising the central limit theorem, it would be better to calculate the 95th percentile from:

$$\exp(\mu + 1.65\sigma)$$

while the fifth percentile would be:

$$\exp(\mu - 1.65\sigma)$$

where μ is the arithmetic mean of the $\log_e$ values and σ is the standard deviation.

Additionally, the WFD will eventually take over the provisions of the Dangerous Substances Directive (European Economic Community (1976): 76/464/EEC), which seeks to eliminate priority hazardous substances (PHSs) from rivers and subsurface water. Such PHSs include cadmium, mercury, nonylphenols and polycyclic aromatic hydrocarbons (Gray, 2005).

8.3 Water-quality sampling

8.3.1 Sampling river water

Choice of sampling site for river water may be governed by an abstraction point or a discharge point associated with an industrial user or waste water treatment works. However, it is often most useful to take water-quality samples at a river gauging station. Ideally, a single sample from the well-mixed waters downstream of a weir would suffice to give a good representation of the water quality of a small river. At a current meter station, the river should be sampled at several points across the channel and in deep rivers (over 3 m) at 0.2 and 0.8 depths. For shallower streams, one sample in the vertical at 0.6 depth should be adequate (Note: These depths correspond to the points giving the mean flow velocity in a vertical section; see Chapter 7). Once the flow characteristics of the river are known, the sampling scheme in Fig. 8.1 can be recommended.

The timing and frequency of sampling also need consideration, particularly if there is a regular pattern of flow control or of effluent discharge from industries above the sampling point. It is often worthwhile having a concentrated sampling period when the river regime is steady to establish a regular norm. Then anomalous conditions of flood flows with their increased load of suspended solids, or of unusual influxes of pollutants from accidental spillages, when sampled, can be related in perspective to average water-quality values. Seasonal changes must be identified in any water-quality variations. In addition to the establishment of the average water-quality characteristics of a river, it is important for regulatory agencies, in the interests of environmental conservation, to be vigilant at all times in maintaining satisfactory river water-quality values. Hence there has been a rapid development of automated monitoring of water quality (Section 8.5).

For taking single samples of river water, standardised instruments have been devised. The *displacement sampler* is recommended for the collecting samples for DO analysis,

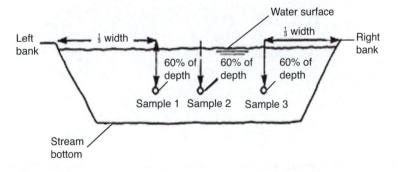

Fig. 8.1 Selected sampling points for a deep wide stream. (Adapted from Nemerow, N. L. (1974) *Scientific Stream Pollution Analysis*, Hemisphere Publishing Corporation.)

but can be used for general sampling in rivers. The inlet is opened when the container is at the required depth and the water is fed into the bottom of a bottle. When the whole container is full, the water flows from the exit, and should continue until the bottle contents have been changed several times before the sampler is removed.

Numerous sampler designs have been adopted in conjunction with studies of suspended solids. The US DH48 is one such sampler and is shown in Fig. 8.2. The container holds a glass or polythene bottle.

When obtaining samples for chemical analysis, great care must be taken against contaminating the water sample; all containers, even the simple bucket dipping into a turbulent well-mixed stream, must be washed out with the flowing river water before being used for a sample. Sample bottles should be sealed with *zero headspace*, so that CO_2, volatile organic compounds (VOCs), etc. are not lost from the water sample. The temperature of the water must be taken at the time of sampling. Sampling bottles must be carried in suitable crates and delivered to the laboratory the same day. Delay in carrying out the analyses can result in spurious values, since some of the chemical properties of the water can be altered by the changing conditions in storage. Indeed, measurement of most water characteristics requires special preservation of samples (Table 8.6).

8.3.2 Sampling subsurface water

Subsurface water can be extracted from both saturated ground and unsaturated ground. To extract water from unsaturated ground, *suction lysimeters* (also called

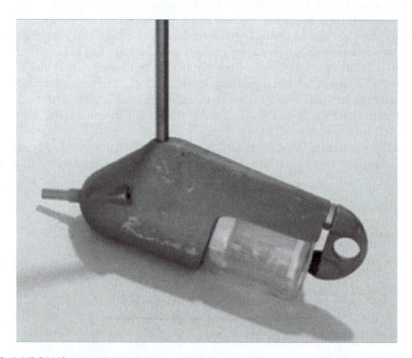

Fig. 8.2 A US DH48 suspended sediment sampler for attachment to a wading rod. (Reproduced by permission of the Ricky Hydrological Company, Columbus, OH, USA.)

Table 8.6 Water sample containers, methods of preservation and minimum sample volumes. (Adapted from Artiola, 2004)

Characteristic	Container	Preservation/storage	Minimum volume (cm^3)
pH, EC, alkalinity, (major anions)*	High-density polyethylene bottle	Keep cool at $4°C$	100 (200)*
Metal cations (except Hg, CrVI)	High-density polyethylene bottle	Add nitric acid (pH<2)	200
Pesticides, phenols	Clear/amber glass bottle	Add reducing agent (sodium thiosuphate).** Keep cool at $4°C$. Adjust pH<2	>1000
Volatile organics	Clear glass vial	Add reducing agent (sodium thiosuphate).** Keep cool at $4°C$. Adjust pH<2	5–25 per group
Coliforms	Plastic bottle	Add reducing agent.** Keep cool at $4°C$	

vacuum samplers) can be used. These devices consist of a porous cup attached to a polyvinyl chloride (PVC) sample accumulation chamber and two tubes that extend to the ground surface (Fig. 8.3). The porous cups are typically made of ceramic, stainless steel or polytetrafluoroethylene (PTFE), and are best installed within an auger-hole with silica flour packed around the cup. The shorter of the two tubes entering the lysimeter (vacuum tube) is used apply a vacuum to the accumulation chamber, with a manual or electric pump. Hours to days of applied vacuum may be necessary to extract sufficient volume of sample into the accumulation chamber. The second tube within the lysimeter (sample tube) extends down to the porous cup, and is used for extracting the sample. A vacuum pump can be used to extract the sample from the sample tube into a sample bottle. Alternatively, a positive pressure can be applied to the vacuum tube to push the sample out through the sample tube. If this alternative system is used, a non-return value needs to be incorporated within the lysimeter to prevent the sample from being pushed through the porous tip.

If subsurface water is to be analysed for trace metals, stainless steel augers coated with nickel, cadmium or zinc metal to reduce oxidation should not be used to install the suction lysimeters.

Where ground is variably saturated, *pan lysimeters* (also called *free-drainage samplers*) and *throughflow troughs* can be used to collect water from the soil or underlying regolith. Care needs to be exercised when interpreting data from these devices, as (artificial) saturation needs to develop at the sampling point for drainage to occur.

For saturated ground, water samples can be extracted from piezometers or open wells using various devices. The first consideration is the installation of the sampling point. Traditional piezometers can be drilled, but consideration needs to minimise the effects of drilling fluids, redistribution of subsurface materials along the drilled

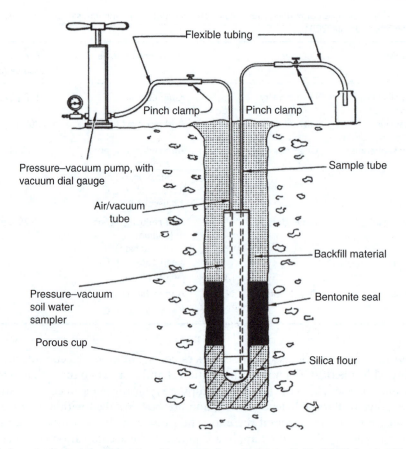

Fig. 8.3 A suction lysimeter. (Reproduced by permission of Nielsen, D. M. (2006) *Practical Handbook of Environmental Characterisation*. Taylor & Francis, Boca Raton.)

hole, and the effect of opening a deep subsurface system to atmospheric air. An alternative is to use *direct-push sampling devices* (e.g. HydroPunch®), which involve fewer disturbances compared to conventional drilling methods (McCall *et al.*, 2006). Consideration also needs to be given to the piezometer construction materials. Within low pH value environments, steel and stainless steel can corrode to add Fe, Mn, Cu, Pb, Cd, Ni, Cr and Mo to subsurface waters. PVC and PVC-cementing agents also degrade. The filter pack around the piezometer screen similarly needs to be comprised of only inert materials, e.g. clean, well-rounded silica sand.

Physico-chemical changes can occur to water samples as a result of the purging and extraction methods. Pressure changes during extraction typically cause CO_2 degassing, which increases sample pH (by between 0.5 and 1.0 pH units), which may then cause trace metals to precipitate. This is particularly the case with *suction-lift pumps*, such as peristaltic pumps (Nielsen and Nielsen, 2006). Some sampling devices also cause temperature increases in samples that can change pH, redox state and the precipitation of carbonates. This can be a particular issue for *electric submersible centrifugal pumps*.

Some sampling devices disturb the water column or mobilise fine material from the filter pack or formation; *bailing devices* and high-flow pumps are particularly prone to these effects.

Analysis of published studies by Nielsen and Nielsen (2006) indicates that water contained within the screened section of piezometers is constantly flushed with natural subsurface water, and poorly connected with the stagnant water higher up within the piezometer. As a consequence, they recommend minimising disturbance of the stagnant water (during purging or sampling) and sampling directly from the screened section. In summary, they recommend the following sampling approaches to minimise changes to water samples:

1   Using low flow rates during purging and sampling.
2   Placement of pump intake with the screened section of the piezometer.
3   Minimising disturbance to the stagnant water column above the screen.
4   Monitoring water-quality indicator characteristics (e.g. EC) during purging.
5   Minimisation of atmospheric contact with samples.
6   Collection of unfiltered samples for metals analysis.

They also provide a full analysis of the operation and disadvantages of the different types of sampler available from *grab samplers* (i.e. *bailers, thief samplers* and *syringe samplers*) to suction-lift devices, *electric submersible centrifugal pumps, positive-displacement pumps* (e.g. a *piston pump*; Fig. 8.4) and *inertial-lift pumps*.

8.4 Laboratory water-quality analyses

Although there are now instruments to measure some physico-chemical properties directly in the field, for example, electrical conductivity monitors, most measures of the physico-chemical content of water are still made by laboratory analyses of samples. It is far beyond the scope of this chapter to describe the analytical techniques for all substances likely to be found in river water, but a knowledge of the main types of analysis used would benefit the hydrologist. The analytical methods may be grouped under the headings: gravimetric, titrimetric, colorimetric, chromatographic, mass spectrometry and electrode-based.

Gravimetric analysis is used to determine the suspended sediment (SS) concentration. With this method, a known volume of water sample is filtered through a pre-weighed filter paper with a pore-size of $0.45 \, \mu$m. The SS concentration is given by the increase in weight at drying at $105°$C. Just prior to both weighing stages, the filter papers should be placed within a desiccator.

Titrimetric analysis, the well-known balancing of reactions using coloured indicators can give satisfactory results down to $1 \, \text{mg} \, \text{L}^{-1}$ in determining, for example, DO or alkalinity. This approach is also described as volumetric analysis, as it depends on the measurement of volumes of liquid reagent of known strength.

Colorimetric analysis involves the measurement of colour intensity when a reagent is combined with a water sample. There are several experimental techniques used in colorimetry that are reliable for measuring such chemicals as ammonia, phosphorus and chlorine as long as they are not mixed with other compounds with similar coloration. The colour can be measured by visual methods of comparison (or Nessler) tubes or

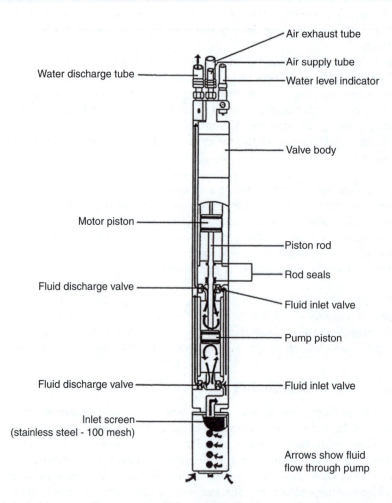

Fig. 8.4 A double-active piston pump used for extracting water samples from piezometers. (Reproduced by permission of Nielsen, D. M. (2006) *Practical Handbook of Environmental Characterisation*. Taylor & Francis, Boca Raton.)

colour papers/discs, but more commonly, instrumental methods employing a colorimeter or spectrophotometer are used. A *colorimeter* uses a photoelectric cell to detect a light source passed through the sample. In contrast, a prism producing monochromatic light of a specific wavelength is used within a *spectrophotometer*. With an *atomic absorption spectrophotometer*, the colour of metal ions burnt within a flame is measured and compared with the light intensity from a known standard solution of the metal ions. With this approach, metals such copper, lead, and zinc are measured.

Chromatographic analysis involves the separation of analytes by their interaction with a microscopic layer of chemicals on the walls of a chromatographic column. *Gas chromatography* (GC) or strictly gas-liquid chromatography is used for measuring phenols and VOCs within water. With this method, the sample is first vaporised and a carrier gas (e.g. helium, nitrogen, argon) is used to transport the sample through the column coated with a microscopic layer of a liquid or polymer. Different analytes within the sample produce eluates that emerge at different times and these are detected with for example a flame-ionization detector. With column chromatography, notably *high-performance liquid chromatography* (HPLC), a water sample is mixed with a solvent and passed through the HPLC column, and the resultant eluates are commonly detected with a ultra-violet absorption detector. HPLC can be used to measure pesticides and polycyclic aromatic hydrocarbons (PAHs) within water.

Mass spectrometry analysis, notably *inductively coupled plasma mass spectroscopy* (ICP-MS) involves the decomposition of water samples to neutral elements in a high-temperature argon plasma, with analysis based on mass to charge ratios. This method is used for measuring trace metals within waters samples. Mass spectroscopy is also used for determining the relative proportions of the stable isotopes of water (^1H, ^2H, ^{16}O, ^{17}O and ^{18}O; DeGroot, 2004) for tracing water pathways or calculating the residence times of different water sources (Kendall and McDonnell, 1998; McGuire and McDonnell, 2007).

Electrode-based analysis is routinely used to measure dissolved oxygen and pH in the laboratory, but can be used to measure ions such as nitrate, potassium, ammonia, cadmium and lead. Details of these analyses are given in the following section on continuous field monitoring. Within laboratory, *dissolved oxygen probes* can be used to measure the biochemical oxygen demand (BOD). This is undertaken by measuring the dissolved oxygen before and after incubation of water samples held at 20°C for 5 days within sealed glass bottles.

Nollet (2007) explains the details of these and other methods of water analysis in considerably more depth.

8.5 Automated field monitoring

Some physico-chemical measurements can be undertaken directly in the field (in rivers or subsurface waters) using data-logged monitors. These characteristics include temperature, turbidity, electrical conductivity, dissolved oxygen, redox potential and pH. *Temperature* is typically measured with a thermocouple. *Turbidity* can be measured with several different types of optical sensor. Typically, a photodiode is used to measure the degree of (coherent) light either penetrating the water or backscattered by particles in the water. The OBS-3+ turbidity probe (D&A Instrument Company, Washington, USA) uses the latter principle. Ideally, the light source and photodiode are cleaned prior to measurement, using an automatic wiper (Fig. 8.5). These turbidity measurements can be calibrated to suspended sediment concentration determined from spot samples by the *gravimetric method* (Section 8.4).

The *electrical conductivity* of water is determined by measuring the potential difference across a couple of electrodes in response to an applied current. The TDS can then be derived from these data using the calibration described in Section 8.1.1. With a *dissolved oxygen* probe, the oxygen diffuses through a gas-permeable plastic membrane to a platinum cathode, and it is reduced:

$$\frac{1}{2}O_2 + H_2O + 2e^- \rightarrow 4OH$$

Fig. 8.5 An OBS-3+ turbidity probe fitted with a Hydrowiper cleaning brush. The installation shown is mounted on a steel support to protect from impact with debris during flood events.

The oxidation taking place at the reference electrode (a silver/silver chloride anode in KCl electrolyte) is:

$$Ag + Cl^- \rightarrow AgCl + e^-$$

A current will then flow that is proportional to the rate of diffusion of the oxygen. Devices to measure *redox potential* (Eh) measure the ability of the water to transfer electrons to and from a reference electrode comprising a silver wire within a silver chloride solution relative to a platinum electrode (platinum because it does not readily oxidise). The *pH* of water requires the use of a glass electrode (which permits the passage of hydrogen ions, but no other ions) coupled to a reference electrode. The potentials created by these two electrodes connected in the same circuit gives a measure that is proportional to the H^+ activity of the solution via the Nernst equation. *Ion-selective electrodes* (ISEs) operate on similar principles, though the ion-selective membranes are not entirely ion specific. ISEs are available to measure ions including nitrate, potassium, ammonia, cadmium and lead.

Several multi-probe sondes are now available that measure several of these characteristics *in situ* within rivers or subsurface wells. An example is the 6-series sonde produced by YSI Inc., which can be configured to measure eight probes (Fig. 8.6).

Fig. 8.6 A 6-series multi-probe sonde. (Reproduced by permission of YSI Inc., Yellow Springs, USA.)

8.6 Bank-side sample analyses (manual and automatic)

Water samples can be extracted manually from rivers or piezometers and analysed in the field using reagent test kits (Nielsen *et al.*, 2006), and portable spectrophotometers in addition to the electrode-based techniques just described. Portable spectrophotometers are available that measure characteristics such as cadmium, colour, hardness, mercury and phosphorus (e.g. Hach DR2800). Spectrophotometer-based systems are also available that can automatically extract samples from rivers (or piezometers) and automatically analyse the sample in the field (Hargesheimer *et al.*, 2002). Within the UK community of water-quality researchers there is a particular focus on the bank-side analysis of phosphorus with these spectrophotometer-based systems (e.g. Jordan *et al.*, 2007; Fig. 8.7).

8.7 Load estimation, discharge consents and compliance

As noted within the introduction to this chapter, the calculation of the *mass flux* or *load* of a substance within rivers rather than just its concentration is important for hydrological research and in the assessment of water quality by regulatory agencies. Load, *L*, is defined over a time period of length, *T*, as:

$$L = \int_{t=0}^{t=T} CQ\,dt$$

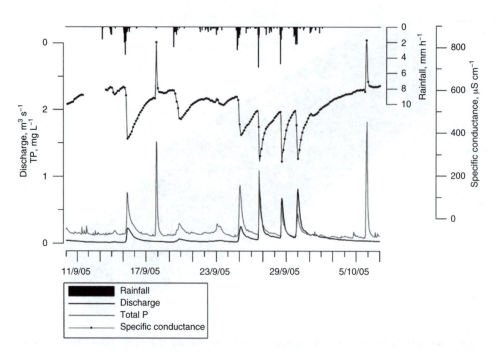

Fig. 8.7 A time series of total phosphorus (total P or TP) measured with a bank-side spectrophotometer. These data are shown with rainfall, river discharge and specific conductance data for a tributary of the Loch Neagh basin, Northern Ireland, UK. (Reproduced from Jordan *et al.*, 2007 with permission of Phil Jordan.)

where C is the measured concentration and Q is the river discharge (or discharge of an effluent) being considered. When continuous samples of discharge and concentration are not available, then this can be expressed in the discrete increment form:

$$L = \sum_{i=0}^{i=N} C_i Q_i$$

where i is a time step index in a period of N time increments. Ideally, monitoring of river discharge (via sub-hourly monitoring of river stage; Section 7.3) and monitoring of concentration (Section 8.5) or sampling of concentration (Section 8.6) is at a very high intensity (i.e. sub-hourly). Loads calculated from such intensive sampling (e.g. Fig. 8.7) have a high degree of accuracy (Littlewood, 1995). It is, however, more common for UK regulatory authorities to collect river stage and hence discharge at such a high intensity, but for the water quality to be sampled only once a week or once every 3 weeks. To obtain the concentration data on the same time step as the discharge data (for subsequent load estimation), a relationship between the observed concentration values and discharge values at the same time is established. This *rating curve* approach does, however, introduce a substantial error into the load calculations. These errors arise because the C–Q relationship is different on the rising and falling stages of storm hydrographs (a phenomenon called *hysteresis*), and because different hydrographs in a time series have different C–Q relationships (see e.g. Sivakumar, 2006; Jordan *et al.*, 2007)

Such load estimates are particularly important in setting licences for discharge consents for point sources of effluents, e.g. an industrial effluent discharge or discharge from a waste water treatment works. Consents are issued as a way of controlling effluent discharges into designated rivers to ensure that the water-quality standards for that river reach can be maintained. This is achieved by taking account of how upstream river and effluent discharges might affect downstream water quality. For a simple conservative substance, this involves simple mixing calculations (see Section 7.5 on dilution gauging and Section 11.3 on mixing models) given information about the upstream load in a river (QC_u) and the load from the effluent (qC_e). Thus, under steady discharge conditions and after downstream mixing has taken place, the downstream concentration (C_d) is calculated:

$$C_d = \frac{QCu + qCe}{(Q+q)}$$

where Q is the upstream river discharge, q is the effluent discharge, C_u is the upstream concentration in the river and C_e is the concentration in the effluent. This type of calculation can be repeated for successive reaches downstream, and at different river discharges and then modelled, e.g. within the Environment Agency's SIMCAT modelling system (WRc, 2006). In setting licensing conditions, this can be used to estimate whether water-quality standards will be not met at any point downstream.

For many water-quality variables, e.g. BOD and DO, conservative mixing of the effluent with the river water will not apply, and chemical reactions will need to be simulated. One of the first water-quality models proposed was the Streeter–Phelps BOD–DO model (Streeter and Phelps, 1925). The Streeter–Phelps model uses simple representations of de-oxygenation and re-oxygenation in a steady discharge to estimate BOD and DO downstream of an effluent input. These relationships are included in the SIMCAT model, which treats discharge in the river network in terms of a succession of steady discharges in each reach. This Streeter–Phelps model is based on the simple analytical solutions:

$$BOD(T) = BOD_o e^{-k_1 T}$$

$$DOD(T) = \frac{k_1 BOD_o}{k_2 - k_1}(e^{-k_1 T} - e^{-k_2 T}) + DOD_o e^{-k_1 T}$$

where *BOD(T)* ($kg\,m^{-1}$) and *DOD(T)* ($kg\,m^{-1}$) are the BOD and DO-deficit below oxygen saturation at a travel time, T, downstream of an effluent discharge point with an initial BOD and DO deficit of BOD_o and DOD_o, respectively. Symbol k_1 is a BOD decay coefficient (d^{-1}) and k_2 is a re-aeration coefficient (d^{-1}), and both k_1 and k_2 are known to vary with temperature and river discharge, and a number of empirical formulae have been proposed for both.

The Streeter–Phelps equations are very simplified representations of the complex physical, chemical and biological processes affecting oxygen in real river reaches. There are more complex water-quality models that have been proposed that deal with a wider range of water-quality variables (e.g. Debele *et al.*, 2006). Such models do, however, incorporate larger numbers of model parameters and input variables, making the predictions much more uncertain (e.g. Dean *et al.*, 2005; see also Section 12.9).

Many river water samples are collected by the Environment Agency for England and Wales are to satisfy to test for compliance with the statutory requirements for quality. When setting consents, the Environment Agency has normally chosen either 95 per cent (percentile limit) compliance or 100 per cent (absolute limit) compliance. The latter means that if any single sample is over the limit set then the site is not compliant. Because of the small number of samples, however, it is still possible that all the samples could be below the limit set but the site might not be compliant. A similar issue arises in testing for 95 per cent compliance. With typical three-weekly sampling producing only 20 samples per year, this would strictly allow only one sample to be over the limit (i.e. 95 per cent $\approx$ 19 out of 20), but because of the small number of samples, it is possible that more samples are over the limit and the site is actually compliant, or for all the samples to be below the limit and the site to actually be non-compliant (e.g. because of lack of samples during storms). Thus there is the possibility in compliance testing of making two types of error:

- *Type I*: the error of judging that the determinant fails to comply when a continuous measurement would have shown that the site was compliant.
- *Type II*: the error of judging that the determinant complies with the set limits when a continuous measurement would have shown that the site was non-compliant.

Given a certain number of samples, and knowing the limits for a particular substance, we can estimate the probabilities of making a *Type I* or *Type II* errors (see McBride, 2005). By convention, we would want both probabilities to be less than 5 per cent before being sure of drawing a conclusion that might lead to a prosecution for failure to comply. For example, even with 365 samples taken over a year all following below the admissible limit, there is still a small probability of making a Type II error (i.e. at an error level of 5 per cent, true compliance is expected in 99.2 per cent of cases). For weekly sampling, for all samples falling below the admissible limit, this probability increases (i.e. at an error level of 5 per cent, true compliance is expected in only 94 per cent of cases). These statistics do depend on the assumption that all the samples are taken from a single underlying distribution. This may not be the case, as there may be storm-event, diurnal or seasonal variations in concentrations that make assessing water quality and compliance to standards much more difficult for those quantities that cannot be monitored or sampled at sub-hourly intensities. This further underlines the need for more monitoring and sampling water-quality characteristics at sub-hourly intervals, as permitted by *in situ*, electrode-based monitors (Sections 8.5) and bank-side analyses (Section 8.6).

References

Artiola, J. F. (2004) Monitoring surface waters. In: Artiola, J. F., Pepper, I. L. and Brusseau, M. L. (eds) *Environmental Monitoring and Characterisation*. Elsevier, Amsterdam. Pp. 142–163.
Bosch, A. (1998) Human enteric viruses in the water environment: a minireview. *International Microbiology* 1, 191–196.

Council of European Communities (1998) Directive 98/83/EC Novemeber 1998 on the Quality of Water Intended for Human Consumption. EC Official Journal L330/41.

Dean, S., Beven, K., Whitehead, P. and Butterfield, D. (2005) Uncertainty assessment of a process-based phosphorus model: INCA-P. *Geophysical Research Abstracts* 7, 02635.

Debele, B., Srinivasan, R. and Parlange, J. Y. (2006) Coupling upland watershed and down-stream waterbody hydrodynamic and water quality models (SWAT and CE-QUAL-W2) for better water resources management of complex river basins. *Environmental Model Assessment* 13, 135–153.

DeGroot, P. A. (2004) *Handbook of Stable Isotope Analytical Techniques.* Amsterdam, Elsevier.

European Commission (2000) Directive 2000/60/EC of the European Parliament and of the Council of 23rd October 2000 establishing a framework for Community action in the field of water policy, Official Journal 22 December 2000 L 327/1 European Commission, Brussels.

European Economic Community (1976) Council Directive of 4 May 1976 on pollution caused by certain dangerous substances discharged into the aquatic environment of the Community (76/464/EEC). Office Journal of the European Communities No. L 129, 18.5.1976.

Gleick, P. (2008) *The World's Water 2008–2009: The Biennial Report on Freshwater Resources.* Island Press, Washington, 432 pp.

Government Printing Office (1996) Safe Drinking Water Act, 1996 Amendments. Public Law 104-182. GPO, Washington.

Gray, N. F. (2005) *Water Technology: An Introduction for Environmental Scientists,* 2nd edn. Elsevier, Oxford, 643 pp.

Grove, A. T. (1972) The dissolved and solid load carried by some West African rivers: Senegal, Niger, Benue, and Shari. *Journal of Hydrology* 16, 277–300.

Hargesheimer, E. E., Conio, O., Popovicova, J. and Proaqua, C. (2002) *Online Monitoring for Drinking Water Utilities.* American Water Works Association, Denver.

Her Majesty's Stationery Office (1989) Water Supply [Water Quality] Regulations 1989. HMSO, London.

Jordan, P., Arnscheidt, J., McGrogan, H. and McCormick, S. (2007) Characterising phosphorus transfers in rural catchments using a continuous bank-side analyser. *Hydrology and Earth Systems Science* 11, 372–381.

Kendall, C. and McDonnell, J. J. (1998) *Isotope Tracers in Catchment Hydrology.* Elsevier Science B.V., Amsterdam.

Littlewood, I. G. (1995). Hydrological refimes, sampling strategies, and assessment of errors in mass load estimates for United Kingdom rivers. *Environment International* 21, 211–220.

Nemerow, N. L. (1974) *Scientific Stream Pollution Analysis.* Scripta, Washington, DC.

McBride, G. B. (2005) *Using Statistical Methods for Water Quality Management: Issues, Problems and Solutions.* John Wiley & Sons Inc., Hoboken, 344 pp.

McCall, W., Nielsen, D. M., Farrington, S. P. and Christy, T. M. (2006) Use if direct-push technologies in environmental site characterisation and groundwater monitoring. In: Nielsen, D. M. (ed.) *Practical Handbook of Environmental Site Characterisation and Ground-Water Monitoring,* 2nd edn. Taylor and Francis, Boca Raton, pp. 345–472.

McGuire, K. and McDonnell, J. J. (2007) Stable isotope tracers in watershed hydrology. In: Lajtha, K. and Michener, W. (eds) *Stable Isotopes in Ecology and Environmental Sciences,* 2nd edn. Blackwell Publishing, Oxford.

Nielsen, D.M. (2006) *Practical Handbook of Environmental Characterisation.* Taylor and Francis, Boca Raton.

Nielsen, D. M. and Nielsen, G. L. (2006) Ground-water sampling. In: Nielsen, D. M. (ed.) *Practical Handbook of Environmental Site Characterisation and Ground-Water Monitoring,* 2nd edn. Taylor & Francis, Boca Raton, pp. 959–1112.

Nielsen, D. M., Nielsen, G. L. and Preslo, L. M. (2006) Environmental site characterisation. In: Nielsen, D. M. (ed.) *Practical Handbook of Environmental Site Characterisation and Ground-Water Monitoring*, 2nd edn. Taylor & Francis, Boca Raton, pp. 35–206.

Nollet, L. M. L. (2007) *Handbook of Water Analysis*, 2nd edn. Marcel Dekker, New York, 784 pp.

Sivakumar, B. (2006) Suspended sediment load estimation and the problem of inadequate data sampling: a fractal view. *Earth Surface Processes and Landforms* 31, 414–427.

Streeter, H. W. and Phelps, E. B. (1925) *Study of the Pollution and Natural Purification of the Ohio River*. Bulletin No. 146, US Public Health Service, Washington, DC.

Stumm, W. and Morgan, J. J. (1996) *Aquatic Chemistry, Chemical Equilibria and Rates in Natural Waters*, 3rd edn. John Wiley & Sons, New York.

Tebbutt, T. H. Y. (1998) *Principles of Water Quality Control*, 5th edn. Butterworth-Heinemann, Oxford, 280 pp.

Twort, A. C., Ratnayaka, D. D. and Brant, M. J. (2000) *Water Supply*. 5th Edition. Arnold, London, 558pp.

United Kingdom Technical Advisory Group of the Water Framework Directive (UTAG; 2008) *UK Environmental Standards and Conditions (Phase 1)*. Final Report, April 2008. UTAG, UK (www.wfduk.org).

WRc (2006) *Production of SIMCAT model structures for England and Wales: Final Report*. WRc ref: UC7189 September 2006. WRc, Swindon.

World Health Organisation (1993) Guidelines for Drinking-Water Quality, 2nd Edition. WHO, Geneva.

Zaitlin, B. and Watson, S. B. (2006) Actinomycetes in relation to taste and odour in drinking water: myths, tenets and truths. *Water Research* 40, 1741–1753.

Chapter 9

Precipitation analysis

The previous chapters have been concerned with the measurement of different hydrological variables. This chapter is concerned with the next stage in the preparation and analysis of information on rainfall and snowfall. It is assumed that the data for computation have been produced by standard observation methods and have been subjected to basic quality assurance checks (see Chapter 3 and World Meteorological Organization, 2008).

9.1 Precipitation extremes

The hydrological extremes of floods and droughts, with their acute effects on human affairs, inevitably provide a large proportion of the work of engineering hydrologists. The simple causes of these notable events are a great surplus of water on the one hand and a great dearth on the other. The first, producing floods, can result from extreme rainfalls, the rapid thawing of a large accumulations of snow, or a combination of both, depending on location and season. Rainfall maxima will be considered here, but the characteristics of flood discharges will be dealt with in the analysis of river flows (Chapter 11). Droughts may also be considered by analysing low river flows, but in many areas, droughts result in dry river beds. However, the study of rainfall deficiencies can give more generally meaningful measures of drought, especially to agriculturalists, and some of these drought studies will be described in this chapter (see Section 9.7).

9.1.1 Maximum observed rainfalls

The world's highest recorded rainfalls for a wide range of durations were assembled by Jennings (1950), and these formed the basis for what has become a well-known log-log plot of rainfall against duration from 1 min to 24 months. The records in the higher durations stemmed from Cherrapunji, India, with a maximum 2-year total of 41 000 mm in 1860–61. New records have recently been set by measured falls from tropical cyclones over the volcanic crater of La Reunion, an island in the Indian Ocean (Quetelard *et al.*, 2008). A revised graph with the values in millimetres is given in Fig. 9.1 The points define an envelope curve for a relationship between the maximum values, R, and duration, D. Higher falls may have occurred, of course, but not been observed. However, because this represents a sample over a period of more than 100 years and over many thousands of gauges, it is expected that the envelope curve would be exceeded only very rarely.

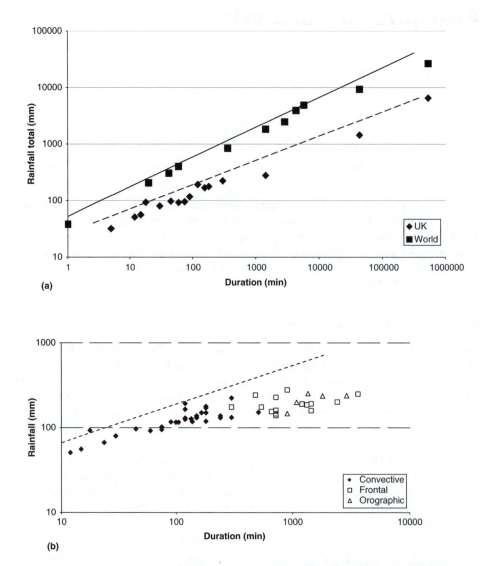

Fig. 9.1 (a) Extreme rainfalls and straight-line envelope curves for the UK and world rainfalls with durations from 1 min to 1 year (data from Tables 9.1 and 9.2). (b) Extreme event rainfalls in the UK classified by event type (data from Hand *et al.*, 2004).

A similar envelope curve can also be constructed for the UK where there are a total of more than 6000 gauges, mostly recording daily catches. Maximum rainfall totals for different durations have been abstracted by the UK Met Office and are also shown in Fig. 9.1a. The UK shows a similar pattern of increasing totals with increasing duration but at much lower values (note the logarithmic scale for the rainfall totals). These extremes are also compared in Table 9.1 for the world and Table 9.2 for the UK.

A useful study of heavy daily falls in the UK was made by Bleasdale (1963) for the years 1893–1960. He listed the 142 occasions on which more than 127 mm

Table 9.1 World rainfall extremes for different durations

Duration	Rainfall (mm)	Place	Date
1 min	38	Barot, Guadeloupe	26 Nov 1970
20 min	206	Curtea-de-Arges, Romania	7 Jul 1889
42 min	305	Holt, USA	22 Jun 1947
60 min	401	Shangdi, Nei Monggol, China	3 Jul 1975
360 min	840	Muduocaidang, China	1 Aug 1977
1 day	1 825	Foc Foc, La Réunion	7–8 Jan 1966
2 days	2 467	Aurere, La Réunion	7–9 Apr 1958
3 days	3 929	Cratère Commerson, La Réunion	24–27 Feb 2007
4 days	4 869	Cratère Commerson, La Réunion	24–28 Feb 2008
10 days	6 028	Cratère Commerson, La Réunion	18–27 Jan 1980
1 month	9 300	Cherrapunji, India	1–31 Jul 1861
1 year	26 461	Cherrapunji, India	Aug 1860–Jul 1861
2 years	40 768	Cherrapunji, India	1860–1861

Table 9.2 UK rainfall extremes for different durations (data from *Hydrology in Practice*, 3rd Edition and Hand *et al.*, 2004)

Duration	Rainfall (mm)	Place	Date
5 min	32	Preston, Lancashire	10 August 1893
12 min	51	Wisbech	27 June 1970
15 min	56	Bolton	18 July 194
18 min	93	Hindolveston (Norfolk)	11 July 1959
30 min	80	Eskdalemuir, Dumfries and Galloway	26 June 1953
45 min	97	Orra Beg (Antrim)	1 August 1980
60 min	92	Maidenhead, Somerset	12 July 1901
75 min	102	Wisley (Surrey)	16 July1947
90 min	117	Dunsop Valley, Lancashire	8 August 1967
120 min	193	Walshaw Dean Lodge, West Yorkshire	19 May 1989
155 min	169	Hampstead, Greater London	14 August 1975
180 min	178	Horncastle., Lincolnshire	7 October 1960
24 h	279	Martinstown, near Dorchester, Dorset	18 July 1955
24 h	314	Seathwaite, Cumbria[a]	18 November 2009
1 month	1 436	Llyn Lydaw, Gwynedd	October 1909
1 year	6 527	Sprinkling Tarn, Cumbria	1954

[a] This event occurred soon after this manuscript was delivered to the publishers. Initial information suggests that this winter synoptic event has now replaced the summer convective event at Martinstown 1955 (see Fig. 9.12) as the highest recorded 24-h rainfall in the UK.

(5 in) had been recorded, and noted that all the falls greater than 102 mm (4 in) numbered over 450. From this study, Bleasdale concluded that no part of the UK could be completely immune from a daily fall of at least 102 mm. Rodda (1973) analysed 121 records with more than 50 years of data and showed how daily rainfall extremes could be related to mean annual average rainfalls across the country. Hand *et al.* (2004) also analysed a variety of extreme rainfall events in the UK and concluded that different types of extreme event (convective, convective with frontal forcing, orographic and frontal) occupied different parts of the rainfall-duration

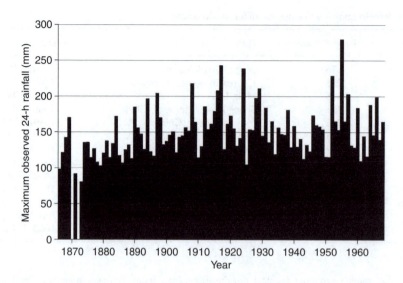

Fig. 9.2 Maximum annual daily rainfalls recorded across all sites in the UK as recorded in British Rainfall 1866 to 1968 (from Rodda *et al.*, 2009, with kind permission of John Wiley & Sons).

plot (Fig. 9.1b). Rodda *et al.* (2009) have produced a digital archive of all the UK rainfall extremes reported in the publications of *British Rainfall Organisation* from 1866 to 1968 (see e.g. the graph of the annual recorded extreme in Fig. 9.2). In the Flood Estimation Handbook (Institute of Hydrology; IoH, 1999), the occurrence of daily totals for given return periods can be derived from the values of the parameters in the depth-duration-frequency (DDF) distribution described below in Section 9.6.1, which are available across the UK on a 5-km grid.

9.1.2 Hydrological impact of an extreme precipitation event: the Boscastle August 2004 flood

The rainfall that led to the flood in the villages of Boscastle and Crackington Haven on the north Cornwall coast on the 16 August 2004 was an example of localized, intense rainfall, falling over an already wet catchment area. It is similar, in this conjunction of very intense rainfalls and wet antecedent conditions, to the flood generating rainfall that occurred in Lynmouth, Devon, 80 km further north-east along the coast, also on 16 August in 1952. In the Lynmouth flood, a maximum of 229.5 mm of rainfall were recorded in 24 h, but the flood that was generated occurred at night and 37 people were killed. Fortunately in Boscastle nobody was killed, but 104 people trapped by the flood were lifted to safety by helicopter from the roofs of buildings and vehicles. Fifty-eight properties were damaged and four had to be demolished. The estimate of total damages was put at £2 m. It is not the first time that Boscastle has been damaged by a floods; events are also recorded in 1827, 1847, 1957, 1958, 1963 and 1996. Even since 2004 there has been another 'mini-flood' in Boscastle in 2007.

The weather conditions on 16 August 2004 show a low-pressure area centred to the west of Ireland. Winds on the Cornish coast were from the south-west with convective cells being triggered along squall lines aligned with the wind. The intensity of these cells was increased by the topography as they crossed the coast, leading to very local high precipitation. Of the ten nearest raing auges to Boscastle, four showed less than 3 mm of rain that day. In Boscastle itself, 89 mm was recorded in 60 min, and at Lesnewth 4 km up the valley, 24 mm in 15 min and 184.9 mm in the day (Fig. 9.3).[1] The highest

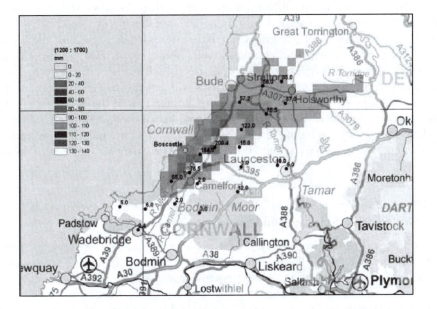

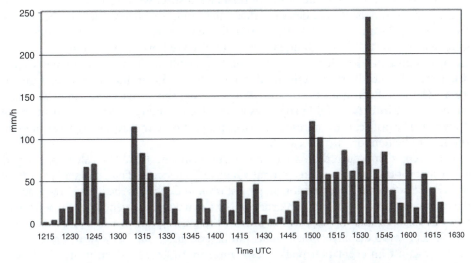

Fig. 9.3 Accumulated radar rainfall totals (after Golding *et al.*, 2005, with kind permission of John Wiley & Sons) and 5-min tipping bucket rainfall record at Lesnewth, Boscastle flood event, 16 August 2004 (after Burt, 2005, with kind permission of John Wiley & Sons).

daily rainfall observed in the area was 200.4 mm at Otterham. A similar event, in Maidenhead, Somerset, in July 1901, provided the largest recorded 60-min rainfall in the UK (see Table 9.2), while similar or more extreme daily falls have been recorded in July 1955 at Martinstown, Dorset (279.4 mm; Table 9.2); in June 1917 at Bruton, Somerset (242.8 mm); in August 1924 at Cannington, Somerset (238.8 mm); and in June 1957 at Camelford, Cornwall (203.2 mm in 24 h).

This type of event is very difficult to predict, both in location and in intensity of the rainfall. Post-event analyses using numerical weather prediction (NWP) models, run at a high-resolution 1-km grid scale, have been able to reproduce the recorded patterns of rainfall in the event reasonably well (Golding *et al.*, 2005), but such models are not yet run at such fine resolution operationally. It is also difficult to predict the runoff produced by such an event because this will depend strongly on the antecedent wetness of the catchment area involved (see Chapter 12). In the Boscastle event, the damages caused seem to have been made worse because of small bridges becoming blocked with debris (including in one case a car), with consequent failure leading to a steep flood wave, up to 3 m high, flowing down the valley at velocities of the order of 4 m s^{-1}.

9.1.3 Hydrological impact of an extreme precipitation event: the Carlisle January 2005 flood

The flooding of Carlisle in January 2005 was part of a much wider event that resulted from very heavy rainfalls across the whole of Cumbria. The rainfall was caused by a flow of unusually warm air, forced northwards ahead of an Atlantic cold front. The rainfall was initially enhanced by strong orographic effects over the mountains of the Lake District and later by strong frontal uplift and convection, as a depression centre passed further north. The maximum observed rainfall during the 7/8 January was 213 mm at Honister in the centre of the Lake District with over 200 mm recorded further east within the River Eden catchment that drains through Carlisle to the Solway Firth (see Fig. 9.4 and Table 9.4). Peak rainfall intensities recorded over 15-min periods were over 20 mm h^{-1} but high intensities were observed over long periods. In fact the initial flooding at Carlisle was pluvial flooding resulting from the drainage system being overwhelmed by the sustained rainfall intensities. Later, fluvial flooding resulted from the River Eden and its tributaries in Carlisle, the Petteril and Caldew, going out of bank. More than 1800 properties were affected and two elderly residents were drowned. The police, fire and rescue services properties in the city centre were flooded and had to be temporarily relocated.

This event also occurred on an already wet catchment area. The calculated runoff coefficient for this event was exceptionally high at 73 per cent. The estimated peak flow of 1520 m^3 s^{-1} at the Sheepmount gauging station in Carlisle is the highest recorded in the catchment (and the highest recorded in the Environment Agency's archive of river flows in England and Wales). At the time of the event, plans for new flood defences for Carlisle were displayed for a final public consultation. They were designed to protect against an event that might occur once in 100 years. Even if they had been built, however, the water levels achieved in this event were higher than the level of protection in the design. The plans have now been revised and new defences built to a higher level of protection.

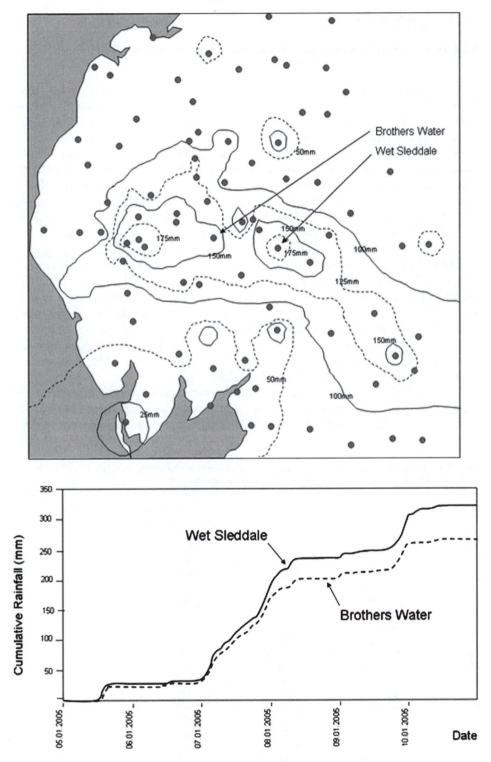

Fig. 9.4 The January 2005 Carlisle flood event: (a) rainfall totals for 6–8 January 2005; (b) 15-min rainfalls recorded at Wet Sleddale and Brothers Water in the period 5–10 January 2005 (redrawn from Environment Agency, 2006).

9.1.4 Hydrological impact of an extreme precipitation event: the UK Summer 2007 floods

The early summer of 2007 in the UK was, in many places, the wettest on record (Marsh, 2008) though there is some dispute as to whether it was really the wettest May to July period in England and Wales since records began in 1776 as widely reported at the time (Eden, 2009). Table 9.3 shows the 20 most extreme 3-month rainfall totals recorded since records began. There is some uncertainty about what the total average over England and Wales should be, but 2007 ranks no higher than 26th in this list. In June, two spells of heavy rainfall, on the 15th and 25th, resulted in 'pluvial' flooding of properties in Hull in Yorkshire as drainage systems were overwhelmed. Fluvial flooding occurred in Sheffield and Doncaster as the River Don overtopped its banks. Totals of over 110 mm were recorded in the Sheffield and Hull areas. July was particularly unsettled and, on the 20th, an active frontal system resulted in daily totals of over 120 mm falling on already wet ground (Fig. 9.5a). The highest recorded rainfall total over 24 h was 142.6 mm at Pershore College, Worcestershire (Fig. 9.5b). This resulted in widespread flooding in several major river catchments in England, such as the Severn in Gloucestershire and Thames in Oxfordshire. In Gloucester, water treatment plants were affected and 400 000 people were left without drinking water. Power was also cut when one electricity substation was flooded, while another was kept in operation by the deployment of temporary flood defences. The total damages in this period in 2007 were of the order of £2bn.

What made this event particularly unusual was the occurrence of such widespread heavy rainfalls and precipitation in summer. In the past, most notably in 1947 but also more recently in 2000 and 2002, such widespread flooding has resulted from synoptic

Table 9.3 The 20 most extreme 3-month periods of rainfalls in the entire England and Wales rainfall record (1776–2007; data from Eden, 2006)

Rank	Period	Year	Rainfall (mm)
1	Oct–Dec	2000	512
2	Sep–Nov	2000	503
3	Nov–Jan	1929/1930	500
4	Oct–Dec	1929	499
5	Aug–Oct	1799	491
6	Jul–Sep	1799	487
7	Oct–Dec	2002	468
8	Sep–Nov	1852	456
9	Oct–Dec	1852	450
10	Nov–Jan	1876–77	449
11	Aug–Oct	1903	444
12	Sep–Nov	1960	439
13	Oct–Dec	1960	438
14	Oct–Dec	1770	433
15	Jul–Sep	1775	431
16	Oct–Dec	1872	429
17	Sep–Nov	1935	424
18	Dec–Feb	1914/1915	423
19	Nov–Jan	1852/1953	423
20	Dec–Feb	1989/1990	419

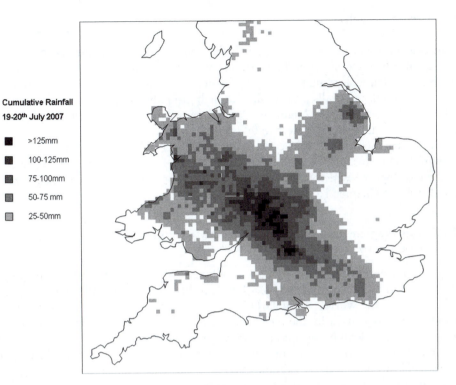

Pershore College, 19-21st July 2007

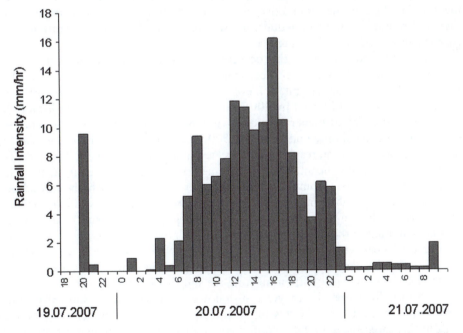

Fig. 9.5 (a) Pattern of total recorded rainfall in southern England, 19–20 July 2007; (b) hourly rainfalls at Pershore College, Worcester, 19–21 July 2007 (redrawn from Marsh and Hannaford, 2007).

scale heavy rainfalls in winter. The particular conditions in 2007 have been linked to a weakened El Niño effect and an extreme southerly position of the jet stream over the North Atlantic, leading to strong depressions tracking further south than normal. The high summer rainfalls also resulted in high groundwater levels in many aquifers, although the second driest August to October period in the record also allowed levels in both rivers and ground waters to fall again.

These extreme rainfalls and consequent flooding resulted in a number of official enquiries. The most significant of these, commissioned by the Government, resulted in the Pitt Report,[2] which made 97 recommendations to improve the management of flood risk (see Chapter 16 for further discussion).

9.1.5 Hydrological impact of an extreme precipitation event: a rain on snow flood

The extreme events discussed above have all been different in the patterns of precipitation and the scale of their impacts but have been the result of rainfalls alone. In other parts of the world, particularly in mountain areas, some of the most extreme flood events are caused by the melting of snow, which may be exacerbated when a warm weather system brings rain to a snow pack that is already close to melting. The additional heat provided by turbulent advection of warm air over the cold pack, and by the infiltration of warmer rain into the snowpack can result in rapid melting.

However, the analysis of such events can be complex, particularly where the temperatures are such that at higher elevations there may be snow during the event while rain is falling at lower elevations and where the snow is not yet close to zero degrees so that heat is required to raise its temperature as well as providing the latent heat of melting. In addition, the ground may or may not be frozen; or may be frozen where there was pasture but not frozen under forest cover. Thus, the runoff generation during such events can be difficult to predict.

This is illustrated by an event that occurred in the Vallée des Ormonts, a tributary to the upper Rhone in the Valaisian Alps, Switzerland (Schoeneich, 1992). On the 10–20 February 1990, a major precipitation event ocurred that initially fell as 30–40 mm of snow but which was followed by 300–350 mm falling as rain at lower altitudes but with further snowfall at higher altitudes (Fig. 9.6). The total precipitation input was estimated as expected only once in about 500 years. In addition, the soil was mostly frozen to depths of 80 cm, resulting in decreased infiltration into the soil surface. This event resulted in a significant, but not extreme flood discharge at Aigle at the outlet of the valley (Fig. 9.6). From the statistics, it was only the tenth highest flood peak recorded at this site. It seems that, in this case, although the inputs to the catchment were extreme, the limited magnitude of the flood peak was a result of the effective area contributing to the runoff generation being relatively small. Only 15 per cent of the rain (or 13 per cent of the total of the initial snowfall plus the rain) contributed to the volume of discharge at Aigle. The rest either fell as snow at higher altitude or was retained in the snow pack without melting the snow or infiltrated into the soil where it was not totally frozen. There were 'slush flows' of rain-saturated snow that moved downslope in the catchment, but without contributing to the discharge peak. It is clear that, in this case, under slightly warmer conditions, the peak discharges could have been much higher.

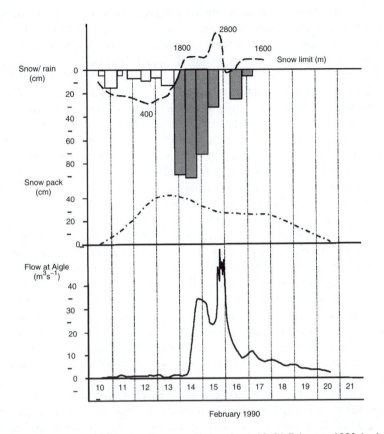

Fig. 9.6 Rain-on-snow event, Vallée des Ormonts, Switzerland, 10–21 February 1990 (redrawn from Schoeneich, 1992).

9.2 Spatial variation in precipitation

9.2.1 Determination of areal rainfall

In describing the measurement of precipitation in Chapter 3, it has been noted that hydrologists still rely heavily on point rain-gauge measurements (Sections 2.3, 3.1 and 3.2); rainfall over a catchment area then has to be estimated from these point measurements. In the UK and elsewhere, these point measurements can now be supplemented by direct areal estimates by rainfall radar (Section 3.6). Radar rainfall, however, does not measure precipitation rates at ground level and is subject to numerous sources of error. Thus in estimating the areal rainfall over a catchment, there will be either uncertainty from the interpolation of point measurements to the catchment area or uncertainty in the integration of radar estimates over the catchment. For either method, the depths of rainfall or the water equivalent of snowfall are expressed in terms of the volume of precipitation falling over a unit area with units of depth. Over a catchment, this value is usually referred to as the areal rainfall or areal precipitation and the term 'average rainfall' is restricted to long-term average values.

The areal rainfall is required for many hydrological studies, and it is most important to have the limits of the catchment carefully defined so as to estimate the volume of input into the catchment area as accurately as possible. For the drainage area down to the river gauging station the 'water parting' (watershed) or boundary of the catchment must be determined and plotted on a topographical map. Now that digital elevation models (DEMs) are widely available, many computer programs have been developed to allow this to be carried out automatically for any point on the river network. One such program is provided for example as part of the *UK Flood Estimation Handbook* discussed in Chapter 13. However, the results from looking only at surface topography are not always correct, particularly where the terrain is flat, and a proper determination of catchment boundaries may require investigations in the field, paying attention to man-made water-courses, such as drainage ditches and leats that may cross a natural topographic watershed boundary. Problems with catchment boundary definitions arise in marshy areas with indeterminate drainage and there may be seasonal differences for such areas in some climates. Knowledge of the geology of the catchment is also necessary since the topographical divide may not be the true water parting for subsurface waters. Part of the surface topographic area of the Pang catchment in the Thames Basin drains as groundwater directly to the Thames; while in north-west England, part of the drainage divide between the Eden and Lune catchment is in limestone geology but overlain by relatively impermeable drumlins deposited during the Pleistocene. In such cases the actual subsurface divide may change with wetting and drying of the catchment, but it must always be remembered that the determination of the catchment area may be a possible source of error in assessing water resources.

There are many ways of deriving the areal precipitation over a catchment from rain gauge measurements. The standard methods and their simpler modifications will be outlined first and a selection of the more sophisticated techniques will follow.

9.2.2 The arithmetic mean

This is the simplest objective method of calculating the average rainfall over an area. The simultaneous measurements for a selected duration at all gauges are summed and the total divided by the number of gauges. The rainfall stations used in the calculation are usually those inside the catchment area, but neighbouring gauges outside the boundary may be included if it is considered that the measurements are representative of the nearby parts of the catchment. A similar approach can be used with radar rainfall estimates, averaging the estimated rainfalls over the pixels and parts of pixels overlaying the catchment in the relevant time period.

The arithmetic mean calculated from gauge measurements gives a very satisfactory measure of the areal rainfall under the following conditions.

(a) The catchment area is sampled by many uniformly spaced rain gauges.
(b) The area has no marked diversity in topography, so that the range in altitude is small and hence variation in rainfall amounts is minimal. The arithmetic mean is readily used when short-duration rainfall events spread over the whole area under study, and for monthly and annual rainfall totals.

If a long-term average for the catchment area is available, then the method can be improved by using the arithmetic mean of the station values expressed as percentages of their annual average for the same long period, and applying the resultant mean percentage to the areal average rainfall (mm) to calculate a weighted average (see next section).

If accurate values of the areal rainfall are obtained first from a large number of rainfall gauge stations, then it may be found that measurements from a smaller number of selected stations may give equal satisfaction. In the Thames Basin of 9981 km^2, it was found that the annual areal rainfall could be determined by taking the arithmetic mean of 24 well-distributed and representative gauges, to within ± 2 per cent of the value determined by a more elaborate method using 225 stations (Institution of Water Engineers, 1937).

9.2.3 The Thiessen polygon

Devised by an American engineer (Thiessen, 1911), this is also an objective method of calculating a weighted average. In this case, the rainfall measurements at individual gauges are weighted by the fractions of the catchment area represented by the gauges as:

$$\overline{R} = \frac{1}{A} \sum_{i=1}^{n} a_i R_i \tag{9.1}$$

where R_i are the rainfall measurements at n rain gauges and A is the total area of the catchment. To calculate the areal weights, given a map of the catchment with the rain-gauge stations plotted, the catchment area is divided into polygons by lines that are equidistant between pairs of adjacent stations. A typical configuration for well-distributed gauges is shown in Fig. 9.7. The polygon areas, a_i, corresponding to the rain-gauge stations are then measured. In the illustrated example, there are nine measurements contributing to the calculation even though gauge 1 is outside the catchment boundary. The area a_1 within the catchment is however nearer to gauge 1 than to the neighbouring gauges 2, 3 and 8, and is therefore better represented by measurements at gauge 1.

The area fractions a_i/A are called the Thiessen coefficients and, once they have been determined for a stable rain gauge network, the areal rainfall is very quickly computed for any set of rainfall measurements. Thus the Thiessen method lends itself readily to computer processing. If there are data missing for one rain-gauge station, it can then be simpler to estimate the missing values and retain the original coefficients rather than to redraw the polygons and evaluate fresh Thiessen coefficients. If, however, a rain-gauge network is altered radically, then the Thiessen polygons have to be redrawn, the new areas measured and a new set of coefficients evaluated.

The simplicity of the Thiessen polygon method derives from the assumption that the rainfall in areas between the gauges can be interpolated linearly. This is only one possible assumption, however. Other methods involve different types of interpolation, such as the annual average rainfall weighting mentioned in the previous section and inverse distance weighting (e.g. Creutin and Obled, 1982; Dirks *et al.*, 1998; Tomczak, 1998; Garcia *et al.*, 2008).

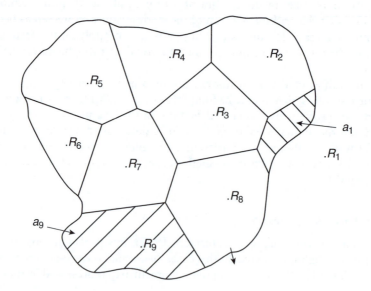

Fig. 9.7 Thiessen polygon method of estimating areal rainfall.

The Thiessen method for determining areal rainfall is sound and objective, but it is dependent on a good network of representative rain gauges. It is not particularly good for mountainous areas, since altitudinal effects are not allowed for by the areal coefficients, nor is it useful for deriving areal rainfall from intense local storms. To overcome some of the shortcomings, investigations into the use of height-weighted polygons combining altitudinal and areal effects on the rainfall measurements have been made. This modification will lead to improved estimates of areal rainfall, relative to the methods based on distance from gauge measurements alone, where there is a strong correlation of rainfall with elevation (annual average rainfall weighting can also be used to reflect such correlations). Previous editions of *Hydrology in Practice* have also outlined the hypsometric method of relating rainfalls to the area in a catchment in a particular elevation band for use where there are multiple gauges at different elevations and such a correlation is particularly strong but statistical interpolation methods are now generally preferred (see Section 9.2.5 below).

9.2.4 The isohyetal method

This is considered one of the most accurate methods, but it is subjective and dependent on skilled, experienced analysts having a good knowledge of the rainfall characteristics of the region containing the catchment area. The method is demonstrated in Fig. 9.8. At the nine rain-gauge stations, measurements of a rainfall event ranging from 26 to 57 mm are given. Four isohyets (lines of equal rainfall), are drawn at 10-mm intervals across the catchment interpolated between the gauge measurements. Areas between the isohyets and the watershed (catchment boundary) are measured. The areal rainfall is calculated from the product of the inter-isohyetal areas (a_i) and the corresponding mean rainfall between the isohyets (R_i) divided by the total catchment area (A).

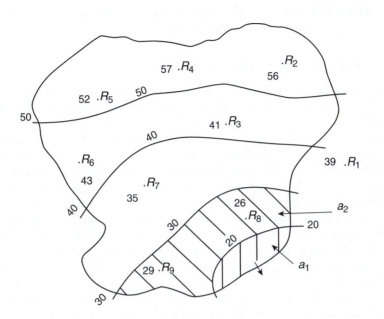

Fig. 9.8 Isohyetal method of estimating areal rainfall.

In the illustration (Fig. 9.8), there are five subareas and the areal rainfall is given by:

$$\overline{R} = \frac{1}{A} \sum_{i=1}^{5} a_i R_i \qquad (9.2)$$

In drawing the isohyets for monthly or annual rainfall over a catchment, topographical effects on the rainfall distribution can be incorporated. The isohyets are drawn between the gauges over a contour base map taking into account exposure and orientation of both gauges and the catchment surface. It is in this subjective drawing of the isohyets that experience and knowledge of the area are essential for good results. The isohyetal method is generally used for analysing storm rainfalls, since these are usually localized over small areas with a large range of rainfall amounts being recorded over short distances.

The UK Met Office has used the isohyetal method for many years and has also developed a modified version in which isopercental lines are drawn instead of isohyets. For a required areal rainfall for a given event, the measurements at each station are plotted as a percentage of the station standard long-term average annual rainfall. Then the lines of equal percentages (isopercentals) divide the catchment into areas that are measured, applied to the mean percentages between the isopercentals, and an overall areal percentage is obtained. This is applied to the standard long-term average annual areal rainfall for the catchment, previously derived, to give the required areal rainfall for the event. This improved technique is more reliable and objective, and can be carried out by computer. However, although it is readily applicable in countries with many long homogeneous rainfall records from which reliable long-term average annual areal rainfall values for all catchments can be evaluated, it is not so useful in developing countries where records are limited.

9.2.5 The multi-quadric surface method

The principles inherent in the isohyetal method of areal rainfall determination can be carried a step further by defining a three-dimensional mathematical description of the rainfall surface. In Fig. 9.9, the diagram represents the volume of precipitation falling on part of a catchment area. The element shown in the x–y plane represents part of the plan area of the catchment and the R values are the depths of rainfall for a given time period. Thus the coordinates (x, y, R) define the position on the rainfall surface of the rainfall measurement R by rain gauge at plan position (x, y). Given a network of rainfall observations, the (x, y, R) surface can be described mathematically in several ways. Two of the most common techniques are by fitting polynomials or harmonic (Fourier) series to the given (x, y, R) points. A third method using multi-quadrics has also been applied successfully to rainfall surfaces and the areal rainfalls subsequently obtained for quantities ranging from 2-min falls to annual totals (Shaw and Lynn, 1972).

Using the latter method, the equation for the rainfall surface is given by summing the contributions from right circular cones placed at each rain-gauge station. Thus the pattern of R can be represented as:

$$R = \sum_{i=1}^{n} c_i \left[(x - x_i)^2 + (y - y_i)^2 \right]^{0.5} = \sum_{i=1}^{n} c_i a(x_i, y_i) \qquad (9.3)$$

where n is the number of data points (x_i, y_i). There are n coefficients c_i that require estimation. For any time interval, the coefficients are obtained from the n equations for the measured rainfalls R_1 to R_n, which written in matrix form are:

$$\underline{R} = \underline{CA} \qquad (9.4)$$

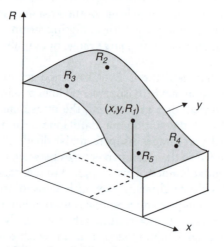

Fig. 9.9 Multi-quadric surface fitting of a rainfall surface.

where $\underline{R}$ and $\underline{C}$ are row vectors, and $\underline{A}$ is a square matrix with elements $a(x_i, y_i)$. Inverting, given the inverse of A gives the coefficients:

$$\underline{C} = \underline{R}A^{-1} \tag{9.5}$$

The equation for R conditioned by these c_i fits all the data points exactly. When the equation for the rainfall surface has been so defined, the volume of rainfall is obtained by integration over the area of the catchment and the areal rainfall results from dividing this value by the catchment area. In practice, the catchment area is subdivided into several rectangles over which the integrations are made and the volumes summed to give the total catchment volume of rainfall.

The method can be used efficiently for all time periods and the computer program can be linked to a contour package to produce machine plotted isohyetal maps of the rainfall event. Versions of multi-quadric interpolation have been used by Wood *et al.* (2000) to interpolate between closely spaced raingauges in the HYREX experiment on the Brue catchment in Somerset and Garcia *et al.* (2008) on the semi-arid Walker Branch catchment in Arizona. The multi-quadric method can also be used for rain-gauge network design by trials, varying the number of gauges and their distribution over a catchment. An alternative to this type of functional interpolation is the thin plate spline method of Hutchinson (1995) that can also incorporate other variables such as elevation in the interpolation.

9.2.6 The kriging and co-kriging methods

Other forms of mathematical interpolation can also be used, including the methods of geostatistics such as kriging, which gives the best linear unbiased statistical interpolation of a set of measurements (e.g. Goovaerts, 1997). This method takes account of the way in which correlation between the catches in individual rain gauges will generally decrease with distance between pairs of gauges. This correlation is important in estimating the areal average (and also controls the uncertainty in the interpolation). It is described by a function called the variogram. The variogram is a representation of the way in which the variance in observed rainfalls between gauges increases with distance between the gauges (it is directly related to the spatial correlation between gauges). Thus, an experimental variogram, $\hat{\gamma}(h)$, can be defined as:

$$\hat{\gamma}(h) = \sum_{i=1}^{N(h)} \left[R(x) - R(x+h) \right]^2 \tag{9.6}$$

where $R(x)$ is the rainfall at position x, and $N(h)$ is the number of rain gauges at a spacing between stations, h. In calculating the experimental variogram from a number of irregularly spaced gauges, it is normal to group the available stations into groups that represent different ranges of spacings. With small numbers of observation points, experimental variograms are often rather erratic, so it is common to fit a mathematical function to the estimated variances so as to smooth the changes (such as the spherical function shown in Fig. 9.10). A number of functions can be used but the form is generally constrained by the value of the variance tending to be rather constant at long distances. The distance at which the variogram reaches a near-constant value

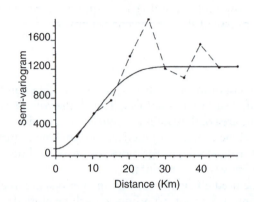

Fig. 9.10 Experimental and fitted spherical variograms for 36 monthly December rainfall observation stations in a 5000 km² area in southern Portugal (after Goovaerts, 2000, with kind permission of Elsevier).

is called the range of the variogram (about 25 km in the case of Fig. 9.10). Some functions allow a non-zero intercept at a distance of zero (this can be seen in Fig. 9.10, though in this case it is small). This is called the nugget variance (it gets this name because a mining engineer call Krige first made the method popular in estimating gold concentrations from point assays in mining operations in South Africa). A large nugget variance is an indication of small-scale variability relative to the spacing of the observation points.

The kriging method has been used for the interpolation of rainfalls at larger scales, e.g. by Goovaerts (2000) in Portugal (Fig. 9.11), and Clark and Slater (2006) in Colorado. The simplest kriging method has one important requirement to be a good interpolation technique: that the variable to be interpolated should be statistically stationary in space. This means that the rainfalls should not be subject to strong trends in the mean values (or in the variance, or other statistical moments). For precipitation variables, of course, this is often the case, particularly where precipitation is strongly related to the topography. In this case, techniques such as kriging with an external drift or co-kriging can take account of other variables, such as elevation, in defining the pattern of rainfalls and integrating it to get the areal average over an area.

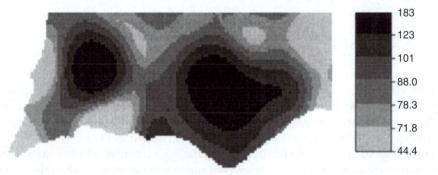

Fig. 9.11 Predicted map of December monthly rainfall amounts in southern Portugal interpolated using ordinary Kriging (after Goovaerts, 2000, with kind permission of Elsevier).

Goovaerts (2000) suggests that this can be a much more reliable estimator than simpler univariate interpolation methods based on distance alone, such as Thiessen polygons or multi-quadric interpolation.

One advantage of kriging as an interpolation technique is that it can provide estimates of uncertainty in the interpolated rainfall at both point and areal average scales. The general use of kriging, however, is limited by the fact that the shape of the variogram is likely to change for every storm or time period (e.g. Bigg, 1991), and requires a large number of gauges to define it well (Fig. 9.10 also illustrates that the observed variogram may not be smooth when derived from only a small number of gauges).

9.2.7 Comparisons of different interpolation methods

There have been many comparisons of rainfall interpolation methods. There are no clear conclusions about which method might be best, partly, of course because we do not generally have enough observational data to say what the 'right' answer might be. Creutin and Obled (1982), for example, concluded that kriging methods generally gave better results; while Dirks *et al.* (1998) suggested that there was no clear advantage over other methods. In a study based on a dense network of rain gauges in the semi-arid Walker Branch catchment in Arizona (100 gauges in $150\,km^2$), Garcia *et al.* (2008) compared multi-quadric and inverse distance weighting methods and suggest that the multi-quadric interpolation is generally superior to the commonly used inverse square distance weighting but that inverse distance cubed weighting can be an improvement on both.

What is clear is that when only a small number of gauges are available to estimate the inputs to a catchment, the errors can be significant, even in UK conditions. Wood *et al.* (2000a) for example, in a study arising from the HYREX experiment with 49 rain gauges in the $135\,km^2$ Brue catchment, suggest that a single gauge will have a standard error of 35 per cent in estimating the true inputs to a $2\,km$ square, and 65 per cent in estimating the inputs to the whole catchment. This compares to 50 per cent and 55 per cent for radar rainfall estimates over the same area. Dynamic calibration of the radar using a single rain gauge can outperform both single rain gauge and radar data alone in estimating the catchment inputs (Wood *et al.*, 2000b). In semi-arid catchments, dominated by small convective rainstorms, the errors might be larger still and can have significant effects on modelling runoff from a catchment (e.g. the results of Faurès *et al.*, 1995, on the Walker Branch catchment).

9.3 Missing data

It is often the case that, for whatever reason, some rainfall data are found to be missing. This can be due to breakdown of a gauge, or the elimination of artefacts, such as bright-band in radar rainfall images (see Section 3.6). Missing data give rise to problems for any technique for interpolating areal rainfalls that rely on weighting coefficients such as in equations 9.1 and 9.2. If there are missing data, then the user can take two options: either recalculate the weighting coefficients with the remaining gauges, or interpolate a value for the missing gauge from the observations at the remaining gauges. Both approaches are used.

Interpolation of data to a missing site will often use one of the techniques outlined in Section 9.2, in particular using the multi-quadric interpolation or one of the kriging interpolators.

9.4 Areal reduction factors and depth–area–duration analysis

In designing hydraulic structures for controlling river flow, a hydrologist needs to know the areal rainfall of the area draining to the control point. Sometimes it is only the average river flow being considered, but more often the works are intended to control flood flows and knowledge of heavy rainfalls is required. There is then an issue that the rainfall depths at a point will not be the same as the average depths over a catchment area estimated by one of the interpolation techniques described above. Estimates of peak rainfall rates for points will then need to be reduced by an areal reduction factor for a catchment. Storm-centred areal reduction factors for single events can be quite variable. In hydrological analyses, fixed area reduction factors are generally used in a region. These do not refer to any particular event but are more statistical in nature, referring to the average ratio of point rainfall to areal rainfall for a given return period.

For an individual storm for which measurements are available, areal reduction factors can be analysed as follows. From the measurements made at all the rain gauges in the area, the pattern of the storm is plotted by drawing the isohyets. The areas enclosed by the isohyets are calculated by planimeter or by digitizing software. Although the average rain between isohyets is taken as the arithmetic mean, the average rain enclosed by the top isohyet has to be estimated. In the case of the extreme Martinstown storm of 1955 shown in Fig. 9.12, there were two 'unofficial' observations in the peak rainfall area (marked at H and G in Fig. 9.12) that helped in this estimation (Clark, 2005). The areal rain for each enclosing isohyet is then plotted against area. The depth–area relationship for the duration of the selected storm is thus obtained. If necessary, several similar duration storms experienced over the area can be used to provide many more data points on the depth–area graph and maximum depth values for the range of areas can be read from an enveloping curve to give design data. Short-duration storms tend to have steep rainfall gradients and hence cover smaller areas than storms of longer duration. This is well recognized in the UK, where intense thunderstorms affect limited areas, whereas prolonged heavy rainfall from an occluded front, for example, usually covers a wide area (as in the Cumbria and Dumfries floods in November 2009).

The analysis of the relationship between areal rainfall depths and area over many storms gives depth–area relationships for different specific durations. Hence in a region where particular types of storms are experienced, the areal rainfall expected from a given catchment area for the catchment response time can be taken from those depth–area relationships for that region.

In the UK, storm patterns are very variable and design rainfalls for different durations and different regions over the whole country were compiled by the Met Office for the original Flood Studies Report (Natural Environment Research Council, 1975), together with estimates of fixed area reduction factors. The resulting areal reduction factors, *ARF*, were also adopted for the *Flood Estimation Handbook* (IoH, 1999)

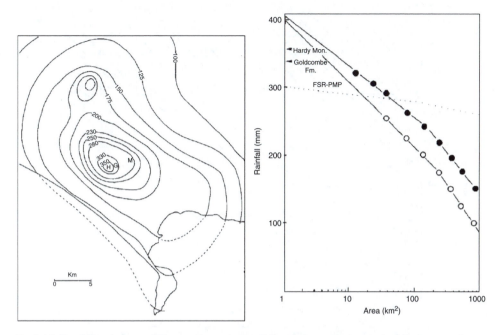

Fig. 9.12 Rainfall isohyets and change in average rainfall with increasing area: the Martinstown Storm of 18 July 1955 that gave the highest ever recorded 24-h point total recorded in the UK until the November 2009 event in Cumbria, see Table 9.2 (after Clark, 2005, with kind permission of John Wiley & Sons).

expressed in mathematical form as

$$ARF = 1 - bD^{-a} \tag{9.7}$$

where D is duration in hours and the coefficents a and b are functions of catchment area given in Table 9.4 (Keers and Wescott, 1977). A graphical representation of these reduction factors is shown in Fig. 9.13.

This type of analysis can be extended to combine considerations of areal rainfall depths over a range of areas and varying durations of heavy falls. The technique of depth-area-duration analysis (DAD) determines primarily the maximum falls for different durations over a range of areas. Analysing the depth–area plots for different

Table 9.4 Flood Estimation Handbook (IoH, 1999) area reduction factor coefficients

Area A (km²)	A	B
$A \leq 20$	$0.40 - 0.0208\ln(4.6 - \ln A)$	$0.0394A^{0.354}$
$20 < A \leq 100$	$0.40 - 0.00382\ln(4.6 - \ln A)^2$	$0.0394A^{0.354}$
$100 < A \leq 500$	$0.40 - 0.00382\ln(4.6 - \ln A)^2$	$0.0627A^{0.254}$
$500 < A \leq 1000$	$0.40 - 0.0208\ln(\ln A - 4.6)$	$0.0627A^{0.254}$
$1000 < A$	$0.40 - 0.0208\ln(\ln A - 4.6)$	$0.1050A^{0.180}$

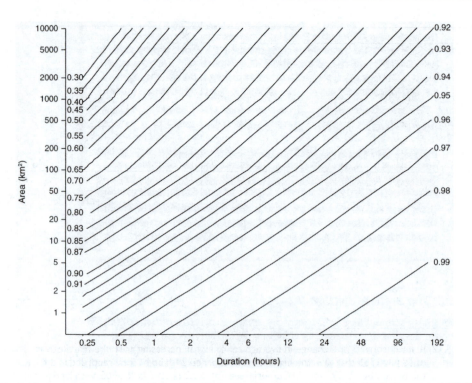

Fig. 9.13 Area reduction factors (after *Flood Estimation Handbook*, Vol. 2, IoH, 1999, Copyright NERC CEH).

durations provides data to compile the highest maximum DAD values from envelope curves drawn for each duration.

9.5 Temporal variation in rainfalls

9.5.1 Checking for consistency at a site

Rainfall amounts are very variable from year to year, even in a humid environment like the UK (e.g. Fig. 9.14). Monthly totals can also show high variability from year to year, totals for individual days even more so. This means that it can be quite difficult to detect inconsistencies in rainfall measurements, for example, where it has been necessary to change the position of a rain gauge. One technique of checking for consistency over time is to use a method called double mass curves. To plot a double mass curve, accumulated rainfall totals from one or more rain gauges are plotted against the accumulated total for the gauge being checked. Fig. 9.15a shows the accumulated rainfalls for three sites in the upper River Eden catchment, the Mallerstang valley. The Castlethwaite site was being closed down by the Environment Agency, with a new site being established at Aisgill at higher elevation, close to the southernmost divide of the catchment. The two sites were run together during the period 1999/2000. Fig. 9.15b shows the double mass plots for both sites plotted against the Scalebeck site for the

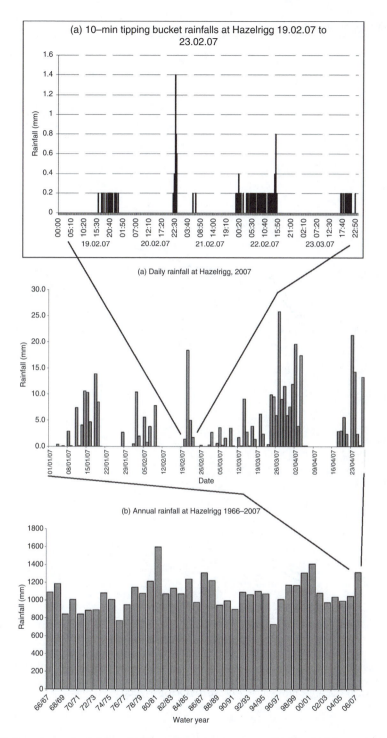

Fig. 9.14 Daily and annual rainfall variability at Hazelrigg, Lancaster: (a) 10-min 0.2 mm tipping bucket rainfalls during part of February 2007; (b) daily rainfalls observed during part of 2007; (c) annual rainfalls observed for the water years (1 October to 30 September) from 1966–7 to 2006–7, mean annual rainfall 1072.7 mm, standard deviation 170.6 mm, mean daily rainfall 2.94 mm day^{-1}.

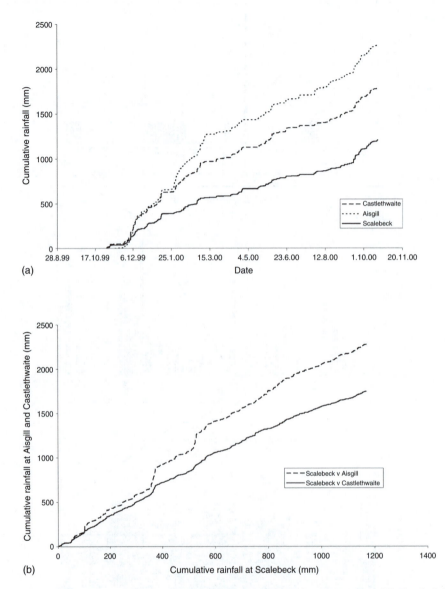

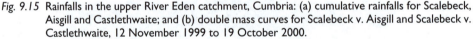

Fig. 9.15 Rainfalls in the upper River Eden catchment, Cumbria: (a) cumulative rainfalls for Scalebeck, Aisgill and Castlethwaite; and (b) double mass curves for Scalebeck v. Aisgill and Scalebeck v. Castlethwaite, 12 November 1999 to 19 October 2000.

period of overlapping data. An examination of the double mass plot shows that the Aisgill site receives on average more rainfall (as would be expected at higher elevations in this region), particularly in some events. Castlethwaite also receives slightly more rainfall than Scalebeck but there is no evidence for either site for significant long-term changes relative to the Scalebeck site.

9.5.2 Long-term changes in rainfalls

There is much speculation about the effects of climate change on rainfall amounts and frequencies. Such changes need to be put into the context of an analysis of historical variability, because we know there has been a lot of variability in rainfalls in the past. Detection of possible changes over time in short duration rainfall extremes is difficult because of the variability (see Section 9.6 on frequency analysis). Analysis of longer term totals can sometimes suggest that there has been change, but it has not all been recent change. The first analyses of this type date back to well before climate change was a real issue (e.g. the study by Nicholas and Glasspoole, 1931).

Fig. 9.16 shows a more recent analysis of seasonal rainfall totals in Scotland taken from Jones and Conway (1997). There appears to be a distinct upward trend in rainfall

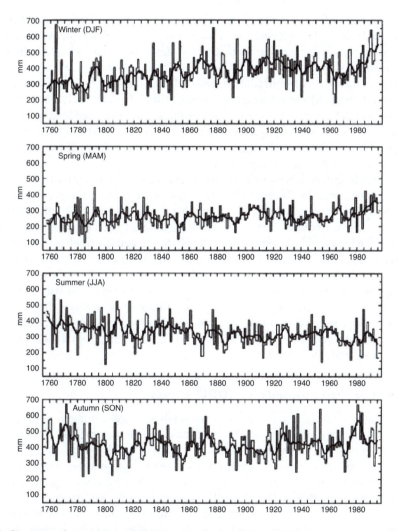

Fig. 9.16 Changes in 3-monthly rainfall totals over Scotland, 1760–1995. The smooth line represents a decadal Gaussian filter centred on each year of data (after Jones and Conway, 1997, with kind permission of John Wiley & Sons).

in the winter months and a downward trend in the summer months, but with a high degree of variability from year to year. In both cases, an analysis for a purely linear trend suggests that the changes are statistically significant. Other regions of the UK show different seasonal patterns of change, not always with significant trends (such as the analysis of data since 1838 at Armagh Observatory in Butler *et al.*, 1998) and with complicating factors such as rain shadow effects (Malby *et al.*, 2007). There is no reason, of course, why we should expect any trends in the data to be linear and there are also other methods for analysis of trends in data that do not assume a purely linear change over time, such as the dynamic harmonic regression method of Young *et al.* (1999) and the general linear models of Chandler and Wheater (2002).

9.6 Rainfall frequency analysis

The analysis of rainfall occurrence depends fundamentally on the length of the rainfall duration for which the information is required. In describing the measurement of precipitation, it has been emphasized that most data are provided by daily gauges, but it is the recording gauges (and radar) that identify the shorter time-scale incidence of rain and give measures of rainfall quantities related to time. The daily storage gauges give rainfall totals without information as to its time of occurrence. Thus, it is logical to consider rainfall frequencies for periods of a day and longer separately from shorter term duration falls derived from recording gauges. Shorter duration frequency analyses still depend heavily on recording gauges since errors in estimating rainfall intensities from radar tend to be high at high intensities.

Considerable differences in rainfall accumulations and the pattern of their occurrence are experienced in different climatic regimes. In general, the greater the annual rainfall amounts the less variable they are from one year to the next. In the semi-arid regions of the world with very limited total amounts, the rainfall is irregular and unreliable. The seasonal pattern of rainfall also has a significant effect on the analysis of its frequency. For the highly seasonal rains of the Tropics or Mediterranean regions it may be advisable to omit considerations of the dry months. When precipitation occurs all the year round, frequency analysis is more straightforward, and attention may be focused more readily on the frequency differences between the rainfall measured over different time periods. The frequency of occurrence of rainfall totals for the day, a month and the year are shown in Fig. 9.17 for a rainfall station close to Lancaster University in the UK.

9.6.1 Depth-duration-frequency (DDF) curves

The analysis of rainfall frequencies is somewhat more complex than the analysis of river discharge frequencies to be considered in Chapter 11. In the case of rainfall extremes we have to consider not only the frequency of a single variable (e.g. flood peaks) but occurrences for which both magnitude and duration are important. Rainfall frequency clearly has an effect on discharge frequency (for both high and low flows) but the relationship between the two will depend on other variables, in particular the antecedent conditions and catchment scale. Small catchment floods, for example, will generally result from small intense rainstorms on already wetted ground. The 2004 Boscastle event in Cornwall was of this type. But for larger catchments, it is the magnitude of rainfall over longer periods that

Daily rainfalls > 0 at Hazelrigg 1966–2007

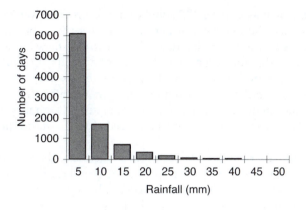

July rainfall totals 1967–2007

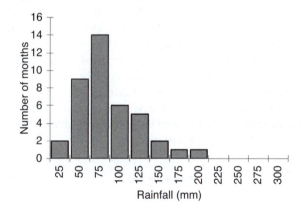

Annual rainfall totals at Hazelrigg 1966–2007

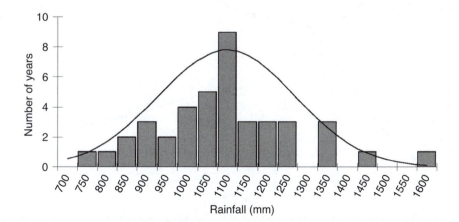

Fig. 9.17 Distributions of daily, July monthly and annual totals shown as histograms for the Hazelrigg site, Lancaster University (see Fig. 9.14 for time series). For the daily totals 40 per cent days with zero rainfall have been excluded from the plot. For the annual totals, a fitted normal distribution is shown. Both daily and monthly data show highly skewed distributions.

will be important, such as in the 2007 summer floods on the River Severn. Thus the concept of a DDF distribution is important in the analysis of recorded rainfalls.

Frequency analysis of any variable depends on certain statistical assumptions. In particular we assume that the occurrence of events of a given magnitude (here rainfall depths) over a long period of time can be represented as a statistical distribution function. Assuming a statistical distribution for extremes also (for the distributions usually used in the analysis of extremes, see Chapter 11) implies an assumption that any particular extreme value might be exceeded in the future (albeit with low probability).

There is evidence that on decadal timescales there has been variability in the frequency of heavy rainfalls. There is also speculation that these frequencies will change with future changes in climate (the statistics may not be stationary). We can, however, only fit a distribution to the data that is available at the current time. As many years of measurements as possible should be used to determine rainfall frequencies. With a record of only 20 years for example, the determination of a representative frequency pattern for annual rainfall, or for the annual maximum series for different durations, will be rather uncertain. The sample is simply too small. Similarly, the analysis of extreme rainfalls lasting for periods of less than a day, entails the abstraction of rainfall depths over specified durations, often from even shorter periods of record. We should expect, therefore some uncertainty in estimating frequencies of occurrence. The general form of the relationship of maximum accumulated rainfall, R, with duration, t, extracted from a recording rain-gauge record during a storm is shown in Fig. 9.18 for the extreme Martinstown Storm event (see Table 9.2 and Fig. 9.12 above). The shape of the average rainfall intensity curve shows how the average maximum rainfall intensity will decrease with increasing duration.

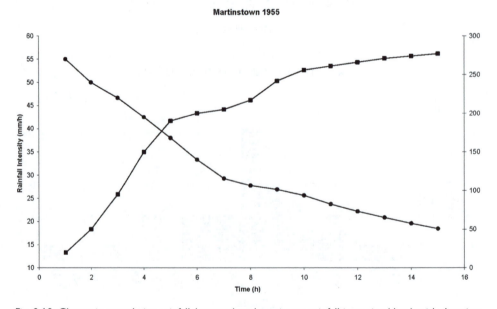

Fig. 9.18 Change in cumulative rainfall (squares) and maximum rainfall intensity (dots) with duration. Storm of 18 July 1955 at Martinstown, Dorset (data taken from Clark, 2005).

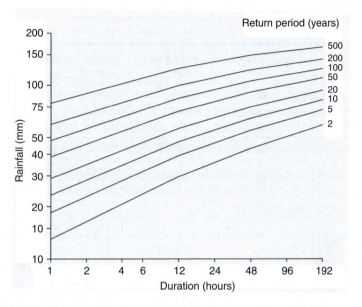

Fig. 9.19 Rainfall depth-duration-frequency curves for Leicester (from *Flood Estimation Handbook*, Vol. 2, IoH, 1999, Copyright NERC CEH).

We can extend the analysis of a single storm to look at how rainfall amounts vary with duration over a large number of storms. This is shown in Fig. 9.19 for a recording rain gauge site at Leicester.

Fig. 9.19 represents the DDF in terms of contours of rainfall depths for different durations for different return periods. The return period is a way of expressing the expected frequency of a statistical variable. Any statistical distribution can be expressed in terms of a cumulative probability of occurrence. Thus for a distribution of rainfall depths r over a duration D, cumulative probability to a value of R is given by the integral

$$F(R|D) = \int_{r=0}^{R} p(r|D)dr \qquad (9.8)$$

where $p(r|D)$ is the probability of r given the duration D.

As hydrologists, we are most often interested not in cumulative probability itself but in the probability of exceedance, which is $P(R|D) = \{1 - F(R|D)\}$. This tells us how often we would expect a event larger than or equal to R to occur for that duration. The probability of exceedance is then directly related to return period, T as:

$$T = \frac{1}{1 - F(R|D)} = \frac{1}{P(R|D)} \qquad (9.9)$$

Thus, the return period is an estimate of the average time between the exceedances of R over a long period of data. For example, if a certain magnitude of daily rainfall has

Table 9.5 Estimated rainfall return periods for different durations at Wet Sleddale: January 2005 Carlisle flood event (data taken from Environment Agency, 2006)

Period	Duration (h)	Rainfall total (mm)	Return period (years)
00:00 Jan 7 – 12:00 Jan 8	36	206.9	173
19:00 Jan 7 – 01:00 Jan 8	6	64.4	21.7
03:15 – 04:15 Jan 7	1	17.6	3.4

an annual probability of exceedance of 0.01, then its return period will be 100 years. A table of estimated return periods for the Wet Sleddale site shown in Fig. 9.4 is given in Table 9.5.

Note that the return period concept does not suggest that if a 1 in 100 year event occurs this year it will be another 100 years before it will occur again. We are treating occurrences of R as a statistical variable, which means that, whether or not there is an occurrence this year, there is finite probability that it will be exceeded next year (0.01 in the case of an event of 100-year return period; 0.1 in the case of a 10-year return period event). There is more discussion of frequency distributions for extreme events in Chapter 11.

In evaluating the frequency of intense rainfalls, mention should be made of two classic studies that established the framework for DDF analyses in the UK. The best known study of the frequency of short-period continuous rainfall in the UK is that by Bilham published in 1936 and reissued by the UK Meteorological Office in 1962. Bilham, who was a civil engineer, assembled 10 years of autographic rain-gauge data from 12 stations representative of lowland England and Wales, and from their analysis developed a formula to relate rainfall depths, durations and frequency of occurrence. A later study by Dillon (1954), then took 35 years of autographic recordings at Cork and identified the relationship of intensity and frequency of occurrence. Dillon proposed expressing the DDF as:

$$R = cD^{1-d}T^f \tag{9.10}$$

in which D is duration, T is return period and c, d and f are constants for a particular data set. This general form reflects the expectation that the higher the frequency of occurrence of a storm of given duration, the smaller the average intensity.

This approach has been extended in the UK *Flood Estimation Handbook* (IoH, 1999), building on the extensive analyses carried out for the earlier Flood Studies Report, by treating the DDF as a number of straight-line segments for each return period on a plot of ln$R$ v. ln$D$ (Fig. 9.20). The straight-line segments run from durations of 1–12 h, 12–48 h and above 48 h. The full DDF is then defined by six parameters, values of which are tabulated for every catchment area in the UK greater draining an area of greater than 0.5 km^2. The DDF distributions have been fitted to rainfall data with durations from 1 h to 8 days and are extrapolated to return periods of up to 1000 years (probability of exceedance 0.001). Thus, the hydrologist in the UK is provided with the frequencies or return periods of rainfalls for any time period from

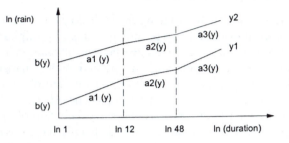

Fig. 9.20 *Flood Estimation Handbook* depth-duration-frequency model showing different linear segments. (Copyright NERC CEH.)

Table 9.6 Factors to convert fixed period rainfall totals to sliding duration equivalents (data taken from Dwyer and Reed, 1995)

Fixed duration (days)	Multiply by	Fixed duration (h)	Multiply by
1	1.16	1	1.16
2	1.11	2	1.08
4	1.05	4	1.03
8	1.01	8	1.01
		≥ 12	1.00

1 min to greater than 8 days for use in engineering design, though the more extreme return period estimates are necessarily somewhat uncertain.

In applying the DDF functions defined in this way, it is important to remember that they have been derived from an analysis of maximum rainfalls in sliding time periods determined from continuous recordings rather than fixed time increments. If the return period for a measured rainfall in a particular fixed time period is then required, the amount must be adjusted to account for the discretisation inherent in fixed period measurements (Dwyer and Read, 1995). The relevant adjustment factors are given in Table 9.6.

9.6.2 The concept of the design event

One of the applications of a rainfall frequency analysis is to provide an input to a design problem, such as determining the capacity of a culvert or storm water detention pond. Depth-duration-frequency analysis provides some standard tools, e.g. those in the UK *Flood Estimation Handbook* described above, for estimating the total storm volume and duration of rainfalls for different probabilities of exceedance. The way in which these are used to produce estimates of discharges with the required return period is explained in Section 13.5. We will only note here that, because of the complexity of runoff generation processes, particularly the effects of antecedent wetness of a catchment on runoff generation, it cannot be assumed that the return period of a design rainstorm will be the same as the discharge resulting from that event (see Fig. 13.4).

9.6.3 Changes in rainfall frequencies

One of the important current issues facing the hydrologist is whether climate change will have a significant impact on the type and amounts of precipitation falling over a catchment area. Climate models do make predictions of changes in precipitation but these are thought to be less reliable than the predictions for changes in temperature and pressures. We can also, however, look at precipitation observations to examine whether there have been changes in the past and there have been a number of studies of changes in rainfall frequencies, particularly of heavy daily falls of rain. It appears that there have been periods of more frequent heavy daily falls in records dating back to the nineteenth century (e.g. Walsh *et al.*, 1982); and analyses suggest that there have been recent changes to both the seasonal timing and frequency of multi-day heavy rainfall events. Fowler and Kilsby (2003) suggest that in southern and eastern parts of the UK, rainfall events that had previously been experienced with return periods of the order of 25 years, since the 1990s are occurring every 6 years on average. Changes have been less in the north and west, although in these regions seasonal winter rainfalls seem to be increasing on average as a result of changing weather patterns (Jones and Conway, 1997).

What we cannot say for sure is whether these changes are a matter of climate variability or climate change. We can speculate about the nature of potential future changes in rainfalls as a result of climate change using global circulation models, but so far, in many parts of the world, these models do not do a good job of reproducing changes in the historical record even for large-scale averages. Thus, while it may be necessary to use such predictions to plan for possible future changes, they should be used with care. In 2009, for the first time, the Hadley Centre is producing projections of regional climate change based on ensembles of climate model runs.[3] While it is difficult to assign real probabilities to the different members of the ensemble, this does represent an important attempt to try to associate some uncertainty with the climate change predictions in a more rigorous way than simply comparing the results from competing global circulation models.

9.6.4 Probable maximum precipitation

Most of the statistical distributions used in the type of frequency analyses described above assume that there is no upper limit to the depth of precipitation that might be recorded at a site in a given duration (the distributions have infinite upper tails). It is certainly the case that, as more and more rainfall measurements become available throughout the world with the increasing number of rain gauges and with the lengthening of existing series of observations, records of maximum rainfall for the various durations continue to be broken. The question arises, however, as to whether records will always continue to be superseded or whether there is a physical upper limit to rainfall. This concept of a finite limit has been named the probable maximum precipitation (PMP), the existence of which has aroused much controversy. A definition of PMP has been given by Wiesner (1970) as 'the depth of precipitation which for a given area and duration can be reached but not exceeded under known meteorological conditions'. The PMP will of course vary over the Earth's surface according to the climatic or precipitation regime. The envelope curves of Fig. 9.1, which show the extremes in both time and space, suggest values that would be expected to be exceeded only rarely at any particular location in the UK or the world, but do not necessarily imply any upper limit.

The idea of a physical upper limit for precipitation comes from meteorological concepts of the maximum precipitable water-holding capacity of a column of air under given meteorological conditions in a particular climatic regime (e.g. Hershfield, 1961). This can be estimated from models of a column of air of different degrees of complexity (recently extreme rainfalls have been studied using fine-resolution convection resolving general circulation models). While such models can be based on the physics of the atmosphere and rainfall generation processes, the resulting rainfall rates will depend on whether the particular parameterisations used are appropriate and on assumptions about the local boundary conditions. In essence, the question about whether PMP is an appropriate concept is replaced by the need to define assumptions about the likely advection of heat and water vapour into the boundaries of the column of air being considered. Clark (2005), for example, in discussing the Martinstown, Dorset storm that is still the record for a 24-h rainfall total in the UK, suggests that the assumptions made in the estimation of PMP in the Flood Studies Report mean that the resulting estimates are too low. Collinge *et al.* (1992) came to similar conclusions in their study of the Hewenden Reservoir storm-event of 1956.

Thus, while the PMP is an attractive concept in engineering design situations requiring protection against catastrophic failures (such as in assessments of dam safety for structures sited upstream of centres of population, e.g. Institution of Civil Engineers, 1996), it is difficult to justify any method of estimating a value for a likely PMP over a catchment area. It might be better to estimate the uncertainty associated with the event of a chosen extreme return period for the design and verge on the side of caution in taking a higher quantile estimate within that range of uncertainty.

9.7 Droughts

Periods of zero or low rainfall over a catchment area are easily observed, but the significance of a lack of rainfall, and what constitutes a 'drought', will vary in the different climatic regimes in the world. In some of the arid zones, for example, there may be several years in which no measurable precipitation occurs and the flora and fauna are adapted to these (normal) conditions. The shortage of rainfall in normally humid parts of the world can, however, result in serious water deficiencies, as for example in a failure of the monsoon in parts of India. Thus drought has widely different connotations according to location and consequences.

A meteorological drought is usually considered to be a period in which the rainfall consistently falls short of the climatically expected amount, such that the natural vegetation does not flourish and agricultural crops fail. Since such extremes are rarely experienced in the UK, low rainfalls more seriously affect water supplies for industry and domestic purposes (e.g. Murray, 1977; March, 2004). In the UK, with regular rainfall all the year round, the occasions of shortages were strictly defined in quantitative terms by the former British Rainfall Organization as an absolute drought where a period of at least 15 consecutive days occurs, none of which has recorded as much as 0.01 in (0.25 mm) of rain, and a partial drought when there is a period of at least 29 days during which the average daily rainfall does not exceed 0.01 in (0.25 mm);

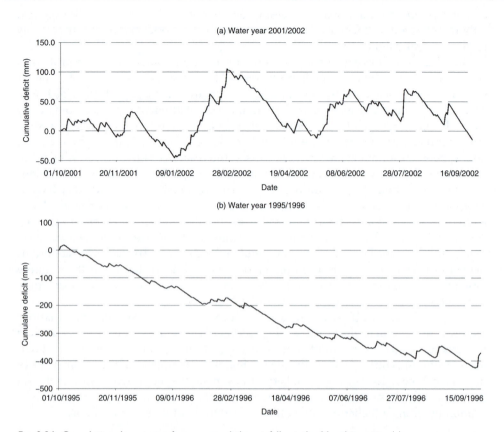

Fig. 9.21 Cumulative departures from mean daily rainfalls at the Hazelrigg site: (a) an average water year 2001/2002 with total rainfall 1080 mm; (b) a very dry water year 1995/1996 with average rainfall with total rainfall 727 mm. Note the difference in deficit scale.

but these criteria are no longer used in UK Met Office publications. Other forms of indexing for the severity of drought have also been used, such as the Herbst method used by Shaw (1979; and earlier editions of *Hydrology in Practice*) and the indices used in the Catalogue of European Droughts.

A good way of visualising rainfall deficiencies is shown in Fig. 9.21 in the form of accumulated departures from mean daily rainfall over a year. A very dry year (WY 1995/1996) with an overall deficit of 400 mm is compared with an average year (WY 2000/2001). This type of diagram was published in the publication *British Rainfall* for selected stations.

The evaluation of water resources requires the appraisal of the incidence of rainfall quantities over periods longer than daily sequences in order to determine storage capacities to meet demands. Hence more importance is normally attached to the identification of rainfall deficiencies in the long term. It has been a statutory requirement since 2003 in the UK that water supply companies produce drought contingency plans. Some droughts, of course, will be more severe than others (e.g. Table 9.7). Thus, the frequency characteristics of droughts of different durations are of interest in a similar way to extreme rainfall amounts. The maximum annual deficiencies can be analysed

Table 9.7 Minimum February–October rainfall totals for the UK (data taken from Marsh, 2004)

Rank	Year	Rainfall (mm)	% of 61–90 mean
1	1921	520	70.1
2	2003	536	72.2
3	1959	557	75.0
4	1955	576	77.6
5	1929	598	80.6
6	1975	604	81.4
7	1911	606	81.6
8	1972	614	82.7
9	1919	620	83.5
10	1902	629	84.7

in a similar way to maximum annual peaks, but it is generally the case that there is greater interest in the frequencies of deficiencies in river flows than of rainfalls (see Chapter 11). Where there is not a flow-gauging site, information about rainfalls can help water engineers assess the potential safe yield of upland catchment areas; a typical assumption is that about 80 per cent of normal annual rainfall will arrive over the driest three consecutive years.

9.8 Stochastic rainfall models

A relatively recent innovation in rainfall analysis has been the use of stochastic rainfall models as a way of generating realistic rainfall sequences in different rainfall regimes. These can then be used with rainfall-runoff models to produce estimates of peak flows and extreme low flows. They are also being used as a way of investigating the impact of future predictions of climate change on hydrology as part of weather generators (e.g. Kilsby *et al.*, 2007). This type of work stems from the seminal papers of Eagleson (1972) who used a simple storm by storm representation of rainfalls as an input to an analytical method for deriving a flood frequency distribution. Eagleson assumed that mean rainfall intensities, storm durations and the arrival times between storms could be described by independent exponential probability distributions. He needed therefore to specify only three parameters.

As computer power has increased and Monte Carlo experiments have tended to replace analytical solutions, stochastic rainfall models have proliferated and become more sophisticated, but with more and more parameters to be defined (e.g. Srikanthan and McMahon, 2001; Cowpertwaite *et al.*, 2002; Wheater *et al.*, 2005). Most models of this type are now based on different representations of rainfall cells, with randomly chosen arrival times, durations and intensities. The sum of all the cells occurring in a time period then produces the final rainfall realization (Fig. 9.22). The parameters of these models are often allowed to vary month by month and are calibrated by fitting to the statistics of the rainfalls (mean, variance, extremes) in each month.

Once the parameters are fitted, arbitrary lengths of rainfall records can be generated to provide statistics and sequences for use in other studies (e.g. the 10 000-year

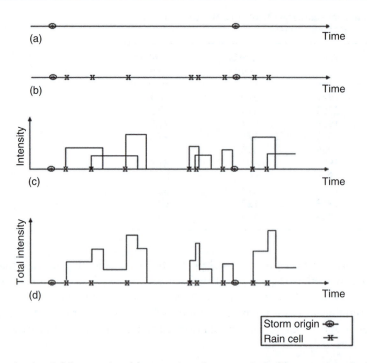

Fig. 9.22 Stochastic rainfall generation: (a) generation of storm arrivals; (b) generation of cell initiations; (c) generation of cell durations and intensities; (d) summation of cell intensities (after Kilsby *et al.*, 2007, with kind permission of Elsevier).

realizations used in estimating extreme flood peak probabilities for dam safety assessment in a Czech catchment by Blazkova and Beven, 2004). The models differ in the assumptions made about the rain-cell-generating mechanisms. In the UK, stochastic rainfall model parameters have been mapped nationally on a 5-km grid basis for both current conditions and modified climate conditions as part of the UK Climate Impacts Program.

Care must be taken with the outputs of such models. In particular, when long realizations are generated from distributions with infinite tails there is always a finite (even if very small) probability of generating very large rainfall amounts that might be considered meteorologically unrealistic (even if there are difficulties in estimating possible maximum rainfalls, see the discussion of the PMP concept in Section 9.6.3). Ways of constraining such extreme values in a way consistent with the upper envelope of Fig. 9.1b have been considered, e.g. by Cameron *et al.* (2001).

For hydrological applications, there is a limit to the information on extreme events that can be gleaned from precipitation records alone. The consequences of excess or deficient rainfalls depend so much on the nature and antecedent condition of the ground that other hydrological processes must be investigated. The solving of engineering problems caused by extreme events only begins with the analysis of the precipitation records.

9.8.1 The design of rainfall networks

One other interesting application of stochastic rainfall models is in the design of rainfall networks. Any design problem requires some assumptions about the nature of the variable being sampled and about the applications for which the network will be used. The design to estimate the mean rainfall over an area to a certain specified uncertainty might be different to that required to provide inputs to a rainfall-runoff model to obtain good estimates of a flood peak. In designing a rain-gauge network it is necessary to make some statistical assumptions about the expected variability and correlation structure in space and time. In doing so, it is somewhat difficult to make assumptions that will be consistent for all rainfall events, since we know that events vary significantly in both variability and correlation structure. A review of the design problem can be found in Bras and Rodriguez-Iturbe (1985). A specific application to the HYREX rainfall experiment on the Brue catchment area in Somerset is provided in Moore *et al.* (2000) and Wood *et al.* (2000a,b).

9.9 Other forms of precipitation

As noted in Chapter 3, there are other forms of precipitation (particularly condensation, occult precipitation and snow) that add to the catchment water balance but which are more difficult to measure and analyse than rainfalls. The principles outlined above for rainfall analysis, in terms of spatial variability and temporal variability, will be the same but the data available for analysis will generally be much more sparse or non-existent. Snow accumulations, in particular, can vary significantly from year to year because of both input variability and redistribution by wind, while occult precipitation is notoriously difficult to measure (although recent work by Bruijnzeel suggests that even in island cloud forests it may not be a significant input to the catchment water balance). As pointed out in Chapter 3, it is also not sufficient to have only snow depth information; snow density is also required to estimate the water equivalent input to the catchment water balance. Snow accumulation and melt are considered further in Chapter 10.

Notes

1 A colour picture of the cumulative radar rainfall estimates may be found at http://www.metoffice.gov.uk/climate/uk/interesting/20040816.html
2 See http://archive.cabinetoffice.gov.uk/pittreview/thepittreview/final_report.html
3 See UK Climate Projections site at http://www.ukcip.org.uk/index.php

References

Bigg, G. (1991) Kriging and intraregional rainfall variability in England. *International Journal of Climatology* 11, 663–675.

Bilham, E. G. (1936) Classification of heavy falls in short periods. *British Rainfall* 1935, 262–280.

Blazkova, S. and Beven, K. J. (2004) Flood frequency estimation by continuous simulation of subcatchment rainfalls and discharges with the aim of improving dam safety assessment in a large basin in the Czech Republic. *Journal of Hydrology* 292, 153–172.

Bleasdale, A. (1963) The distribution of exceptionally heavy daily falls of rain in the United Kingdom, 1863–1960. *Journal of the Institute of Water Engineering* 17, 45–55.

Bras, R. L. and Rodríguez-Iturbe, I. (1985) Random *Functions and Hydrology*. Addison-Wesley Publishing Company, Reading, MA.

Burt, S. (2005) Cloudburst on Hendraburnick Down: The Boscastle storm of 16 August 2004. *Weather* 60, 219–227.

Butler, C. J., Coughlin, A. D. S. and Fee, D. T. (1998) Precipitation at Armagh Observatory 1838–1997. *Proceedings of the Royal Irish Academy, Biology and Environment* 98B, 123–140.

Cameron, D., Beven, K. J. and Tawn, J. (2001) Modelling extreme rainfalls using a modified random pulse Bartlett–Lewis stochastic rainfall model (with uncertainty). *Advances in Water Resources* 24, 203–211.

Chandler, R. E. and Wheater, H. S. (2002) Analysis of rainfall variability using generalized linear models: a case study from the west of Ireland. *Water Resources Research* 38, 1192, doi:10.1029/2001WR000906.

Clark, C. (2005) The Martinstown storm 50 years on. *Weather* 60, 251–257.

Clark, M. P. and Slater, A. G. (2006) Probabilistic quantitative precipitation estimation in complex terrain. *Journal of Hydrometeorology* 7, 3–22.

Collinge, V. K., Thielen, J. and McIlveen, J. F. R. (1992) Extreme rainfall at Hewenden Reservoir, 11 June 1956. *Meteorological Magazine* 121, 166–171.

Cowpertwait, P. S. P., Kilsby, C. G. and O'Connell, P. E. (2002) A space-time Neyman–Scott model of rainfall: empirical analysis of extremes. *Water Resources Research* 38, doi:10.1029/2001WR000709.

Creutin, J. D. and Obled, C. (1982) Objective analysis and mapping techniques for rainfall fields. An objective comparison. *Water Resources Research* 18, 413–431.

Dillon, E. C. (1954) Analysis of 35-year automatic recordings of rainfall at Cork. *Transactions of the Institution of Civil Engineers, Ireland* 80, 191–283.

Dirks, K. N., Hay, J. E., Stow, C. D. and Harris, D. (1998) High-resolution studies of rainfall on Norfolk Island Part II: Interpolation of rainfall data. *Journal of Hydrology* 208, 187–193.

Dwyer, I. J. and Reed, D. W. (1995) *Allowance for Discretisation in Hydrological and Environmental Risk Estimation (ADHERE)*, Report No. 123, Institute of Hydrology, Wallingford, UK.

Eagleson, P. S. (1972) Dynamics of flood frequency. *Water Resources Research* 8, 878–898.

Eden, P. (2009) The Government's response to the summer floods of 2007. *Weather* 64, 18–22.

Environment Agency (2006) *Cumbria Floods Technical Report: Factual report on Meteorology, Hydrology and Impacts of Flooding January 2005 in Cumbria.*

Faurès, J.-M., Goodrich, D. C., Woolhiser, D. A. and Sorooshian, S. (1995) Impact of small-scale spatial rainfall variability on runoff modeling. *Journal of Hydrology* 173, 309–326.

Fowler, H. J. and Kilsby, C. G. (2003) Implications of changes in seasonal and annual extreme rainfall. *Geophysical Research Letters* 30(13), 1720, doi:10.1029/2003GL017327.

Garcia, M., Peters-Lidard, C. D. and Goodrich, D. C. (2008) Spatial interpolation of precipitation in a dense gauge network for monsoon storm events in the southwestern United States. *Water Resource Research* 44, W05S13, doi:10.1029/2006WR005788.

Golding, B., Clark, O. and May, B. (2005) The Boscastle Flood: meteorological analysis of the conditions leading to flooding on 16 August 2004. *Weather* 60, 230–235.

Goovaerts, P. (1997) *Geostatistics for Natural Resources Evaluation*. Oxford University Press, Oxford.

Goovaerts, P. (2000) Geostatistical approaches for incorporating elevation into the spatial interpolation of rainfall. *Journal of Hydrology* 228, 113–129.

Hand, W. H., Fox, N. I. and Collier, C. G. (2004) A study of twentieth-century extreme rainfall events in the United Kingdom with implications for forecasting. *Meteorological Applications* 11, 15–31.

Hershfield, D. M. (1961) Estimating the probable maximum precipitation. *Journal of Hydraulics Division, American Sociey of Engineers* 87, HY5, 99–116.

Hutchinson, M. F. (1995) Interpolating mean rainfall using thin plate smoothing splines. *International Journal of Geographical Information Science* 9, 385–403.

Institute of Hydrology (IoH) (1999) *Flood Estimation Handbook* (5 vols). IoH, Wallingford.

Institution of Civil Engineers (1996) *Floods and Reservoir Safety: an Engineering Guide*, 3rd edn. Thomas Telford, London.

Institution of Water Engineers (1937) Report of Joint Committee to consider methods of determining general rainfall over any area. *Transactions of the Institution of Water Engineers* XLII, 231–259.

Jennings, A. H. (1950) World's greatest observed point rainfalls. *Monthly Weather Review* 78, 4–5.

Jones, P. D. and Conway, D. (1997) Precipitation in the Bristish Isles: an analysis of area-average data updated to 1995. *International Journal of Climatology* 17, 427–438.

Keers, J. F. and Wescott, P. (1977) *A Computer-based Model for Design Rainfall in the United Kingdom*. Met Office Scientific Paper No. 36. HMSO, London.

Kilsby, C. G., Jones, P. D., Burton, A., Ford, A. C., Fowler, H. J., Harpham, C., James, P., Smith, A. and Wilby, R. L. (2007) A daily weather generator for use in climate change studies. *Environmental Modelling and Software* 22, 1705–1719.

Malby, A. R., Wgyatt, J. D., Timmis, R. J., Wilby, R. L. and Orr, H. G. (2007) Long-term variations in orographic rainfall: analysis and implications for upland catchments. *Hydrological Sciences Journal* 52, 276–291.

Marsh, T. J. (2004) The UK drought of 2003: a hydrological review. *Weather* 58, 224–230.

Marsh, T. J. (2008) A hydrological overview of the summer 2007 floods in England and Wales. *Weather* 63, 274–279.

Marsh, T. J. and Hannaford, J. (2007) *The Summer 2007 Floods in England & Wales. National Hydrological Monitoring Programme*. Centre for Ecology and Hydrology, Wallingford.

Moore, R. J., Jones, D. A., Cox, D. R. and Isham, V. S. (2000) Design of the HYREX raingauge network. *Hydrology and Earth System Sciences* 4, 523–530.

Murray, R. (1977) The 1975/76 drought over the United Kingdom hydrometeorological aspects. *Meteorological Magazine* 106, 1258, 129–145.

Natural Environment Research Council (NERC) (1975) *Flood Studies Report, Vol. IV. Hydrological Data*. NERC, Wallingford, 541 pp.

Nicholas, F. J. and Glasspoole, J. (1931) General monthly rainfall over England and Wales, 1727 to 1931. *British Rainfall* 1931, 299–306.

Quetelard, H., Bessemoulin, P., Cerveny, R. S., Peterson, T. C., Burton, A. and Boodhoo, Y. (2008) World record rainfalls (72-hour and four-day accumulations) at Cratère Commerson, Réunion Island, during the passage of Tropical Cyclone Gamede. *Bulletin of the American Meteorological Society*, DOI: 10.1175/2008BAMS2660.1.

Rodda, H. J. E., Little, M. A., Wood, R. G., MacDougal, N. and McSharry, P. E. (2009) A digital archive of extreme rainfalls in the British Isles from 1866 to 1968 based on British rainfall. *Weather* 64, 71–75.

Rodda, J. C. (1973) A study of magnitude, frequency and distribution of intense rainfall in the United Kingdom. *British Rainfall* 1966, Part III, 204–215.

Schoeneich, P. (1992) Rain on snow, a specific type of event. Examples from the western Alps. *Internationales Symposion INTERPRAEVENT, Bern, 1992*. VHB Tagundspublikation Band 1, 182–192.

Shaw, E. M. (1979) The 1975/76 drought in England and Wales in perspective. *Disasters* 3, 103–110.

Shaw, E. M. and Lynn, P. P. (1972) Area rainfall evaluation using two surface fitting techniques. *Bulletin of the International Association of Hydrological Sciences* XVII(4), 419–433.

Srikanthan, R. and McMahon, T. A. (2001) Stochastic generation of annual, monthly and daily climate data: a review. *Hydrology and Earth Systems Science* 5, 653–670.

Thiessen, A. H. (1911) Precipitation for large areas. *Monthly Weather Review* 39, 1082–1084.

Tomczak, M. (1998) Spatial interpolation and its uncertainty using automated anisotropic inverse distance weighting (IDW) – cross-validation/jackknife approach. *Journal of Geographic Information and Decision Analysis* 2, 18–30.

Walsh, R. P. D., Hudson, R. N. and Howells, K. A. (1982) Changes in the magnitude-frequency of flooding and heavy rainfalls in the Swansea Valley since 1875. *Cambria* 9, 36–60.

Wheater, H. S., Chandler, R. E., Onof, C. J., Isham, V. S., Bellone, E., Yang, C., Lekkas, D., Lourmas, G. and Segond, M.-L. (2005) Spatial–temporal rainfall modelling for flood risk estimation. *Stochastic Environmental Research and Risk Assessment* 19, 403–416, doi:10.1007/s00477-005-0011-8.

Wiesner, C. J. (1970) *Hydrometeorology*. Chapman and Hall, London.

World Meteorological Organization (WMO) (2008) *Guide to Meteorological Instruments and Methods of Observation*. WMO-No. 8, 7th edn. WMO, Geneva, 681.

Wood, S. J., Jones, D. A. and Moore, R. J. (2000a) Accuracy of rainfall measurement for scales of hydrological interest. *Hydrology and Earth Systems Science* 4, 531–543.

Wood, S. J., Jones, D. A. and Moore, R. J. (2000b) Static and dynamic calibration of radar data for hydrological use. *Hydrology and Earth Systems Science* 4, 545–554.

Young, P. C., Pedregal, D. J. and Tych, W. (1999) Dynamic harmonic regression. *Journal of Forecasting* 18: 369–394.

Chapter 10

Energy budget analysis, evapotranspiration and snowmelt

Hydrological processes depend not only on the water budget but also on the surface energy budget, which affects, in particular, the transfer of water back to the atmosphere as evaporation and transpiration and also the way in which precipitation as snow builds up into a snow pack and later melts. Evapotranspiration requires energy to change the phase state of water from liquid to vapour; snowmelt requires energy to change the phase state of water from ice to liquid (or sometimes directly to vapour, a process called *sublimation*). In both cases, the amount of energy required to change the phase of water is known (these are the latent heat of melting and latent heat of vaporisation). The difficulty is in estimating how much energy will be available from the energy budget in different circumstances. The energy budget involves several variables that are difficult to estimate by measurement, particularly at the catchment scale.

10.1 The energy budget and evapotranspiration

The factors governing evaporation from open water and from a vegetated surface have been described in Chapter 4, where methods of measurement of water losses are outlined. The practical approach of those methods considers the evaporation process from the liquid phase as the loss rate from a surface, with the estimate of evapotranspiration being given in the form of an equivalent depth of water lost over a selected time period. In energy terms this can also be expressed as a flux of latent heat. The evaluation of evapotranspiration into the gaseous phase considers the gain of water vapour by the air above the open water or vegetation (i.e. the absorption of the water vapour by the air measured over a period of time). To complete the energy balance, estimation of the fluxes of sensible heat is also important. The instrumentation required to make such measurements are described in Chapter 4. Here the related analysis and computations using the measured meteorological variables are set out. Such estimates of latent and sensible heat fluxes are important not only to the hydrologist, but also to the meteorologist, since they provide a lower boundary condition for atmospheric models, such as those used for weather forecasting and climate change predictions. The representation of the surface and its hydrology in such models is often called a *land surface parameterisation*.

With the exception of direct estimation of evaporation from a large open water body, the estimation of actual evapotranspiration rates is difficult, since it depends on the prevailing meteorological conditions, the physical characteristics of the surface including the angle and aspect of the local hillslopes, the physiological characteristics of any vegetation cover, and the availability of water both at the surface and via the roots of the vegetation. These factors interact. For example, vapour absorbed by the atmosphere at one point in space may affect evapotranspiration rates downwind or a lack of available water may cause plants to reduce their transpiration rates. There is also the problem of heterogeneity in the landscape. Real landscapes very often show rapid changes in vegetation cover. A change of physical roughness, or albedo of the surface or water supply to the vegetation will affect local evapotranspiration rates and through the transport and mixing of vapour in the lower boundary layer of the atmosphere may also have effects downwind. Actual evapotranspiration rates may therefore depend on wind direction and fetch, which is why measurement sites for evapotranspiration fluxes are often chosen where there is a fairly homogeneous vegetation cover in all directions. Such sites are, however, unusual in many landscapes. Certain measurement techniques, such as laser scintillometry (see Section 4.6), can make estimates of integrated fluxes over a long path length across a heterogeneous surface, but it is not yet well understood how such measurements are affected by the complex mixing processes over such surfaces so that the accuracy of such measurements may be uncertain.

This is an important issue to the hydrologist who will be primarily interested not in the estimation of evapotranspiration at a point, but in estimates over a catchment area with its complex of open water and different vegetation surfaces. The difficulty of measuring catchment scale evapotranspiration directly has meant that the hydrologist has found it difficult to close the water balance equation (i.e. to get direct measurements of all the input and output terms). Indeed, the only way of getting catchment-scale evapotranspiration estimates has been to use the water balance equation in the form:

$$E_t = P - Q - G \pm [\Delta S]$$

where E_t is the catchment average actual evapotranspiration flux, P is the estimated catchment average precipitation input, Q is the measured stream discharge, G is any groundwater discharge across the basin divides, and ΔS is the change in storage (in brackets because it is often assumed negligible over of long time period such as the water year).

This approach, however, assumes that accurate estimates of the precipitation inputs and discharge outputs are available, and that the change of storage can be either considered to be negligible or estimated. Unfortunately, estimating changes of storage at the catchment scale is equally fraught with difficulty. The result is that any errors in the terms on the right-hand side of the water balance equation will result in error in the estimate of actual evapotranspiration. Thus, in what follows, we will consider ways of estimating actual evapotranspiration more directly.

Water balance estimates of evapotranspiration can, however, be useful. In 1954, an engineer for the Fylde Water Board, Frank Law, started lysimeter and water balance experiments at the Stocks Reservoir site in the Forest of Bowland, Lancashire to investigate the hypothesis that tree-covered catchments lost more water to evapotranspiration than grassland catchments. His conclusion that planting forest on water supply reservoir catchment areas might greatly reduce water yield was controversial and led to the more detailed studies in the paired catchment experiments at Plynlimon, mid-Wales. These experiments, at least initially, showed that Frank Law was right, with up to 60 per cent greater evapotranspiration at Plynlimon in the early stages of forest growth on the forested River Severn catchment compared with the grassland River Wye (Fig. 10.1). This simple conclusion can, however, be complicated by other effects. At the small Coalburn catchment in the Kielder Forest in northern England, Robinson (1998) shows how forest-ditching associated with the planting of trees resulted in drier soils and initially caused a *reduced* actual evapotranspiration loss and that it was some 20 years before evapotranspiration rates returned to pre-plantation levels (Fig. 10.2).

To obtain more direct estimates of evapotranspiration, we need to consider the energy budget of a surface. In Fig. 10.3, we distinguish water and land surfaces. Over a water surface, water is always available, but energy from the Sun can penetrate the surface and act to heat the water body as well as evaporate water from the surface as latent heat. Sensible heat energy can also be supplied directly from (or lost to) the lower layer of the atmosphere, depending on the difference in temperature between the water and the air. The transfer of sensible heat will be much more efficient with a higher wind speed even though a water surface is relatively smooth (has a relatively high aerodynamic resistance to transport). Over a land surface, on the other hand,

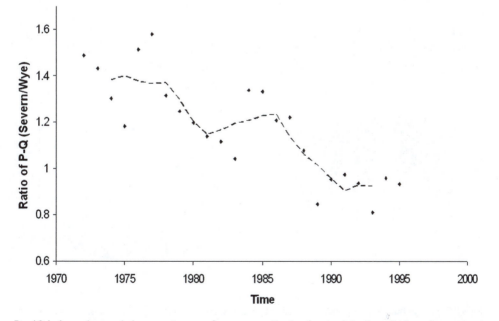

Fig. 10.1 Annual water balance estimates of evapotranspiration (as precipitation–discharge) expressed as a ratio of totals from the 70 per cent forested River Severn to the grassland River Wye catchments at Plynlimon, mid-Wales. The dotted line is a 5-year moving average. Data taken from Hudson *et al.* (1997). Forest harvesting in parts of the River Severn started in 1985.

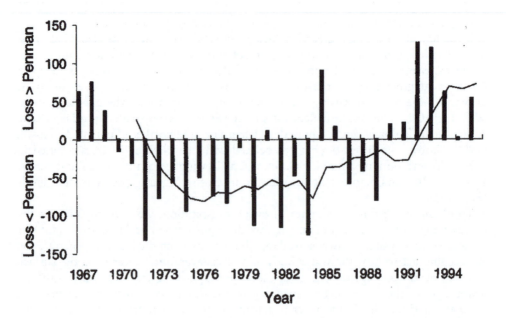

Fig. 10.2 Changes in actual evapotranspiration (as precipitation–discharge in millimetres) in comparison with Penman potential evapotranspiration (see Section 10.2.3) at the Coalburn experiment catchment in northern England. Drainage and planting of conifer plantations started in this catchment in 1972. The solid line is a moving average (from Robinson, 1998).

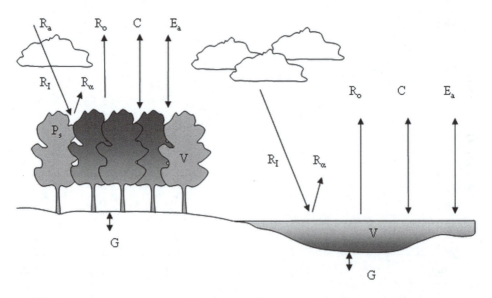

Fig. 10.3 Energy budget components of vegetation and water surfaces. R_I is incoming clear-sky short-wave radiation; R_α is reflected short wave radiation; R_o is outgoing long-wave radiation; C is sensible heat flux; λE_a is actual latent heat flux; V is storage of energy in vegetation or water body; G is energy exchange with ground; P_s is energy used in photosynthesis. Note that all terms are expected to vary in space and time.

such transport can be much more efficient over a rough tree canopy (which has a low aerodynamic resistance) but evaporation and transpiration may be limited by the availability of water, particularly after extended dry periods. The structure of a vegetated surface may also be much more complicated, with multiple leaf layers at different levels contributing to the use of energy in photosynthesis and loss of water by transpiration. A wet soil surface, if energy is available to provide the latent heat of vaporisation, might also contribute to the total latent heat flux through evaporation. The soil can also act as a heat store, generally heating up during the day and releasing heat during the night. Evaporation from a bare soil surface is somewhat simpler, but can still be limited by water availability once the surface starts to dry out.

In the diagram, R_I is the incoming short-wave solar radiation, R_α is the reflected short-wave radiation, and R_o is the net outgoing long-wave radiation from the land or water surface. These terms can be measured by a net radiometer as net radiation R_N. C is the sensible heat transfer to the air, V is the change in stored energy in the vegetation canopy or water body, and G is the energy transfer between canopy or water and the underlying soil or lake bed. P_s is the energy used in photosynthesis. This is normally small relative to the other terms and usually neglected. Then the energy balance equation can be written as follows:

$$\lambda E = R_N \pm C \pm V \pm G - [P_s] \tag{10.1}$$

where λ is the latent heat of vaporisation of water ($\sim$2470 kJ kg^{-1}; see Appendix Table A4). If all the terms are expressed as energy per unit surface area as W m^{-2}, then E has units of kg m^{-2} s$^{-1} \approx$ mm s^{-1} (assuming a density of water of 1000 kg m^{-3}). Note that if the right-hand side (RHS) of (10.1) sums to a negative value, it suggests that latent heat of vaporisation is being released. This can occur when the meteorological conditions are such that there is condensation of vapour from the atmosphere to the surface.

10.2 Calculation of open water evaporation E_o

As noted in Chapter 4, we can distinguish between potential evapotranspiration when water is not limiting and the actual rate of evapotranspiration from a surface, E_t. By far the easiest case to consider is evaporation directly from an open water body, E_o. This might be a lake, or the types of evaporation pan considered in Section 4.2.2. In either case, water availability is not an issue. There are two major approaches that may be adopted in calculating evaporation from open water, E_o. The *energy budget* method considers all the heat sources and sinks on the RHS of (10.1) and calculates the energy available for the evaporating process by closing the energy balance. The second or *mass transfer* method, sometimes called the *vapour flux* method, calculates the upward flux of water vapour from the evaporating surface in the lower boundary layer of the atmosphere. A third method uses a combination of the two physical approaches.

10.2.1 Energy budget method

Evaporation from a lake or reservoir may be calculated on a weekly or monthly basis by taking into consideration the heat or energy required to effect the evaporation.

The measurements required to evaluate the full-energy budget formula are described in Section 4.4. While the measurement of net radiation over a water surface is relatively straightforward, estimating the other terms in the energy balance involves multiple wind speed and temperature measurements of the air and water surface to estimate the sensible heat flux and temperature measurements at various depths to estimate changes in total heat storage (including in the bed for a shallow water body). Consequently the data processing involved is consequently extensive and time consuming. Modern sensors and recording equipment, although simplifying the work, make it an expensive method to use. In reservoirs, installation of the instrumentation on a raft can provide relatively reliable estimates of E_o by this method. It is sometimes used in a specific reservoir for a limited period until satisfactory mass transfer coefficients have been determined. Then the empirical mass transfer calculations of the next section can provide estimates of the reservoir evapotranspiration while the expensive energy budget equipment is used at other locations.

10.2.2 Mass transfer methods

Mass transfer methods depend on determining the rates at which vapour can be carried away from a wet surface in the turbulent lower boundary layer of the atmosphere. This will depend on the humidity of the air above the water surface, the roughness of the surface and the wind speed, which both affect the turbulent eddies in the boundary layer. The mass transfer of water vapour away from the surface results in a humidity gradient in the atmosphere, while the transfer of sensible heat results in a temperature gradient that can be used to estimate the sensible heat flux in (10.1). It is not normally possible, however, to measure the full gradients of humidity and temperature (except in some research projects); it is more usual to make measurements at a single reference height. There are then different ways of calculating the vapour flux by this method. In Fig. 10.4, three sets of measurements are shown diagrammatically and each gives rise to a separate method of calculation of E_o. The minimum requirements are to measure wind speed at some reference height, with dry and wet bulb temperatures to give the humidity deficit in the boundary layer. It is customary to represent humidity of the air in terms of vapour pressure, e. Then the humidity gradient can be represented as the difference between e_s, the saturated vapour pressure at the water surface temperature, and e_d the saturated vapour pressure of the air at T_d the dew point or wet-bulb temperature at the reference height, which is equivalent to the vapour pressure at the local air temperature T_a (see Fig. 10.4).

Then the most straightforward method uses the bulk aerodynamic equation originating with John Dalton in the early nineteenth century:

$$\lambda E = f(u)(e_s - e_d) \tag{10.2}$$

where E is the evaporation rate, $f(u)$ is a function of wind speed, and $(e_s - e_d)$ is the vapour pressure deficit. Thus the evaporation is related to the wind speed and is proportional to the vapour pressure deficit, the difference between the saturated vapour pressure at the temperature of the water surface and the actual vapour pressure of the air above.

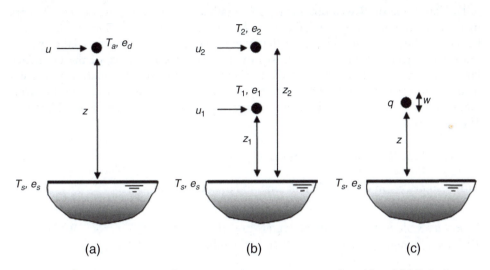

Fig. 10.4 Measurement strategies for mass transfer over a water surface. (a) and (b) T_s is the temperature of the surface, T_a is the air temperature at height z (T_1 and T_2 at two heights z_1 and z_2), e_s is the saturated vapour pressure of air at water surface, e_d is the vapour pressure of air (e_1 and e_2 at heights z_1 and z_2), and u is the horizontal wind velocity at height z (u_1 and u_2 at heights z_1 and z_2). (c) w is the vertical wind velocity at height z, and q is the specific humidity at height z.

The function $f(u)$ will depend on the stability of the lower boundary layer as well as surface roughness and wind speed but is usually assumed to take one of two forms: either $a(b+u)$ or Nu, where a, b and N are empirical mass transfer coefficients. Many investigations have produced values of a, b and N for various conditions and measurements. Special care is needed in noting the height of the anemometer and thermometers above the surface. In the UK, the detailed studies of Penman (1948) using the first form of $f(u)$, resulted in:

$$E_o = 0.467(0.5 + 0.1862u_2/1000)(e_s - e_d) \qquad (10.3)$$

for which the air measurements are made at 2 m above the surface. The vapour pressures are in hPa, wind speed in m s^{-1} and E_o is in mm day^{-1}.

In the USA and Australia, the second form of the wind term, Nu, is more commonly used, and again values of the mass transfer coefficient N are dependent on the height and units of the air measurements. Harbeck and Meyers (1970) give values of N from three famous studies. Thus in the general equation:

$$E_o = Nu_2(e_s - e_d) \qquad (10.4)$$

With units converted to SI units of wind speed at 2 m in m s^{-1}, vapour pressures in hPa and E_o in mm day^{-1}. Harbeck and Meyers (1970) found:

$N = 0.0120$ for Lake Hefner

$N = 0.0118$ for Lake Mead and

$N = 0.01054$ for Falcon Reservoir.

The last value of N was obtained from detailed measurements made over 2 years by equating the mass transfer estimate of E_o with E_o values calculated by the energy budget method. We should expect that there will also be an effect of lake size on evaporation since, when dry air moves over a lake, the humidity of the boundary layer will increase with distance across the lake. From a study of numerous reservoirs of different sizes up to 12 000 hectares, a factor of surface area can be incorporated into the equation to determine evaporation loss from a reservoir (Harbeck, 1962). Thus:

$$E_o = 0.291A^{-0.05}u_2(e_s - e_d)\,\text{mm day}^{-1} \tag{10.5}$$

with A in m^2, u_2 in m s^{-1} at height 2 m, and e_s and e_d in hPa.

Example: Calculate the annual water loss from a $5\,\text{km}^2$ reservoir, when u_2 is $10.3\,\text{km h}^{-1}$, and e_s and e_d are 18.9 and 14.6 mb, respectively.

$$A = 5\,\text{km}^2 = 5 \times 1000^2\,\text{m}^2$$

$$u_2 = 10.3\,\text{km h}^{-1} = 10.3 \times 1000/(60 \times 60) = 2.86\,\text{m s}^{-1}$$

$$e_s = 18.9\,\text{mb} = 18.9\,\text{hPa}$$

$$e_d = 14.6\,\text{mb} = 14.6\,\text{hPa}$$

$$E_o = 0.291(5 \times 1000^2)^{-0.05} \times 2.86(18.9 - 14.6)\,\text{mm day}^{-1}$$

$$= 1.66\,\text{mm day}^{-1}$$

$$= 606\,\text{mm year}^{-1}(\text{assuming constant loss rate})$$

Estimate of total loss from reservoir in a year $= E_o A = 0.606 \times 5 \times 1000^2 = 3.03\,\text{mm}^3$

The second *vapour flux method* of calculating evaporation uses air temperature measurements at two fairly close levels above the water surface to estimate the vapour pressure gradient and considers the turbulent transfer of water vapour through the small height difference (Fig. 10.4b). An equation for E_o due to Thornthwaite and Holtzman (1939) takes the form:

$$E_o = \frac{538.27\kappa^2\rho\,(u_2 - u_1)\,(e_1 - e_2)}{p\,(\ln z_2/z_1)^2}\,\text{mm day}^{-1} \tag{10.6}$$

where e_1 and e_2 are vapour pressures (hPa) at heights z_1 and z_2 (m), u_1 and u_2 are wind speeds (m s^{-1}) at heights z_1 and z_2 (m), p is atmospheric pressure (hPa), ρ is the density of air (kg m^{-3}), and κ is von Karman's constant $= 0.41$. This equation is valid only for neutral stability in the boundary layer when the mass transfer is dominated by eddies due to frictional turbulence. Under these conditions the Prandtl theory of turbulence in a boundary layer gives a logarithmic profile of wind speed with height above the surface. With greater heating of the ground, the vapour flow is also affected by convective turbulence. Under such conditions, and also in temperature inversions when turbulence is suppressed, the logarithmic relationship of wind speed with height used in the Thornthwaite–Holtzman equation does not hold. Much research has been undertaken by meteorologists on this approach to evaporation but it remains difficult to compensate for non-neutral conditions for general applications of mass transfer methods.

The direct eddy-flux or eddy-transfer method calculates energy for evaporation from measurements of vertical wind velocity and vapour content of the air at a single point above the evaporating surface (Fig. 10.4c; see also Section 4.5):

$$E_o = \lambda \overline{(\rho\omega)' q'} \qquad (10.7)$$

where ρ is the air density (kg m^{-3}), ω is the vertical wind speed (m s^{-1}), q is the specific humidity of the air (ratio of mass of water vapour to mass of air in kg kg^{-1}) and λ is the latent heat of vaporization of water as before.

In (10.7) the primes denote the departure of an instantaneous value from a mean value and the over-bar signifies a mean value over a specific time period. Thus $(\rho\omega)'$ represents an instantaneous fluctuation in the rate of upward air flow at the point of measurement and q' is the associated moisture content fluctuation. The evaporation rate is the mean of the products of such measurements during a given time period. In the last decade, methods for measuring short-term fluctuations in wind speed and humidity have become cheaper and much more reliable such that this is now the approach used in a worldwide network of evaporation measurement stations called FLUXNET,[1] including 15 sites in the UK and 150 in the USA (Section 4.5). Most of these sites are also equipped to measure fluctuations in temperature and carbon dioxide concentrations which also allow estimates of sensible heat and carbon flux away from vegetated surfaces.

10.2.3 Penman formula (combination method)

In a classical study of natural evaporation, Howard Latimer Penman (1948) developed a formula for calculating open water evaporation based on fundamental physical principles, with some empirical concepts incorporated, to enable standard meteorological observations to be used. This latter facility resulted in the Penman formula being enthusiastically acclaimed and applied the world over, especially by practising engineers seeking a relatively simple way of estimating evaporation from water bodies.

The physical principles combine the two previous approaches to evaporation calculation, the mass transfer method and the energy budget method. The basic equations are modified and rearranged to use meteorological constants and measurements of variables made regularly at climatological stations.

In a simplified form of the energy balance (10.1):

$$H = \lambda E_o + C \qquad (10.8)$$

where H is the available energy (= $R_N \pm V \pm G$), λE_o is latent energy flux from evaporation and C is sensible heat energy flux (all normally measured in W m^{-2}).

During daytime conditions, and when averaged over daily or longer periods, H is normally dominated by the net radiation term, but for estimates of actual evapotranspiration for shorter (e.g. hourly) time steps, the storage and release of energy in the ground, vegetation or water body (V and G) may become important.

The values of E_o and C can be defined by the aerodynamic equations:

$$\lambda E_o = f(u)(e_s - e_d) \tag{10.9a}$$

and

$$C = \gamma f_1(u)(T_s - T_a) \tag{10.9b}$$

where γ is called the *psychrometric constant* ($\sim 0.67\,\mathrm{hPa\,K^{-1}}$; see Appendix Table A4). It is generally assumed that $f(u) = f_1(u)$. The problem with using these equations directly is that the temperature of the surface is not generally known from standard measurements. We do, however, know the air temperature T_a, so that an alternative expression for latent heat flux can be written as:

$$\lambda E_a = f(u)(e_a - e_d) \tag{10.10}$$

where e_a is the saturated vapour pressure at air temperature T_a (which is tabulated in Appendix Table A4), and thus $(e_a - e_d)$ is the saturation deficit (e_d, the vapour pressure of the air, is the saturated vapour pressure at the dew point, T_d).

If Δ represents the slope of the known curve of saturated vapour pressure plotted against temperature (see Appendix Table A4), then:

$$\Delta = \frac{de}{dT} \approx \frac{e_s - e_d}{T_s - T_d} \approx \frac{e_a - e_d}{T_a - T_d} \tag{10.11}$$

Going back to (10.9), this can then be used to eliminate the unknown surface temperature T_s. Thus:

$$
\begin{aligned}
C &= \gamma f(u)\left[(T_s - T_d) - (T_a - T_d)\right] \\
&= \gamma f(u)\left[\frac{(e_s - e_d)}{\Delta} - \frac{(e_a - e_d)}{\Delta}\right] \\
&= \frac{\gamma \lambda E_o}{\Delta} - \frac{\gamma \lambda E_a}{\Delta}
\end{aligned}
\tag{10.12}
$$

Then substituting for C in the energy balance equation (10.7):

$$\lambda E_o = H - \frac{\gamma \lambda E_o}{\Delta} - \frac{\gamma \lambda E_a}{\Delta} \tag{10.13}$$

or

$$\Delta \lambda E_o + \gamma \lambda E_o = \Delta H + \gamma \lambda E_a$$

so that

$$\lambda E_o = \frac{\Delta}{\Delta + \gamma} H + \frac{\gamma}{(\Delta + \gamma)} \lambda E_a \ (\text{W m}^{-2}) \tag{10.14}$$

This final equation is the basic Penman formula for open water evaporation. It requires values of H and E_a as well as Δ for its application.

The heat available for evaporation is often approximated by the net radiation. If net radiation measurements are available, then H, the available heat may be obtained directly. If not, H must be calculated from incoming (R_I) and outgoing (R_o) radiation determined from sunshine records, temperature and humidity, using:

$$H = R_I(1 - \alpha) - R_o \tag{10.15}$$

where α is the *albedo*, the ratio of reflected short-wave radiation to incoming short-wave radiation for a given surface (about 0.05 for water). R_I is a function of R_a, the clear sky solar radiation (fixed by latitude and season) modulated by a function of the ratio, n/N, of measured to maximum possible sunshine duration. Using $r = 0.05$ gives:

$$R_I(1 - \alpha) = 0.95 R_a f_a(n/N) \tag{10.16}$$

Penman used $f_a(n/N) = 0.18 + 0.55 n/N$ in the original work, but later studies have shown that the function $f_a(n/N)$ depends on the clarity of the atmosphere and latitude.

The term R_o in (10.14) is given by:

$$R_o = \sigma T_a^4 \left(0.56 - 0.12 e_d^{0.5}\right)(0.1 + 0.9 n/N) \tag{10.17}$$

where σT_a^4 (W m^{-2}) is the theoretical black body radiation at T_a (in K) from the *Stefan–Boltzmann law*, which is then modified with functions of the humidity of the air (e_d in hPa) and the cloudiness (n/N). Thus H in (10.14) is obtained from values found via (10.16) and (10.17) inserted into (10.15). The value of the Stefan–Boltzmann constant, σ, has the value 5.67040×10^{-8} W m^{-2} K^{-4}.

Next, for a water surface, E_a in (10.14) is calculated using the empirical coefficients derived by experiment (10.3):

$$E_a = 0.467(0.5 + 0.1862 u_2/1000)(e_a - e_d) \ (\text{mm day}^{-1}) \tag{10.18}$$

Finally a value for Δ is found from the curve of saturated vapour pressure against temperature corresponding to the air temperature, T_a (see Appendix).

The equations given are those originally published by Penman with the coefficients changed to SI units. The four measurements required to calculate the open water evaporation are thus:

T_a mean air temperature in the required period, normally a week, 10 days or a month (K);

e_d mean vapour pressure for the same period (hPa);

u_2 mean wind speed at 2 m above the surface (m s^{-1});
n bright sunshine over the same period (h day^{-1}).

R_a and N can obtained from standard meteorological tables (see Appendix Tables A1 and A2).

With meteorological observations often recorded in various units, care is needed in converting measurements into the SI units (see Appendix Table A5). The evaporation from open water E_o is finally in mm/day.

The Penman combination formula was later extended for use in estimating potential evapotranspiration over short grass with unlimited water supply (Penman, 1950a,b). Based on empirical data from field experiments, Priestley and Taylor (1972) suggested that over large areas the potential evapotranspiration could be estimated as 1.26 times the first term in equation (10.14) (which led to other values of the multiplier being suggested based on other datasets, e.g. Bastiaanssen *et al.*, 1996). The more general Penman–Monteith equation for evapotranspiration is essentially based on a similar combination of energy balance and mass transfer equations, and can be used to estimate actual evapotranspiration directly (see next section).

Example: A calculation of E_o using the original formula and empirical equations is to be made from data assembled at the Lancaster University climatological station at Hazelrigg (latitude 54°2'N, elevation 94.1 m above mean sea level) for the month of August 2003. Temperature and vapour pressure values come from thermometer measurements in a Stevenson screen, wind speed from an anemometer at 2 m above the ground and sunshine hours from a Campbell–Stokes recorder (see Section 4.4.2). August 2003 was a particularly dry hot month (a total of 14 mm rainfall).

Mean temperature for the month is 16.9°C (290.1 K) giving $e_a = 17.04$ hPa and $\Delta = 1.01$ hPa K^{-1}

Mean wet bulb temperature is 14.78°C (287.98 K) giving $e_d = 15.27$ hPa

Mean wind speed is 5.3 m s^{-1}

Mean daily bright sunshine, n, is 6.15 h day^{-1}, and N the day length in August from Table A2 is 14.80 h so that $n/N = 0.415$.

The mean monthly potential R_a radiation input interpolated to 54°N from Table A1 is 379.1 W m^{-1}.

To calculate the energy available for evapotranspiration we need to apply equations (10.14)–(10.17). Thus:

$$R_I(1 - \alpha) = 0.95 \times 379.1 \times [0.18 + 0.55 \times 0.415] = 147.11 \, \text{W m}^{-2}$$

$$R_o = (5.67040 \times 10^{-8}) \times (290.1)^4 \times [0.56 - 0.12(17.04)^{0.5}]$$

$$\times [0.1 + 0.9 \times 0.415] = 12.80 \, \text{W m}^{-2}$$

$$H = R_I(1 - \alpha) - R_o = 134.31 \, \text{W m}^{-2}$$

Then, from (10.18)

$$E_a = 0.467(0.5 + 0.1862 \times 2.6/1000) \times (17.04 - 15.27) = 0.415\,\text{mm day}^{-1}$$

Finally, we can apply (10.14) to get an estimate of E_o

$$\lambda E_o = \left(\frac{1.01}{1.01 + 0.67}\right)131.14 + \left(\frac{0.67}{1.01 + 0.67}\right)2470 \times 0.416$$

$$= 87.27\,\text{W m}^{-2}$$

or

$$E_o = 87.27 \times 60 \times 60 \times 24/2\,470\,000 = 3.05 \times \text{mm} \times \text{day}^{-1}$$

Estimated Penman evaporation for August 2003 $= 3.05\,\text{mm day}^{-1} = 94.64\,\text{mm}$ month^{-1}

A more detailed calculation based on data for every day in the same month produces a Penman evapotranspiration estimate of $95.30\,\text{mm month}^{-1}$. In the case of the Hazelrigg site, this can be compared with a measurement of open water evapotranspiration from a standard evaporation pan, for which the average daily evaporation was $3.30\,\text{mm day}^{-1}$ or $103.20\,\text{mm month}^{-1}$. In this case, the Penman equation underestimates the pan measurement by 8 per cent in this unusually dry and hot summer month. This may be because the Penman equation underestimates the effects of a surrounding dry soil on the pan evaporation, or because the pan is subject to heat-storage effects in this dry period that increase the measured evaporation.

10.3 Calculation of evapotranspiration, E_t

The evaluation of actual water loss from a vegetated land surface by evaporation plus transpiration (E_t) adds further complexities to the processes involved in the evaporation from an open water surface (E_o). Instead of adding directly to the body of water, some of the rainfall (or snowfall) is intercepted by the vegetation and, from the various wetted surfaces, moisture is readily returned to the atmosphere by direct evaporation. However, much of the precipitation eventually reaches the ground, where it is absorbed by the soil or runs off over impervious surfaces. As the soil dries out, transpiration may be limited by the availability of water to the roots of the vegetation. The ways in which plants control transpiration and regulate the passage of water through the plant pores is thought to be reasonably well understood, but remains difficult to quantify or represent in mathematical terms. In drying soils, for example, it has been shown that, because of the way in which soil hydraulic conductivity decreases dramatically with soil water content, the roots of some plants may grow towards water faster than water will move towards the roots under a physical hydraulic gradient.

In fact, the issue of how water is supplied to the transpiring leaf surfaces is intriguing, especially in tall vegetation. Transpiration is a by-product of the need for the plant to exchange carbon dioxide and oxygen with the atmosphere through the openings on leaf surfaces, called *stomata*, as part of the process of photosynthesis. So to raise water from the roots to the leaf surfaces, the plant has to work against the effects of gravity. In Section 6.2, the concepts of capillary potential and capillary rise have been introduced. Water, under the effects of surface tension, will naturally rise in fine capillary tubes until the pressure drop across the curved meniscus is balanced by the height of rise. The finer the tube, the higher the rise. The water passageways in a plant (the *xylem*) are essentially capillary tubes, with diameters of the order of 50–100 μm. This is, however, much too coarse to support the rise to the leaves of a tall tree. The relevant diameters are the air–water interfaces in the cell walls of the leaf stomata where the effective pore diameters are much smaller, of the order of 5–10 nm. Theoretically this would be enough to raise water by several kilometres, but that is without accounting for the fact that the water has to move through the pores. Even though the movement is laminar and of low velocity, this generates a velocity gradient in the xylem and loss of energy due to viscosity, greatly limiting the possible rise. The plant reduces this effect by using the larger xylem to transport water but generating the capillary forces necessary to raise water in the stomata walls at the ends of the flow path. The problem for the plant is that, under these negative pressures, any gas bubble will tend to fill the pathway and cause the flow to break down (a form of embolism). The hydrogen–oxygen bonds in the water molecule are such that this is unlikely to occur from the breakdown of the water itself. It can, however, happen under very dry conditions by air being drawn in through the walls of the xylem. The pathway then becomes useless for water transport and the plant wilts. The woody structures found in plant stems are different strategies for avoiding the formation of bubbles. Moving water across a root membrane into the plant is also not only a result of a physical hydraulic gradient but also on solute concentration in the water. This is usually higher within the root than in the soil, which results in an *osmotic gradient* across the root membrane that causes water to move into the root by a process called *osmosis*.

The methods of estimating E_t are similar to those used for open water evaporation E_o, but have to take account of the additional controls and characteristics of the surface and the availability of water on the energy balance and mass transfers of heat and vapour. As noted above, eddy flux methods may be used to measure fluxes of heat, vapour and carbon directly but, where such measurements are not available, a modification of the Penman combination equation due to John Monteith is often used. The Penman–Monteith equation is based on the concept of the plant–soil–atmosphere system as a sequence of resistances in series as in Fig. 10.5. The land surface parameterisations of meteorological models commonly follow a similar approach, but increasingly with multiple layers in both the vegetation cover and the soil. In the Penman–Monteith approach, a complex vegetated surface is treated as if it acts as a *"big leaf"*, which allows the separation of the soil–plant resistance and the plant to atmosphere resistance.

The starting point is the simplified energy budget of (10.7), where for short (sub-daily) time periods it will be necessary to include the heating and cooling of the canopy and soil, as well as net radiation, in estimating H. Equation (10.7) can be rearranged in the form:

$$\lambda E_t = H - C = \frac{H}{1 + \beta} \tag{10.19}$$

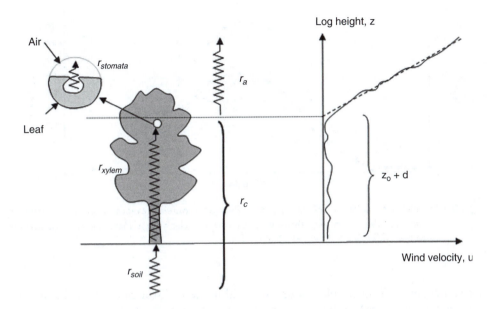

Fig. 10.5 Resistance analogy and logarithmic wind profile in a tall vegetation canopy for the big leaf model. Canopy resistance, r_c, is intended to represent the bulk effect of the soil, xylem and stomatal resistances at canopy scale.

where $\beta = C/\lambda E_t$ is known as the *Bowen ratio*. The value of the Bowen ratio indicates how the available energy is partitioned into that used to heat the atmosphere and that used in evapotranspiration. Experimental studies have shown how, for a vegetated surface, the Bowen ratio can stay fairly constant (at least for clear-sky conditions with no water limitations) (e.g. Crago and Brutsaert, 1996, see also Fig. 10.6).

As for open water, the fluxes C and λE_t are represented by gradient equations, here in the form:

$$C = \frac{1}{r_{a,H}} \rho_a c_p \left(T_o - T_a\right) \tag{10.20}$$

and

$$\lambda E_t = \frac{1}{r_{a,V}} \frac{\rho_a c_p}{\gamma} \left(e_o - e_a\right) \tag{10.21}$$

where T_o and e_o are temperature and vapour pressure at the canopy surface, T_a and e_a are air temperature and vapour pressure at the reference height, c_p is the specific heat of air ($\approx 1.005\,\mathrm{kJ\,kg^{-1}\,{}^\circ C^{-1}}$), and $r_{a,H}$ and $r_{a,V}$ are the *aerodynamic resistances* to the transport of heat and vapour from the surface to the atmosphere. The problem in applying these equations is that the surface temperature and vapour pressure of the 'big leaf' surface approximation cannot be easily measured. Penman and Schofield (1951) and later Monteith (1965, 1973) came up with the idea of using an additional

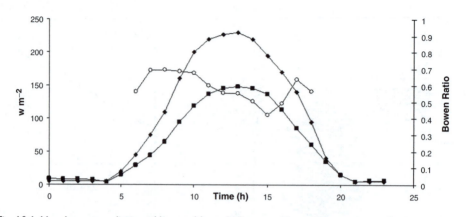

Fig. 10.6 Hourly net radiation (diamonds) and Penman–Monteith evapotranspiration estimate (squares) in $W\,m^{-2}$ for dry, short, grass canopy on a clear day in May. Values of the Bowen ratio ($\beta = C/\lambda E_t$) are shown as open circles.

resistance for the transfer of vapour from within the leaf stomata (where the air can be assumed saturated) to the surface to eliminate these variables from the equations. Thus the vapour flux may then also be written:

$$\lambda E_t = \frac{1}{r_c}\frac{\rho_a c_p}{\gamma}\left(e_s(T_o) - e_o\right)$$

(10.22)

where r_c is called the *canopy resistance*, and $e_s(T_o)$ is the saturated vapour pressure at the canopy temperature. An important feature of the canopy resistance is that it should be zero when the canopy is wet, i.e. when there is intercepted water on the canopy during and after rainfall. It will be shown below how this can make a significant difference to rates of water loss from the canopy, particularly for rough canopies with low aerodynamic resistance. Under dry conditions, we expect the canopy resistance to be related to the stomatal resistance within individual leaves, but under the big leaf assumption it is used here as an effective parameter for the canopy as a whole. It will increase as the plant starts to restrict the loss of water by closing the stomatal openings under dry conditions.

Combining (10.21) and (10.22) gives:

$$\lambda E_t = \frac{1}{r_{a,v} + r_c}\frac{\rho_a c_p}{\gamma}\left[e_s(T_o) - e_a\right]$$

(10.23)

The aerodynamic resistance is often approximated by assuming that both $r_{a,H}$ and $r_{a,v}$ are equal to the equivalent resistance for momentum flux in a neutrally stable atmospheric boundary layer above the canopy. For the resulting logarithmic wind-velocity profile, r_a is given by:

$$r_a = \frac{\ln\left[(z-d)/z_o\right]^2}{\kappa^2 u_z} \tag{10.24}$$

where u_z is the measured wind speed at the reference height z, d is called the zero plane displacement, z_o is called the roughness height and κ is the von Karman constant ($= 0.41$). The parameters d and z_o are introduced to allow for the fact that over a tall, rough, vegetated surface the effective zero wind velocity will be displaced above the ground. We should expect that the resistance for a rough tree canopy will be much lower than that of a smooth grass sward. For a range of vegetation canopies, d and z_o can be very approximately estimated from the canopy height h as $d = 0.67h$ and $z_o = 0.1h$.

Then under the assumption that $r_a = r_{a,V} = r_{a,H}$, we can use the same approximation as in (10.12) to (10.14) above to eliminate T_o from the equations by using the gradient of the saturation vapour pressure curve at the measured air temperature, Δ. The result, after some rearranging, is the Penman–Monteith equation in the form:

$$\lambda E_t = \frac{\Delta H + \rho_a c_p \left[e_s(T_a) - e_a\right]/r_a}{\Delta + \gamma\left(1 + r_c/r_a\right)} (\text{W m}^{-2}) \tag{10.25}$$

Thus, in applying the Penman–Monteith equation, we can proceed in a similar way to the calculation of the Penman E_o, outlined above. To do so, the user must specify:

- the energy available for evapotranspiration, H;
- wind speed at the reference height, u_z;
- zero plane displacement, d;
- roughness height, z_o;
- dry-bulb air temperature at the reference height, T_a;
- wet-bulb air temperature at the reference height (to give vapour pressure, e_a);
- the canopy resistance, r_c.

Fig. 10.6 shows these calculations applied on an hourly basis for a clear day in May for a dry grass canopy (with $r_a = 50\,\text{s m}^{-1}$; Table 10.1), together with the Bowen ratio ($\beta = C/\lambda E_t$). The estimates in energy units (W m^{-2}) can be converted to depth of water (m s^{-1}) by dividing by the latent heat of vaporisation, λ ($2.470 \times 10^{-6}\,\text{J kg}^{-1}$). With such short time steps, storage of energy in the ground and canopy starts to become more important. In this case, the data indicate that there is a source of sensible heat from the atmosphere and ground at night, leading to evapotranspiration rates that are higher than the net radiation, although over the day the energy balance is driven by the net radiation and dominated by the split between sensible and latent heat transfers to the atmosphere. Fig. 10.7 shows the application of the Penman–Monteith equation at a daily time step using the same August 2003 Hazelrigg data as above to estimate daily potential evapotranspiration rates, assuming a dry grass canopy without water limitation. In this case, the actual evapotranspiration total will be much less because

Table 10.1 Typical values of resistances in the Penman–Monteith equation for different surface and wetness conditions

	Aerodynamic resistance (s m^{-1})		Canopy resistance (s m^{-1})		
	r_a ($u = 2\,\mathrm{m\,s}^{-1}$)	r_a ($u = 5\,\mathrm{m\,s}^{-1}$)	r_c (dry canopy surface)	r_c (wet surface)	r_c (soil water limitation)
Open water	140	70	–	0	–
Grass	70	30	50	0	200
Mature wheat	25	10	70	0	250
Pine trees	10	5	100	0	300

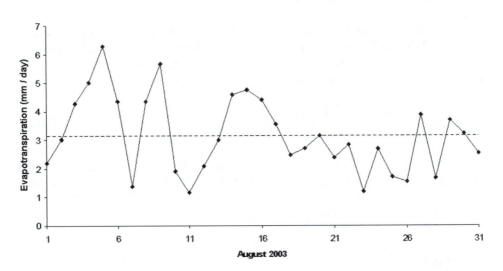

Fig. 10.7 Daily estimates of Penman–Monteith evapotranspiration (in mm day^{-1}) for August 2003 at the Hazelrigg climate station at Lancaster University assuming a short, dry, grass canopy without water limitation. Dotted line is mean daily potential evapotranspiration for the month.

the hot dry conditions will mean that the grass evapotranspiration will be water limited and canopy resistance will increase.

Beven (1979) looked at the sensitivity of evapotranspiration estimates to changes in the different parameters for typical UK summer conditions. His analysis showed quite clearly how the predicted rates increase with low canopy resistance (wet canopies) and low aerodynamic resistance (rough canopies) and that rates can be very high for a rough wet canopy (Fig. 10.8). This is one reason why frequently wetted forest canopies in the UK uplands tend to show greater cumulative volumes of evapotranspiration than grassland (e.g. Fig. 10.1). Typical values of these resistances for different canopies are shown in Table 10.1. Values of canopy resistance are known to exhibit a diurnal variation (e.g. Szeicz and Long, 1969; Stewart and Thom, 1973) and many canopies will also show a seasonal variation (Calder, 1977). We can examine the sensitivities

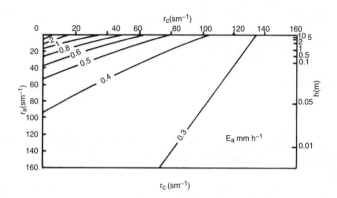

Fig. 10.8 Sensitivity of Penman–Monteith actual evapotranspiration estimates to changes in aerody-namic (r_a) and canopy (r_c) resistance parameters for mean August midday conditions in central England (from Beven, 1979, with kind permission of Elsevier).

directly by calculating the estimates of evapotranspiration for different values of the canopy and aerodynamic characteristics (see example).

Example: Comparison of estimated evapotranspiration for different canopies
For the midday conditions for the plot in Fig. 10.6, the net radiation value is 227 W m^{-2}, the air temperature is 16°C (289.2 K), the wet-bulb temperature is 12.1°C, the wind speed is 2.5 m s^{-1}. At this air temperature, the density of air, ρ_a, is 1.22 kg m^{-3} (Appendix Table A4), the specific heat of air, c_p, is 1.005 kJ kg^{-1} K^{-1}(Appendix Table A4), the saturation vapour pressure is 18.17 hPa (Appendix Table A4), and from the wet-bulb temperature the vapour pressure deficit is 4.16 hPa. The slope of the satu-rated vapour pressure v. temperature curve at this air temperature (Δ) is 1.61 hPa K^{-1} and the psychometric constant, γ, 0.67 hPa K^{-1}.

For a dry, short, grass canopy, assume an aerodynamic resistance at this wind speed of 57 s m^{-1} and a canopy resistance of 50 s m^{-1}.

Substituting into the Penman–Monteith equation (noting that 1 W = 1 J s^{-1})

$$\lambda E_t = \frac{\Delta H + \rho_a c_p \left[e_s(T_a) - e_a\right]/r_a}{\Delta + \gamma \left(1 + r_c/r_a\right)}$$

$$= \frac{1.61 \times 227 + 1.22 \times 1005 \times (4.16/57)}{1.61 + 0.67 \times (1 + 50/57)}$$

$$= 146.26\,\text{W m}^{-2}$$

To convert from energy to an equivalent depth of water, we need to divide by the latent heat of vaporisation λ (2.470×10^{-6} J kg^{-1}) and convert to the required depth and time units, so assuming a density of water of 1000 kg m^{-3} (so that 1 kg m^{-2} = 1 mm m^{-2} water)

$$E_t = 146.26 \times 60 \times 60/(2.470 \times 10^{-6}) = 3.52\,mm\,h^{-1}.$$

Repeating the calculations for a tree canopy with aerodynamic resistance under the same wind conditions of $5.7 \, \mathrm{s \, m^{-1}}$ and canopy resistance of $50 \, \mathrm{s \, m^{-1}}$ gives

$$\lambda E_t = \frac{1.61 \times 227 + 1.22 \times 1005 \times (4.16/5.7)}{1.61 + 0.67 \times (1 + 50/5.7)}$$

$$= 159.09 \, \mathrm{W \, m^{-2}}$$

$$E_t = 3.64 \, \mathrm{mm \, h^{-1}}.$$

The rougher tree canopy in this case enhances the transfers of both sensible and latent heat to the atmosphere such that the Bowen ratio stays about the same giving a similar latent heat flux.

For a grass canopy that is short of water after a long dry period, with the same aerodynamic resistance but with a canopy resistance of $200 \, \mathrm{s \, m^{-1}}$, the loss of latent heat from the canopy is

$$\lambda E_t = 47.97 \, \mathrm{W \, m^{-2}}$$

$$E_t = 1.11 \, \mathrm{mm \, h^{-1}}$$

If we assume that a tree canopy has been wetted by rain during the morning such that the canopy resistance is reduced to zero (but with no changes to temperature or humidity),

$$\lambda E_t = 673.03 \, \mathrm{W \, m^{-2}}$$

$$E_t = 15.36 \, \mathrm{mm \, h^{-1}}$$

Note the large differences in these values under the same midday meteorological conditions.

Thus, the Penman–Monteith estimate of evapotranspiration from a wet canopy under these conditions is actually far higher than the net radiation. This means that evapotranspiration of the intercepted water on the leaf surfaces is being driven by sensible heat provided by aerodynamic transfer from the atmosphere (a negative C in the energy balance of (10.8)). This indicates the potential for the efficient evaporation of intercepted water on a rough canopy (though in reality we would also expect this to increase the humidity above the canopy and change the wet-bulb temperature and calculated vapour pressure deficit).

Since the original appearance of the Penman–Monteith equation, there have been a number of modifications to allow for sparse vegetation (e.g. Shuttleworth and Wallace, 1985), where there may be some evaporation directly from the soil (the so-called *two-source* model); to allow for corrections to the aerodynamic resistance dependent on the stability of the atmosphere; and to take account of the effects of soil water on the effective canopy resistance. Each introduces additional parameters that must be estimated before the equation can be applied.

The most important issue in the application of the Penman–Monteith equation is to know how the canopy resistance changes with changing weather and soil water conditions. It is well known that most plants will act to reduce water losses under dry conditions by regulating the opening of the stomata on the leaf surfaces. To make this as efficient as possible, the stomata should stay open as much as possible, but this then means that water vapour is also lost and must be replenished by water from the roots. When the soil water content is a limiting control on water availability to the plant, then the stomata will tend to close and consequently the canopy resistance will rise. Other factors can also affect the opening of the stomata, including radiation, carbon dioxide concentrations in the air, leaf temperatures and biochemical signalling in the plant itself. The biophysical processes affecting stomata are complex but recent work suggests that it might be possible to approximate the plant response by invoking optimality principles (e.g. Caylor *et al.*, 2004).

10.4 Estimating wet canopy evaporation and interception losses

Table 10.1 shows how low values of canopy resistance are expected for the evapo-ration of intercepted water on a wet canopy. Interception can be an important part of the water balance, with losses of up to 30–40 per cent of the input precipitation where dense canopies are frequently wetted in a windy environment. Sparse canopies and shrub and heather vegetation have much lower interception losses (Calder, 1990). The adjustment of canopy resistance when a canopy is wetted after rainfall can make an important difference to total estimates of actual evapotranspiration. Section 3.8 explains how net precipitation (P_n) is measured so that storm-based (or longer) inter-ception losses can be estimated by subtracting the net from gross precipitation (P_g). The simplest model of interception loss (I_{loss}) relates these data to the gross precipi-tation, i.e. $I_{loss} = I_o + C(P_g - I_o)$, where C is a coefficient and I_o is some initial loss before the rainfall penetrates the canopy. More sophisticated treatments of intercep-tion loss have been developed, including representations that include partially wetted canopies and which allow for drainage of water from the canopy as throughfall and stemflow. The most commonly used interception model was first suggested by Rutter *et al.* (1975). This made use of a storage capacity for the canopy and drainage that was a simple exponential function of storage at any time. Other, more complex, stochastic representations of interception have also been developed (see Calder, 1990).

10.5 Estimating E_t over a landscape area or catchment

One of the problems of estimating E_t is that of estimating the total flux over an area, including the combined effects of evaporation from water surfaces and the soil, evaporation of intercepted water on the vegetation canopy and transpiration from leaf surfaces. This is similar to the problem of estimating precipitation over an area based on point rain-gauge measurements, but with the difference that the number of sites with the measurements necessary to estimate evapotranspiration will in general be much smaller. Thus, *de facto*, hydrologists will often assume that actual evapotranspiration is relatively conservative over an area, so that a measurement at one site will provide a good estimate of the flux over a larger area. This is probably not a bad assumption when rates of actual evapotranspiration are limited by the energy inputs; it can be a very bad assumption when rates of actual evapotranspiration are limited by water availability.

A very obvious example occurs in areas that are seasonally dry, e.g. in summer in regions with a Mediterranean climate. As the soils dry out, progressively more and more of the landscape will be subject to water limitations (something that might also depend on the flow of water downslope and depths of soil and rooting in the landscape). An extreme case is where there is a landscape of dry hillslopes where evapotranspiration is small and wet valley bottoms that are still transpiring freely (see example calculation).

Example: *Evapotranspiration over a catchment area with wet and dry surfaces*
Consider a situation where, at the end of a dry period, 80 per cent of a catchment is effectively water limited with actual evapotranspiration close to zero, while the remaining 20 per cent in the valley bottoms or irrigated agricultural fields are still transpiring freely at, say 5 mm day^{-1}. Thus the actual average flux rate from the catchment would be:

$$\overline{E_t} = 0.8 \times 0 + 0.2 \times 5$$
$$= 1 \text{ mm day}^{-1}$$

Note that in this case, the air moving over the landscape would be warmed, and its specific humidity reduced, by the specific heat flux from the dry hillslopes where we would expect the Bowen ratio of $C/\lambda E_t$ to be very high. Thus evapotranspiration over the valley bottoms would be expected to be enhanced as a result of advection of energy in the air moving from the upwind slopes.

However, if the average over the area was based only on the measurements in the dry area, then it would be estimated as zero; if it was based on measurements over the wet area, then it would be estimated as 5 mm day^{-1}. Both would be quite wrong.

Thus the pattern of actual evapotranspiration over a landscape will have an important effect on catchment average actual evapotranspiration rates and therefore closure of the water balance for a catchment. In this case, the estimate could be improved by having two *tiles* with different characteristics (an approach now often used in the land surface parameterisations of atmospheric circulation models; see Section 10.6) but the mix of different rates of E_t over different parts of the landscape will usually be more complex. As noted earlier, latent heat losses over one part of the landscape will affect the humidity of the air downwind. This is, in fact, the basis for a quite different approach to estimating E_t based on a concept of complementarity.

10.5.1 The complementary concept

In 1963 Bouchet suggested a complementary concept for estimating large-scale evapotranspiration rates. His idea, since developed by Fred Morton and others, was that advection of air over a surface would mean that the humidity of the air would come into a dynamic equilibrium with the latent heat fluxes from the surface. What is measured as local evaporation and transpiration locally will therefore depend on what is happening up-wind. Thus, if the surface was dry, the humidity of the air would be relatively low, and the apparent potential evapotranspiration would be very high. If, on the other hand, the air mass was already humid and moving with low velocity, such that the vapour pressure deficit was low and the aerodynamic resistance was high, the apparent potential evapotranspiration would be low, even if energy and water were available. Based on this idea, both theoretical and practical studies have developed a relationship between actual evapotranspiration over an area, E_t, to potential evapotranspiration, E_P, resulting in acceptable operational estimates of areal evapotranspiration. A full account of the development of the method is given in Morton (1983).

Potential evapotranspiration is generally defined as the local evapotranspiration under given atmospheric conditions when water supply is non-limiting. The use of evaporation pans as a way of estimating potential evapotranspiration (see Section 4.2.2) is essentially based on this idea. However, when considering a larger catchment or landscape scale, the complementary concept recognises that there is a feedback mechanism whereby changes in E_t alter the temperature and humidity of the over-passing air which in turn changes E_P.

Thus instead of using E_P as a causal agent for estimating E_t, Morton treats E_P as an *effect* of changes in E_t caused by changes in the availability of water for evaporation from a larger area. This is demonstrated, under conditions of constant energy supply, in Fig. 10.9. E_t increases from zero when there is no water available for evaporation from the surrounding area to a constant rate of E_w when there are no limitations on the availability of water. In contrast, E_P decreases from $2E_w$, when $E_t = 0$ and the air is hot and dry, down to a constant rate of E_w, when $E_t = 0$ and the air is cool

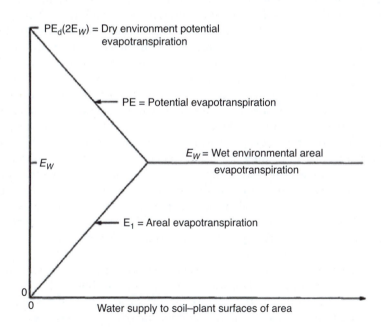

Fig. 10.9 Complementarity of areal actual evapotranspiration and potential evapotranspiration with change in water supply.

and humid. Thus Morton postulates that, where conditions give near-steady fluxes, E_P and E_t are complementary such that:

$$E_t = 2E_w - E_P \tag{10.26}$$

where the actual values of E_w and E_P must reflect changes in energy supply.

This equation provides the basis for the complementary relationship areal evapotranspiration model (CRAE). In conjunction with the CRAE model, a complementary relationship lake evaporation model (CRLE) provides estimates of lake evaporation from routine measurements of temperature, humidity and sunshine duration in the land environment (Morton, 1986). While evidence suggests that such a complementary relationship cannot strictly hold (LeDrew, 1979; Lhomme and Giulioni, 2006), Morton's results and modifications of the method (Crago and Crawley, 2005; Szilagyi and Jozsa, 2008), suggest that this can be a simple and useful way of estimating catchment-scale actual evapotranspiration, given estimates of potential evapotranspiration. Nash (1989) and Szilagyi and Jozsa (2008) point out that the complementary concept was not incompatible with the Penman approach to estimating potential evapotranspiration. It has been used with remote sensing information to estimate spatial patterns of evapotranspiration (e.g. Venturini *et al.*, 2008) and as the basis for estimating evapotranspiration in a number of hydrological models.

10.5.2 Energy balance closure using remote sensing

Another approach to obtaining areal estimates of actual evapotranspiration is to make use of the spatial information in remote sensing images. This was an approach that was pioneered in the UK by Holwill and Stewart (1992) and in the Netherlands by Bastiaanssen *et al.* (1998, 2005) whose SEBAL program is used on a routine operational basis there. In this approach, multispectral images can be used to identify different classes of vegetation, multidirectional sensing can be used to estimate the effective albedo of the surface under different conditions, and thermal infrared wavelength images can be used to estimate the temperature of the surface. Then, given some information about incoming radiation, air temperatures and wind speed in an area, and a number of parameters, the various flux terms of the energy balance can be estimated so that patterns of evapotranspiration can be derived. The effects of slope angle, aspect and horizon shading can also be taken into account in accounting for the net radiation. Cloudiness can also be determined from thermal and visual wavelength images. In some circumstances it is possible to use active and passive microwave sensing to estimate the near-surface soil moisture content, but microwaves do not penetrate more than a few centimetres into the soil and so this is not always a strong control on the effective water availability in the full profile, which may also be affected by downslope subsurface water flows.

Studies of this type have revealed some very interesting patterns of estimated actual evapotranspiration rates in the landscape (e.g. Fig. 10.10), but studies of the uncertainties associated with the energy balance closure have suggested that the absolute values of E_t predicted in this way may be rather uncertain (e.g. Franks *et al.*, 1999). Patterns of wind speed, in particular, which are important in controlling the local

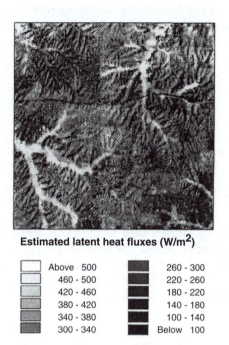

Estimated latent heat fluxes (W/m^2)

☐	Above 500	■	260 - 300
☐	460 - 500	■	220 - 260
☐	420 - 460	■	180 - 220
☐	380 - 420	■	140 - 180
☐	340 - 380	■	100 - 140
☐	300 - 340	■	Below 100

Fig. 10.10 Derived actual evapotranspiration estimates for the FIFE site in Kansas at 1500 on 15 August 1987 (after Franks and Beven, 1999, reproduced by permission of American Geophysical Union).

aerodynamic resistance can be difficult to predict over a complex landscape, especially in mountain areas.

10.6 Empirical formulae for E_t and E_P

Previous editions of *Hydrology in Practice* have summarised a number of different empirical formulae for estimating potential and actual evapotranspiration. The main reason for using such formulae is when the data required for using, say, the Penman–Monteith equation as set out above, are not available (Xu and Singh (2006), e.g. compare the performance of seven different simplified formulae). In data-sparse situations, e.g. it may be possible to obtain data on daily maximum and minimum temperatures, but there may not be a nearby meteorological measurement station (or FLUXNET site) that will give wind speed and humidity data (or direct estimates of E_t). However, the cost of obtaining the relevant variables has, with the availability of modern automatic weather stations, been greatly reduced. There are even recent experiments where large numbers of weather measurements are being made using cheap sensors linked by wireless network technology, such as the SensorScope program at EPFL, Lausanne.[2] Thus, it is suggested that, wherever possible, the Penman–Monteith equation now be used for point estimates of actual and potential evapotranspiration in preference to any of the empirical equations.

It has been shown how the Penman–Monteith equation can be used to estimate E_t directly. Over a long period, however, and in particular a long dry period, this does require information about how canopy resistance for a given type of surface will change with water availability. There have been many studies with different crops and natural vegetation canopies to suggest the nature of this relationship, but it is also possible to use the Penman–Monteith equation to estimate potential evapotranspiration and then use a functional relationship between the ratio of actual to potential evapotranspiration and soil moisture to estimate E_t. One thing should be noted in taking this approach. From the derivation of the Penman–Monteith equation, it is expected that the canopy resistance r_c will only be zero for a canopy that is wet. When the canopy is dry but transpiring at the potential rate for the prevailing conditions, r_c may be greater than zero even if the water supply is non-limiting (Table 10.1).

For this reason, the Food and Agricultural Organisation of the United Nations (FAO) use the term *reference crop evapotranspiration*, E_{To}, when water is not limiting rather than potential evapotranspiration (Allen *et al.*, 1998, 2005). The FAO now recommend the Penman–Monteith equation for estimating E_{To} but has taken an empirical approach to determining E_t from E_{To} and has tabulated crop coefficients relating E_t to E_{To} for a wide variety of crops under different cropping patterns during a production season. E_t for a given crop is then calculated from estimates of E_{To} as:

$$E_t = K_c \, E_{To} \qquad\qquad (10.27)$$

Some examples of crop coefficients are shown in Table 10.2.[3] They will vary with the stage of growth of the crop (see Allen *et al.* (1998), who outline procedures for constructing seasonal crop coefficient curves). Given some information about E_{To} in

Table 10.2 Single (time-averaged) mid-growing season crop coefficients, K_c, for non-stressed, well-managed crops in sub-humid climates (relative humidity >45%, wind speed >2 m s) for use with the FAO Penman–Monteith E_{To}

Crop	K_c
Small vegetables	1.05
Roots and tubers	1.10
Legumes (*Leguminosae*)	1.15
Perennial vegetables (with winter dormancy and initially bare or mulched soil)	1.00
Fibre crops	1.15
Oil crops	1.15
Cereals	1.15
Turf grass – cool season	0.95
Sugar cane	1.25
Banana	1.05
Coffee – bare ground	0.95
Rubber trees	1.00
Berries (bushes)	1.05
Apples cherries pears active ground cover	1.20
Citrus no ground cover (70% cover)	0.65
Citrus with active ground cover or weeds (70% canopy)	0.70
Conifer trees	1.00

a given region and these crop coefficients for typical cropping patterns, the water requirements for different crops can be estimated.

10.7 Soil moisture deficit

The calculation of potential evaporation (E_P) from readily available meteorological data is seen to be a much simpler operation than the computation or measurement of actual evapotranspiration (E_t) from a vegetated surface. However, water loss from a catchment area does not always proceed at the potential rate, since this is dependent on a continuous water supply. When the vegetation is unable to abstract water from the soil, then the actual evaporation becomes less than potential. Thus the relationship between E_t and E_P depends upon the soil moisture content. Early attempts to quantify actual evapotranspiration and crop water use by agricultural hydrologists made use of the concepts of *soil moisture deficit* and *field capacity* in defining this relationship.

When the soil is saturated, it can hold no more water. After rainfall ceases, saturated soil relinquishes water and becomes unsaturated until it can just hold a certain amount against the forces of gravity; it is then said to be at 'field capacity'. A consideration of the principles of soil physics suggests that this is not a fixed capacity, but practically after a day or two of drainage, hydraulic conductivities and rates of gravity drainage in most soils become rather slow so that further drying will be affected as much by evapotranspiration as by drainage. Under these conditions, the negative pressure potential may be approximately $-100\,\text{hPa}$ (see Section 6.3).

When the soil is close to field capacity, $E_t = E_P$, evapotranspiration occurs at the maximum possible rate determined by the meteorological conditions. If there is no rain to replenish the water supply, the soil moisture gradually becomes depleted by the demands of the vegetation to produce a soil moisture deficit (SMD), defined as the amount of water required to restore the soil to field capacity. As SMD increases, E_t becomes increasingly less than E_P. The values of SMD and E_t vary with soil type and vegetation, and the relative changes in E_t with increasing SMD have been the subject of considerable study by botanists and soil physicists. Penman (1950b) introduced the concept of a '*root constant*' (RC) that defines the amount of soil moisture (millimetre depth) that can be extracted from a soil without difficulty by a given vegetation. Thereafter, E_t becomes less than E_P as moisture is extracted with greater difficulty as shown in Fig. 10.11. As the SMD increases further, the vegetation wilts and E_t becomes very small or negligible. Before the onset of wilting, vegetation will recover if the soil moisture is replenished, but there is a maximum SMD for each plant type at a 'permanent wilting point' (typically $-15\,000\,\text{hPa}$) from which the vegetation cannot recover and dies. The total water available to the vegetation is called the *available water capacity* (AWC).

There have been a variety of SMD models of this type reported in the literature. One of the most interesting is a comparison of different formulations by Calder *et al.* (1983). Fig. 10.12 shows the sequence of SMD predicted for two sites in the UK in comparison with measured SMDs based on neutron probe soil moisture profile measurements. In both cases, the best fitted models are shown. It is clear that these simple concepts can reproduce the observations quite well, especially when it is remembered that, for the Thetford Forest site, the years 1975/1976 were among the driest on record. In fact, this

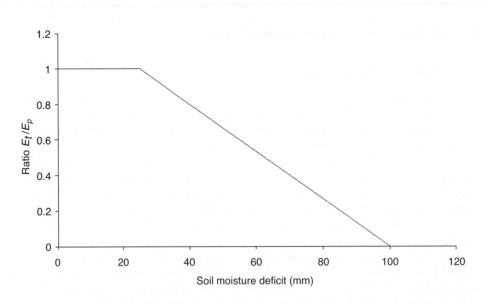

Fig. 10.11 Representative decline of E_t/E_p with soil moisture deficit (storage less than field capacity).

study also showed that the very simplest approaches could do well. One of the model structures that they tried used only a one parameter annual sine wave to estimate daily potential evapotranspiration such that:

$$E_P = \overline{E_P}\left[1 + \sin\left(\frac{360i}{365} - 90\right)\right] \tag{10.28}$$

where i is day number in the year. Application of this equation to estimate E_p therefore requires only the estimation of the mean annual daily evapotranspiration $\overline{E_p}$. Soil moisture deficits can be applied to estimate evapotranspiration rates over a catchment by allowing for the different rooting characteristics and available water capacities of the soils. It is, however, a one-dimensional approach. Every point is treated independently, with no explicit account taken of the effects of any drainage to replenish soil moisture deficits downslope or in valley bottoms.

10.7.1 UK Met Office rainfall and evaporation calculating system

A version of soil moisture deficit modelling is still used by the UK Met Office in their UK Met Office Rainfall and Evaporation Calculating System (MORECS) system. They produce predictions of potential and actual evapotranspiration rates and soil moisture deficits updated on a daily basis for different crop types mapped over the UK on a 40 by 40 km grid. The system uses measured rainfalls and weather variables as inputs to predict potential evapotranspiration using the Penman combination approach and uses a simple soil moisture deficit model to allow for the effects of soil moisture on actual evapotranspiration. Since the system was first introduced in 1978, it has been

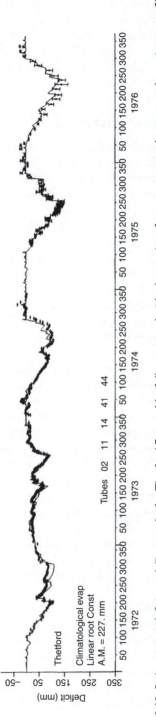

Fig. 10.12 Soil moisture deficit modelling results for Thetford Forest, Norfolk compared with observations from neutron probe access tube moisture profile measurements (from Calder *et al.*, 1983, with kind permission of Elsevier). A.M., Available Moisture (Available Water Capacity).

gradually modified to improve the estimates (Hough and Jones, 1997), including the inclusion of interception, and improvements to the representation of albedo. However, while it is still used as an operational tool to produce predictions for customers of the Met Office, it is in the process of being replaced by more sophisticated *land surface parameterisations* as a result of developments in representing the surface boundary conditions for atmospheric models.

10.8 Land surface parameterisations

In the last few decades, there have been dramatic developments in models of the atmospheric circulation, driven by the needs for weather forecasting and predicting climate change. Hydrology is important in these models; it controls the latent and sensible heat fluxes that are needed as the lower boundary condition for the atmosphere over land surfaces. Early representations of hydrology in these land surface parameterisations (LSPs) were limited by computational constraints (and still are, even with implementations of global and regional circulation models on current supercomputers). They were therefore of similar form to the very simple soil moisture deficit models outlined in the previous section. The advantages of such an approach were that few parameters were required when applied globally and the low computational requirements. The disadvantages were that such simple models did not properly reflect understanding of the physics. This became particular apparent in higher latitudes where it was necessary to implement multiple layers to allow for heat transfer, storage and freezing in the soil.

Thus the next generation of LSPs were still one dimensional but started to incorporate more information about the soil and vegetation, and more explicit calculation of soil water transfers in multiple soil layers, including root density profiles to allow for root extraction with depth. Evapotranspiration calculations are often based on Penman–Monteith formulations but multiple vegetation layers have also been incorporated in some LSPs with resistance-based calculations of fluxes between layers. A typical example of a LSP is the MOSES scheme used in the UK Met Office Unified Model of the atmosphere (Fig. 10.13). The second-generation MOSES scheme allows for multiple independent *tiles* of different vegetation/land-use types making up fractions of any grid square of the atmospheric model. The fluxes calculated for each tile are integrated to give the total boundary fluxes for that grid square. The same approach is used in the latest development of MOSES, the Joint UK Land Environment Simulator (JULES[4]), which allows nine different surface types: broadleaf trees, needleleaf trees, C3 (temperate) grass, C4 (tropical) grass, shrubs, urban, inland water, bare soil and ice.

As well as being used as an LSP for an atmospheric model, MOSES and JULES can also be used as stand-alone representations of the surface hydrology and will replace MORECS as a way of calculating the patterns of soil moisture status across the UK. As an example of the use of this type of land surface model, Smith *et al.* (1994) have combined MOSES with high resolution analyses of precipitation, cloud and near-surface atmospheric variables (available from the Met Office NIMROD product), and the Probability Distributed (PDM) runoff model (see Chapter 12) to provide hourly updates of soil moisture, snowmelt, runoff and evaporation. Some results of this model in time and space are shown in Figs 10.14 and 10.15.

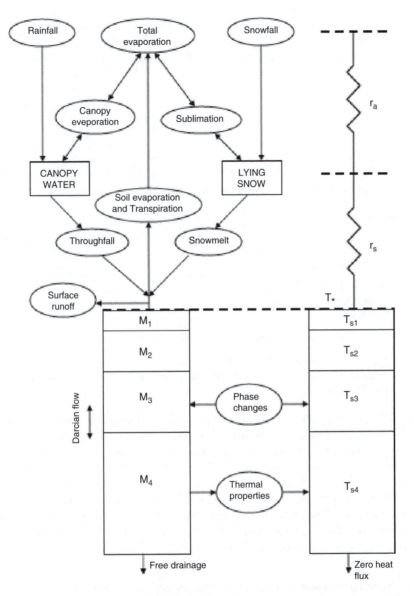

Fig. 10.13 A representation of the MOSES land surface parameterisation scheme (after Cox et al., 1999, with kind permission of Springer Science and Business Media).

10.9 Snow and the energy budget

In many parts of the world snow forms an important part of the hydrological cycle. Where it is the major input of water in the annual water balance, it can be critical in water resources assessment, while the snowmelt period can be the most critical periodfor flooding in some areas. In California, for example, up to 80 per cent of the annual discharge of some rivers is generated by snowmelt. Thus the build-up and

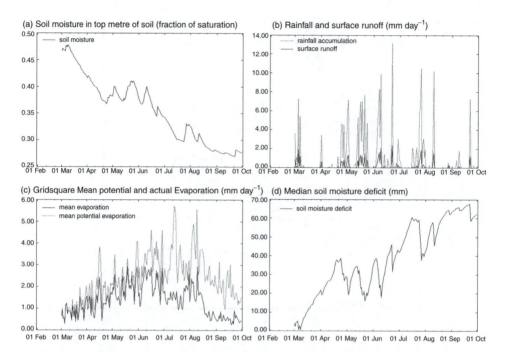

Fig. 10.14 Predictions with the MOSES-PDM model for a grid square in southern England, March to September in the dry summer of 2003 (after Smith *et al.*, 2004, © Crown copyright 1994, the Met Office).

melting of the snowpack is then an important hydrological problem. It depends on both the amounts of input precipitation but once a snowpack has developed, also on the energy budget of the surface. Rapid snowpack melting can be an important cause of flooding in some parts of the world. Warm air masses and heavy rainfalls on to a snowpack can cause such rapid melting. A good example was the Red River Flood in April and May of 1997 on the US–Canada border. This flood was the result of a heavy winter snowfalls followed by a period of extremely warm temperatures in the melt season. The Red River reached flood levels of 16.6 m at Grand Forks, North Dakota, and caused the evacuation of 50 000 people and \$3.5bn damages.

10.9.1 Estimating snowmelt by the energy budget method

The principles are much the same as for evapotranspiration. In this case, a simplified surface energy budget at a point can be expressed as

$$\lambda_M M = R_N \pm C \pm V \pm G + Ic_p \left(T_P - T_S\right) \tag{10.29}$$

where λ_M is the latent heat of melting (334 kJ kg^{-1}), M is the melt rate in water depth per unit area per unit time (mm s^{-1}), R_N is net radiation, C is sensible heat flux from the atmosphere, V is energy storage in the snowpack, G is heat exchange with the underlying soil, I is the precipitation input rate as a depth of water equivalent per unit

Fig. 10.15 Soil moisture deficit predictions with the MOSES-PDM model over western Europe for 20 September 2003. White: 0 mm; dark blue: 0–20 mm; green: 60–80 mm; yellow: 80–100 mm; orange: 100–140 mm (after Smith *et al.*, 2004, © Crown copyright 1994, the Met Office).

time, c_p is the specific heat of water, T_P is the temperature of the precipitation and T_S is the temperature of the snow. Under some circumstances, water can also be lost from a cold snowpack directly to the atmosphere as vapour, a process called *sublimation*. There may also be additional small transfers of water from a humid atmosphere to a cold snowpack as *condensation*.

The energy storage term, V, is important in snowmelt, since melt will not occur until the snowpack is *'ripe'*, i.e. at or close to 0°C (273.2 K). Thus in following the accumulation and melt of a snowpack over time it is necessary to keep track not only of the water equivalent of the pack but also of its temperature. While the pack is still at temperatures below freezing, the change in temperature over time will depend on the inputs of energy from net radiation, and exchanges with the atmosphere and the soil. Losses and gains of heat in the pack will also depend on the thermal characteristics of the snow, which can change as the pack ripens and the structure of the snow grains changes due to compaction and diurnal freeze/thaw processes. The evolution of the temperature of the pack will also depend on the temperature at which new precipitation is added to the pack. Snow can fall and add to the pack at temperatures well below zero degrees, but this is particularly important when warm rain falls on to a ripe snowpack. Under these conditions, often found, for example, during spring melt periods on the west coast mountains of the USA, the additional input of energy from the rain and turbulent transfers of sensible heat in (relatively) warm air moving over the snow surface can lead to periods of rapid melting and consequent snowmelt floods. Northern California is also particularly susceptible to such events (locally called a *'pineapple express'*) when warm tropical air masses move inland from the Pacific during winter. One such period occurred in the early January 1997 in California, when torrential rains on to a deep snowpack produced some of the highest discharges recorded in some Californian rivers and filled reservoirs close to overtopping. It is worth noting that a number of such floods have been recorded, with that in January 1862 producing the highest recorded river levels.

In the various terms of (10.29), net radiation and sensible heat flux can be either measured or estimated as before, given the surface temperature of the pack and the air, together with information about the albedo and aerodynamic roughness of the pack and wind speed (using equations 10.15, 10.16, 10.17, 10.20). The storage term and ground heat flux term can be estimated by knowing the surface temperature of the pack and the thermal properties of the snow and ground. The input of energy from precipitation can be estimated from knowing the temperature and intensity of the input water equivalent.

The difficulty of implementing a full energy budget for a snowpack in this way comes from the fact that where snow accumulation and melt is most important, which is in mountain areas, it can be very difficult to estimate many of the different quantities required. The net radiation varies with slope and aspect. The temperature of incoming precipitation (and whether it is rain or snow) can depend on elevation. Air temperature will also vary with elevation, but with lapse rates that vary with the prevailing weather system. The accumulation of snow in different parts of a catchment can depend on drifting due to changing wind patterns so that there can be a wide distribution of snow depths in a given range of elevation. The albedo of snow will normally decrease over time as the snowpack ages and ripens, unless there is a fall of fresh snow. All of these factors can make it difficult to predict spatial patterns of snow accumulation and quantify rates of melt. An example of this is given in Section 9.1.5 where the pattern of rain and snow in different parts of a catchment in Switzerland proved to be so important in the magnitude of a flood peak.

10.9.2 Estimating snowmelt by the degree-day method

There are now systems based on geographical information system (GIS) databases that allow such predictions based on the energy budget to be made (see e.g. Blöschl *et al.*, 1991) but the predictions will necessarily be uncertain when not all the information required about snow depths, albedo, aerodynamic roughness and other factors can be known accurately everywhere. Thus some simpler methods are used in routine practice: one older method that depends only on temperature data called the degree-day method; one modern method that depends on regular remote sensing imaging to follow the snow-covered area in a catchment during a melt period.

In the degree-day method, the transfer of heat to and from the snowpack is assumed to be a empirical function only of the difference between the average pack temperature and the air temperature. In this approach, melt is predicted only if air temperature,

Table 10.3 Typical values of degree-day factor, M_f (from Hock, 2003)

Type of site	M_f snow (mm day^{-1} °C^{-1})	M_f ice (mm day^{-1} °C^{-1})
Glacier sites	2.7–11.6	5.4–20
Non-glaciated sites	2.5–5.5	–

T_a, is above some base temperature, T_b, such that

$$M = M_f \left(T_a - T_b \right) \tag{10.30}$$

where M_f is an empirical *melt factor* (normally expressed in units of mm day^{-1} °C^{-1}). If the base temperature is taken as zero degrees, then the melt factor is also known as the *degree-day factor*.

Appropriate melt factor values have been widely studied, and estimates are available for seasonal changes for different regions (e.g. Table 10.3). Equation (10.30) suggests that no melt will take place when the air temperature is below the base temperature, commonly taken as zero (or slightly less to make some allowance for radiation and ground heat sources of energy). The air temperature is also used to decide on whether precipitation falls as rain or snow in following the build-up of a snow pack to know how much water equivalent is available to melt. There is still the problem in complex mountainous terrain of knowing what the temperatures might be at different points in a catchment, but if measurements are available at one elevation, then this is normally approximated by simply assuming a temperature *lapse rate*. The ambient lapse rate will vary with weather conditions, but where there is no information about changes in temperature with elevation, it is common to assume that the dry adiabatic lapse rate holds everywhere (approximately 10°C km^{-1} of elevation). In this way, rates of snow accumulation and melt can be extrapolated across the landscape or catchment area.

Example: A warm humid air mass moves over a ripe snowpack with average depth of 80 cm, density of 120 kg m^{-3}, and a temperature of 0°C, and produces a rainfall total of 25 mm at an average air temperature of 10°C. Assuming a melt factor of 3 mm day^{-1} °C^{-1} and base temperature of 0°C, estimate the melt rate from the air temperature alone, and the additional melt resulting from the rain falling on the pack, relative to the snow water equivalent of the snow pack.

The snow water equivalent, *SWE*, of the snow pack can be calculated from the depth and density (assuming a water density of 1000 kg m^{-3})

$$SWE = 800 \times 120/1000 \, \text{mm}$$

$$= 96 \, \text{mm}$$

From (10.30), melt due to the warm air will be:

$$M_{air} = 3 \times (10 - 0) = 30 \, \text{mm}$$

In calculating the additional heat available from rain falling on the pack, we will assume that the temperature of the rainfall equilibrates in the pack to zero degrees, so that the heat available for melting snow per unit area will be

$$H = \text{specific heat of water}(kJ\,kg^{-1}\,K^{-1}) \times \text{mass of water per unit area (kg)}$$

$$\times \text{temperature change (K)}$$

$$= 4.184 \times 25 \times (10 - 0)$$

$$= 1046\,kJ$$

The melt resulting from this energy can be calculated by dividing by the latent heat of melt ($334\,kJ\,kg^{-1}$) and converting it to millimetres ($1\,kg\,m^{-2}$ is equivalent, assuming a water density of $1000\,kg\,m^{-3}$, to 1 mm per unit area).

$$M_{rain} = 1046/334 = 3.13\,mm$$

Total melt is therefore estimated at 33.13 mm, some 35 per cent of the snow water equivalent of the pack, which must be added to the 25 mm of water falling as rainfall in calculating the local input.

Clearly, this is a gross simplification of the processes involved, and a variety of modifications to the basic degree-day method have been suggested. One is not only to keep track of the accumulation of water equivalent in the pack but also the *cold content* of the pack by using a similar equation to (10.30) to keep track of exchanges of heat with the atmosphere even when no melt is taking place. Knowing the current water equivalent of the pack, these exchanges can then be used to calculate the evolving temperature of the pack so that no melt is allowed until it is ripe (when the cold content reaches zero).

This still takes no account of the role of radiation in the melt process, however (except indirectly through its effect on air temperatures), and another modification has been suggested that adds a radiation term to (10.30) so that:

$$M = M_f\left(T_a - T_b\right) + M_R R_N \tag{10.31}$$

where M_R is a conversion factor for energy flux density to millimetres of snowmelt ($kg\,W^{-1}$) and R_N is the net radiation acting on the snowpack ($W\,m^{-2}$) (e.g. Ambach, 1988).

10.9.3 Estimating snowmelt using remote sensing information

The degree-day method has been an important tool in estimating snowmelt in the past, but a modern technique using remote sensing images is becoming more routinely used in areas where snow accumulation and melt is important to water resources. Remote sensing can most easily reveal the area of snow-covered ground and the surface temperature of the snow pack; it cannot so easily determine the total snow water equivalent in a pack. Passive microwave sensors have been used to estimate snow water equivalent (deeper packs have lower microwave brightness temperatures) but available satellite systems have too coarse a resolution to be useful other that on large areas of relatively flat terrain (Schmugge *et al.*, 2002). Estimates of snow-covered area can still be operationally useful, however, during the melt season. This is because the changes in area can be superimposed over a digital terrain map of an area and extrapolated patterns of air temperature to estimate how much water equivalent has melted between successive images. The snowmelt runoff model (SRM) combines the changes in area with the degree-day method to provide operational forecasts of snowmelt runoff (Seidel and Martinec, 2003).

The MODIS sensors on the AQUA and TERRA satellites of the Earth Observation Program provide both visible wavelength and thermal infra-red images and allow repeat imaging every 1–2 days at a resolution of 500 m (e.g. Fig. 10.16). Snow can be differentiated from cloud cover by using both visible and infra-red images,

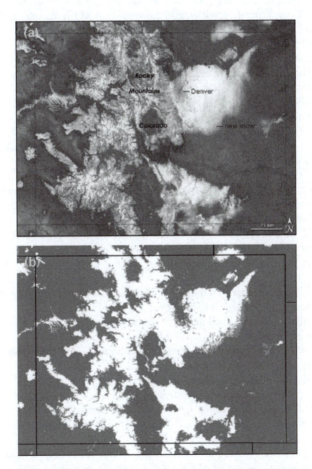

Fig. 10.16 (a) Moderate-resolution imaging spectroradiometer true-colour image acquired on 12 April 2005, showing the result of a new snowfall in Colorado; (b) the corresponding swath snow map, with pink areas indicating a classification as possibly either snow or cloud (after Hall and Riggs, 2007, with kind permission of John Wiley & Sons).

although cloud cover can prevent full coverage of an area being retrieved for all images. Operational snowmelt prediction systems based on satellite data are now being used in several countries including the USA, Canada and Spain.

Notes

1 See http://www.fluxnet.ornl.gov/fluxnet/index.cfm
2 http://sensorscope.epfl.ch/index.php/Main_Page
3 For a much wider range of crops and information on growing season, see http://www.fao.org/docrep/x0490e/x0490e0b.htm#crop%20coefficients
4 http://www.jchmr.org/jules/

References

Allen, R. G., Pereira, L. S., Raes, D. and Smith, M. (1998) *Crop Evapotranspiration – Guidelines for Computing Crop Water Requirements*. FAO Irrigation and drainage paper 56, FAO – Food and Agriculture Organization of the United Nations, Rome, 1998. Available at http://www.fao.org/docrep/X0490E/X0490E00.htm

Allen, R. G., Pereira, L. S., Smith, M., Raes, D. and Wright, J. L. (2005) FAO-56 dual crop coefficient method for estimating evaporation from soil and application extensions. *Journal of Irrigation and Drainage Engineering, ASCE* 131, 2–13.

Ambach, W. (1988) Interpretation of the positive degree-day factor by heat balance characteristics in West Greenland, *Nordic Hydrology* 19, 217–224.

Bastiaanssen, W. G. M., Menenti, M., Feddes, R. A. and Holtslag, A. A. M. (1998) A remote sensing surface energy balance algorithm for land (SEBAL): 1. Formulation. *Journal of Hydrology* 212–213, 198–212.

Bastiaanssen, W. G. M., Noordman, E. J. M., Pelgrum, H., Davids, G., Thoreson, B. P. and Allen, R. G. (2005) SEBAL model with remotely sensed data to improve water-resources management under actual field conditions. *Journal of Irrigation and Drainage Engineering ASCE* 131, 85–93.

Bastiaanssen, W. G. M., Pelgrum, H., Menenti, M. A. and Feddes, R. A. (1996) Estimation of surface resistance and Priestley–Taylor α-parameter at different scales. In: Stewart, J. *et al.* (eds) *Scaling Up in Hydrology Using Remote Sensing*. Wiley, Chichester, pp. 93–111.

Beven, K. J. (1979) A sensitivity analysis of Penman–Monteith actual evapotranspiration estimates. *Journal of Hydrology* 44, 169–190.

Bloeschl, G., Kirnbauer, R. and Gutknecht, D. (1991) A spatially distributed snowmelt model for application in Alpine terrain. *IAHS Publication* 205, 51–60.

Bouchet, R. J. (1963) Evapotranspiration réelle, evapotranspiration potentielle, et production agricole. *Annals of Agronomy* 14, 543–824.

Calder, I. R. (1977) A model of transpiration and interception loss from a spruce forest in Plynlimon, central Wales. *Journal of Hydrology* 33, 247–265.

Calder, I. R. (1990) *Evaporation in the Uplands*. Wiley, Chichester.

Calder, I. R., Harding, R. J. and Rosier, P. T. W. (1983) An objective assessment of soil moisture deficit models. *Journal of Hydrology* 60, 329–355.

Caylor, K. K., Scanlon, T. M. and Rodriguez-Iturbe, I. (2004) Feasible optimality of vegetation patterns in river basins. *Geophysical Research Letters* 31, article L13502.

Cox, P. M., Betts, R. A., Bunton, C. B., Essery, R. L. H., Rowntree, P. R. and Smith, J. (1999) The impact of new land surface physics on the GCM simulation of climate and climate sensitivity. *Climate Dynamics* 15, 183–203.

Crago, R. D. and Brutsaert, W. (1996) Daytime evaporation and the self-preservation of the evaporation fraction and the Bowen ratio. *Journal of Hydrology* 178, 241–255.

Crago, R. D. and Crowley, R. (2005) Complementary relationships for near-instantaneous evaporation. *Journal of Hydrology* 300, 199—211.

Franks, S. W. and Beven, K. J. (1999) Conditioning a multiple patch SVAT model using uncertain time-space estimates of latent heat fluxes as inferred from remotely-sensed data. *Water Resources Research* 35, 2751–2761.

Gash, J. H. C. (1979) Analytical model of rainfall interception by forests. *Quarterly Journal of the Royal Meteorological Society* 105, 43–55.

Hall, D. K. and Riggs, G. A. (2007) Accuracy assessment of the MODIS snow products. *Hydrological Processes* 21, 1534–1547.

Harbeck, G. E. (1962) *A Practical Field Technique for Measuring Reservoir Evaporation Utilizing Mass Transfer Theory.* US Geological Survey Professional Paper 272-E.

Harbeck, G. E. and Meyers, J. S. (1970) Present day evaporation measurement techniques. *Proceedings of the American Society of Civil Engineers* HY7, 1381–1389.

Hock, R. (2003) Temperature index melt modelling in mountain areas. *Journal of Hydrology* 282, 104–115.

Holwill, C. J. and Stewart, J. B. (1992) Spatial variability of evapotranspiration derived from aircraft and ground-based data. *Journal of Geophysical Research* 97(D17), 18673–18680.

Hough, M. N. and Jones, R. J. A. (1997) The United Kingdom Meteorological Office rainfall and evaporation calculation system: MORECS version 2.0 – an overview. *Hydrology and Earth System Sciences* 1, 227–239.

Hudson, J. A., Crane, S. B. and Blackie, J. R. (1997) The Plynlimon Water Balance 1969–1995: the impact of forest and moorland vegetation on evaporation and streamflow in upland catchments. *Hydrology and Earth System Sciences* 1, 409–427.

LeDrew, E. (1979) A diagnostic examination of a complementary relationship between actual and potential evapotranspiration. *Journal of Applied Meteorology* 18, 495–501.

Lhomme, J. P. and Giulioni, L. (2006) Comments on some articles about the complementary relationship. *Journal of Hydrology* 323, 1–3.

Monteith, J. (1965) Evaporation and environment. In: *The State and Movement of Water in Living Organisms.* Cambridge University Press, London.

Monteith, J. L. (1973) *Principles of Environmental Physics.* Edward Arnold, London.

Morton, F. I. (1983). Operational estimates of areal evapotranspiration, and their significance to the science and practice of hydrology. *Journal of Hydrology* 66, 1–76.

Morton, F. I. (1986) Practical estimates of lake evaporation. *Journal of Climate and Applied Meteorology* 25, 371–387.

Nash, J. E. (1989) Potential evaporation and 'the complementary relationship'. *Journal of Hydrology* 111, 1–7.

Penman, H. L. (1948) Natural evaporation from open water, bare soil and grass. *Proceedings of the Royal Society, London* 193, 120–145.

Penman, H. L. (1950a) Evaporation over the British Isles. *Quarterly Journal of the Royal Meterological Society* LXXVI, 330, 372–383.

Penman, H. L. (1950b) The water balance of the Stour catchment area. *Journal of the Institution of Water Enginers* 4, 457–469.

Penman, H. L. and Schofield, R. K. (1951) Some physical aspects of assimilation and transpiration. *Symposia of the Society for Experimental Biology* 5, 115–129.

Priestley, C. H. B. and Taylor, R. J. (1972) On the assessment of the surface heat flux and evaporation using large scale parameters. *Monthly Weather Review* 100, 81–92.

Robinson, M. (1998) 30 years of forest hydrology changes at Coalburn: water balance and extreme flows. *Hydrology and Earth System Sciences* 2, 233–238.

Rutter, A. J., Morton, D. J. and Robins, P. C. (1975) A predictive model of rainfall interception by forests. II. Generalisations of the model from observations in some coniferous and hardwood stands. *Journal of Applied Ecology* 12, 367–380.

Sceicz, G. and Long, I. F. (1969) Surface resistance of crop canopies. *Water Resources Res.* 5, 622–633.

Schmugge, T. J., Kustas, W. P., Ritchie, C., Jacskon, T. J. and Rango, A. (2002) Remote sensing in hydrology. *Advances in Water Resources* 25, 1367–1385.

Seidel, K. and Martinec, J. (2003) *Remote Sensing in Snow Hydrology: Runoff Modelling, Effect of Climate Change.* Springer, Berlin.

Shuttleworth, W. J. and Wallace, J. S. (1985) Evaporation from sparse crops – an energy combination theory. *Quarterly Journal of the Royal Meteorological Society* 111, 839–855.

Smith, R. N. B., Blyth, E. M., Finch, J. W., Goodchild, S., Hall, R. L. and Madrey, S. (2004) Spoil state and surface hydrology diagnosis based on MOSES in the Met Office Nimrod nowcasting system, Forecasting Research Technical Report No. 428, Met Office: Blackwell.

Stewart, J. B. and Thom, A. S. (1973) Energy budgets in pine forests. *Quarterly Journal of the Royal Meteorological Society* 99, 154–170.

Szilagyi, L. and Jozsa, J. (2008) New findings about the complementary relationship based evaporation estimation methods. *Journal of Hydrology* 354, 171–186.

Thornthwaite, C. W. and Holzman, B. (1939) The determination of evaporation from land and water surfaces. *Monthly Weather Review* 67, 4–10.

Venturini, V., Islam, S., Rodriguez, L. (2008) Estimation of evaporative fraction and evapotranspiration from MODIS products using a complementary based model. *Remote Sensing of Environment* 112, 132–141.

Xu, C.-Y. and Singh, V. P. (2001) Evaluation and generalization of temperature-based methods for calculating evaporation. *Hydrological Processes* 15, 305–319.

Chapter 11

River flow analysis

The ultimate aim of many computational techniques in engineering hydrology is the derivation of river discharges, and it might appear that, once these are obtained, the hydrologist's work is done. However, whether they are gained indirectly from considerations of other hydrological variables (to be described in following chapters) or directly from river discharge measurements, the discharge data are only samples in time of the behaviour of the river. The hydrologist then must assess the utility of the data and their representativeness over the period for which the information is required, usually the expected life of a water engineering project.

The basic river flow data that are normally readily available from the responsible agencies[1] are daily mean discharges and instantaneous peak discharges. These are usually derived from river stages (Section 7.3) and complementary velocity–area measurements (Section 7.4) to give a stage–discharge rating curve (Section 7.7), or from discharges obtained with a calibrated structure (Section 7.6). They will normally have been quality controlled, but for higher peak discharges may be associated with some uncertainty, because of the need to extrapolate the rating curve beyond the range for which measurements are available. For greater detail, the hydrologist may also be able to obtain hourly or 15 mm interval stage recordings and equivalent discharges.

Two examples of a year's record of daily mean discharges from rivers in the UK are shown in Fig. 11.1 from the gauging station at Sheepmount on the River Eden in the Environment Agency north-west region and the Panshanger Park gauging station on the River Mimram in the Thames region (note that the discharge scale is logarithmic). The daily mean discharges are for the year 2007 and are shown in relation to the extreme daily mean flows, the maximum and minimum discharges on corresponding days in the complete period of record. Although 2007 was a generally average year overall, the extremely high flows in June and July on the River Eden show quite clearly in relation to the long-term extremes (Fig. 11.1a). June and July in 2007 was the period of extensive flooding in Yorkshire and further south in the Severn and Thames catchments (see Section 9.1.4). The great irregularity shown by the sequence of daily mean flows during the wet months is indicative of a catchment responding rapidly to rainfall and the steep recessions in January/February and March/April emphasizes lack of storage in the drainage basin. A contrasting record is seen in Fig. 11.1b. The daily mean discharges for the Mimram, one of the headwaters of the River Lea north of London, for the same year of 2007, show a much more even pattern. The heavy rainfalls in the summer period have produced a few peaks on the hydrograph but the

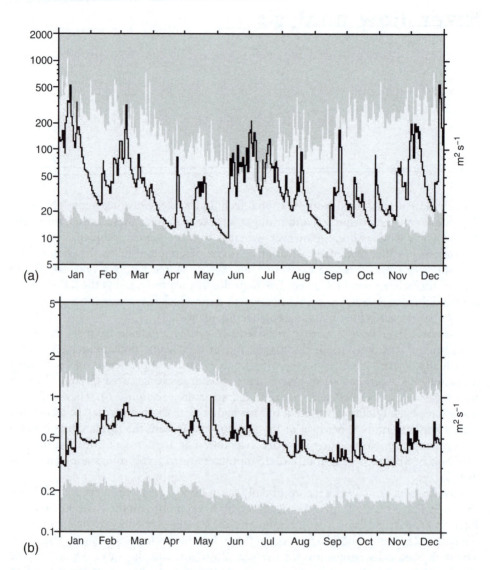

Fig. 11.1 Mean daily discharge record for 2007 and daily extremes (darker grey shading) recorded in the period of record: (a) Eden at Sheepmount (076007), 1967–2008; and (b) Mimram at Panshanger Park (039019), 1952–2008. (Reproduced from the UK National River Flow Archive, Centre for Ecology and Hydrology.[2] Copyright NERC CEH.)

flow is generally well within the long-term extreme values. Lack of variation in the daily flows is a notable feature of a catchment with a large storage. The Mimram drains a part of the Chiltern Hills, an area predominantly composed of chalk, which dampens the effects of minor irregularities in daily rainfalls, but a clay subcatchment responds quickly to rainfall and is largely responsible for the summer peaks.

The records that are shown here are gauged daily mean discharges calculated directly from the measurements made at a gauging station. However, some rivers may be

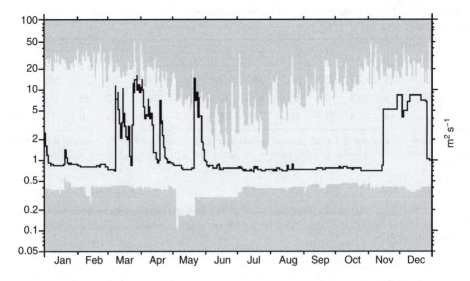

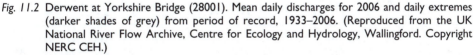

Fig. 11.2 Derwent at Yorkshire Bridge (28001). Mean daily discharges for 2006 and daily extremes (darker shades of grey) from period of record, 1933–2006. (Reproduced from the UK National River Flow Archive, Centre for Ecology and Hydrology, Wallingford. Copyright NERC CEH.)

affected by large abstractions upstream of the gauging station or the river flow may be controlled totally by regulated reservoir releases. The hydrologist must pay attention to potential modifications such as those illustrated in Fig. 11.2 by the hydrograph for 2006 for the River Derwent, a tributary of the River Trent, at Yorkshire Bridge, immediately downstream of series of reservoirs built for the water supply of Sheffield. The effect of the reservoirs, in both maintaining low flows and controlling high flows, is evident throughout the summer and in November/December when the reservoirs were refilling. For some types of application it is important to account for abstractions to give estimates of naturalized or gross flows (see also Section 17.6.1).

Abstractions from a river, although taken regularly each day for domestic or other water supplies, are usually quantified on a monthly basis. Hence a very useful statistic of river flow is the average of the daily mean flows over a month (the monthly mean flow). Table 11.1 shows the monthly mean gauged and naturalized discharges for the River Thames at Teddington for 1973. Teddington Weir is the tidal limit of the Thames, and thus the difference between the gauged and naturalized flows gives a measure of the demands of London and other towns abstracting water supplies from the river on the freshwater resources of the river.

Further monthly statistics for a selection of rivers are given in Table 11.2. The highest instantaneous peaks recorded are essential for assessing regulation requirements and the maximum and minimum daily mean discharges indicate the range of water availability. The value of these statistics is enhanced with each year of record: the longer the record at a gauging station, the more reliable can be the evaluation of water resources and the estimation of extreme events either of dangerous floods or of harmful droughts (see sections below).

Table 11.1 Monthly mean gauged and naturalized discharges (m^3 s^{-1}) River Thames at Teddington 1973 (drainage area 9870 km^2)

Flow	J	F	M	A	M	J	J	A	S	O	N	D
Gauged	50.1	42.9	25.4	22.0	33.7	21.7	22.6	11.7	11.6	9.8	10.0	20.3
Naturalized	69.5	60.3	43.5	42.1	50.7	40.4	41.3	29.6	27.5	28.2	27.3	41.4
Difference	19.4	17.4	18.1	20.1	27.0	18.7	18.7	16.9	15.9	18.4	17.3	21.1

Figures rounded off to the first decimal place. (Reproduced from Department of the Environment (1978) *Surface Water: United Kingdom 1971–73*, by permission of the Controller, Her Majesty's Stationery Office. © Crown copyright.)

Table 11.2 Monthly statistics of gauged flows (m^3 s^{-1}; data from National River Flow Archive, Centre for Ecology and Hydrology). Monthly highest instantaneous peaks and extreme daily mean discharges in 1990

River	Area (km^2)	Flow (m^3 s^{-1})	J	F	M	A	M	J	J	A	S	O	N	D
Tay at	4587	Peaks	776	1746	1102	252	137	496	297	84	194	597	213	535
Ballathie		Max.	663	1647	1033	223	119	234	228	81	112	423	201	443
		Min.	98	243	212	107	47	45	47	43	51	51	99	79
Derwent at	1586	Peaks	57.1	59.2	23.5	8.2	9.2	24.4	9.6	4.2	4.7	17.8	28.2	84.5
Buttercrambe		Max.	52.5	57.6	22.4	8.0	8.4	19.8	8.8	3.9	4.3	15.5	27.2	83.2
		Min.	6.8	13.9	7.8	5.3	4.3	3.9	3.2	2.9	2.8	3.4	5.7	6.5
Little Ouse at	699	Peaks	5.6	15.2	9.1	4.0	3.2	2.2	2.1	1.4	2.0	−2.7	3.2	
Abbey Heath		Max.	4.0	13.9	8.4	3.6	3.0	1.8	1.3	1.2	1.3	1.5	2.0	2.2
		Min.	2.3	3.8	3.0	2.6	1.5	1.1	0.9	1.0	1.0	1.1	1.0	1.2
Usk at Chain	912	Peaks	404	627	80	17	16	19	18	7	41	99	74	286
Bridge		Max.	230	472	68	14	12	14	15	6	20	62	58	146
		Min.	35	54	11	8	5	5	4	3	3	5	10	12
Eden at	2286	Peaks	485	705	160	42	103	46	117	21	64	310	192	457
Sheepmount		Max.	333	528	126	38	69	29	77	16	38	207	101	318
		Min.	34	84	21	18	13	13	11	10	10	17	20	18

Example: Converting units of discharge (see also Appendix, Table A5).

Taking the maximum daily flow for the River Eden at Sheepmount for January 1990 of 333 m^3 s^{-1} over a catchment area of 2286 km^2 (Table 11.2), this can also be expressed as:

$$333 \times 24 \times 60 \times 60 = 28\,771\,200\,\text{m}^3\,\text{d}^{-1} = 28\,771.2\,\text{ML}\,\text{day}^{-1}$$

or

$$333 \times 24 \times 60 \times 60 \times 1000/(2286 \times 1000 \times 1000) = 12.59\,\text{mm}\,\text{day}^{-1}$$

11.1 Peak discharges

High river discharges are caused by various combinations of extreme conditions. Heavy rainfalls over short durations, deep snow cover melted by warming rain and moderate rainfalls on frozen ground or saturated soil, can all contribute to a rapid and large runoff. Flood conditions are of great concern, and notable events are always studied in detail (see the discussion of the Boscastle 2004, Carlisle 2005, UK Summer 2007 and Aigle 1991 events in Section 9.1). With the expansion of the river gauging network in the UK, many more floods are now being measured, but when the river has overtopped the gauging control and where a river is ungauged, only estimates of the peak discharges can be made by hydraulic calculations (see Section 7.7.2).

Peak discharges from major runoff events in the UK were first assembled in relation to reservoir practice (Institution of Civil Engineers, 1933). Later this was updated by the Institution of Civil Engineers (1960), the Flood Studies Report (Natural Environment Research Council; NERC, 1975) and the *Flood Estimation Handbook* (Institute of Hydrology, 1999). A database of historical flood events for over 1000 sites in England and Wales has been collated by the UK Environment Agency,[3] while the University of Dundee hosts the British Hydrological Society database of historical floods in the UK[4] (see Section 11.9). In the USA, the USGS National Water Information Service maintains a database of flood peaks for over 27 000 sites.[5]

From a combination of flood discharge measurements and post-event discharge calculations taken from the available records in the UK, a graph of maximum recorded peak discharge against catchment area can been drawn (Fig. 11.3a). Twenty peak discharges for areas ranging from $4 \, km^2$ to nearly $1800 \, km^2$ are plotted, and an envelope curve is drawn. Most of the points represented by catchments less than $25 \, km^2$ pertain to the tributary flows contributing to the disastrous Lynmouth flood in August 1952, where the peak flow was estimated at $650 \, m^3 \, s^{-1}$ from $101 \, km^2$ (Dobbie and Wolf, 1953). The highest measured flow in the UK is the $2402 \, m^3 \, s^{-1}$ on the River Findhorn at Forres in Scotland in 1969 from a catchment area of $782 \, km^2$. The highest measured in England and Wales is the $1516 \, m^3 \, s^{-1}$ on the River Eden at Sheepmount in the 2005 Carlisle flood from a catchment area of $2286 \, km^2$. There are other historic records of flood levels on many rivers in the country,[3] and some of these could possibly provide peak discharges above the curve in Fig. 11.3a. However, changes in the river profiles and cross-sections over the ensuing years makes the conversion of level data into discharges unreliable. A worldwide relationship between maximum floods and catchment area is shown in Fig. 11.3b from data provided by O'Connell and Costa (2004).

The US data come from drainage areas that range from $2 \, km^2$ to over $8000 \, km^2$, with corresponding peak discharges from $144 \, m^3 \, s^{-1}$ to $21\,000 \, m^{-3} \, s^{-1}$ (Costa and Jarrett, 2008). These extreme events in the USA come from a wide range of different climatic regions, some of which experience tropical rainfall intensities. From the enveloping curves of the two plots, a $5 \, km^2$ catchment somewhere in the USA may well produce a peak of $600 \, m^3 \, s^{-1}$, but in the UK a peak of only $150 \, m^3 \, s^{-1}$ may be expected from the same drainage area. (The Findhorn record in Scotland would not be worth plotting on the USA graph!) Thus, when making a general appraisal of peak discharges, such a relationship of peak discharge to catchment area should be made according to climatic region. For the Flood Studies Report in the UK, regional subdivisions were

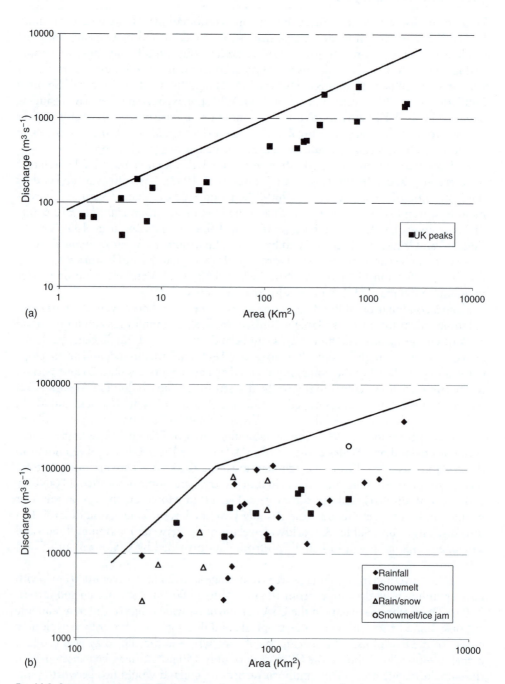

Fig. 11.3 Some extreme peak discharges and upper envelope curves for drainage basins of different area for: (a) the UK (data added to those of Boorman *et al.*, 1990); and (b) world (data from O'Connell and Costa, 2004). Note the different scales.

Table 11.3 Peak discharge per unit area from some documented floods in the UK

River (site)	Area (km²)	Peak discharge $m^3 s^{-1} km^{-2}$	Year
Chulmleigh (Devon)	1.7	40.0	1982
Caldwell Beck (Dumfries)	5.8	32.6	1979
Claughton Beck (Lancs)	2.2	30.2	1967
Red-a-ven (Dartmoor)	4.0	27.6	1917
Hoaroak Water (Lynmouth)	8.1	18.4	1952
Tyne (East Lothian)	4.1	10.0	1948
Alphin Brook (Exeter)	7.2	8.3	1960
Valency (Boscastle)	23.1	6.09	2004
Wye (Pant Mawr)	27.2	6.40	1973
Divie and Dorback (Moray)	365.0	5.31	1970
Muick (Invermuick)	110.0	4.28	1981
Findhorn (Forres)	782.0	3.07	1969
Oykel (Easter Turnaig)	330.7	2.56	1978
Tywi (Dolau Hirion)	231.8	2.30	1979
Rawthey (Brigg Flatts)	200.0	2.24	1982
Dart (Austins Bridge)	247.6	2.22	1979
Tyne (Haydon Bridge)	751.1	1.24	2005
Eden (Sheepmount)	2286	0.66	2005

made, but not only for hydrological reasons. At the time, the regions corresponded to the administrative areas of the Regional Water Authorities. In the Water Act of 1995 parts of the responsibilities of the Water Authorities were subsumed into the national Environment Agency. This is reflected in the *Flood Estimation Handbook* that abandoned the regional analyses in favour of grouping catchments together according to their hydrological characteristics.

Some hydrologists prefer to convert the absolute peak discharges into relative values per unit area for plotting against catchment area, thus giving an inverse relationship and a declining curve with increased catchment area. Sample values from the UK are shown in Table 11.3. These have been taken from Boorman *et al.* (1990). They demonstrate the contrast between the exceptional discharges from small areas ($< 10 \, km^2$) evaluated after the events and the peak flows at formal river gauging stations.

Numerous empirical formulae can be found in the hydrological literature for the relationship between peak discharges and catchment areas, with the coefficients specifically determined for particular countries or climatic regions. In attempting to use such formulae to obtain peaks for ungauged catchments, the hydrologist must guard against applying them to inappropriate conditions and areas.

11.2 River regimes

In Table 11.1, the monthly mean discharges for the River Thames in 1973 exhibit a distinctive seasonal pattern, with the highest values occurring in the winter months. The expected pattern of river flow during a year is known as the *river regime*. Flow records for 20–30 years are required to provide a representative pattern, since there may be considerable variations in the seasonal discharges from year to year. The averages of the monthly mean discharges over the years of record calculated for each month, January to December, give the general or expected pattern: the regime of the river.

The river regime is the direct consequence of the climatic factors influencing the catchment runoff, and can be estimated from knowledge of the climate of a region. The eminent French geographer, Pardé, first identified and classified distinctive river regimes and this can be a great help to engineers faced with unfamiliar conditions and sparse data. The classification is based on an understanding of the role of the main climatic features, temperature and rainfall, in causing river runoff. An illustration of *simple river regimes* resulting from a single dominant factor is given in Fig. 11.4. For each river, the monthly mean discharges from January to December are represented as proportions of the mean of the 12 monthly values. In this way, the graphs are comparable and independent of the absolute values of the monthly mean discharges and catchment areas.

11.2.1 Temperature-dependent regimes

Rivers with a dominant single source of supply, initially in the solid state (snow or ice), produce a simple maximum and minimum in the pattern of monthly mean discharges according to the seasonal temperatures.

- *Glacial.* When the catchment area is over 25–30 per cent covered by ice, the river flow is dominated by the melting conditions. Such rivers are found in the high mountain areas of the temperate regions. There is little variation in the pattern from year to year, but in the main melting season, July and August, there are great diurnal variations in the melt water flows.
- *Mountain snowmelt.* The seasonal peak from snowmelt is lower and earlier than in a glacial stream, but the pattern is also regular each year providing there has been adequate winter snowfall. The low winter flows are caused by freezing conditions.
- *Plains snowmelt.* The regular winter snow cover of the interior regions of the large continents in temperate and sub-Polar latitudes melts quickly to give a short 3-month season of high river flows. The timing of the peak month depends on latitude, with the more southerly rivers (e.g. the River Don in Fig.11.4a) having rapid flood peaks in April, whereas further north there is a slower melt and the lower peak occurring in June is more prolonged (the River Ob, Fig. 11.4a).

11.2.2 Rainfall-dependent regimes

In the equatorial and tropical regions of the world with no high mountains, the seasonal rainfall variations are the direct cause of the river regimes. Temperature effects in these areas are mostly related to evaporation losses, but with these being dependent on rainfall, the overall effect of evaporation is of secondary importance in influencing the river flow pattern.

- *Equatorial.* Drainage basins wholly within the equatorial belt experience two rainfall seasons with the annual migration of the intertropical convergence zone, and these are reflected directly in the river regime (e.g. the River Lobe in Cameroun in Fig. 11.4a).

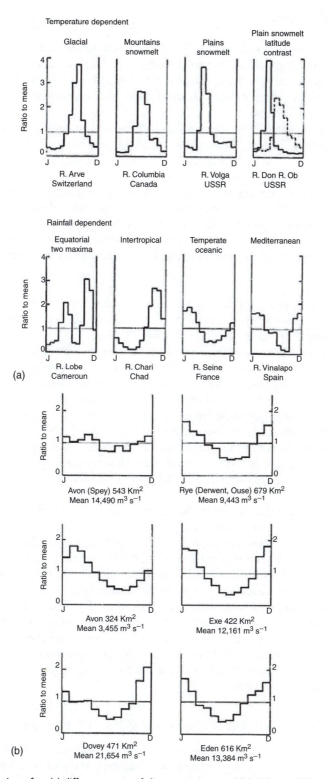

Fig. 11.4 River regimes for: (a) different types of climatic regime; and (b) different UK rivers expressed as a ratio of mean monthly flow to the annual mean flow.

- *Seasonal tropical.* Within the tropics, there are usually marked wet and dry seasons that vary in length according to latitude, and the flow ratios vary according to rainfall quantities. The River Chari in Chad is a clear example. Some catchments have double peaks, as in the equatorial region, but only separated by 1 or 2 months rather than 6 months.
- *Temperate oceanic.* The River Seine in France demonstrates the characteristic regime of these regions. Rainfall occurs all the year round, but the summer evaporation provides the relatively small variation in the seasonal flow pattern, which is however annually irregular.
- *Mediterranean.* The regime resembles that of the temperate oceanic regions, but is more extreme. The dry summers result in very low flows; along the desert margins, rivers dry up completely. Most of the flow results from the winter rains, but occasional very heavy summer storms may produce flood flows. The river regime is also very variable from one year to another, in parallel with the irregular incidence of rainfall.

The flow regime of a river draining a catchment area within a single climatic region may be readily estimated by considering the features demonstrated by the simple regimes. It must be remembered that away from the equatorial region, the patterns will be reversed in the southern hemisphere. However, the conditions in the high mountains in any of the regions will result in modifications to the expected pattern. Rivers with two or more sources of supply have a *mixed regime*. For example, a spring maximum may be identified as snowmelt, whereas early winter rains give a second peak.

More *complex regimes* result from the overlapping of different causes. These are usually characteristic of large rivers, especially those flowing through several climatic zones. The major rivers of the world, the Congo, Nile, Mississippi and the Amazon come into this category. With mixed or complex regimes, the range between the extreme months is usually small, and the annual variability decreases with increase in catchment size.

Examples of river regimes within the UK are given in Fig. 11.4b. The general pattern resembles that for temperate oceanic regions with a relatively small range between the wettest and driest months. The Scottish Avon, a tributary of the Spey, shows the least seasonal variation, the result of persistent rainfall throughout the year. Further south, the effect of summer evaporation losses becomes more apparent, and the Rye in east Yorkshire reflects this tendency. The slight differences in the other four diagrams stem from varying location and catchment characteristics. There are sustained later spring flows above normal from the chalk of the south coast Avon, and the exposure of the Dovey catchment in central Wales causes high flows from the mountains in early winter. These are average conditions calculated from the years of record available for each station, but there are great variations from year to year that result from the irregular incidence of rainfall.

11.3 Mixing models for determining runoff sources

As well as the amount of runoff, the sources of water that make up the stream hydrograph can also sometimes be of interest. In Chapter 1, the idea of separating different sources of water using mixing models was introduced in the discussion of concepts of runoff generation. It was noted there that this has been very important in the development of scientific hydrology since a large number of studies have shown that in many different catchments the hydrograph from a rainfall event is not made up of rainfall falling in that event but of *old* or *pre-event* water stored in the catchment prior to that rainfall falling. This should not be expected to be the case everywhere (the hydrograph from an intense rainfall falling on an arid catchment might be expected to be dominated by event water), but such mixing model analyses conflicted with a generally held perception at the time that storm hydrographs were caused by surface runoff made up of rainfall. It is therefore worth considering the nature of mixing model calculations in more detail.

Inevitably some simplification is necessary since in reality there is certainly a whole spectrum of different sources of storage in a catchment that might contribute to the hydrograph. It is also the case that even the simplest separation will not be possible if the potential sources cannot be differentiated by having different chemical characteristics in some way. A problem that then arises is that such chemical characteristics will not be constant over time, but will change as a catchment wets and dries. Thus any such separations will be, at best, approximate.

11.3.1 Two-component mixing

Consider first a simple two-component mixing model with storm rainfall as one source and all other sources from pre-event storage as the second. A number of chemical characteristics have been used in such separations including silica (normally nearly zero in rainfall, but higher in soil and groundwaters), and the stable isotopes of hydrogen and oxygen. The former has the disadvantage that there is evidence that rainwater might increase in silica concentration quite rapidly once in contact with the soil. The latter have the advantage that hydrogen and oxygen are part of the water molecule and will therefore necessarily follow the flow pathways of water, but the disadvantage that the rainfall inputs do not always have a significantly different concentration from the pre-event water sources (and that different pre-event sources may have rather different concentrations). The analysis then requires that concentration measurements are available for both rainfall and for the pre-event water. The latter is usually achieved by sampling the river water prior to an event. This may represent the mix of concentrations from different sources that is then assumed to remain constant throughout the event. That this is not always a good assumption is demonstrated in the next section where three component mixing is considered.

For this two-component mixing, we can then write down mass balance equations for both the water and the chemical constituents as follows:

for water

$$Q = Q_o + Q_n \tag{11.1}$$

and for the chemical constituent

$$CQ = C_o Q_o + C_n Q_n \tag{11.2}$$

where Q is flow, C is concentration, the subscript o indicates old or pre-event water, and the subscript n represents new or event water. These are two simultaneous equations in two unknowns (Q_o and Q_n) and can therefore be solved easily when it can be assumed that the concentrations for the two sources are constant in time and distinctive. Substituting (11.1) into (11.2) and rearranging gives the result:

$$Q_o = Q(C - C_n)/(C_o - C_n) \tag{11.3}$$

The need for a difference in the old and new water concentrations is then obvious. If the two waters cannot be differentiated, the denominator in (11.3) is zero and the calculation is indeterminate. Given a distinctive difference in the source concentrations, however, (11.3) can then be applied at every time step during a hydrograph to effect the separation between old and new water. An example using isotope concentrations is shown in Fig. 11.5. In effect, the constant concentration assumption is equivalent to assuming that the sources of pre-event water are well mixed before the event and are displaced from storage by the storm inputs.

Example: Oxygen isotope (^{18}O) concentration data are to be used to carry out a two-component hydrograph separation for a storm on 26 July 1995 in the 160-ha catchment of Shelter Creek in the Catskill Mountains of New York State (adapted from Brown *et al.*, 1999). The mean concentration of ^{18}O in the rainfall recorded for the event was -4.83 $\delta^{18}O$ (oxygen isotope concentrations are expressed in δ units, as differences from an international reference standard). The concentration in the stream prior to the event, taken as indicative of the old water component was -9.85 $\delta^{18}O$. Concentrations in the stream water were measured on grab samples taken during the storm.

At the peak discharge of $0.59\,\mathrm{mm\,h^{-1}}$, the ^{18}O concentration in the stream was $-7.06\delta^{18}O$. Applying equation (11.3) gives an old water contribution to the total discharge of:

$$Q_o = 0.59 \times (-7.06 + 4.83)/(-9.85 + 5.83) = 0.33\,\mathrm{mm\,h^{-1}}$$

This is 55 per cent of the total flow at the hydrograph peak, leaving only 45 per cent estimated as being supplied by event water. The 'old' water is being displaced from storage by the infiltration of event water into the soils. This calculation can be repeated at each time step for which an isotope concentration is available. The full results for this storm are shown in Fig. 11.5.

Brown *et al.* (1999) also carried out a three-component mixing (see next section) which suggested that there was an additional component of pre-event soil water contributing to the hydrograph.

A particularly interesting application of two-component mixing has been in the investigation of incremental lateral subsurface inputs into a stream channel. For each reach of the channel, the two components are the flow from upstream and the subsurface lateral input (which again have to have different concentrations). This can also be considered a form of successive dilution gauging (see Section 7.5). An early application by Huff *et al.* (1982) using an artificial tracer added to the Walker Branch stream

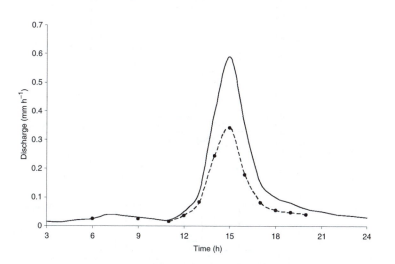

Fig. 11.5 Results of a two-component hydrograph separation using ^{18}O as an environmental tracer for the event of 26 July 1995 in the Shelter Creek catchment, Catskills, New York State. Solid line is total discharge (mm h^{-1}), dotted line is pre-event or old water component (mm h^{-1}). (Data taken from Brown *et al.*, 1999, with kind permission of Elsevier.)

in Tennessee revealed that there were strong spatial differences in subsurface inputs to the stream related to upward dipping bedrock structures. These results were later confirmed at different discharges by Genereux *et al.* (1993; Fig. 11.6). They also investigated the changing patterns of lateral inflows under different discharge conditions. A modern variation on this is to use the spatial pattern of temperatures in a stream to indicate mixing between surface and subsurface sources. This is a much more complicated problem, however, and requires modelling a number of different energy balance components in each reach of a river (Westhoff *et al.*, 2007).

11.3.2 End-member mixing analysis

More complex mixing models are also possible and are referred to under the general name of *end-member mixing analysis* (EMMA). Modifications to the two-component model have been made to allow for temporal variability in the input and end-member concentrations and to include more components. If we wish to separate out more components, then more chemical constituents will be needed. Three components will result in three unknown variables and, assuming that mass balance holds, will require two chemical constituents that have different concentrations for different water sources. For three-component mixing this can be visualised in the form of a mixing diagram such as that of Fig. 11.7. This shows how, for this small catchment in Switzerland, measured concentrations of calcium and silica in rainwater are clearly different from those in soil water and those in deeper groundwater. This is a result of the particular conditions in this catchment, in which the soils are relatively acid, favouring high-silica concentrations, but waters that penetrate to the Tertiary period

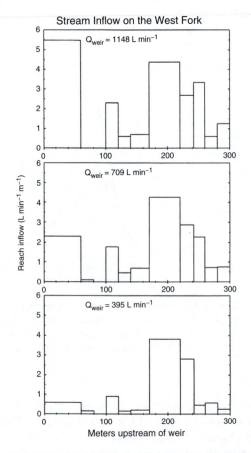

Fig. 11.6 Lateral inflows to the West Fork of Walker Branch, Oak Ridge (TN) determined by a two-component mixing model using salt as a tracer. (Reproduced from Genereux *et al.*, 1993, with kind permission of Elsevier.)

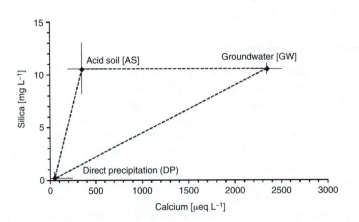

Fig. 11.7 Mixing diagram for separation of sources of stream discharge in the Haute Menthue catchment, Switzerland, using silica and calcium concentration observations (after Iorgulescu *et al.*, 2005, with kind permission of John Wiley & Sons).

Molasse bedrock are alkaline and high in calcium. Thus, in this case, the mixing equations are as follows

for water

$$Q = Q_s + Q_g + Q_n \qquad (11.4)$$

for the silica

$$C_{Si}Q = C_{Si,s}Q_s + C_{Si,g}Q_g + C_{Si,n}Q_n \qquad (11.5)$$

for the calcium

$$C_{Ca}Q = C_{Ca,s}Q_s + C_{Ca,g}Q_g + C_{Ca,n}Q_n \qquad (11.6)$$

The solution to these equations, again under the assumptions that the concentrations stay constant during an event, is shown in Fig. 11.8. Such an assumption is less tenable if we wish to consider a whole sequence of events, since in a second event the input of rainwater from the first event might expect to have an effect on the concentrations in different storages in a catchment. Iorgulescu *et al.* (2005) have attempted to model this, under different hypotheses about mixing of different sources over time for this same catchment. The results suggest that the proportion of soil water sources in the hydrograph increases significantly over time (Fig. 11.9). Fig. 11.7 also shows that the

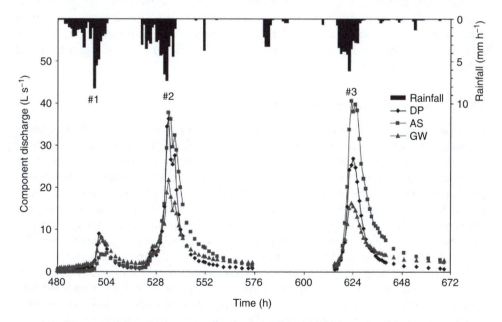

Fig. 11.8 Analysis of contributions to total Stream discharge of different sources in the Haute Menthue catchment using a three-component end-member mixing analysis (after Iorgulescu *et al.*, 2005, with kind permission of John Wiley & Sons). DP, direct precipitation; AS, soil waters; GW, groundwater.

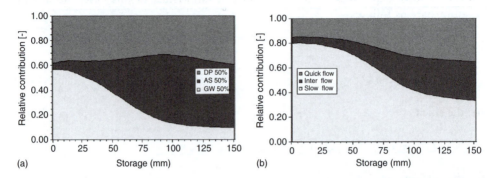

Fig. 11.9 Changing contributions of different sources of stream discharge in the Haute Menthue catchment with changing catchment storage (after Iorgulescu *et al.*, 2005, with kind permission of John Wiley & Sons). DP, direct precipitation; AS, soil waters; GW, groundwater.

concentrations measured for each source also show some variability. Such variability should be expected to result in some uncertainty in the calculated proportions for each component. A number of different methods are available for assessing such uncertainty (see e.g. Joerin *et al.*, 2002).

In some catchments, concentrations may be available for many different chemical constituents in waters from different sources. One way of trying to distinguish end members in this case is to use *principal components analysis* to determine linear combinations of the different concentration observations that provide separation between the end members. Burns *et al.* (2001), for example, used the first two principal components in an analysis of seven different chemical characteristics to allow a three-component separation of runoff sources in the Panola catchment, Georgia (in this case, hillslope runoff water, riparian groundwater and runoff from a bare granite outcrop in the catchment, Fig. 11.10).

11.4 Flow–duration curves

For many problems in water engineering, the hydrologist is asked for the frequency of occurrence of specific river flows or for the length of time for which particular river flows are expected to be exceeded. Thus frequency analysis forms one of the important skills required of a hydrologist. Estimates of the frequencies of floods of a particular magnitude are important in assessing flood risk. Estimates of the frequencies of low flows are important in assessing reservoir yields and the potential for low head hydropower schemes in rivers. In attempting to provide any answer to the questions of frequency, good reliable hydrological records are essential, and these must if possible extend beyond the expected life of the engineering scheme being considered.

From the basic assemblage of river flow data comprising the daily mean discharges and the instantaneous peaks, analysis of the daily mean flows will be considered first. Taking the n years of flow records from a river gauging station, there are $365n +$ leap year days of daily mean discharges. The frequencies of occurrence in selected discharge classes (groups) are compiled, starting with the highest values. The cumulative frequencies converted into percentages of the total number of days are then the basis for the *flow–duration curve*, which gives the percentage of time during which any selected

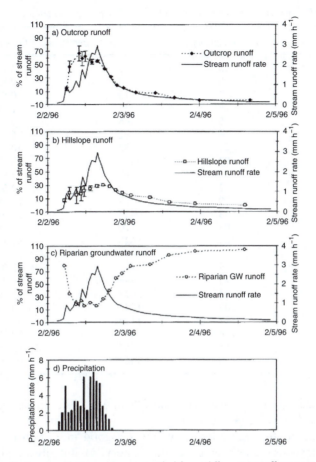

Fig. 11.10 Stream discharge and proportions provided from different runoff sources (hillslope runoff, runoff from a bare granite outcrop and riparian groundwater) in the Panola catchment, Georgia (after Burns *et al.*, 2001, with kind permission of John Wiley & Sons).

discharge may be equalled or exceeded. An example is demonstrated in Table 11.4, in which the daily mean discharges for 4 years for the River Thames at Teddington Weir are analysed. The flow–duration curve plotted on natural scales is seen in Fig. 11.11a. The area under the curve is a measure of the total volume of water that has flowed past the gauging station in the total time considered. For the reliable assessment of water supply, the flow–duration curves for the wettest and driest years of the record should be derived and plotted.

The representation of the flow–duration curve is improved by plotting the cumulative discharge frequencies on log-probability paper (Fig. 11.11b). (The abscissa scale is based on the normal probability distribution; if the logarithms of the daily mean discharges were normally distributed, they would plot as a straight line on the log-probability paper.) From the plot (Fig. 11.11b), it can be readily seen e.g. that for 2 per cent of the 4-year period, flows exceeded $290 \, m^3 \, s^{-1}$. At the other extreme, flows of less than $12 \, m^3 \, s^{-1}$ occurred for the same proportion of the time. Alternatively it can be stated that for 96 per cent of that 4-year period, the flow in the River

Table 11.4 Flow frequencies over a 4-year period: River Thames at Teddington

Daily mean discharge ($m^3 s^{-1}$)	Frequency (days)	Cumulative frequency	Cumulative percentage
Over 475	3	3	0.21
420–475	5	8	0.55
365–420	5	13	0.89
315–365	8	21	1.44
260–315	25	46	3.15
210–260	36	82	5.61
155–210	71	153	10.47
120–155	82	235	16.08
105–120	52	287	19.64
95–105	42	329	22.52
85–95	50	379	25.94
75–85	58	437	29.91
65–75	83	520	35.59
50–65	105	625	42.78
47–50	72	697	47.71
42–47	75	772	52.84
37–42	73	845	57.84
32–37	84	929	63.59
26–32	103	1032	70.64
21–26	152	1184	81.04
16–21	128	1312	89.80
11–16	141	1453	99.45
Below 11	8	1461	100.00

Thames at Teddington is between 12 and $290\,m^3\,s^{-1}$. The 50 per cent time point provides the median value ($45\,m^3\,s^{-1}$).

The shape of the flow–duration curve gives a good indication of a catchment's characteristic response to its average rainfall history. An initially steeply sloped curve results from a very variable discharge, usually from small catchments with little storage where the stream flow directly reflects the rainfall pattern. Flow–duration curves that have a very flat slope indicate little variation in flow regime, the resultant of the damping effects of large storages. Groundwater storages are provided naturally by extensive chalk or limestone aquifers, and large surface lakes or reservoirs may act as runoff regulators either naturally or controlled by man. Examples of some different flow–duration curves are given in Fig. 11.12. Hydrographs for these same sites can be seen in Fig. 11.1. The comparisons are simplified by plotting the logarithms of the daily mean discharges as percentages of the overall daily mean discharge. The Cumbria Eden drains the Eastern Lake district and northern Pennines, whereas the catchment of the Mimram, a tributary of the Lee is nearly all chalk.

The comparison of flow–duration curves from different catchments can also be used to extend knowledge of the flow characteristics of a drainage area that has a very limited short record. At least 1 or 2 years of records are required overlapping with those of a long-term well-established gauging station on a nearby river, whose flow–duration curve for the whole length of its record may be taken to represent long-term flow conditions. For satisfactory results, the two catchments should be in the same hydrological region and should experience similar meteorological conditions.

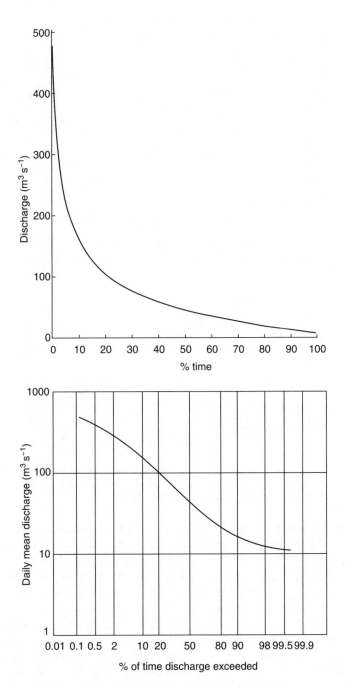

Fig. 11.11 Flow–duration curves for the Thames at Teddington: (a) discharge v. per cent of time exceeded; (b) log discharge v. per cent of time exceeded on normal probability scale.

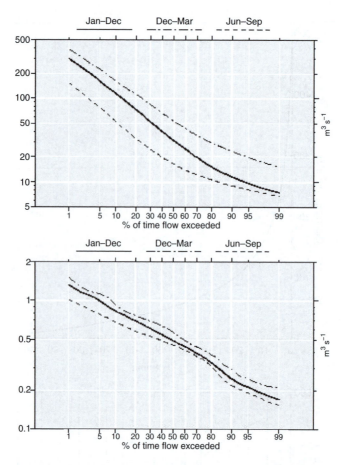

Fig. 11.12 Seasonal flow–duration curves for: (a) Eden at Sheepmount (076007); and (b) the Mimram at Panshanger Park (038003). (Reproduced from the National River Flow Archive, Centre for Ecology and Hydrology. Copyright NERC CEH.)

The method for supplementing the short-term record is to construct its long-term flow–duration curve by relating the overlapping short-period flow–duration curves of both catchments, as shown in Fig. 11.13. The available data are plotted and flow–duration curves (in full lines) are drawn in Fig. 11.13a and b. S_a and S_b are flow–duration curves for the short overlapping records; L_a is the flow–duration curve for the long-term neighbouring record. Selected percentage discharge values from the two short-period flow–duration curves are plotted in Fig. 11.13c, and a straight-line relationship drawn. Time percentage discharge values for L_a are converted to corresponding percentage values for L_b via the S_a and S_b relation. The derived long-term duration curve for the short-period station, L_b, is shown by a broken line in Fig. 11.13. By these means, the variation in the flow characteristics embodied in the long-term flow–duration curve has been translated to the short-period station.

Flow–duration curves from monthly mean and annual mean discharges can also be derived, but their usefulness is much less than those constructed from daily mean

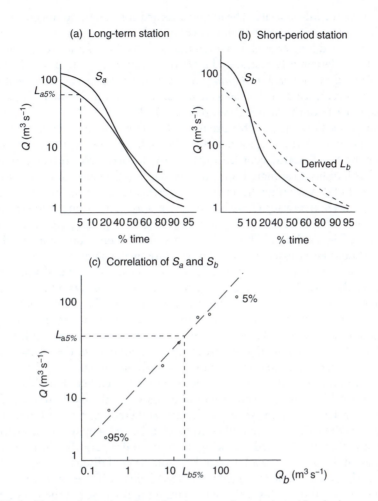

Fig. 11.13 Method for supplementing a flow–duration curve for a station with only short-term record, given a flow–duration curve for a similar station with a long-term record.

flows, since the extreme high and low discharges are lost in the averaging. It should be stressed that no representation of the chronological sequence of events is portrayed or enumerated in flow–duration curves. For assessing water resources, the frequency of *sequences* of wet or dry months can be evaluated, given a suitably long record. Hydrograph sequences can also be modelled, given a suitably calibrated rainfall–runoff model for a site and either measured or model generated rainfalls. Such models are considered in later chapters.

11.5 Flood frequency

One of the most important hydrological analyses is the assessment of the frequency with which discharges of a given magnitude are exceeded at a site and one of the most important data sets is the measured instantaneous flood peak discharges estimated from the stage records at a site. The longer a record continues, homogeneous and with no missing peaks, the more its value is enhanced. Even so, it is very rare to have a satisfactory record long enough to match the expected life of many engineering works required to be designed (which may be 50 years or more) or to assess flood risk for the 0.01 *annual exceedance probability* (AEP, also referred to as 1 per cent AEP or 1 in 100 year) event, used in the UK as a standard for flood risk mapping as required under Section 105 of the Water Resources Act 1995.

As many peak flows as possible are needed in assessing flood frequencies but care should be taken in evaluating estimates of the higher peaks. A detailed review of the most extreme flood peaks recorded in the USA, for example, revealed that the discharges that had been estimated for a number of floods needed some revision (Costa and Jarrett, 2008). Very often flood discharge estimates are based on an extrapolation of the stage–discharge rating curve based on measurements at much lower discharges, if only because of the logistical and safety difficulties of getting measurements of discharge at the highest flows. Thus, where possible, the uncertainty in the flood peak estimates should be assessed as part of a flood frequency analysis.

The hydrologist defines two data series of peak flows: the *annual maximum series* (often called the AMAX series) and the *partial duration or peaks over threshold series*. These may be understood more readily from Fig. 11.15. The annual maximum series takes the single maximum peak discharge in each year of record so that the number of data values equals the record length in years. For statistical purposes, it is necessary to ensure that the selected annual peaks are independent of one another. This is sometimes difficult, e.g. when an annual maximum flow early in January may be related to an annual maximum flow at the end of the previous December. For this reason, it is generally advisable to use the water year rather than the calendar year; the definition of the water year depends on the seasonal climatic and flow regimes. In humid temperate areas, such as the UK, it is often taken to be October to September (in October, soils should normally have rewetted following the summer dry period). The *partial duration series* takes all the peaks over a selected threshold level of discharge. Hence the partial duration series is often called the *peaks over threshold* (POT) series. There will be more data values for analysis in this series than in the annual series, but there is more chance of successive peaks being related such that the assumption of statistical independence of the data might be less valid.

In Fig. 11.14, P_1, P_2 and P_3 form an annual series and P_1, p_1, P_2, P_3 and p_3 form a POT series. It will be noted that one of the peaks in the POT series, p_3, is higher than the maximum annual value in the second year, P_2. For sufficiently long records it may be prudent to consider all the major peaks and then the threshold is chosen so that there are N peaks in the N years of record, but not necessarily one in each year. This is called the *annual exceedance series,* a special case of the POT series.

Flood frequency analysis entails the estimation of the peak discharge, which is likely to be equalled or exceeded on average once in a specified period, T years. This is called the *T-year event* and the peak, Q_T, is said to have a *return period* or *recurrence interval* of T years. The return period, T years, is the long-term average of the intervals between successive exceedances of a flood magnitude, Q_T, and is effectively a shorthand way of referring to a probability with which Q_T might be expected to be exceeded in any 1 year (see next section). It is important not to misunderstand the return period concept. The intervals with which Q_T is exceeded might vary considerably around the

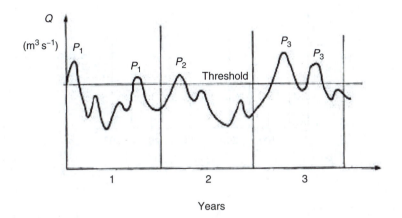

Fig. 11.14 Definition of annual maximum discharges (upper case) and peaks over threshold (upper and lower case) peaks over a 3-year period.

average value T. Thus a very long record could show 10-year events, Q_{10}, occurring at intervals much greater or much less than 10 years, and even in successive years.

The annual series and the partial duration series of peak flows form different probability distributions but, for return periods of 10 years and more, the differences are minimal and the annual maximum series is the one most usually analysed. Both types of analysis can also be used to estimate the probability of exceedance or return period of a particular event in the historical record at a site.

11.6 Flood probabilities

When a series of annual maximum flows is sub-divided by magnitude into discharge groups or classes of class interval ΔQ, the number of occurrences of the peak flows (f_i) in each class can be plotted against the discharge values to give a frequency diagram (Fig. 11.15a). It is convenient to transform the ordinates of the diagram in two ways: with respect both to the size of the discharge classes (ΔQ) and the total number, N, of events in the series. By plotting $f_i/(N.\Delta Q)$ as ordinates, it is seen that the panel *areas* of the diagram will each be given by f_i/N, and hence the sum of those areas will be unity, i.e.

$$\sum_{i=1}^{n}\left(\frac{f_i}{N\Delta Q}.\Delta Q\right) = \frac{N}{N} = 1 \tag{11.7}$$

where n is the number of discharge classes.

If the data series is now imagined to be infinitely large in number and the class intervals are made infinitesimally small, then in the limit, a smooth curve of the *probability density distribution* is obtained (Fig. 11.15b), the area under this curve being unity, i.e.

$$\int_{0}^{\infty} p(Q)dQ = 1 \tag{11.8}$$

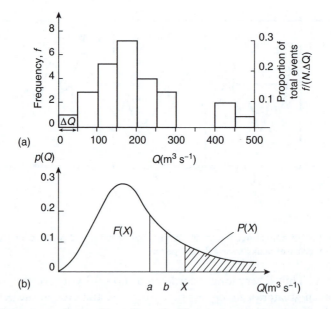

Fig. 11.15 (a) Frequency of annual maxima of different magnitude plotted as a histogram. (b) Continuous probability distribution of annual maximum flood peaks.

The probability that an annual maximum, Q, lies between two values, a and b, is given by:

$$P = \int_a^b p(Q)dQ$$

For any given magnitude, X, the probability that an annual maximum equals or exceeds X, i.e. that $Q \geq X$ is:

$$P(X) = \int_X^\infty p(Q)dQ \qquad (11.9)$$

which is the area shaded under the probability curve (Fig. 11.15b). If $F(X)$ is the probability of $Q < X$:

$$F(X) = \int_0^X p(Q)dQ \qquad (11.10)$$

and clearly:

$$P(X) = 1 - F(X) \qquad (11.11)$$

$P(X)$ is the probability of an annual maximum equalling or exceeding X in any given year, since it is the relative proportion of the total number of annual maxima that have equalled or exceeded X. If X is equalled or exceeded r times in N years (N large), then $P(X) \rightarrow r/N$. The return period for X is, however, $T(X) = N/r$. Thus:

$$P(X) = \frac{1}{T(X)} = \frac{1}{1 - F(X)} \tag{11.12}$$

It follows that

$$F(X) = \frac{T(X) - 1}{T(X)} \tag{11.13}$$

Thus, if $T(X) = 100$ years, $P(X) = 0.01$ and $F(X) = 0.99$. Thus it is seen that return period $T(X)$, annual probability of exceedance $P(X)$, and cumulative probability $F(X)$, for a given flood peak with magnitude X are directly and simply related. In flood frequency it is really the probability of exceedance that is important. The engineering hydrologist is interested in the probability that a flood will be exceeded over the next year, or the next 10 years, or the next 50 years. It is thus better for technical purposes to think in terms of probabilities rather than return periods.

For example, it is interesting to calculate the probability with which a value of X might be exceeded in a certain defined time period. This is given by the formula

$$P_N(X) = 1 - F(X)^N = 1 - (1 - P(X))^N = 1 - \left[1 - \frac{1}{T(X)}\right]^N \tag{11.14}$$

Values for the probability of a flood of a given magnitude during different period lengths are given in Table 11.5. From this it may be seen, for example, that there is only a 63.4 per cent chance of the 0.01 probability (1 in 100-year) event being exceeded in any given period of 100 years. These probabilities are independent of the choice of a statistical distribution chosen to represent flood peak probabilities at any particular site.

These figures are of interest in engineering design since they can be used in the evaluation of how well a structure might be protected over its design life against an event of a particular magnitude. For example, from Table 11.5, if a flood defence

Table 11.5 The probability of different magnitude of events occurring during any given period of length N years

Probability of exceedance, $P(X)$	Return period, T	Length of period (N, years)			
		10	20	50	100
0.1	10	0.651	1	1	1
0.05	20	0.401	0.642	0.923	0.994
0.02	50	0.183	0.332	0.636	0.867
0.01	100	0.096	0.182	0.395	0.634
0.005	200	0.003	0.095	0.222	0.394

is intended to protect against the 100-year event and have a design life of 50 years, then the probability of it being over-topped (failing to provide protection) in that period is 39.5 per cent. This might be considered to be unacceptably high. Even a design to give protection against the 200 year event would be expected to fail over the 50-year design life with a probability of 22.2 per cent. This gives a good idea of the real difficulty of protecting property against major floods that underlies the move away from the concept of 'protection' to the *management of flood risk* (see Chapter 16) that is the basis for legislation such as the EU Floods Directive of 2007.

In the UK, these probabilities are used by the Environment Agency to classify different zones for flood hazard. Zone 1 represents land that is outside the area that would be inundated by a 0.1 per cent AEP event. These areas are defined as having low probability of flooding. Zone 2 is the area between the 1 per cent and 0.1 per cent AEP of inundation for fluvial flooding and 0.5 per cent and 0.1 per cent AEP for coastal flooding. These areas are defined as having a medium probability of flooding. Areas at high risk (Zone 3) are those within the 1 per cent probability of inundation area for fluvial flooding, and the 0.5 per cent probability area for coastal flooding. These different zones are defined has having different appropriate uses under Planning Policy Statement 25: Development and Flood Risk (Cabinet Office, 2006). Defining the different zones requires converting the estimate of discharge with the required probability of exceedance into a flood hazard map. Normally this is done using the type of hydraulic model described in Chapter 14.

11.7 Analysis of an annual maximum series

There are many different statistical distributions that have been used to represent flood peak frequencies in different studies in the past. To facilitate flood frequency analysis for a data sample, special probability graph papers may be used. There are also many different software packages available, such as WINFAP-FEH that was produced as part of the *Flood Estimation Handbook* (*FEH*, Institute of Hydrology, 1999; see also Section 13.3.2). In a graphical analysis, the probabilities, $F(X)$, are made the abscissa, with X the ordinate. For any given probability distribution, the values of $F(X)$ are transformed to a new scale such that the X versus $F(X)$ relationship is made linear for that distribution. An example of such a probability paper, designed for the Gumbel (extreme value type 1 or EV1) probability distribution is seen in Fig. 11.16. Distributions used in the *FEH* are described later in Section 13.3.

In Table 11.6, the procedure for analysing annual maximum flows is given using the 24 annual maximum discharges from a continuous homogeneous gauged river record. The peak flows ($m^3 s^{-1}$) are arranged in decreasing order of magnitude with the second column showing the rank position. The probability of exceedance, $P(X)$, is then calculated for each value, X, according to a plotting position formula devised to overcome the fact that when N is not large, r/N is not a good estimator. Of the several formulae in use, the best is due to Gringorten:

$$P(X) = \frac{r - 0.44}{N + 0.12} \tag{11.15}$$

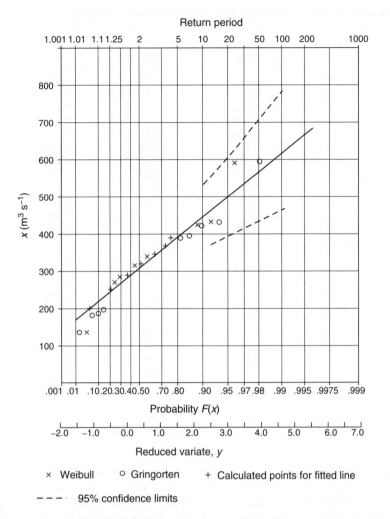

Fig. 11.16 Plot of estimated exceedance probability for the annual maximum peak series of Table 11.6 using both Gringorten and Weibull plotting positions on Gumbel (extreme value type I distribution) probability paper.

where r is the rank of X and N is the total number of data values. The Weibull formula:

$$P(X) = \frac{r}{N+1} \tag{11.16}$$

is also widely used. Both formulae, which generally give similar results, have been applied in the example and the two sets of values of $P(X)$ and $F(X)$ are shown in Table 11.6.

All the Weibull values of $P(X)$ have been plotted for their corresponding values of X(Fig. 11.16) but only the top 5 and bottom 4 Gringorten values have been plotted

Table 11.6 Flood frequency analysis (24 years of an annual maximum series)

$X(m^3 s^{-1})$	Rank r	P(X) Gringorten	F(X) Gringorten	P(X) Weibull	F(X) Weibull
594.7	1	0.023	0.977	0.040	0.960
434.7	2	0.065	0.935	0.080	0.920
430.4	3	0.106	0.894	0.120	0.880
402.1	4	0.148	0.852	0.160	0.840
395.9	5	0.189	0.811	0.200	0.800
390.8	6	0.231	0.769	0.240	0.760
369.5	7	0.272	0.728	0.280	0.720
356.8	8	0.313	0.687	0.320	0.680
346.9	9	0.355	0.645	0.360	0.640
342.9	10	0.396	0.604	0.400	0.600
342.6	11	0.438	0.562	0.440	0.560
321.4	12	0.479	0.521	0.480	0.520
318.0	13	0.521	0.479	0.520	0.480
317.2	14	0.562	0.438	0.560	0.440
300.2	15	0.604	0.396	0.600	0.400
290.0	16	0.645	0.355	0.640	0.360
284.6	17	0.687	0.313	0.680	0.320
283.2	18	0.728	0.272	0.720	0.280
273.3	19	0.769	0.231	0.760	0.240
256.3	20	0.811	0.189	0.800	0.200
196.8	21	0.852	0.148	0.840	0.160
190.3	22	0.894	0.106	0.880	0.120
185.5	23	0.935	0.065	0.920	0.080
138.8	24	0.977	0.023	0.960	0.040

$\overline{X} = 323.5$
$\sigma_X = 97.1$

to show the location of their main divergence from the Weibull points. The distribution of the plotted points on the graph would be linear if the annual maxima came from a Gumbel distribution and N was large, but with small samples, departures from a straight line are always to be expected. It will be noted that the highest peak $(594.7 m^3 s^{-1})$, which has the appearance of an outlier, falls into line better in the Gringorten plotting position. Outliers commonly occur and arise from the inclusion in a short record of one or more events with a probable return period that is much longer than the length of series available.

Although the Gumbel distribution is used for the present example, many other probability distributions have been investigated for application to the extreme values produced by flood peak discharges (see Section 11.8).

11.7.1 Fitting the Gumbel distribution (EVI)

Although subjective graphical curve fitting by eye may be justifiably adequate in the analysis of inaccurately measured or even estimated flood discharges, objective methods of curve fitting are preferred by national agencies seeking uniformity of practice. For example, the straight line drawn through the plotted points in Fig. 11.16 could well have been drawn by eye, especially using the Gringorten plotting positions, but in fact the line has been drawn by fitting the Gumbel distribution to the data by the *method of moments*.

The equation of the Gumbel or EV1 is given by

$$F(X) = \exp\left[-e^{-\left(\frac{X-a}{b}\right)}\right] \qquad (11.17)$$

where $F(X)$ is the probability of an annual maximum $Q \leq X$ as defined previously, and a and b are parameters.

Denoting the first moment (the mean) of the distribution by μ_Q and the second moment (the variance) by σ_Q^2 then the parameters a and b are given by the following expressions:

$$a = \mu_Q - 0.5772b$$

$$\qquad (11.18)$$

$$b = \frac{\sigma_Q \sqrt{6}}{\pi}$$

In (11.17) and (11.18), μ_Q and σ_Q^2 pertain to the whole statistical population of floods at the site; with a finite sample they can only be estimated from the moments of the data sample. Thus, denoting the sample estimates of μ_Q and σ_Q^2 by $\hat{\mu}_Q$ and $\hat{\sigma}_Q^2$:

$$\hat{\mu}_Q = \overline{Q} = \frac{1}{N}\sum_{i=1}^{N} Q_i \qquad \text{(the sample mean)}$$

$$\hat{\sigma}_Q^2 = s_Q^2 = \frac{1}{N-1}\sum_{i=1}^{N} (Q_i - \overline{Q})^2 \qquad \text{(the sample variance)}$$

Example: Fitting the Gumbel distribution to the annual maximum peaks in Table 11.6 using the method of moments.

Sample estimates of the mean and variance for the data in Table 11.6 are:

$$\overline{X} = \overline{Q} = \hat{\mu}_Q = 323.5 \text{ m}^3\text{s}^{-1}$$

$$s_x = s_Q = \hat{\sigma}_Q = 97.1 \text{ m}^3\text{s}^{-1}$$

Substituting for μ_Q and σ_Q^2 in (11.8), we have

$$b = \frac{\sigma_Q \sqrt{6}}{\pi} = 75.75$$

$$a = \mu_Q - 0.5772b = 279.8$$

The equation of the Gumbel distribution is thus given by

$$F(X) = \exp\left[-e^{-\left(\frac{X-279.8}{75.75}\right)}\right]$$

A Gumbel distribution will plot as a straight line on Gumbel probability paper as in Fig. 11.16. To plot the distribution only two points therefore need to be calculated. For $X = 200$, $F(X) = 0.057$ and for $X = 500$, $F(X) = 0.947$.

Examination of Fig. 11.16 reveals that three different scales have been used to describe the horizontal axis: return period, probability and the *reduced variate y*. It will be seen that the latter scale is a linear scale and is specific to the Gumbel distribution used. The reduced variate is related directly to the a and b parameters of the Gumbel distribution to give a linear relationship with peak discharge X as:

$$y = (X - a)/b \tag{11.19}$$

so that

$$F(X) = \exp[-e^{-y}] \tag{11.20}$$

The relationship between return period, probability of exceedance and the Gumbel reduced variate is given in Table 11.7.

The straight-line plot can be used to provide an estimate of the return period of a given peak discharge, but extrapolations beyond the limits of the data must be treated with care. In the example of Fig. 11.16, the return periods of flows up to $500\,\mathrm{m^3\,s^{-1}}$ can be provided with some confidence by this analysis. The validity of the outlying peak needs further checking, and confidence limits to the fitted curve would be recommended before estimates are made of the return periods of higher discharges. Data for which the Gumbel distribution is not a suitable distribution will not plot as a straight line. In Section 11.4, the more flexible generalised extreme value distribution is described, while Section 13.2 introduces the generalised logistic and generalised pareto distribution used in the *Flood Estimation Handbook*.

Table 11.7 Relationship between return period, probability of exceedance and reduced variate for the Gumbel distribution

Return period T	Probability of exceedance P(X)	Reduced variate y	Reduced variate y	Probability of exceedance P(X)	Return period T
2	0.500	0.367	−2	0.9994	1.001
5	0.200	1.500	−1	0.9340	1.071
10	0.100	2.250	0	0.6321	1.582
20	0.050	2.970	1	0.3078	3.249
50	0.020	3.902	2	0.1266	7.900
100	0.010	4.600	3	0.0486	20.59
200	0.005	5.296	4	0.0181	55.10
500	0.002	6.214	5	0.0067	148.9
1000	0.001	6.907	6	0.0025	403.9
			7	0.0009	1097.1

11.7.2 Estimating the T-year flood

The above formulation permits $F(X)$ to be found for a specified annual maximum X. Once $F(X)$ is known, $P(X) = 1 - F(X)$ is known, and therefore the return period $T(X) = 1/P(X)$ is known.

To reverse the procedure, the estimation of the annual maximum for a given return

$$F(X) = \exp\left[-e^{-\left(\frac{X-a}{b}\right)}\right] = \frac{T(X) - 1}{T(X)} \tag{11.21}$$

Taking logarithms of both sides twice:

$$-\frac{(X-a)}{b} = \ln\left\{-\ln\left[\frac{T(X) - 1}{T(X)}\right]\right\} \tag{11.22}$$

and rearranging:

$$X = a - b\ln\ln\left[\frac{T(X)}{T(X) - 1}\right] \tag{11.23}$$

Substituting for the parameters a and b with the sample mean $\overline{Q}$ and standard deviation s_Q as estimates of the population values μ_Q and σ_Q, then estimates of X may be obtained from:

$$\begin{aligned}
\hat{X} &= \overline{Q} - \frac{0.5772 s_Q \sqrt{6}}{\pi} - \frac{s_Q \sqrt{6}}{\pi}\left\{\ln\ln\left[\frac{T(X)}{T(X) - 1}\right]\right\} \\
&= \overline{Q} - \frac{\sqrt{6}}{\pi}\left\{0.5772 + \ln\ln\left[\frac{T(X)}{T(X) - 1}\right]\right\} s_Q \\
&= \overline{Q} + K(T) s_Q
\end{aligned} \tag{11.24}$$

where:

$$K(T) = -\frac{\sqrt{6}}{\pi}\left\{0.5772 + \ln\ln\left[\frac{T(X)}{T(X) - 1}\right]\right\} \tag{11.25}$$

$K(T)$ is called the *frequency factor*, which is a function only of the return period, T. Although it is independent of the parameters a and b, $K(T)$ in (11.25) is specifically for the Gumbel (EV1) distribution. It is tabulated in Table 11.8.

Thus if an estimate of the annual maximum discharge for a return period of 100 years is required, then $T(X) = 100$ years, $K(T) = 3.14$ and $\hat{Q}_{100} = \overline{Q} + 3.14 s_Q$. In the plotted example of Fig. 11.16 with $\overline{Q} = 323.5$ and $s_Q = 97.1$, $\hat{Q}_{100} = 628.4 \, \text{m}^3 \, \text{s}^{-1}$. Thus, assuming a Gumbel distribution and having calculated the mean and standard deviation of a sample of annual maximum flows, an estimate of the peak flow for any required return period can be obtained from (11.24) using the appropriate $K(T)$ value from Table 11.8.

Table 11.8 The *T-K(T)* relationship for the Gumbel distribution

T	K(T)	T	K(T)	T	K(T)
1	$-\infty$	10	1.30	80	2.94
2	-0.16	15	1.64	90	3.07
3	0.25	20	1.86	100	3.14
4	0.52	25	2.04	200	3.68
5	0.72	30	2.20	400	4.08
6	0.88	40	2.40	600	4.52
7	1.01	50	2.61	800	4.76
8	1.12	60	2.73	1000	4.94
9	1.21	70	2.88		

11.7.3 Confidence limits for the fitted data

It is often advisable, particularly with a short data series, to construct confidence limits about the fitted straight-line relationship between the annual maxima and the linearised probability variable. A first step is the calculation of the standard error of estimate for a peak discharge (X) in terms of the return period (T). The expression for the standard error is dependent on the probability distribution used, and for the Gumbel distribution it is given by:

$$SE(\hat{X}) = \frac{s_Q}{\sqrt{N}}\left[1 + 1.14K(T) + 1.10K(T)^2\right]^{0.5} \tag{11.26}$$

where N is the number of annual maxima in the sample. Then the upper and lower limits are calculated from:

$$\hat{X} \pm t_{\alpha,v}SE(\hat{X}) \tag{11.27}$$

where $t_{\alpha,v}$ are values of the t distribution obtained from standard statistical tables with α the probability limit required and v the degree of freedom.

The calculations for the 95 per cent confidence limits for the plotted example in Fig. 11.16 are set out in Table 11.9. The value of the t statistic is 2.06 for $\alpha = 100 - 95$ per cent $= 5$ per cent and $v = n - 1 = 23$. The curves of the 95 per cent confidence limits are plotted on Fig. 11.16 for the range of selected T values.

11.7.4 The generalised extreme value distribution

The Gumbel distribution of the previous section is a special case of the generalised extreme value (GEV) distribution, which is described by the equations for discharge X at any value of $F(X)$

$$X(F) = a + \frac{b}{k}\left\{1 - \left[-\ln F(X)\right]^k\right\} \qquad for\ k \neq 0$$
$$X(F) = a + b\left\{-\ln\left[-\ln F(X)\right]\right\} \qquad for\ k = 0 \tag{11.28}$$

Table 11.9 Calculation of 95 per cent confidence limits for the Gumbel distribution

T (years)	10	20	30	50	100
K(T) (from Table 11.8)	1.30	1.86	2.20	2.61	3.14
$\hat{X}$ (equation 11.14) m^3 s^{-1}	449.7	504.1	537.1	576.9	628.4
SE($\hat{X}$) (equation 11.16) m^3 s^{-1}	41.3	52.2	58.9	67.1	77.8
$t_{5.23}$ SE($\hat{X}$) m^3 s^{-1}	85.1	107.4	121.3	138.3	160.4
$\hat{X}_{97.5\%}$ m^3 s^{-1}	534.8	611.5	658.4	715.2	788.8
$\hat{X}_{2.5\%}$ m^3 s^{-1}	364.6	396.7	415.8	438.6	468.0

The second form with $k = 0$ is the form of the Gumbel (EV1) distribution. If $k \neq 0$, then the sample of floods will not plot as a straight line on Gumbel probability paper such as that shown in Fig. 11.16. If it appears that there is a trend for the more extreme floods to curve upwards above a straight line, then k will be negative and the distribution is called an EV2 distribution. If it appears that there is a trend towards an upper limit below a straight line plot, then the distribution is called an EV3 distribution. Both have been used in past analyses of annual maximum floods, including in the UK Flood Studies Report (NERC, 1975).

One of the techniques used in comparing flood frequency distributions between different catchments is the *growth curve*. The growth curve is defined by normalising the floods observed at a site by dividing by the value of an *index flood*. In the *FEH* (Institute of Hydrology, 1999), the index flood is taken to be the median of the observed series (*QMED*; the value of X when $F(X) = 0.5$). The ratio of the observed discharge to the index flood provides a non-dimensional scale of discharge for comparing catchments but will not change the form of the frequency distribution. Defining $x = X/QMED$, the growth curve for the GEV distribution can then be defined by substitution in (11.28) and rearranging as:

$$x(F) = 1 + \frac{\beta}{k}\left\{\ln(2)^k - \left[-\ln F(x)\right]^k\right\}$$

where (11.29)

$$\beta = \frac{b}{a + \frac{b}{k}\left[1 - \ln(2)^k\right]}$$

The growth curve can also be related directly to return period, T, by substituting for $F(x)$ such that:

$$x_T = 1 + \frac{\beta}{k}\left[\ln(2)^k - \left(\ln\frac{T}{T-1}\right)^k\right]$$

(11.30)

11.8 Other statistical distributions used in flood frequency analysis

The statistical theory of extreme values suggests that the maximum values in independent sample series of fixed length taken from *any* fixed distribution should, as the number of samples increases, asymptotically approach the form of the GEV distribution, while peaks over threshold series should be approximately distributed as a generalised pareto (GP) distribution. A series of annual maximum series of flood discharges might also therefore approach the form of the GEV as the sample size increases. Why, therefore, are other distributions used in flood frequency analysis? This is because the number of samples is generally small, and we cannot be sure that there is a fixed underlying distribution for the occurrences of floods because of climate and hydrological variability.

In the USA, after comparing six different distributions, the log Pearson Type III was selected, although many factors, other than statistical, governed the final choice (Benson, 1968). The log-normal distribution, in which the logarithms of the peak flows conform to the Gaussian or normal distribution, also takes the form of a probability curve with a positive skew, as shown in Fig. 11.15b and is often applied successfully to annual maximum flow series.

In the UK, the GEV distributions were chosen as the basis for analysis of annual maximum flows in the Flood Studies Report (NERC, 1975). This has been replaced by the generalised logistic (GL) distribution in the *FEH* (see Chapter 13). With the longer series available in *FEH*, the GL distribution was selected as better representing annual maximum series. The advantage of the GL distribution over the GEV is that fitting the GEV more often results in a distribution with an upper bound (an EV3 distribution). It is perhaps debatable whether we should expect floods to have some upper bound on the basis of meteorological and hydrological conditions in any region. Reaching such a bound would normally occur at such a long return period that, for practical purposes, the assumption of an unbounded distribution can be taken as reasonable; hence the choice of the GL distribution in the *FEH*. Details of the GL and GP distributions may be found in Chapter 13 on the *FEH* procedures.

Distributions with greater numbers of parameters are also sometimes fitted to flood frequency data, including the four-parameter kappa distribution, and the five-parameter Wakeby distribution. The GEV and GL distributions are both special cases of the kappa distribution. Mixed distributions are also sometimes used where there is evidence that different events in the flood record are generated by different types of mechanism. Hydrological understanding would suggest that we might expect mixed distributions in some cases, such as the synoptic rainfall events and snowmelt or rain-on-snow events that make up the series of extremes in British Columbia (see Woo and Waylen, 1984); or the mixture of rainstorms and rare hurricanes, such as the destructive Hurricane Agnes in 1972, on the east coast of the USA. In all these cases, however, fitting distributions with more parameters will often result in higher uncertainty associated with the parameter estimates and wider confidence limits on estimates at longer return periods, unless very long sequences of extremes are available for analysis. Since this is rarely the case, the two- and three-parameter distributions remain the most commonly used in practice.

11.9 Using historical flood data

Whilst the *FEH* and other formal procedures generally rely on using gauged flow and rainfall records, there is often potential to supplement these estimates by making use of 'historical' flood information. Here the term 'historical' refers to information not contained within the gauged record, but to a mixture of quantitative and qualitative data such as flood marks on historic buildings, contemporary accounts of notable floods, photographs and engravings and records of changes in factors such as catchment land drainage or river channel cross-section shape.

The use of incomplete flood records in statistical approaches based on censored annual maximum series was discussed in the Flood Studies Report (NERC, 1975). There have been studies demonstrating the use of historical information since then, e.g. Acreman and Horrocks (1990) and Archer *et al.* (2007). However, historical flood data has not been incorporated routinely into statistical flood estimation procedures. Floods in the UK in the 1990s prompted a more systematic effort to look back to major historical events for comparison and to attempt to incorporate these into flood risk assessments. Prompted also by investigations into the potential effects of climate and land-use change (see Chapter 19), there has been a growing recognition that the mid-twentieth century might have been unusually 'flood sparse' and that statistical flood estimates might therefore underestimate flood risk. On the other hand, relatively short gauged records that include the floods in the 1990s onwards may overestimate the frequency of these events. The British Hydrological Society (BHS) *Chronology of British Hydrological Events*[6] was launched in 1998 (Black and Law, 2004). It provides a readily accessible searchable database of hydrological events and reports including both floods and droughts.

Bayliss and Reed (2001) published guidance on using historic data in flood risk estimation, and recommended a graphical approach to include historical data on to a plot of the systematic gauged data and fitted frequency distribution based on adjusted plotting position formulae. Stedinger and Cohn (1986) have described an approach to fit probability distributions to a flood peak data series that combines continuous AMAX records from a gauging station with historical records of flooding, which are viewed as analogous to an incomplete AMAX record. The approach adopts a binomial distribution to model the probability of obtaining the number of historical peaks that have been observed. This is combined with a probability distribution for the gauged AMAX series (the GEV distribution above or the GL distribution described in Chapter 13 are suitable choices) and the resulting combined function fitted by the method of maximum likelihood. The method has the useful feature that, even if the magnitude of an historical flood flow is unknown, knowledge of the mere occurrence of a flood that exceeds a given threshold can be used to improve the statistical model, hence qualitative information about past events can be incorporated.

Fig. 11.17 shows a flood frequency analysis using these methods for the River Wansbeck at Morpeth, draining a predominantly lowland catchment in north-east England. The data comprises a combination of continuous records from gauging stations at Highford and Mitford, where the catchment area is 287 km^2, and also historical information from nearby sites. The gauged record from 1963 to 2008 includes significant high-flow events in 2008, 1992, 1982, 1967 and 1963. Flood marks on

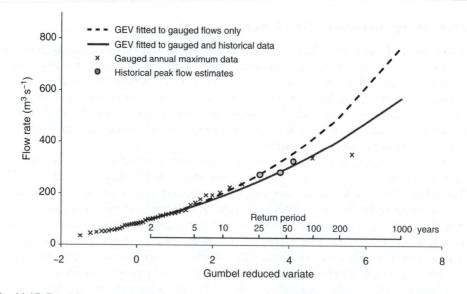

Fig. 11.17 Flood frequency curves for the Wansbeck at Morpeth based on fitting the generalised extreme value (GEV) distribution to gauged annual maximum flows and to a composite record including estimated flows for events in 1878, 1886 and 1898.

nearby buildings at East Mill and Bothal Mill also allow flow rates to be estimated for floods that occurred in 1878, 1886 and 1898, using reconstructed rating curves. The results illustrate that the inclusion of historical events may lead to a reassessment of flood frequency estimates from the gauged record. Although the events of 1878, 1886 and 1898 rank highly in the composite record, the historical information effectively extends the record and places the largest events, including gauged peaks in 1963 and 2008, into a longer historical context (effectively shifting the plotting positions of the two largest AMAX values to the right). Here, the inclusion of this information dating back to the nineteenth century suggests that statistical estimates based on the gauged record alone may overestimate the frequency of high flows. Note that this is not a general result; underestimation may be apparent in other cases and a careful interpretation of the gauged and historical information is required in each individual study.

11.10 Droughts

Droughts as rainfall deficiencies have been considered earlier in Chapter 9 in terms of both long-term precipitation characteristics and in their effect on evaporation and transpiration losses to the atmosphere. Droughts are also important in terms of water supply. When there are deficiencies in river flows, it is the bulk users of water, namely agriculture, industry and the large urban concentrations of domestic consumers that begin to suffer. Water authorities and Water utilities whose duty it is to maintain public supplies of water are always concerned for their resources during periods of low flows and by the increase in demand that dry spells generally stimulate. An analysis of the expected minimum flows in those rivers providing water supplies is therefore essential. It is also important in other engineering design problems, such as low head hydropower installations. The economic viability of a design may be dependent on the expected durations of flows below some minimum threshold for power production.

In the UK, all major rivers are perennial, but some of the headwaters and small tributary streams rising in limestone or chalk country are intermittent or *ephemeral*. Their upper courses dry up in summer, and in drought periods the lack of water in the streams may extend further down the valleys, particularly when the flows are also affected by groundwater abstractions. Increasing attention is being paid to flow deficiencies, and in the perennial streams it is considered desirable to maintain a defined minimum discharge. The sustaining of a minimum discharge is particularly important in all rivers receiving waste water effluents in order to ensure required dilution of pollutants and the habitat conditions necessary for good ecological status within the requirements of the water framework directive (see Section 8.2.2 and Chapter 17).

In the analysis of low-flow conditions, it is preferable to have natural discharge records unaffected by major abstractions or sewage effluent discharge. In drought periods, hydrologists are encouraged to make extra gaugings on small tributary streams as well as at the established gauging stations. In the UK, the Environment Agency has carried out many such low-flow surveys on the major river networks. Such records provide more detailed information on changes in base flow conditions in relation to the geology of the catchment.

At gauging stations, where continuous river discharge records are available, several features of the data sets can be abstracted or computed to give measures of the characteristics of low flows. A variety of different indices have been used in the analysis of low flows derived from the flow–duration curve, consideration of low-flow spells, and the frequency analysis of the annual minimum series of low flows.

The *flow–duration curve* (Figs 11.11 and 11.12) was considered earlier and gives the duration of occurrence of the whole range of flows in the river. Selected points on the lower end of the curve can give measures of flow deficiency. The flow that is exceeded 95 per cent of the time, $Q95$, and the percentage of time that a quarter of the average flow is exceeded, are two suggested indices. In Fig. 11.12, the 95 per cent exceedance flow is $13.5 \, m^3 \, s^{-1}$ and a quarter of the average flow $(44 \, m^3 \, s^{-1}) \, 11 \, m^3 \, s^{-1}$, is exceeded 99.45 per cent of the time. To compare low-flow values between catchment areas, the $Q95$ value may be expressed as a runoff depth over the catchment in millimetres per day. Hence for the Thames curve, $Q95/A = 13.5 \, m^3 \, s^{-1}/9870 \, km^2$, which gives $0.118 \, mm \, day^{-1}$.

Similarly for the other flow–duration curve examples (Fig. 11.11), where the $Q95$ flows are: for the Eden, $9.90 \, m^3 \, s^{-1}$ over a catchment of $2286 \, km^2$, equivalent to $0.374 \, mm \, day^{-1}$ or 30 per cent of the mean daily flow; and for the Miriam $0.218 \, m^3 \, s^{-1}$ over a catchment of $133.8 \, km^2$, equivalent to $0.141 \, mm \, day^{-1}$ or 42 per cent of the mean daily flow. These catchments are therefore rather different in their baseflow characteristics, reflecting the geological differences between the two catchments: hard rock, partly overlain by glacial moraines, and chalk, respectively. The index $Q95/A$ can be correlated with various catchment characteristics and useful regional patterns of drought properties can be identified.

The study of *low-flow spells* seeks to overcome the shortcomings of the flow–duration curve, which gives no indication of *sequences* of low flows. The most serious stress to water supply arises when there are periods of extended low flows. Such periods occurred in the UK, e.g. in 1976, 1995 and 2003. From the continuous record of daily mean discharges, the number of days for which a selected flow is not exceeded defines a low-flow spell. A $Q95$ flow may not be exceeded for a sequence of 10 days, thus giving a low-flow spell of 10 days' duration. The frequency or probability of occurrence of low-flow spells of different durations may be abstracted and assessed from the record. Hence D days, the duration of spells of low-flow $\leq Q95$, can be plotted against the percentage of low-flow spells $> D$. Catchments can again be compared by plotting individual results on the same graph. Where the volume of flow deficit below a threshold is also of interest, such as in assessing inputs to water supply reservoirs, the cumulative deficit can also be used as an index for frequency analysis.

In the UK, the estimation of low flows has been the subject of an extended study by the Institute of Hydrology and Centre for Ecology and Hydrology at Wallingford. This has resulted in a commercial software package, Low-Flows 2000[7] (Young et al., 2003), that can be used to estimate flow–duration curves and the low-flow characteristics for ungauged sites. More details on these studies are given in Chapter 13.

11.11 Frequency of low flows

Although the investigation of flood flows always attracts a great deal of research, the frequency analysis of low flows has not been neglected. In the UK a *low-flows study* was published by the Institute of Hydrology in 1980. Tallaksen and van Lanen (2004) provide a comprehensive overview of low-flow frequency analysis. Guidelines for frequency analysis of low flows in the UK were updated by Zaidman et al. (2002). Low-flow frequency analyses are carried out using a series of annual minimum ('AMIN') flow data. Low-flow periods can be prolonged and hence the annual minimum series should include a definition of the duration of interest, e.g. AMIN(30) refers to the series of annual minimum discharges of 30-day duration. In fitting a statistical distribution to annual minimum discharges, the principal requirements are that the distribution should be skewed, have a finite lower limit ≥ 0 and, as in case of short flood records, have a small number of parameters. The EV3 distribution of the smallest value is one of the most reliable and is relatively simple to compute.

The minimum daily mean flows for 10 years (Table 11.10) are used to demonstrate the method. In practice, at least 20 years of flow data should be obtained in order to constrain uncertainty in the frequency estimates. The annual minimum values are ranked starting with the highest. The probabilities $P(X)$, although calculated from the Gringorten and Weibull plotting positions as before, have a different meaning for the low-flow case. $P(X)$ is now the probability that $(Q_{annual\,minimum} \leq X)$.

All the Weibull points and the two extremes of the Gringorten points are plotted in Fig. 11.18 on Gumbel extreme value paper. Such a plot is sufficient to demonstrate that the Gumbel distribution itself is not an adequate representation of low-flow frequency because it is not bounded. Thus, if a straight line were fitted by eye through the plotted points, it would give a probability of exceedance of zero flow of approximately 0.98 and thus apparently a probability of $Q < 0$ of 2 per cent. Since this is not the case in this catchment, which remains perennial in flow, the EV3 distribution will be more appropriate. Similar arguments would apply to the analysis of the total volume of flow deficit below some specified threshold discharge.

Table 11.10 Frequency of low flows (annual minimum daily mean flows)

$X(m^3 s^{-1})$	Rank R	P(X) Gringorten	P(X) Weibull
0.408	1	0.055	0.091
0.351	2	0.154	0.182
0.315	3	0.253	0.273
0.256	4	0.352	0.364
0.238	5	0.451	0.455
0.222	6	0.549	0.545
0.210	7	0.648	0.636
0.187	8	0.747	0.727
0.152	9	0.846	0.818
0.074	10	0.945	0.909

$\overline{X}_m = 0.241$
$s_m = 0.098$

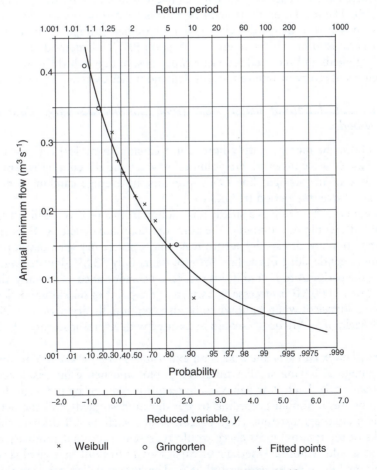

Fig. 11.18 Frequency plot of annual minimum flows for data of Table 11.10.

11.12 Low flows and water yield analysis

An important issue that arises in water supply management is the *water yield* of a catchment area. The utility of a water supply reservoir, for example, will depend critically on the yield under dry conditions. When there is a sufficiently long representative series of discharge records for analysis, the occurrence and frequencies of periods of critically low flows can be determined. Some regard must be paid both to the proposed life of the water resource scheme in accepting the representativeness of the data, and to the possibility of an increased yield being required in the future. The expected loss by evaporation from a reservoir must also be taken into account. Using the runoff time series, there are several straightforward methods of analysis.

Monthly data may be used for evaluating the amount of storage required in a reservoir to guarantee a given demand or supply rate. An initial determination of the reservoir capacity needed to ensure supplies over a low-flow period can be obtained by evaluating the cumulative volumes over that period. In the *mass curve* or *Rippl method*, the sequences of months from the historical record having the lowest flows are abstracted and for each sequence the cumulative amounts plotted against time (see Fig. 11.19). The testing of all drought periods is needed before deciding on a design drought on which to assess the yield given a specified storage. Some of the notable drought periods in the UK were given in Section 9.7. This method gives a good first estimate of yields and required reservoir capacities, but gives little information on the probabilities of reservoir failure to meet demand under extreme dry conditions.

Example: Calculation of water yield from monthly discharge data using the Rippl method

In Fig. 11.19, the mean monthly flows for a catchment of $150 \, \text{km}^2$ for the 5 years 1973–77 have been converted into volumes of water and the cumulative sums plotted.

The slope of the straight line *OA* represents an average catchment flow rate of $74 \, \text{ML day}^{-1}$ over the period from *O* to *A*.

To sustain this flow rate as a steady abstraction from a reservoir full at *O*, a storage of 13 000 ML would be required at the point of maximum deficit, *X*. Over this period, there are two minor dry spells, but from *A* a more severe dry spell develops.

The more gentle slope of the line *AB* represents only $53 \, \text{ML day}^{-1}$, average inflow rate over the period *A* to *B*, and if this rate is to be maintained as a steady abstraction, a storage of 12 000 ML is necessary, as seen at the point of maximum deficit *Y*.

On the evidence of what is known to be an exceptionally dry period 1975–76 *(AB)*, a reliable yield of $53 \, \text{ML day}^{-1}$ would be assured with reservoir storage of 12 000 ML.

The residual mass curve method is an extension of the mass curve technique with the advantage of having smaller numbers to plot and hence increased accuracy on the ordinate scale. Each flow value in the record is reduced by the mean flow (mean monthly or mean annual according to the duration studied) and the accumulated residuals plotted against time (Fig. 11.20). A line such as *AB* drawn tangential to the peaks of the residual mass curve would represent a residual cumulative constant yield that would require a reservoir of capacity *CD* to fulfil that yield starting with the reservoir full at *A* and ending full at *B*. The largest deficit between this residual yield line and the residual mass curve gives the minimum storage required to maintain

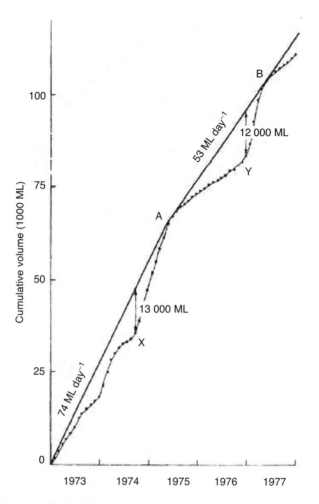

Fig. 11.19 Cumulative volume of monthly discharge in ML for the years 1973–1977. See example for explanation of lines 0–A and A–B.

the yield. In Fig. 11.20, the residual mass curve has been plotted for the same data used in Fig. 11.19. The mean rate of flow over the 5-year period is 1855 ML per month. The slope of the residual yield line *(AB)* is -240 ML month^{-1}, which when adjusted by the mean rate gives $1855 - 240 = 1615$ ML month^{-1} or 52 ML day^{-1} for the actual yield rate for the period $A \rightarrow D \rightarrow B$. The storage required is then 12 000 ML given by the line *CD*. This result compares favourably with the answer obtained in Fig. 11.19 for the necessary storage to sustain a 52 ML day^{-1} yield over the particularly dry period represented by the period $A \rightarrow D \rightarrow B$ in Fig. 11.20.

The minimum flows over various durations (e.g. 6, 12, 18, 24, 30, etc. consecutive months) can be abstracted from the runoff record and their flow volumes plotted against the corresponding durations (Fig. 11.21). A required yield line at the relevant slope is drawn from the origin of the plot, and a parallel line tangential to the plotted

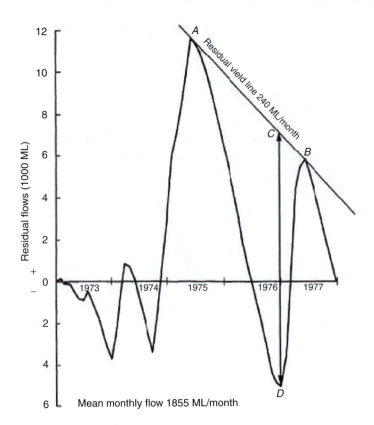

Fig. 11.20 Residual mass curve analysis.

droughts curve identifies both the critical drought duration and the amount of storage, S, needed to provide the yield. In practice, it is recommended to add a year's required supply to the storage as a safety factor. An example of the yield assessment of Vyrnwy Reservoir in North Wales was given in Shaw (1989).

11.13 Some concluding remarks

This chapter has set out various methods for the analysis of river discharge data, including river regime plots, mixing models, flow–duration curves, flood frequency analysis, the analysis of low flows and yield estimation. It is often the case that such analyses reveal some data anomalies and missing data. These might be apparent errors of timing relative to recorded rainfalls, flows that are larger than the measured rainfalls, apparently flat-topped peaks, or strange sudden steps in the recorded flows. The hydrologist always needs to bear two things in mind when analysing flow data. The first is that, at most river flow-gauging sites, flow is not measured directly. Stage or water level is measured and then converted to flow using a rating curve. This introduces the potential for error, both in extrapolation beyond the range of available discharge measurements used to infer a rating curve and also for changes in the rating over time

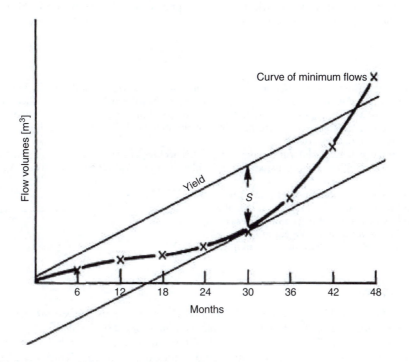

Fig. 11.21 Reservoir yield from analysis of minimum flows.

(such as after major floods). Where possible, try to check the original measurements for the rating curve and any other available gaugings at a site to check for such anomalies, particularly at sites that do not have a proper flow measurement structure. Flood peak discharge estimates in particular should be looked at with some care (e.g. Costa and Jarrett, 2008).

The second issue is that very few rivers are now unaffected by the influence of man, either through abstractions of flow for water supply and agriculture, the addition of effluents, the operation of weirs or the storage of water behind dams and reservoirs. Sometimes a series of reconstructed 'naturalized' discharges are available from agencies, but the assumptions on which the naturalization has been carried out might vary from agency to agency or region to region. The effects of abstractions, effluent discharges and storages are expected to be greatest at low flows but in some catchments, flood detention basins and controls on flood plain storage might affect flood peaks. The lesson is always to examine available discharge data with some care.

A third issue is that new projects requiring estimates of river flows are rarely planned close to stream gauges with an observed discharge available. The vast majority of sites, e.g. for new low-head hydropower projects, for the design of a new bridge, or for the mapping of flood plain inundation, will be ungauged. Thus, procedures are required to estimate the flow characteristics at such sites, informed by the type of analyses of gauged sites described above. In hydrology, the process of estimating flows at ungauged sites is called *regionalisation*. In the UK, the FEH (Institute of Hydrology, 1999) and Low Flows 2000 (Young et al., 2003) provide the basis for the regionalisation of flood frequency and flow–duration curves, respectively. The use of the FEH methods is described later in Chapters 13 and 17.

Notes

1 Such as the Environment Agency for England and Wales (EA), the Scottish Environmental Protection Agency (SEPA), the Rivers Agency of Northern Ireland, the Office of Public Works (OPW) in Eire, or the United States Geological Survey, which makes its data freely available in real time on the internet through the National Water Information System; see http://waterdata.usgs.gov/nwis.
2 http://www.ceh.ac.uk/data/nrfa/index.html (This archive contains over 50 000 years of data from 1300 gauging stations using data provided by the EA, SEPA or the Rivers Agency of Northern Ireland.)
3 http://www.environment-agency.gov.uk/hiflows/91727.aspx
4 http://www.dundee.ac.uk/geography/cbhe/
5 http://nwis.waterdata.usgs.gov/usa/nwis/peak
6 http://www.dundee.ac.uk/geography/cbhe/
7 See http://www.hydrosolutions.co.uk/lowflows.html

References

Acreman, M. C. and Horrocks, R. J. (1990) Flood frequency analysis for the 1988 Truro floods. *Journal of Institute of Water and Environmental Management* 4, 62–69.

Archer, D. R., Leesch, F. and Harwood, K. (2007) Assessment of severity of the extreme River Tyne flood in January 2005 using gauged and historical information. *Hydrological Sciences Journal* 52, 1–12.

Bayliss, A. C. and Reed, D. W. (2001) *The Use of Historical Data in Flood Frequency Estimation.* Report to Ministry of Agriculture, Fisheries and Food, Centre for Ecology and Hydrology, Wallingford.

Benson, M. A. (1968) Uniform flood frequency estimating methods for federal agencies. *Water Resources Research* 4, 891–908.

Black, A. R. and Law, F. M. (2004) Development and utilization of a national web-based chronology of hydrological events. *Hydrological Sciences Journal* 49, 237–246.

Boorman, D. B., Acreman, M. C. and Packman, J. C. (1990) *A Review of Design Flood Estimation Using the FSR Rainfall-runoff Method.* Report No. 111. Institute of Hydrology, Wallingford.

Brown, V. A., McDonnell, J. J., Burns, D. A. and Kendall, C. (1999) The role of event water, a rapid shallow flow component and catchment size in summer stormflow. *Journal of Hydrology* 217, 171–190.

Burns, D. A., McDonnell, J. J., Hooper, R. P., Peters, N. E., Freer, J. E., Kendall, C. and Beven, K. J. (2001) Quantifying contributions to storm runoff through end-member mixing analysis and hydrologic measurements at the Panola Mountain Research Watershed (Georgia, USA). *Hydrological Processes* 15, 1903–1924.

Cabinet Office (2006) *Planning Policy Statement 25: Development and Flood Risk.* HMSO, London.

Costa, J. E. and Jarrett, R. D. (2008) *An Evaluation of Selected Extraordinary Floods in the United States Reported by the U.S. Geological Survey and Implications for Future Advancement of Flood Science.* USGS Scientific Investigations Report 2008–5164, Washington, DC. (available at http://pubs.usgs.gov/sir/2008/5164/pdf/sir20085164.pdf)

Dobbie, C. H. and Wolf, P. O. (1953) The Lynmouth flood of August 1952. *Proceedings of the Institution of Civil Engineers*, Pt. III, 2, 522.

Genereux, D. P., Hemond, H. F. and Mulholland, P. J. (1993) Spatial and temporal variability in streamflow gauging on the West Fork of the Walker Branch watershed. *Journal of Hydrology* 142, 137–166.

Huff, D. D., O'Neill, R. V., Emmanuel, W. R., Elwood, J. W. and Newbold, J. D. (1982) Flow variability and hillslope hydrology. *Earth Surface Processes and Landforms* 7, 91–94.

Institute of Hydrology (IoH) (1980) *Low Flow Studies*. Research Report and addenda. IoH, Wallingford.

Institute of Hydrology (1999) *Flood Estimation Handbook* (5 vols). IoH, Wallingford.

Institution of Civil Engineers (ICE) (1933) *Floods in Relation to Reservoir Practice*. Interim Report of the Committee on Floods. ICE, London.

Institution of Civil Engineers (1960) *Floods in Relation to Reservoir Practice*. Reprint of 1933 Report with Additional Data, 66 pp. ICE, London.

Iorgulescu, I., Beven, K. J. and Musy, A. (2005) Data-based modelling of runoff and chemical tracer concentrations in the Haute-Menthue (Switzerland) research catchment. *Hydrological Processes* 19, 2557–2574.

Joerin, C., Beven, K. J., Iorgulescu, I. and Musy, A. (2002) Uncertainty in hydrograph separations based on geochemical mixing models. *Journal of Hydrology* 255, 90–106.

Matalas, N. C. (1963) *Probability Distribution of Low Flows*. US Geological Survey Professional, Paper 434-A, 27 pp.

Natural Environment Research Council (NERC) (1975) *Flood Studies Report,* Vol. IV. Hydrological Data 541 pp. NERC, Swindon.

O'Connell, J. E. and Costa, J. D. (2004) *The World's Largest Floods, Past and Present: Their Causes and Magnitudes*. USGS Circular 1254, US Department of the Interior, Washington, DC.

Shaw, E. M. (1989) *Engineering Hydrology Techniques in Practice*. Ellis Horwood, Chichester, 349 pp.

Stedinger, J. R. and Cohn, T. A. (1986) Flood frequency analysis with historical and paleoflood information. *Water Resources Research* 22, 5, 785–793.

Tallaksen, L. M. and van Lanen. H. A. J. (eds) (2004) *Hydrological Drought, Vol. 48: Processes and Estimation Methods for Streamflow and Groundwater*. Elsevier Science, Amsterdam, 579 pp.

Westhoff, M. C., Savenije, H. H. G., Luxemburg, W. M. J., Stelling, G. S., van de Giesen, N. C., Selker, J. S., Pfister, L. and Uhlenbrook, S. (2007) A distributed stream temperature model using high resolution temperature observations. *Hydrology and Earth System Sciences*, 11, 1469–1480.

Woo, M. K. and Waylen, P. R. (1984) Areal prediction of annual floods generated by 2 distinct processes. *Hydrological Sciences Journal* 29, 75–88.

Young, A. R., Grew, R. and Holmes, M. G. R. (2003) Low Flows 2000: a national water resources assessment and decision support tool. *Water Science and Technology* 48, 119–126.

Zaidman, M. D., Keller, V. and Young, A. R. (2002) *Low Flow Frequency Analysis: Guidelines for Best Practice*. Environment Agency R&D Technical Report W6-64/TR1, Environment Agency publications catalogue product code SW6-064-TR1-E-E. Environment Agency, Bristol.

Chapter 12

Catchment modelling

12.1 The essentials of a catchment model

The derivation of relationships between the rainfall over a catchment area and the resulting flow in a river is a fundamental problem for the hydrologist. In most countries, there are usually plenty of rainfall records, but the more elaborate and expensive streamflow measurements, which are needed for the assessment of water resources or of damaging flood peaks, are often limited and are rarely available for a specific site under investigation. Modelling the way in which rainfall becomes river discharge has stimulated the imagination and ingenuity of hydrologists for over a century, but has been an important area of research in the last 40 years as digital computers have become more widely available. Catchment models are now routinely used in flood forecasting, the design of flood defences and urban drainage systems, water resources assessment, and predicting the response of ungauged catchments. The use of models in hydrology is widespread in the applications illustrated in the remaining chapters of this book.

As well as these practical applications, however, catchment models are also an important way of doing science in hydrology, since they provide a mechanism for formalising hydrological understanding. We can take the type of mathematical representations of hydrological processes that have been described in previous chapters, and implement them as computer programs that will produce predictions of the consequences of assuming that those representations are correct. Those predictions can then be compared with what we know about the response of a catchment to see whether they are satisfactory as hypotheses about how the hydrological system is functioning.

To facilitate comparisons it is usual to express values for rainfall and river discharge in similar terms. The amount of precipitation (rain, snow, etc.) falling on a catchment area is normally expressed in millimetres (mm) depth per unit area over a specified time period, but may be converted into a total volume of water, cubic metres (m^3) falling on the catchment by multiplying by the catchment area. Alternatively, the river discharge (flow rate), measured in cubic metres per second ($m^3\,s^{-1}$ or cumecs) for a comparable time period may be converted into total volume (m^3), or expressed as an equivalent depth of water (in mm) by dividing by the catchment area. The discharge per unit area, often termed *runoff*, is then easily compared with rainfall depths over a defined time period. To make it easier to assess the water balance in a catchment, other variables such as storages and evapotranspiration flux can also be expressed as millimetres over the catchment area.

Estimating runoff or discharge from rainfall measurements is very much dependent on the timescale being considered. For short durations (hours) the complex interrelationship between rainfall and runoff is not easily defined, but as the time period lengthens, the connection becomes simpler until, on an annual basis, a straight-line correlation between the cumulative depths or volumes of rainfall and runoff may often be obtained. The purpose of an application will define what type of data time step is needed in developing relationships between rainfall and runoff. Relating a flood peak to a heavy storm will require continuous or short time-step records, but determining water yield from a catchment can be accomplished satisfactorily using relationships between totals of monthly or annual rainfall and runoff (noting, however, that accurate estimates of monthly or annual totals might also require continuous records, particularly in small or flashy catchments).

Naturally, the size of the area being considered also affects the relationship. For very small areas of a homogeneous nature – a stretch of impermeable motorway, say – the derivation of the relationship could be fairly simple; for very large drainage basins and for long time periods, differences in local rainfalls and runoff production are smoothed out giving relatively simple rainfall–runoff relationships. However, in general and for short time periods, great complexities occur when spasmodic rainfall is unevenly distributed over an area of varied topography, soil and geology characteristics with spatially heterogeneous antecedent soil water storage. For catchments with locally variable surface characteristics affected by a single severe storm – say catchments up to about 200–300 km^2 in humid temperate climates – the direct relationship between specific rainfalls and the resulting discharge or runoff is extremely complicated and often quite non-linear (e.g. doubling the rainfall input will more than double the discharge output).

The non-linearity arises because, at intermediate scales of both area and time, other physical and hydrological factors, such as evaporation, infiltration, groundwater flow, and the processes of wetting and drying over a sequence of events, are very significant and thus any direct relationship between rainfall alone and runoff is not easily determined. In this chapter, models for estimating the runoff resulting from a rainstorm or continuous sequence of rainstorms will be discussed. These methods range from simple procedures derived by engineers for immediate practical use to complex computer models.

Any catchment model needs to address two fundamental characteristics of the relationship between rainfall and runoff. The first is the proportion of the volume of rainfall represented by the storm *hydrograph* (or that part of the hydrograph defined as *stormflow* over and above the *baseflow* that would have occurred if the storm had not happened, Fig. 12.1). We know that, given an accurate estimate of rainfall over the catchment, this proportion will be less than one, but may vary in non-linear ways with the rainfall pattern in time and space and the antecedent wetness of the catchment. The runoff generation process must therefore be treated non-linearly. The second characteristic is the distribution of the runoff generated in time to make up the shape of the storm hydrograph. In this case, hydraulic theory suggests that flow velocities should vary non-linearly with discharge rate, but many hydrograph estimation procedures make the assumption that the time distribution does not change very much with the nature of the storm, i.e. that the runoff routing process can be treated linearly.

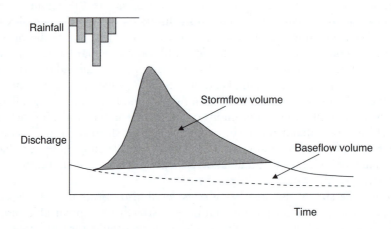

Fig. 12.1 Definition of stormflow and baseflow.

This is an assumption that seems to work quite well in practice, the reasons why will be discussed below.

There is now an enormous variety of catchment models available, ranging from the very simple to complex distributed models that make predictions in space as well as time. Two points are worth making before going on to consider the different types of models that might be used in practice. The first is that even the most complex model will still be only an approximate representation of the concepts of runoff generation described earlier in Section 1.6. The predictions will necessarily therefore be to some degree approximate or uncertain and it will be worthwhile considering carefully whether the assumptions of a particular modelling approach are appropriate for a particular application. The second is that the predictions of a model can only be as accurate as the input data used to drive the model. Poor input data will result in poor predictions. In some types of application it is possible to compensate for data deficiencies (see e.g. Section 12.4) but it is also worthwhile considering carefully the possible errors in the input data when using a model. A much more complete discussion of catchment models is provided by Beven (2001) while details of some specific models are given in the books edited by Singh (1995) and Singh and Frevert (2002a,b).

12.2 A simple catchment model

In practice, the choice of a method of modelling catchment response might depend on the nature of the application. There are very many applications, for example, that could not justify the expense of applying a complex distributed catchment model. This is nowadays not so much a matter of the expense of running the model, but rather the cost of the time necessary to assemble the data sets needed (if they exist at all). Thus simpler methods may often still be useful. The simplest methods of predicting runoff aim to predict only the hydrograph peak discharge. These derive from the so-called '*rational method*', which is also the oldest form of catchment model, dating back to the work of Thomas Mulvaney in Ireland and Emil Kuiching in the USA in the nineteenth century. A modern implementation of the rational method, still used in the UK, is the ADAS345 method, so called as it is set out in Report 345 produced by the Agricultural Development and Advisory Service (ADAS) for the then UK Ministry of Agriculture, Fisheries and Food (MAFF, 1981), now Defra. The method is intended to predict the peak discharge from an area underlain by field drains and was based on an analysis of the response of small rural catchments. It is only recommended for use in small catchment areas less than 30 ha.

The ADAS345 model is the simple equation

$$Q = S_T FA \qquad (12.1)$$

where Q is the peak flow in $L s^{-1}$; S_T is the soil type factor, which ranges between 0.1 for a very permeable soil to 1.3 for an impermeable soil; F is a factor which is a function of catchment average slope, maximum drainage length and average annual rainfall; and A is the area of the catchment being drained in hectares.

Guidance on the values of the above variables is given in the ADAS report, together with a nomograph that can be used to estimate the flow. The return period of the peak flow varies with the coefficient F. The ADAS345 model is based on empirical analysis rather than hydrological theory and, while the results would be expected to be rather uncertain in any particular application, effectively this provides a simple and low cost but accepted convention for the estimation of the runoff expected from a small area, perhaps a potential development site in a previously undeveloped rural area. In many applications, more sophisticated catchment models will be justified that attempt to predict both the amount of rainfall that becomes runoff and the timing of the hydrograph. However, it is worth noting that even the most complex hydrological models that are described later in this chapter depend, to some greater or lesser extent, on empirical representations of hydrological processes.

12.3 Estimating the proportion of rainfall equivalent to the stormflow hydrograph

At first sight this would appear to be a simple problem. A rainfall event occurs, the stream responds, discharge increases during the '*rising limb*' to a peak and then falls during the '*falling limb*' or '*recession*' before the next event. This is the storm hydrograph shown in Fig. 12.1. By calculating the volume of rainfall input, and the volume represented by the stormflow, it should be easy to calculate that part of the rainfall represented in the stormflow hydrograph, otherwise known as the '*effective rainfall*'.[1] The proportion of the rainfall represented by the stormflow is also known as the event '*runoff coefficient*'.

Unfortunately it is not that simple because, except in some arid and semi-arid environments, where discharges are often zero between events, the storm discharges from individual events are not easily separated. The rising limb from one event always starts somewhere on the falling limb from the previous event. Thus, 'hydrograph separation' requires distinguishing between what will be called *stormflow* and what will be called *baseflow* (Fig. 12.1). The only really objective way to estimate the volume of discharge from any single event would be to try to estimate what would have happened if no new events had occurred (as suggested by the dotted line in Fig. 12.1; this was tried by Reed *et al.*, 1975). But this will then include baseflow discharges that could have long timescales, much longer than what might be called '*stormflow*'. Thus, in the past, many pragmatic methods of estimating the stormflow component in a hydrograph have been proposed (such as the solid separation line in Fig. 12.1) but it is important to remember that they are all rather arbitrary; in general there is no easily separated 'stormflow' component, particularly where much of the stormflow may be water displaced from pre-event storage by the event inputs so that the stormflow is *not* the same water as that which fell on the catchment during the storm (see discussion of runoff processes in Sections 1.2 and the hydrograph separations based on tracer concentrations in Section 11.3).

Having said that, this type of hydrograph separation has been rather important in the history of rainfall–runoff analysis because of the way it allowed estimates of input (the effective rainfall) and output (the stormflow) to be matched. Effectively, after baseflow separation has been carried out (by whatever method), the volume of effective rainfall (and therefore the storm runoff coefficient) required is known precisely. Mass balance is then assured in the analysis, even if not all the water is accounted for. Having matched the input and output volumes for many events for a catchment, the way in which the effective rainfall varies between storms, and the way in which the flow processes modify the time distribution of the hydrograph could be studied. It turned out that understanding how much of the rainfall became effective rainfall was rather difficult, but that, for many catchments, the time distribution stayed fairly constant (see the next section). In fact, we do not need to invoke any process interpretation for the generation of the 'stormflow'. We only need to insist that the stormflow is defined in a consistent way since the matching effective rainfall will be conditional on how stormflow is defined.

The simplest and perhaps most widely used method of determining the effective rainfall to match a given stormflow is the *phi-index* approach. Phi (the Greek letter Φ) is a coefficient that is used as a threshold on each time increment of rainfall in a storm. Any rainfall above the threshold is assumed to be effective rainfall (a modification is to allow some initial deficit to be satisfied before implementing the threshold but this introduces an additional variable parameter, since the initial loss might be different for wet and dry antecedent conditions). The value of Φ is chosen such that the total volume of effective rainfall is equal to the total volume of stormflow. In this way, the greatest contribution to effective rainfall is given by time increments with high rainfall intensities (Fig. 12.2a). The value of Φ is chosen on an event by event basis in analysis. The difficulty then remains to choose a value for Φ in predicting the effective rainfall in predicting a new event.

12.4 Estimating the time distribution of runoff

12.4.1 The time–area method

The time–area method of obtaining runoff or discharge from rainfall can be considered as the first attempt to create a hydrological model that is distributed in space. The concept was (possibly) first introduced by Imbeaux (1892) in the Durance River in France, and later used by Ross, Zoch and Clark in the USA and Richards in the UK (see Beven, 2001). The method attempts to divide a catchment into areas based

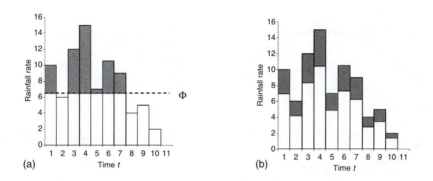

Fig. 12.2 Different ways of assigning a volume of effective rainfall based on total rainfall hyetograph: (a) phi index method; (b) proportional rainfall method. The volume of effective rainfall is the same in both cases.

on how long it will take runoff generated on that area to reach the catchment outlet. Once this areal discretisation of the catchment has been made, estimates of the runoff generated on each fractional part of the catchment can be delayed by the appropriate time delay and then added to calculate the storm hydrograph. The normal assumption is that the delays do not change over time. This implies that this method of routing runoff is linear, i.e. that if the estimate of the runoff generation is doubled, the result will be a simple doubling of the output hydrograph (see Sections 12.5 and 12.6 below).

Thus the storm hydrograph, Q_T is the sum of flow-contributions from subdivisions of the catchment defined by time contours (called *isochrones*). The method is illustrated in Fig. 12.3a. The flow from each contributing area bounded by two isochrones $(T - \Delta T, T)$ is obtained from the product of the mean runoff generated on that area (i) from time $T - \Delta T$ to time T and the area (ΔA). Thus Q_4, the flow at X at time $t = 4$ h is given by:

$$Q(t) = i_4 A_1 + i_3 A_2 + i_2 A_3 + i_1 A_4 \tag{12.2}$$

i.e.

$$Q(T) = \sum_{k=1}^{NT} i_{(NT+1-k)} A_k \tag{12.3}$$

where NT is the number of time steps of length ΔT equivalent to the time of concentration T_c and i_k is the effective rainfall generated on a fractional area A_k. Note that different sets of units can be used for this calculation. If runoff generation is expressed as a depth per unit time step and area as a fraction, then the predicted flows will also have units of depth per unit time (e.g. mm over the time step). If runoff generation is expressed as m s^{-1} (equivalent to mm h^{-1}/(1000×3600)) and area in m^2, then the predicted flows will have units of m^3 s^{-1}.

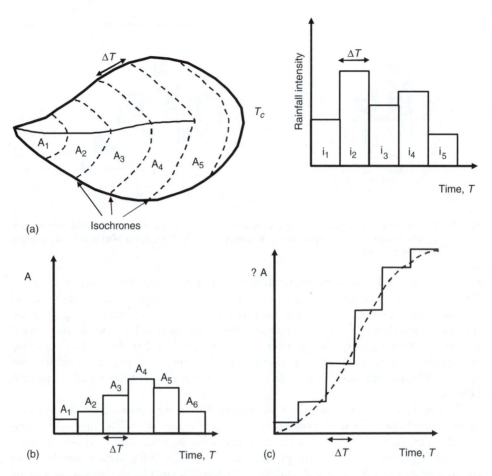

Fig. 12.3 The time–area method: (a) division of storm rainfalls and catchment area into areas with equal travel time ΔT to the outlet. (b) time–area–concentration curve. (b) cumulative time–area curve.

The whole catchment is taken to be contributing to the flow after T equals T_c, which is called the *time of concentration* of the catchment. Hence, in deriving a flood peak for design purposes, a design storm with a critical sequence of runoff generation intensities can be used for the maximum intensities applied to the contributing areas of the catchment that have most rapid runoff.

To fix the isochrones considerable knowledge of the catchment is required, so that the times of overland flow and flow in the river channels may be determined. This is most easily achieved by deriving a representation of the time–area histogram from an analysis of storm data. The simple discrete form of the time–area concept can be generalised by making ΔT very small and considering increases in the contributing area to be continuous with increasing time. Thus, in Fig. 12.3c, the plot of catchment area against time is shown as a dashed line and this is known as the *time–area curve*. Its limits are the total area of the catchment and the time of concentration. For any value of T, the corresponding area A gives the maximum flow at the river outfall caused by a rainfall of duration T. The derivative of the time–area curve shown in Fig. 12.3b gives the rate of increase in contributing area with time, and is called the *time–area–concentration curve*, since the length of the time base is equal to the time of concentration of the catchment. It is a form of linear *impulse-response function* or *transfer function* for the catchment. A generalisation of the time–area transfer function that has been very widely used in hydrological analysis is the *unit hydrograph*.

12.5 The unit hydrograph

A major step forward in hydrological analysis was the concept of the *'unitgraph'* introduced by the American engineer Leroy K. Sherman (1932). Sherman (1942) later called this the *unit hydrograph*, which is the name that is normally used today in hydrology. The unit hydrograph was actually a rather neat idea since it was a way of generalising the time–area curve in a way that did not require the rather difficult task of estimating which parts of the catchment actually fell within each time increment. Sherman defined the unit hydrograph as the hydrograph of stormflow resulting from effective rainfall falling in a unit of time such as 1 h or 1 day and produced uniformly in space and time over the total catchment area. By averaging the runoff generation as a uniform effective rainfall over the catchment, and generalising the transfer function as the unit hydrograph, a useful predictive tool was provided.

In practice, a T-hour unit hydrograph is defined as resulting from a unit depth of effective rainfall falling in T h over the catchment. The magnitude chosen for T depends on the size of the catchment and the response time to major rainfall events. The standard depth of effective rainfall was taken by Sherman to be 1 in, but now 1 mm is more normally used. The definition of an effective rainfall–runoff relationship is shown in Fig. 12.4a, with 1 mm of uniform effective rainfall occurring over a time T producing the hydrograph labelled TUH. The units of the ordinates of the T-hour unit hydrograph are $m^3 s^{-1}$ per mm of rain. The volume of stormflow is given by the area under the hydrograph and is equivalent to the 1-mm depth of effective rainfall over the whole catchment area.

The unit hydrograph method makes several assumptions that give it simple properties that make it easy to apply.

(a) There is a direct proportional relationship between the effective rainfall and the stormflow. Thus in Fig. 12.4b, two units of effective rainfall falling in time T produce a stormflow hydrograph that has its ordinates twice the TUH ordinates, and similarly for any proportional value. For example, if 6.5 mm of effective rainfall fall on a catchment area in T h, then the hydrograph resulting from that effective rainfall is obtained by multiplying the ordinates of the TUH by 6.5.

(b) A second simple property, that of superposition, is demonstrated in Fig. 12.4c. If two successive amounts of effective rainfall, R_1 and R_2, each fall in T h, then the stormflow hydrograph produced is the sum of the component hydrographs due to R_1 and R_2 separately (the latter being lagged by T h on the former). This property extends to any number of effective rainfall blocks in succession. Once a TUH is available, it can be used to estimate design flood hydrographs from design storms.

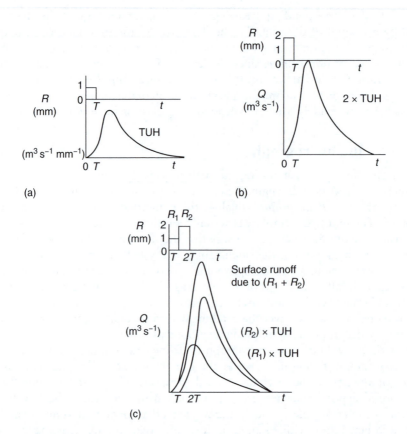

Fig. 12.4 The unit hydrograph concept. (a) The *T* hour unit hydrograph (TUH) arising from 1 mm of effective rainfall in a time step of *T* h. (b) The linearity assumption: 2 mm of effective rainfall results in a time step of length *T* in a predicted hydrograph of 2 × the TUH. (c) The superposition principle: linear addition of the contributions of effective rainfall in successive time steps.

(c) A third property of the TUH assumes that the effective rainfall–stormflow relationship does not change with time, i.e. that the same TUH always occurs whenever the unit of effective rainfall in *T* h is applied. Using this assumption of invariance, once a TUH has been derived for a catchment area, it could be used to represent the response of the catchment whenever required.

The assumptions of the unit hydrograph method must be borne in mind when applying it to natural catchments. In relating total rainfall to stormflow, the amount of effective rainfall will depend on the state of the catchment before the storm event. If the ground is saturated or the catchment is impervious, then a high proportion of the rain would be expected to be effective in producing stormflow. On unsaturated ground, however, the soil will have a certain capacity to take up rainfall before stormflow is generation. Only when any storage deficits have been made up and the rainfall becomes fully effective will extra rainfall in the same time period produce proportionally more runoff. The first assumption of proportionality of response to effective rainfall conflicts with the observed non-proportional behaviour of river flow. In a second period of effective rain, the response of a catchment will be dependent on the effects of the first input, although the second assumption makes the two component contributions independent (Fig. 12.4c). The third assumption of time invariance implies that, whatever the state of the catchment, a unit of effective rainfall in T h will always produce the same TUH. However, the response hydrograph of a catchment might be expected to vary according to the season: the same amount of effective rainfall will be longer in appearing as stormflow in the summer season when vegetation is at its maximum development and the hydraulic behaviour of the catchment will be 'rougher'. In those countries with no marked seasonal rainfall or temperature differences and constant catchment conditions throughout the year, then the unit hydrograph would be a much more consistent tool to use in deriving stormflow from effective rainfall.

Another weakness of the unit hydrograph method is the assumption that the effective rainfall is produced uniformly both in the time step T and over the area of the catchment. The areal distribution of rainfall within a storm is very rarely uniform, and, as discussed in Section 1.2, we expect stormflow contributing areas to expand and contract in space as the catchment wets and dries. For small or medium-sized catchments (say up to 100 km^2), a significant rainfall event may extend over the whole area and, if the catchment is homogeneous in composition, a fairly even distribution of effective rainfall may be produced. More usually, storms causing large river discharges vary in intensity in space as well as in time, and the consequent response is often affected by storm movement over the catchment area. However, rainfall variations are damped by the integrating action of the catchment, so the assumption of uniformity of effective rainfall over a selected period T is less serious than might be supposed at first. The effect of variable rainfall intensities in time can be reduced by making T smaller. The effect of variable rainfall intensities in space can be reduced by developing TUH for different sub-catchments where the data are available to do so or developing TUHs for storm type of different origins showing different characteristic patterns.

Despite these conceptual limitations, the unit hydrograph method has the advantage of great simplicity. Once a unit hydrograph of specified duration T has been derived for a catchment area (and/or specific storm type), then for any sequence of effective rainfalls in periods of T, an estimate of the stormflow can be obtained by adopting the assumptions and applying the simple properties outlined above. The technique has been adopted and used worldwide over many years. It is therefore very useful for the hydrologist to understand the unit hydrograph concept and its simplifying assumptions, since it is still the basis for many analysis and estimation procedures in hydrology. Previous editions of *Hydrology in Practice* have dealt with aspects of unit hydrograph theory in some detail, including the derivation of the TUH from simple storms and multiple storms, changing the time step T by using the S-curve technique, and the *instantaneous unit hydrograph* (IUH) that is the (theoretical) response from an instantaneous unit input of effective rainfall (Nash, 1957). For the IUH, the summation of equation 12.3 is replaced by an integral and it is no longer necessary to assume that the effective rainfall is uniform over long time steps (though it is still necessary to assume that it is uniform in space). Discharge at any time t is then predicted as:

$$Q(t) = \int_0^t R(t-\tau).H(\tau)\, d\tau \qquad (12.4)$$

This is called a *convolution integral*, analogous to the discrete summation of (12.3), in which the time series of effective rainfalls $R(t)$ is convolved with the transfer function that is the IUH, $H(\tau)$, where time t is now treated as a continuous variable rather than a series of discrete time steps. This was not easy to evaluate for any arbitrary form of the IUH in the days before digital computers were widely available; hence the discrete TUH remained popular. Now, however, it is relatively easy to program a numerical solution to the convolution integral (12.4).

12.5.1 The unit hydrograph as the routing component of a catchment model

We have already seen how the unit hydrograph is essentially a linear model for routing runoff generation (or effective rainfall) to the catchment outlet. In the late 1950s, through the work of the Irish engineering hydrologists Jim Dooge and Eamonn Nash, it was realised that this allowed a link to be made to general linear storage models (Dooge and O'Kane, 2003, provide a detailed summary of this work). This allowed the unit hydrograph to be represented in terms of some simple functional forms with just a small number of parameters. Nash (1960), for example, proposed a model of the unit hydrograph that consisted of n linear stores in series, each of which has a mean residence time of K. The equation for a single storage element is then that outflow

$$Q = S/K \qquad (12.5)$$

where S is storage and K is a time constant. This has an impulse response or unit hydrograph of the form of an instantaneous rise followed by an exponential decline, or

$$H(\tau) = \frac{1}{K} \exp(-\tau/K)$$

and for n stores in series

$$H(\tau) = \left(\frac{\tau}{K}\right)^{n-1} \frac{\exp(-\tau/K)}{K\Gamma(n)} \qquad (12.6)$$

This is the equation of a gamma distribution, with $\Gamma(n)$ being the tabulated gamma function. It has the advantage that it can take on a variety of forms depending on the values of n and K (Fig. 12.5). In the general case, n need not be an integer value. Jim Dooge (1959) also showed how the unit hydrograph could be represented by a general class of linear models with combinations of fast and slow storages, how a constant time delay could also be incorporated into the model in cases where the rising limb did not start immediately, and how the same concepts could be used for routing flood waves in large rivers (see also Dooge and O'Kane (2003) and the discussion of the data-based mechanistic modelling concepts in Section 12.8.3 below).

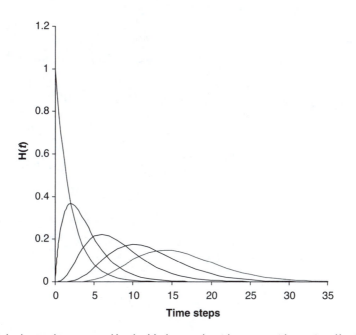

Fig. 12.5 Unit hydrographs generated by the Nash cascade with mean residence time $K = 2$ time steps in each store and number of stores $n = 1,2,4,6,10$.

12.5.2 The geomorphological unit hydrograph and channel-width function

A more recent interesting development of the unit hydrograph concepts has been to try to relate the catchment response in the form of the IUH to the characteristics of the channel network. There have been two ways of doing this, one based on characterising the network in the form of statistics of stream links of different order, the other by representing the network as a distance histogram or '*channel network width function*'.

The geomorphological instantaneous unit hydrograph (GIUH) depends heavily on the concept of stream order introduced by Robert Horton in his classic 1945 paper on the development of catchment geomorphology. The ordering system most commonly used is now that of Arthur Strahler (1956; Fig. 12.6). In Strahler's system, first-order streams are those segments of the channel network that originate in stream sources; second-order segments occur below the junction between two first-order segments; third-order segments below the junction of two second-order segments, and so on. Junctions with lower order streams do not change the order of a segment. Horton noticed that in a larger order network of channel segments the numbers, lengths and catchment areas of different order streams had ratios that were approximately constant over a range of orders, and proposed that these ratios could be considered as laws at the catchment scale (in fact, the very nature of a space-filling dendritic (tree-like) network ensures that the ratios will be approximately constant; see the detailed discussion of the properties and development of channel networks in Rodriguez-Iturbe and Rinaldo, 1997).

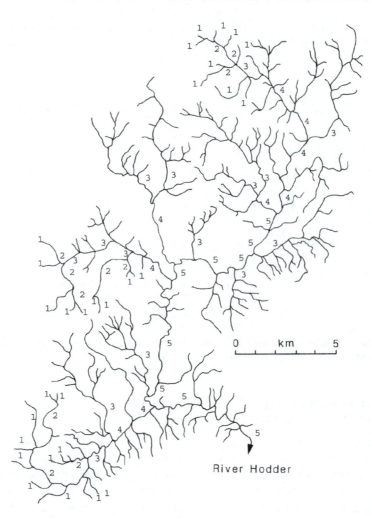

Fig. 12.6 Strahler ordering system for channel links in the $\Omega = 5$ River Hodder stream network. The River Hodder is a tributary of the River Ribble in northern England, draining the southern slopes of the Forest of Bowland. Not all network links are labelled for clarity. Stream order increases when two streams of equal lower order join.

The GIUH concept, as originally introduced by Ignacio Rodriguez-Iturbe and Juan Valdes (1979), makes use of these laws and a probabilistic argument that a drop of effective rainfall could be produced randomly anywhere in the catchment. It will then flow to the catchment outlet in the segments of the channel network. This can be viewed as a form of transition probability matrix between states in the system. In any time step there is a probability that a drop of effective rainfall on a hillslope (zero-order state) will reach a first or higher order stream; a probability that a drop in a first-order stream will reach a second or higher order stream; and so on for all the possible transitions until a drop reaches the catchment outlet which acts as a sink or trapping state. The model is completed by assuming that the distribution of waiting times in each channel state is an exponential distribution with a single parameter that depends on the order of the segment. This is equivalent to assuming that each state acts as a linear store as in the Nash model (Chutha and Dooge, 1990); more complex distributions could be used, but this will both increase the number of parameters that need to be defined and the difficulty of the mathematics (see Gupta et al., 1980). It can then be further assumed that the mean residence time in each channel state is related to the mean length of segments of that order and a constant routing velocity (see Section 14.4 for a discussion of the difference between routing velocity and actual flow velocity of the water).

It was found that the resulting unit hydrograph expressions could be well represented by approximations for the peak, q_p, and time to peak, t_p as:

$$q_p = 0.36R_L^{0.43} v L_\Omega^{-1}$$
$$t_p = 1.58\left[\frac{R_B}{R_A}\right]^{0.55} R_L^{-0.38} L_\Omega v^{-1} \tag{12.7}$$

where v is the routing velocity, L_Ω is the length of the stream of highest order Ω, R_L is the ratio of mean stream lengths in different orders, R_B is the ratio of stream numbers in different orders, and R_A is the ratio of mean stream catchment areas in different orders.

Renzo Rosso (1984) also shows how these functions can be used to define a full unit hydrograph in the form of a gamma distribution (similar to that in the Nash cascade of (12.6) and Fig. 12.5) so that

$$H(\tau) = \left[\frac{\tau}{k}\right]^{\alpha-1} \frac{\exp(-\tau/k)}{k\Gamma(\alpha)} \tag{12.8}$$

where the two parameters, α and k are given by

$$\alpha = 3.29\left[\frac{R_B}{R_A}\right]^{0.78} R_L^{0.07}$$
$$k = 0.70\left[\frac{R_A}{R_B R_L}\right]^{0.48} L_\Omega v^{-1} \tag{12.9}$$

The GIUH provides a unit hydrograph that, with some simple assumptions, is related directly to the form of the channel network as generalised by Horton's 'laws' of stream numbers, lengths and areas. It has been used quite widely and, with some adjustments of the velocity parameter by calibration, can produce good predictions given a good estimate of the effective rainfall.

It is not, however, necessary to simplify the channel network in this generalised way. In developing the GIUH Rodriguez-Iturbe and Valdez (1979) did so for two reasons. The first was to simplify the mathematical treatment in moving from one stream order to another; the second was to continue Horton's attempt to provide a general synthesis of catchment hydrology and geomorphology. In doing so, however, some

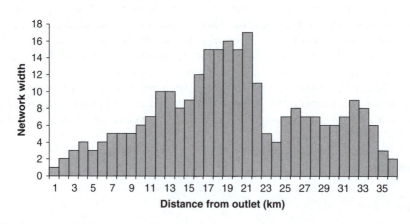

Fig. 12.7 Network width function for the River Hodder network shown in Fig. 12.6.

information about the detailed form of the channel network is lost, in fact unnecessarily, since the actual channel network (as represented by the blue lines on a map) is generally one of the very easiest pieces of information to obtain for any catchment. By making the same constant routing velocity assumption, travel times in the network can then be determined directly from mapping the pattern of distances to the outlet along the channels. This is similar to the time delay histogram of Section 12.4.1 above, but now only considering the channel network. At any given distance, there may be one or more channel segments (see Fig. 12.6) and the histogram of numbers of channel segments against distance is known as the channel network width function (Fig. 12.7). This has been used directly as a linear flow routing algorithm (e.g. Kirkby, 1976; Beven, 1979) and with the addition of a dispersive component for each network link as a simple routing algorithm (e.g. Franchini and O'Connell, 1996), including at continental scale (Naden *et al.*, 1999). It has been incorporated into several catchment rainfall–runoff models, including Topmodel (see Section 12.8.4). Other ways of deriving routing models from network geomorphology have also been considered (see, e.g. Shamseldin and Nash, 1998; Cudenec *et al.*, 2004, Nourani *et al.*, 2009). Routing of flood waves along the main channel of a large catchment is considered later in Chapter 14.

12.5.3 A review of unit hydrograph methods

The unit hydrograph has proven to be a very useful technique in hydrological analysis and prediction. It has been widely used since its introduction by Sherman in the 1930s. As noted earlier in this chapter, it actually requires two components: one to determine how much rainfall becomes effective rainfall, and one (the unit hydrograph itself) to distribute that effective rainfall in time to predict the resulting hydrograph. In the 1930s, the combination of Sherman's unitgraph with the Horton infiltration excess concept of stormflow generation to predict the effective rainfall, meant that a practical rainfall–runoff model at the catchment scale was available. In fact it was so useful that later work often concentrated on improvements to ways of deriving the unit hydrograph, or relating the unit hydrograph to catchment characteristics, or improving the prediction of effective rainfalls, than on querying the rather strong assumptions on which it is based. The method is still used extensively in the UK as part of the FEH flood hydrograph method (see Section 13.5).

There is no doubt that, in some catchments, these assumptions will not be valid. The study of Minshall (1960), for example, is well known for showing that, in a small catchment, the derived unit hydrograph might change with the magnitude of the peak flow (see Beven, 2001). If this is the case, then the assumptions of linearity, superposition and invariance are not tenable. Such an analysis will be dependent on the rules that have been used to separate stormflow from baseflow. The derived unit hydrograph will vary, depending on what technique of hydrograph separation has been used. Many studies have suggested that it is difficult to carry out such a separation objectively, even if once a set of rules is chosen they are applied consistently. A general framework for relating both the unit hydrograph response and the actual travel times of water in a catchment in a way that properly reflects the time variability of the responses has been provided by Botter et al. (2010).

The main problem with the unit hydrograph approach, however, has always been in calculating the effective rainfall for an event, both in analysis and prediction. As noted above, in analysis this problem is minimised by setting the volume of effective rainfall to the volume of stormflow after hydrograph separation to impose a mass balance. In prediction, it is somewhat more problematic. We have the measured rainfalls, we know that the effective rainfall should depend on both the antecedent state of the catchment and the pattern of rainfalls, but we have not had very good ways of predicting just how much of the rainfall becomes effective rainfall in any particular event.

In fact, it would be better to avoid hydrograph separation and effective rainfall calculations completely by trying to predict the whole hydrograph. Young and Beven (1994), e.g. have shown how a general discrete time linear transfer function model could be used to represent both the storm and baseflow components of the hydrograph, in combination with a non-linear function on the input that would provide an estimate of the total effective rainfall for both components. This provided a complete catchment model with a structure inferred directly from the data, what Peter Young calls a *data-based mechanistic model* (see Section 12.8.3). This approach has since been developed further for both rainfall–runoff simulation and flood forecasting (Young, 2001, 2002, 2003, 2009; Romanowicz et al., 2008). Unit hydrographs in the form of linear transfer functions also provide the routing components for many different conceptual catchment rainfall–runoff models, as will be seen in the descriptions of some of the available models below. Routing of flood waves along the main channel of a large catchment is considered later in Chapter 14.

12.6 Conceptual catchment rainfall–runoff models

Unit hydrograph models of catchment response are examples of conceptual rainfall–runoff models. Although relatively simple, they reflect the two most important aspects of the hydrograph: a loss function to account for the fact that not all the rainfall in an event becomes streamflow, and a time distribution to account for the delays between runoff generation and the hydrograph measured at a catchment outlet. Rainfall–runoff modelling has become one of the most important tools available to the hydrologist. Many different models have been developed in the last 50 years, once digital computers became available to researchers and, later, practitioners. There are now so many, in fact, that a number of texts have appeared to provide detailed information about modelling techniques and particular models (see Beven, 2001; Abbott and Resfsgaard, 1996; Singh and Frevert, 2002a,b; Wagener *et al.*, 2004). Here we will take an overview of the types of models used in practice and give some examples of their use.

The best way to classify hydrological models is by purpose. There are *distributed* models for predicting the spatial pattern of flow processes, there are *lumped* models for predicting the hydrograph response without worrying too much about what is happening in space, and there are *real-time forecasting* models that are used for flood forecasting or drought management within a *data assimilation* framework with the aim of improving the predictions as a flood or drought event unfolds. Models are also sometimes classified as to whether they are *deterministic* or *stochastic*. A deterministic model run has only one possible outcome; a stochastic model takes account of uncertainties in the representation of the system and might have a distribution of outcomes. Today this distinction has become somewhat blurred, however, since deterministic models can be run many times with randomly chosen inputs or parameter values in *Monte Carlo simulation* to produce distributions of outcomes (see Section 12.7 below).

The first widely recognised catchment model on a digital computer was the Stanford watershed model, developed by Norman Crawford during his Ph.D. with Ray Linsley at Stanford University in the early 1960s. This was a lumped, deterministic model that was later developed commercially and applied worldwide by their company called Hydrocomp. Versions of the model are still being used and can be freely downloaded as part of the US Environmental Protection Agency (EPA) software suite as HSPF (hydrologic simulation package fortran).[2] Recent versions include components for non-point pollutant sources, including transport and modification in river channels. It has been extensively used in the EPA's Chesapeak Bay programme.

The Stanford watershed model set the framework for many of the rainfall–runoff models that followed: it consisted of a set of storage elements linked in series and in parallel, to represent different parts of the catchment system (surface runoff, the upper soil zone store, lower soil zone store, a ground water store, channel routing, etc. The use of digital computers meant that the equations describing the operation of each store did not have to be simple and did not have to be linear. The computations could be easily programmed in a high-level language such as Fortran. The Stanford watershed model had some 25 parameter values (34 if a snowmelt component was included). Some could be estimated on the basis of catchment characteristics, others had to be calibrated by fitting observed discharges.

There are many models of this type. They vary in the numbers of stores and the equations used for each store. They all, however, have a similar problem in applications. Many of the equations used will contain parameters that must be specified before a run of the model can be made. But the parameter values are not generally easily determined on the basis of catchment characteristics. Therefore, it is often the case that the model user tries to *calibrate* the parameters by adjusting the values until a good fit is obtained to an observed discharge time series. In the early days of using such models, this was done by trial and error, with the results assessed by eye. More recently, automatic optimisation methods have been developed, with goodness of fit being assessed by one or more numerical performance measures, of which perhaps the most widely used in hydrology is the Nash and Sutcliffe (1970) model efficiency measure, *NSE*, defined by:

$$NSE = 1 - \frac{\sum (q_{obs} - q_{sim})^2}{\sum (q_{obs} - \overline{q_{obs}})^2} = 1 - \frac{\sigma_e^2}{\sigma_o^2} \qquad (12.10)$$

where q_{obs} is observed discharge and q_{sim} is simulated discharge at each time step, $\overline{q}_{obs}$ is the mean observed discharge, the summations are over all time steps, σ_e^2 is the variance of the model residuals, and σ_o^2 is the variance of the observations. This measure has a range from – infinity to 1. When the value is zero, the model will have no more predictive power than a simple 'model' that is the mean of the observations. Negative values mean that the fit is worse than this. When the value is 1, the fit is perfect.

Although widely used, the *NSE* is not an ideal measure (e.g. Beran, 1999; McCuen *et al.*, 2006; Schaefli and Gupta, 2007). In particular, timing errors and, more generally, series of residuals showing temporal autocorrelation will reduce the value even though visually the shape of the hydrograph can appear to be good. A feature of the Nash–Sutcliffe efficiency measure is that the value is determined relative to the variance of the observed discharges. This means that, if a catchment has a relatively low observed variance (e.g. if it is dominated by a slowly changing baseflow component) the model fit must be very much better in absolute terms than if the catchment is flashy with high observed variance, to get an efficiency close to 1.

It is obvious that this form of calibration can only be carried out if there is an observed discharge record. It also soon became clear that, the simpler a model, and the fewer the parameters that had to be calibrated, the easier it was to fit the model. Rather early in the history of hydrological modelling Dave Dawdy and Terrence O'Donnell (1965) produced a model that was much simpler than the Stanford Watershed Model with a view to making automatic calibration easier, while Mike Kirkby (1975) suggested that the information content of a rainfall–discharge record was only sufficient to calibrate perhaps five or six parameters. It is also worth thinking about why it is worth calibrating a model at gauged sites when we already have some information about the catchment response. As we will see in the next section, in real applications models are most needed to predict catchment responses where we do not have data (the ungauged basin problem), but then we cannot calibrate parameter values and must try to make use of values determined at gauged sites.

12.7 The application of rainfall–runoff models

A basic distinction in the applications of rainfall–runoff models is between applications at gauged sites and those at ungauged sites. We will assume that we have some measurements of inputs to a model and want to predict the catchment responses to those inputs. At gauged sites we can calibrate the parameters of a model against the observed discharges; at ungauged sites we need to estimate the parameter values in order to make predictions.

Why would we need a model at a gauged site when we already have observed discharges? There are a number of reasons:

(a) We may wish, particularly for research purposes, to test a particular representation of the processes controlling how a catchment response works before using a model structure for a wider range of applications. This is often called model *validation* (though the word validation suggests an element of 'truthfulness' that may not be justified, model *evaluation* is preferable). Such model evaluation should be done with care, it is not enough to simply calibrate a model and show that it gives good results after adjusting the parameters; it is important to show how it gives good results for an independent period of data without further adjustments.

(b) We may wish to use the model to extend the period of discharge measurements in some way. It is quite often the case in the UK (and widely elsewhere) that there are longer periods of rainfall available (at least for daily totals) than there are discharge measurements. After fitting the model to the available discharges, a longer period of discharge can be generated using the observed rainfalls. There is an implicit assumption in doing so that the catchment hydrological response has not changed during the additional period. The longer record could then be used, for example, to improve flood frequency estimates.

(c) We may wish to predict the effects of some future change to the catchment. This could be a climate change (there have been many studies of this type) in which case some scenarios of future changes to precipitation and evapotranspiration can be used to provide a new sequence of inputs to a model with parameters fitted to a period of observed discharges. Somewhat more difficult is to predict the impacts of land-use change, including urbanisation. This would require a change in the model parameters but it is not always clear how to relate changes in the catchment (which may affect only part of the catchment) to effective values of model parameters. See Chapter 19 for examples in which hydrological models have been applied in this way.

(d) We may wish to calibrate the parameters on many gauged catchments with a view to using the calibrated parameter values in a regionalisation exercise to estimate the parameters required to run the model for other ungauged catchments.

(e) We may wish to use a calibrated model in real-time flood forecasting applications with a view to providing timely flood warnings or control of flood defences.

An important issue in all types of applications is how to define the values of the model parameters. Can they be calibrated against some measured responses; can they be estimated on the basis of catchment characteristics alone; can they be measured directly in the field; or can they be estimated from some regionalisation exercise (as in point (d) above)? All of these approaches have been tried; all result in significant uncertainties in predicting the response of a catchment. This should not be unexpected. The processes controlling the form of the hydrograph in a catchment are very complex, and there is no reason why they should be necessarily well represented by a simple model construct. In addition, the inputs to a catchment are often not very well measured, and we should not expect a model to perform better than we can define the inputs. There is an increasing appreciation of such uncertainties and a recognition that model predictions should be provided with an estimate of the associated uncertainties (see Section 12.9 below).

To illustrate the use of different types of hydrological models, in what follows we will describe several different rainfall–runoff models that span this range of applications. The different models are representative of the different generic types. Examples of *lumped* models are the PDM (probability-distributed model) developed in the 1980s by Moore and Clarke (1981) at the Institute of Hydrology at Wallingford, UK and the DBM (data-based mechanistic) modelling approach of Young (2001, 2003). A number of the applications illustrated in later chapters of this book have used the PDM. An example of a model that is lumped for calculation purposes but where the predictions can be mapped back into space, is Topmodel (topography derived model) developed in the 1970s by Beven and Kirkby (1979) at the University of Leeds but widely used elsewhere. Examples of models that are *distributed* are the SHE model (Système Hydrologique Européen) of Abbott *et al.* (1986) and the InHM (Integrated Hydrologic Model) of VanderKwaak and Loague (2001).

There are also a variety of models available developed for more specialised uses, including models of hydraulic models for predicting flood inundation (see Chapter 14); groundwater systems (see Chapter 15); and for predicting the response of urban drainage systems (see Chapter 18).

12.8 Examples of rainfall–runoff models

12.8.1 The probability-distributed model (PDM)

The PDM, developed by Bob Moore and Robin Clarke (1981) at the Institute of Hydrology in Wallingford (now the Centre for Ecology and Hydrology, CEH), represents one of the earliest attempts to allow for the spatial heterogeneity of runoff generation (though it is worth noting that the Stanford Watershed Model also used an 'infiltration' function that effectively allowed for a range of infiltration capacities uniformly distributed throughout a catchment). In doing so, Moore and Clarke, recognised that, in any application to a real catchment, it would not be possible to go out and measure the heterogeneity of the soil characteristics and that the heterogeneity of runoff generation might be produced in different ways or by different mechanisms. Accordingly, to keep things simple, they suggested representing the local storage deficits that needed to be satisfied before fast runoff generation would occur in a catchment conceptually as a probability distribution. Different catchments might then be represented by different forms of distribution. The characteristics of the distribution would then need to be calibrated, but they also had the idea that the model could be formulated so as to simplify the calibration process (Moore, 1985, 2007). A form of the PDM model is implemented in the Imperial College rainfall–runoff modelling MATLAB toolbox (see Wagener *et al.*, 2004).

Essentially, the PDM assumes that at any point and at any time step, stormflow will be produced whenever any local storage deficit is filled so that

$$q(t) = r(t) - D(t) - e_a(t) \tag{12.11}$$

where $r(t)$ is the rainfall at time t, $e_a(t)$ is the actual evapotranspiration, and $D(t)$ is the local storage deficit at time t. It is expected that the maximum possible local storage capacity (when dry) will vary throughout a catchment area while local deficits

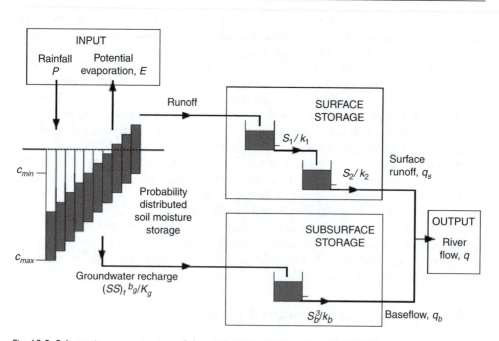

Fig. 12.8 Schematic representation of the probability-distributed model (PDM).

will vary as the catchment wets and dries. The approach taken by Moore and Clarke was to assume that the spatial variation takes the form of a statistical distribution function. They gave equations for the exponential, gamma and Weibull distributions, for all of which there is no maximum capacity, while Moore (1999) suggests that, from experience in the UK, a suitable function is the Pareto distribution, which does have a maximum. This has a cumulative density function of the form:

$$F(c) = 1 - \left(1 - \frac{c}{c_{max}}\right)^b; \qquad 0 < c < c_{max} \tag{12.12}$$

where c is the local storage capacity and c_{max} is the maximum storage capacity in the catchment (Fig. 12.8). For $b = 1$ the stores are uniformly distributed from zero to c_{max}; for $b = 0$, the storage capacity is the same everywhere in the catchment. This form allows the changing contributing area for fast runoff as the catchment wets and dries to be derived analytically (Moore, 2007).

At any time step, incoming rainfall will saturate part of the catchment and produce stormflow. The remaining net rainfall will decrease the available local storage deficit over the unsaturated part of the catchment, while any actual evapotranspiration will increase the deficits everywhere in the catchment. Integrating the local changes at each time step produces an estimate of the contributing area, stormflow and change in total storage. This allows the runoff generation in the model to be a non-linear function of the rainfall and antecedent wetness of the catchment.

The model is completed by a routing component. This has been implemented in a variety of different ways. In the original paper the runoff generated from the distribution of stores was routed into an additional linear store. Later, a constant split into a fast flow pathway and a slow flow pathway was used. The two pathways were both treated as linear stores but with different mean residence time parameters. Later still, input into the slow pathway has been calculated as a function of bulk storage in the distribution of stores, while the stormflow has been routed directly into the fast pathway (Fig. 12.8).

A recent development of the PDM model has involved developing relationships between the model parameters and gridded soil and topographic information. Maximum soil storage capacity in a grid square is related to regional maximum gradient and storage capacities while the exponent b in (12.16) is treated as a function of the mean slope angle in a grid square (Bell and Moore, 1998; Cole and Moore, 2008). The analytical relationships of the PDM still hold at the grid scale level. The resulting grid to grid (G2G) model has been applied to predict runoff across the whole of the UK.

There have been some other models of this type, using functional forms to represent the non-linearity of stormflow generation based on an interpretation of spatially distributed storage deficits. Perhaps the most widely used has been the Xinanjiang/Arno/Variable Infiltration Capacity (VIC) model. This was originally proposed by Zhao in China in the 1970s, then used for flood forecasting on the River Arno by Todini in Italy, and later adopted for use as a representation of land surface hydrology in large-scale grid elements of general circulation models (see Beven, 2001). Topmodel (see Section 12.8.3) is also based on predicting a pattern of storage deficits in a catchment, but in that case the pattern can be mapped back into the catchment, so this will be discussed later in this chapter.

12.8.2 Data-based mechanistic (DBM) modelling

The DBM modelling concepts were developed by Peter Young and his collaborators as a use of his CAPTAIN time series analysis program (which is now implemented as a MATLAB toolbox). CAPTAIN provides routines for calibrating linear transfer functions to input–output data. It can be used both for rainfall–runoff modelling and routing flood waves from upstream to downstream gauging stations (see Chapter 14). The idea is to let the data give an indication of the structure of the transfer function needed by fitting many different linear model structures and choosing the structure that gives the best compromise between goodness-of-fit and simplicity. The family of models from which the structures are chosen is essentially one or more linear storages, linked in either series or parallel, and with a suitable time delay. A general linear transfer function model can be written in the form:

$$Q_t = a_1 Q_{t-1} + a_2 Q_{t-2} + \ldots a_n Q_{t-n} + b_o U_{t-1-d} + b_2 U_{t-m-d}$$
$$+ b_2 U_{t-2-d} + \ldots b_m U_{t-m-d} \tag{12.13}$$

where U_t is the input at time t (here effective rainfall), Q_t is the predicted variable at time t (here discharge), d is a time delay, and the a and b coefficients define the model. Introducing the backwards difference operator (z^{-1}), which is defined by

$Q_{t-1} = z^{-1}Q_t$, then the transfer function can be written more simply as:

$$Q_t = \frac{b_o + b_1 z^{-1} + b_2 z^{-2} + \ldots b_m z^{-m}}{-a_1 z^{-1} - a_2 z^{-2} + \ldots a_n z^{-n}} U_{t-d}$$

or

$$Q_t = \frac{B(z)}{A(z)} U_{t-d} \qquad\qquad (12.14)$$

where $A(z)$ and $B(z)$ are polynomials in the backward difference operator with n a and $m+1$ b parameters, respectively.

All transfer functions that are modelled as one or more linear stores (12.5) in series or in parallel are of this general type. The Nash cascade mentioned earlier in the discussion of the unit hydrograph is a specific example of a such a model, although DBM models generally have only an integer number of elements (the model *order* or number of *a* coefficients), do not constrain the time constants to be equal and include the possibility of including the time lag d. A further requirement is that the model structure should have a mechanistic interpretation. In the rainfall–runoff case this means that the transfer function should have a sensible time delay and have ordinates that are everywhere positive (see also the discussion of the Muskingum flood routing model as a linear transfer function in Section 14.2.5). One of the features of the DBM modelling approach is that application of the general form (12.14) allows many different model structures to be tried very quickly. Thus, the model structure appropriate to a particular data set can be determined rather than a specific model structure set beforehand. However, in doing so it is necessary to be careful to avoid *over-parameterisation*. By adding more *a* and *b* coefficients, it will often be the case that model fit will increase but, as with fitting any polynomial function, more coefficients does not necessarily mean that the model performs better in prediction. Too many coefficients (over-parameterisation) may mean that the model is not robust in prediction. Thus, the principle of *parsimony* should be followed (the simplest model that will provide acceptable predictive accuracy should be chosen). Parsimonious models should be more robust in prediction in that they will capture the dominant modes of the response even if they do not give the highest values of performance measures such as (12.10) in calibration. Over-parameterisation is an issue for many hydrological models, since trying to include more process understanding into models often means introducing more parameters. The DBM approach tries to let the data suggest how complex a model is needed.

In the same way as the unit hydrograph, this type of modelling approach is limited in its application to rainfall–runoff predictions by its assumption of linearity. Catchment rainfall–runoff relationships are not linear, so an additional model component is required to obtain a good fit to the data over a wide range of conditions. There are two ways of doing this. The first is to assume a conceptual structure for the non-linearity. This is the approach taken by the PDM model above, and the IHACRES model which simultaneously fits the parameters of a simple 'soil water store' and the transfer function parameters. IHACRES is an acronym of the Institute of Hydrology, Wallingford, and the Centre for Research on Environmental Systems at the Australian

National University in Canberra who jointly developed the model (see Jakeman *et al.*, 1990). It is interesting because, by fitting the model to many different gauged catchments, it has been used as the basis for some regionalisation studies in both the UK and Australia (see Sefton and Howarth, 1998; Post and Jakeman, 1996).

A second approach takes advantage of the fact that linear models can be used to both estimate outputs from inputs and to estimate inputs from outputs. Thus, having fitted an approximate DBM transfer function to rainfall–runoff data under wet conditions, that transfer function can then be used to estimate the effective rainfall inputs that would have produced the measured discharge outputs. In the DBM approach this is achieved by recursive (time step by time step) updating of a gain parameter on the rainfall inputs. When the catchment is wet, the gain should be high, when the catchment is dry, the gain should be low. This is one way of estimating a series of effective rainfall inputs, which gives more stable results than directly inverting the transfer function. Given a functional description of how the gain varies, the transfer function can be refitted, and so on. Early application of this approach revealed that, since the most readily available index of catchment wetness is the discharge itself, an effective rainfall function of the form

$$U = RQ^n \tag{12.15}$$

provided a good representation of the non-linearity (Young and Beven, 1994). This type of approach has now been used in a variety of contexts, including real-time flood forecasting and flood routing. It has also been generalised to allow more flexible forms of the non-linearity to be used, where the shape appears to be more complex (see Young, 2003). Again, the idea is to let the data suggest the correct form of non-linearity that should be used for that particular catchment, with a realistic interpretation.

In assessing model structures derived in this way, an important part of the methodology is to make an assessment of the model residuals to see whether there is any remaining structure that might be represented by an additional model component. Ideally the model residuals at the end of a modelling exercise should be '*white*', that is they should be random with a variance that does not change with time or magnitude of the prediction, and with no clear autocorrelation from time step to time step. If this is not the case, then there is some feature of the response that has yet to be explained. This can be demonstrated by the application of the DBM methodology to one of the Coweeta experimental catchments reported in Young (2001) shown in Fig. 12.9. Having fitted the basic DBM model to the (daily time step) data (Fig. 12.9a,b), it seemed that there was a long-term seasonal component remaining in the residuals (Fig. 12.9c). This was then fitted by using mean daily temperature as an input to an additional linear transfer function component, improving the predictive capability of the model. The mechanistic interpretation is that there is an additional effect of evapotranspiration on effective rainfall that is not totally accounted for using the simple non-linear function of (12.15). Clearly, the additional component means that the model has more parameters, but the parameters are justified by the data (as revealed in the residuals) in this case (see Fig. 12.9c). The residuals of the more complex model were much closer to being 'white'. A Matlab toolbox[3] is available to fit DBM models to data, while the basic concepts are included in the Lancaster University TFM[4] program or the Imperial College rainfall–runoff modelling toolbox (Wagener *et al.*, 2004).

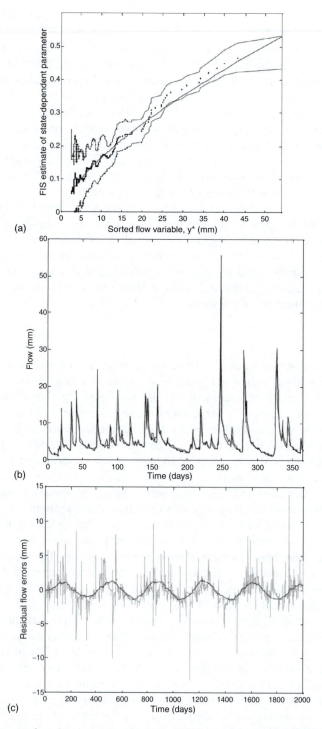

Fig. 12.9 Application of a data-based mechanistic (DBM) model to modelling daily discharges from the small catchment at Coweeta, N. Carolina: (a) fitted power law non-linearity; (b) a section of the deterministic flow prediction in comparison with the observed discharges; (c) temperature-dependent seasonal component fitted to residual errors. (From Young, 2001, with permission of John Wiley & Sons.)

The DBM approach to rainfall–runoff modelling is well suited to real-time flood forecasting (see Section 12.10 below). It has also been used to model the transport of pollutants in rivers as the aggregated dead zone (ADZ) model (Young and Wallis, 1993), and for flood routing along mainstream river channels (see Section 14.2.8).

12.8.3 Topmodel

Topmodel is a TOPography-based MODEL designed to simulate the runoff from hillslopes and source areas of gauged and ungauged catchments with inputs of rainfall. It was first suggested by Beven and Kirkby (1979) and has been widely used since. In Topmodel, the catchment is sub-divided into relatively homogeneous sub-catchment units based on the channel network and the separate outflows are routed downstream using a channel width function based, constant velocity, time-delay histogram to give the final catchment discharge.

The essential feature of the model is the prediction of saturated contributing areas based on the distribution of a topographic index in the catchment, previously suggested by Kirkby (1975), and the mean soil water deficit as it changes over time. As in the other models based on distribution functions described above, this greatly simplifies the calculations but, in the case of Topmodel, the topographic index can be mapped for the catchment, so that the predictions of saturated areas can be checked against any field information.

In making such a match, the assumptions are, however, critical. The three most important assumptions are as follows.

(1) The water table is nearly parallel to the soil surface, so that the hydraulic gradient is locally equal to the surface slope. This implies that the soil should not be too deep, that there should be a lower impermeable layer that is also near parallel to the surface, and that the slopes should not be flat or very steep.
(2) The water table takes up a configuration *as if* the storage at any point on a hillslope was being maintained by a constant uniform recharge rate over the upslope area draining through that point. This is treating the saturated zone storage as a succession of steady-state equivalent forms.
(3) The downslope transmissivity of the saturated zone can be represented as a simple function of the local storage deficit. In the original model, an exponential decline with increasing storage deficit was used, but other functions are also possible (Ambroise *et al.*, 1996; Iorgulescu and Musy, 1997). Different transmissivity functions lead to different forms of topographic index that should be used in the Topmodel equations.

It is evident that such assumptions will not be applicable everywhere. A schematic representation of the sub-catchment model is given in Fig. 12.10.[5]

The essence of Topmodel lies in one equation that can be derived from the assumptions above. This gives the relationship between a local storage deficit resulting from gravity drainage, the sub-catchment mean storage deficit, and the distributions of topographic index and soil transmissivity in the catchment. Keeping with

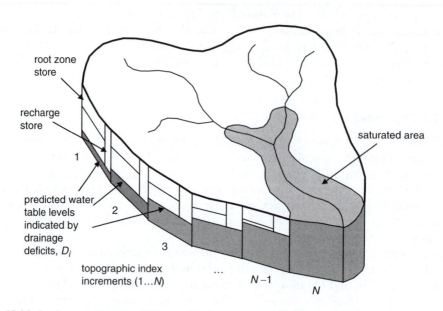

root zone store

recharge store

saturated area

1

predicted water table levels indicated by drainage deficits, D_i

2

3

topographic index increments (1...N)

...   N−1

N

Fig. 12.10 A schematic diagram of Topmodel showing how calculations made for each topographic index increment (here ranked from low to high values, 1:N) can be mapped back into space.

the assumption of an exponential decline of transmissivity with deficit, i.e.

$$T = T_o \exp(-D/m) \tag{12.16}$$

where T_o is the transmissivity when the soil is just saturated, D is storage deficit, and m is a parameter controlling the steepness of the decline in transmissivity, then it can be shown that:

$$\frac{D_i - \overline{D}}{m} = \left[\lambda - \ln\left(\frac{a}{\tan\beta}\right) \right] + \left[\ln T_o - \overline{\ln T_o} \right] \tag{12.17}$$

Both sides of this equation represent deviations within distribution functions. The left-hand side is the deviation of a local storage deficit, D_i, from the sub-catchment mean value, $\overline{D}$, scaled by the parameter m of the exponential transmissivity function, while the right hand-side represents sum of a (negative) deviation of the local topographic index, $\ln(a/\tan\beta)$ from the mean sub-catchment value of the index, λ, and the deviation of the local value of T_o from the mean sub-catchment value. In the topographic index, a is the area drained per unit contour length at a point and $\tan\beta$ is the local slope gradient at that point, and

$$\lambda = A^{-1} \int \ln\left(\frac{a}{\tan\beta}\right) dA \tag{12.18}$$

where A is the total sub-catchment area. In most applications it is not possible to know how the transmissivity varies in space and so both T_o and m are normally treated as

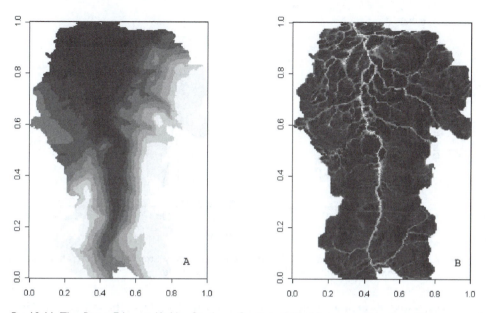

Fig. 12.11 The River Eden at Kirkby Stephen, Cumbria, UK. (a) topographic data; (b) pattern of ln(a/tan β) topographic index (light values are high).

constants and the last bracketed term in (12.17) is assumed equal to zero (the local value is everywhere equal to the mean). The variation in local deficit is then only a function of the topographic index and the parameter m. In particular, for any value of $\overline{D}$ we can calculate where the local drainage deficit D_i is zero. This represents the saturated area. Saturation will be more likely when the area draining though a point is high (e.g. in convergent hollows) and where the slope gradient is relatively low (e.g. Fig. 12.11). The model will predict that the saturated area will expand and contract as the catchment wets and dries so that the mean deficit $\overline{D}$ gets smaller and larger.

The mean soil moisture deficit will increase with drainage of the saturated zone and evapotranspiration and decrease with rainfall (and snowmelt). Under the exponential transmissivity function assumption, it can be shown that drainage from the saturated zone, q_b, is given by

$$q_b = q_o \exp(-\overline{D}/m) \qquad (12.19)$$

where the value of q_o is a function of T_o and λ, and m is the same constant as before.

The model has been well tested on several UK catchments over wet and dry periods and for different vegetation types with the best performance demonstrated in the humid and temperate climate and topographical conditions for which it was developed (e.g. Fig. 12.12). It compares very favourably in goodness of fit with other notable deterministic computer models which have a greater number of parameters, even using only measured and estimated parameters (Beven *et al.*, 1984). However, the model assumptions will not be appropriate to all catchments, even if the parameters can be calibrated to give a good fit to a discharge record. Some relaxations of the assumptions

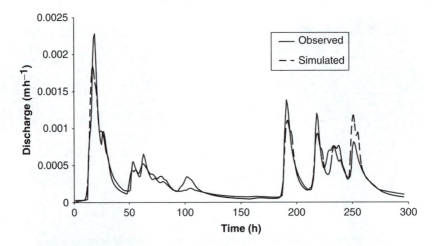

Fig. 12.12 Application of Topmodel to the River Eden at Kirkby Stephen, Cumbria, UK. Part of calibration period, with Nash–Sutcliffe efficiency = 0.896.

have been tested. By changing the saturation reference level from the soil surface, Piñol *et al.* (1997) show how the concepts can be applied to fast subsurface responses and Quinn *et al.* (1991) show how the model concepts can be applied to deeper water tables. Beven and Freer (2001a) have also shown how the steady-state assumption can be relaxed in a dynamic version of the model in a way that allows other definitions of similarity of response to be used in addition to topography while retaining computational efficiency. It is the computational efficiency of the model that has made it a very useful research tool in exploring uncertainties in model predictions (e.g. Beven, 1993; Beven and Freer, 2001b) and in using continuous simulation for flood frequency estimation (see Beven, 1987; Cameron *et al.*, 2000).

The ability to check the spatial pattern of the model predictions has also proven to be interesting. Because of the simplifying assumptions used in the model, we would not expect the predictions to be correct everywhere and in going from the storage deficits used in the simplest form of the model to a measured water table depth, it is necessary to introduce an additional assumption and parameter. What has been found in a number of studies is that the broad pattern of modelled contributing areas is often well predicted (e.g. Beven and Kirkby, 1979; Seibert *et al.*, 1997; Güntner *et al.*, 1996), but that there are areas where saturation is not obviously related to the surface topography (Ambroise *et al.*, 1996) and that local water table predictions can be improved if local transmissivities are calibrated (e.g. Lamb *et al.*, 1998; Blazkova *et al.*, 2002; Freer *et al.*, 2004), although this does not necessarily result in improved discharge predictions. A review of Topmodel applications is provided by Beven (1997).

12.8.4 The Système Hydrologique Européen model

The Système Hydrologique Européen (SHE) model is a physically based distributed model accounting directly for spatial variations in hydrological inputs and catchment responses. Originally developed jointly by hydrologists in the UK, France and Denmark

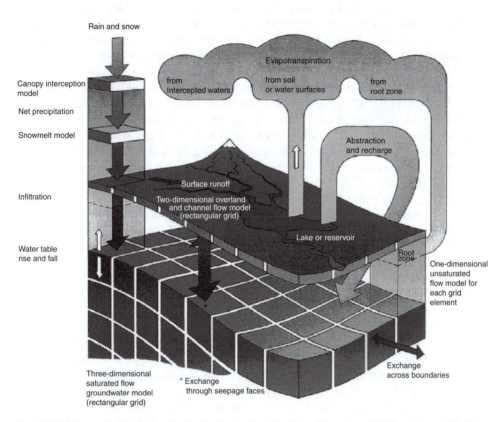

Fig. 12.13 Schematic diagram of the Système Hydrologique Européen (SHE) model (with kind permission of the Danish Hydraulics Institute).

starting in 1977, it has now been commercialised by the Danish Hydraulic Institute as MIKE-SHE and developed into a research tool as SHETRAN in the UK. In SHE, the catchment is discretised using a square grid. Past applications have used grid sizes ranging from 50 m or less in small catchments to 4 km in a large catchment in India (Jain *et al.*, 1992). Finite difference methods are used to obtain the solutions of the non-linear flow equations representing overland and channel flow, unsaturated and saturated subsurface flow. The model structure is shown in Fig. 12.13 (Abbott *et al.*, 1986; Refsgaard and Storm, 1995). In the unsaturated zone, the system is simplified with a one-dimensional vertical flow component used to link the two-dimensional surface flow and saturated groundwater components. The saturated zone can interact directly with the channel network along designated river segments.

The computational sequence is as follows. The *precipitation rate* is the data input to the interception model. A layered snowmelt model component may be applied next (different snow accumulation and melt models of different complexity are included). The *evapotranspiration loss model* operating from several vertical zones requires four meteorological inputs or a specified potential evapotranspiration rate. The *overland-channel flow model* requires boundary flow and initial flow depth conditions, topography of the overland flow plane and channels and particulars of any man-made discharge alterations. The *unsaturated flow model* is based on Darcy's law and controls infiltration and recharge and allows different soil layers and vegetation root patterns. It requires an initial water content profile for each grid square. The *saturated flow model* requires input data on boundary conditions and topography of the base aquiclude, initial water table levels or saturated thicknesses and any man-made interference to natural conditions (such as pumped wells or drainage tunnels) must be defined. It requires an initial depth of saturation for each grid square. In the first versions of the model the saturated zone was represented as a two-dimensional depth-averaged solution of the Darcy equation, but both MIKE-SHE and SHETRAN now have fully three-dimensional options, and MIKE-SHE also has a simple conceptual lumped store option. The different components can use different, automatically controlled, time steps. The operation is made flexible by a frame structure controlling the coupling of the individual component models so that they may be applied to a variety of catchment conditions.

The distributed nature of the model means that it can take account of the variability of rainfall and other meteorological variables over a catchment. An interface to a geographical information system (GIS) can also be used to facilitate the specification of the topography and soil and vegetation types. Model predictions can also be presented in a form compatible with GIS systems with on-screen animations of the distributed predictions.

There have been a number of interesting applications of the SHE model, including attempts to make predictions based only on physically based estimates of the vegetation, soil and groundwater parameters needed. If this were possible, then it would be important for two reasons. One is that with such a large number of parameters required for each grid square, and very many grid squares, calibration of the individual parameter values required by the model is not really possible (though limited calibration has been tried, e.g. Bathurst, 1986). A second is that, if the model can be shown to perform well using only estimated parameter values, then we might be able to have more belief in predictions of future conditions with, for example, parameters representing changed land use. Some of the earlier applications using SHE were greatly limited by computer time, which meant that the periods of simulations used were relatively short, so that results might have been unduly influenced by the specified pattern initial conditions. Later studies have shown some success in reproducing both discharges and internal state variables (Parkin *et al.*, 1996; Bathurst *et al.*, 2004; Thompson *et al.*, 2004) but it is not yet clear that this type of model has real advantages over other rainfall–runoff models except where highly detailed local data are available (e.g. Ebel and Loague, 2006) and even then it is not clear that the Darcian descriptions of flow in the unsaturated and saturated zones on which these models are based are adequate to describe the complex flow pathways in reality (see Beven, 1989; and discussions in Abbott and Refsgaard, 1996). The model has been extended to include the transport of soluble contaminants, sediment transport and landslides.

12.8.5 Other grid-based models

There are a number of other distributed-based models that do not aim to be as 'physically based' as the SHE model but which were developed to try to take account of the spatial patterns of inputs (including snow) and soil and vegetation characteristics in large catchments. A good example of this is the SLURP model developed by Kite and Kouwen (1992) in Canada where it has been used to model the Mackenzie and other large basins. SLURP is based on the concept of *hydrological response units* (HRUs) that have similar soil, vegetation and elevation characteristics, and are therefore expected to be similar in their hydrological response. In SLURP a single grid square may contain a number of different HRUs in an arbitrary pattern, with the areas of each being grouped together for calculation purposes. A similar approach, based on hexagonal units, has recently been applied across a range of scales in Russia (Semenova and Vinogradova, 2009). The HRUs can be defined automatically using overlays of GIS and remote sensing information. A somewhat similar approach has been taken with the LISFLOOD model of de Roo *et al.* (2000). This is now being used operationally at a 5-km grid scale at the European Union Joint Research Centre in Ispra, Italy, as the basis for a European Flood Warning System. The aim is to use ensemble numerical weather prediction (NWP) forecasts from the European Medium Range Weather Forecasting Centre to provide up to 10-day ahead forecasts of potential flood situations for all the major river basins in Europe. As noted in Section 12.8.2 above, a similar gridded approach in the grid-to-grid (G2G) model has been applied nationally at the Centre for Ecology and Hydrology, Wallingford.

Gridded models of this type are ideally suited to take advantage of the facilities of GIS, which can be used to store different layers of spatial information and to facilitate the presentation of distributed model outputs in graphical form. GIS systems have not traditionally been good at dealing with time-dependent spatial patterns needed in hydrological simulation (with some exceptions, such as the PC-Raster system that was used to create the LISFLOOD model). Some distributed models, such as SHE, have developed their own GIS interfaces. A general hydrological GIS toolkit has been developed in the Arc Hydro system (Maidment, 2002). Arc Hydro makes use of the proprietary GIS software *ArcGIS*, making use of spatial data layers for streams and hydrography, catchment drainage areas, channel sections, surface topography, soil and land use, and vegetation, together with time series information from rainfall and stream gauging sites. The system makes use of a Visual Basic interface to ArcGIS to allow flexibility in the programming of hydrological models.

12.8.6 Irregular element distributed models

Square grids are not the only way of discretising a distributed catchment model. In fact, the use of square or rectangular grids comes more from the ease of implementing approximate finite difference solutions to the subsurface flow equations than from any process considerations. Hillslope flow processes are not readily represented by a square grid in plan, even if that grid is rather fine, while sloping soil horizons and complex geological structures are not easily represented in the vertical by rectangular elements. Thus, it might be more realistic to have a more flexible discretisation.

Two main approaches have been followed. There are other solution techniques to the subsurface flow equations that are not based on a square or rectangular grid. Models based on the finite element solution method, that use triangular or quadrilateral elements to discretise the flow domain have also been available since the earliest days of distributed hydrological models. A recent finite element based model is the Integrated Hydrological Model (InHM) of VanderKwaak and Loague (2001). This has been used in a series of papers that have modelled the small R5 catchment at Coshocton, Ohio, gradually making the representation of the processes more complicated. The original perception of the hydrological response of this catchment was that it was dominated by surface runoff, but the final papers in the series have recognised that there may be an important subsurface component (Loague *et al.*, 2005). The InHM model has been recently extended to include sediment and chemical transport.

The other approach starts off with a non-raster description of the topography. While most of the readily available digital maps of topography are in raster (grid) form, many of these have been derived from vector (contour) or point elevation data. In fact, some approximation is introduced in going from point and vector data to a raster grid. It is, however, possible to create spatial discretisations directly from point elevation data, e.g. using a triangular irregular grid (TIN) based on Delauny triangulation. There have been a number of models based on the TIN approach. One is still being developed at the Massachusetts Institute of Technology and now includes complex representation of vegetation processes as affected by interactions with the hillslope flow processes (see Ivanov *et al.*, 2008).

12.9 Uncertainty in rainfall–runoff modelling

The models described above are typical examples of those used in practice. It is clear that even the most complex distributed models can only approximate the complex response of a real catchment and the parameter values can only approximate the complex characteristics of a real catchment. In addition, in any practical application to a real catchment, we will only have approximate estimates of the rainfall input and evapotranspiration output boundary conditions for any model that is applied. It should therefore be *expected* that the predictions of any chosen model will be necessarily approximate. This implies that the predictions will be associated with some error. Traditionally, this issue was addressed by practitioners only by trying to minimise the error in model calibration, i.e. to find the best or optimal model possible in some sense. This was achieved either by a visual comparison of plots of observed and predicted variables, or by a maximising a performance measure, such as the Nash–Sutcliffe efficiency measure (12.10).

Recently, however, the uncertainties of model predictions have started to be addressed more explicitly. This is because there is now more computer power available to enable the multiple model runs that are needed for the prediction uncertainties to be estimated. The simplest form of uncertainty analysis makes assumptions about the nature of the sources of uncertainty (typically rainfall inputs and model parameters) and then carries out a *forward uncertainty analysis* using either analytical (for linear models) or Monte Carlo simulation techniques to propagate the assumed uncertainty through the model. The outcomes from such an analysis will be stochastic, but will clearly depend totally on the assumptions made about the nature of the uncertainties and their interactions.

A more interesting case arises when there are some observations available with which to try and constrain the uncertainties. In this case the performance of different runs of one or more models can be compared and those that give better performance can be given more weight (or greater likelihood) in assessing a distribution of predicted variables. Studies of this type have shown that there is not generally a clear optimal model, and that models with different structures will often result in similar performance. This is, at least in part, a result of the fact that we normally do not know too much about the uncertainties in the input data provided to a model. This is often a limiting factor in model performance and the first stage in any modelling study should be to check the hydrological consistency of the data set.

When model predictions are being compared to observed discharges, a similar issue arises. We should not expect the performance of any model to be better than the accuracy of the observations with which it is being compared. Remember that it is rare that discharge is measured directly. More often water level is measured and converted to discharge by means of a rating curve. This introduces uncertainty in the discharge measurements, particularly when the rating curve has to be extrapolated to flood discharges well above the measured values on which the rating curve is based. Thus, where possible, it is also worth checking the rating curve before carrying out any model runs to assess the range of discharges for which the observations could be considered reliable.

Given the potential for error in both model inputs and discharge observations, there is a possibility of *over-fitting* a model in calibration, particularly when only an optimal model is sought. This means that the fit of the model is partly a result of the errors in the calibration data. When such a model is used for prediction (where the input errors can be expected to be different), over-fitting may result in poorer predictions. Over-fitting is more likely with a complex model with many parameters to be calibrated (more *degrees of freedom*). One way of checking for over-fitting is to make a *split record test*. This is where only part of the data available is used in calibrating the model. The rest is retained for use in evaluating the predictions. If model performance deteriorates badly in the evaluation period, then the model may have been over-fitted. Split record tests should be used whenever possible. They can be used to assess model predictions both with and without an estimation of prediction uncertainties in the calibration period.

There are many different methods for uncertainty estimation that have been used in hydrological modelling. More detail on the techniques available can be found in Beven (2009).

12.10 Real-time forecasting models

Uncertainty of predictions is also an important issue in real-time forecasting because decisions about flood warnings or water resource management might have important economic and human consequences. In this respect, forecasts need to be both accurate and timely in the sense of providing adequate 'lead time' for any decision to be implemented. Accuracy and adequate lead time are conflicting requirements. One hour ahead predictions of flood flows in a river might well be accurate, but would not provide a sufficient lead time of potential over-topping of flood defences to be useful for flood warning purposes. There would just not be time for those who would be most affected to respond to any warning. A much less accurate 6- or 12-hour ahead forecast might be much more useful in preventing loss of life and allowing those who might be affected to try to protect their property. A 24-hour ahead forecast, but still less accurate forecast, might be useful in deciding about where to commit resources in deploying temporary or demountable flood defences.

The ability to make accurate forecasts with adequate lead time depends very heavily on the scale of the catchment. In large catchments such as the Rhone, Rhine or Danube, it will take several days for a flood wave generated in the mountain headwaters to move downstream. Even for the much smaller catchments in the UK, at catchment scales of $1000 \, \text{km}^2$ and above, given rain-gauge and radar rainfall estimates in real time, can usually provide warnings to downstream towns with a lead time of 6–12 h or more. At smaller scales, with much shorter times of concentration in a catchment, then this is not possible as was seen in the Boscastle event on 16 August 2004 (see Section 9.1.2). On the basis of numerical weather prediction (NWP) warnings of severe rainfalls were issued by the UK Met Office, but they could not be precise about either the location or the rainfall amounts (one grid square in the regional NWP model is much larger than the catchment area draining to Boscastle). In the event, up to about 200 mm of rain fell upstream of the village, but the flood wave occurred during the day and while many people had to be rescued by helicopter, no-one was killed. Some 58 properties were affected by the flood and some £2m of damage was caused, including cars that were swept out to sea. Fifty years before, in the night of the 15 August 1952, in a flood in the village of Lynmouth, further up the Devon coast, a flood wave caused by up to 300 mm of rain in the catchments of the East and West Lyn rivers, swept through the village at night, killed 34 people, and deposited 200 000 tons of debris. In these types of flash flood events on small catchments, it will not be possible to provide warnings with adequate lead times until the accuracy of short-term NWP rainfall predictions improves significantly. Despite the continuing improvements in NWP, this still seems to be some way off for this type of event even though post-event local predictions using an atmospheric model with a 1-km grid have shown some success in reproducing the nature of the Boscastle event (e.g. Golding et al., 2005).

Forecasting during droughts is not quite so time critical. Things happen much more slowly during drought periods, and water resource management decisions can consider the 'worst-case' scenario of no rainfall into the future. It is still, however, the case that any model predictions might start to deviate from what is actually being observed in real time and one of the features of the forecasting problem is to try to take account of whether a model is currently under- or over-predicting in improving the future forecasts. This is called *adaptive forecasting* or *data assimilation*. It is used routinely in weather-forecasting models but, until recently, has not been used widely in hydrological forecasting.

There are two basic approaches to adaptive forecasting for real-time applications. In both, the starting point is a model that has been calibrated and shown to work well for some historical datasets. The first approach is to update the state variables in a model (e.g. values of storage) that has already been calibrated on historical datasets, to try to match the current observations on the basis that the updated model will provide better predictions into the future. For linear models and near-linear models, a useful robust methodology for state updating is the *Kalman filter* (KF). The KF can make use of one or more observables in updating the model. At each update step or *innovation* it uses the latest set of residuals between observed and predicted variables (or *innovations*) to update the states of the model (e.g. storages) *recursively* (i.e. the estimates at the last time step provide the starting point for the updating at this time step). The amount of updating depends on the magnitude of the residual and an estimate of the covariance of the state estimates. Thus, not only are the predictions improved by the updating, but the covariance matrix can be used to give estimates of the uncertainty in the predictions. It is also worth noting that the model parameter values can also be included in the updating, so that both storages and parameters can be updated. However, there will only be a limited amount of information in the innovations, so that, if it is attempted to update too many variables, then the estimates may fluctuate rapidly without great benefit in improving the forecasts (this is equivalent to the over-fitting problem in model calibration). The very simplest form of updating in this way is to use only the gain parameter in the updating algorithm (see Young, 2002).

For the non-linear models that are more usual in hydrological modelling, the *extended Kalman filter* (EKF) is more appropriate. The EKF uses a form of linearisation at each time step in doing the state and parameter updating. This means that the EKF can be subject to stability problems for highly non-linear models or strongly interacting states in the model. To avoid the approximation arising due to the linearisation, a new method called the *ensemble Kalman filter* (EnKF) has been introduced recently. This uses an estimate of the covariance of the states and parameters at a time step to sample multiple realisations of the model. These are then used to provide an ensemble of predictions at the required lead time, including all effects of non-linearities in the model. As a new set of observables is received, a form of the KF filter is used to update the covariance matrix. Then a new ensemble of models is sampled and used to project the predictions forward to the required lead time and so on. Further details on the various forms of Kalman filter and other data assimilation methods may be found in Beven (2009).

The second approach, called *error correction*, is to try to model the residual between the current observations and a calibrated model to try to improve the predictions into the future. This second approach requires that there be some structure in the residuals that allows projection into the future. If the residuals were purely random, then there would be no additional predictability, but this is not often the case with hydrological models, which will often under-predict or over-predict for successive time steps (the residuals can exhibit *bias* and *autocorrelation*). This might, of course, be so only because the inputs to the model are under- or over-estimated, but this still requires some adaptive capability to improve the forecasts. The advantage of this approach is that while the model might be highly non-linear, it might be possible to model the residuals with a linear model. Techniques such as the Kalman Filter might then be used to update the model of the residuals recursively as new observables are received, although the nature of the residuals of rainfall–runoff models does not normally conform to simple statistically assumptions such that it may be difficult to specify an adequate model.

There is one important feature of real-time forecasting that makes it different from other applications of hydrological simulation models. In forecasting, it is not so very important that we represent the processes controlling the response in any explicit way. The only requirement is to produce a forecast at the required lead time with good accuracy and minimum uncertainty. Hence, the usefulness of adaptive forecasting where, for example, the adaptation may mean that the model does not necessarily maintain mass balance. Indeed, we would not want to maintain mass balance in the model if, for example, an underestimate of the true rainfall given the available rain-gauges, would lead to an under-prediction in forecasting. An interesting example of this is provided by the study of Romanowicz *et al.* (2008). In developing a flood forecasting

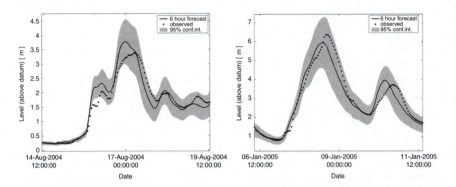

Fig. 12.14 Real-time adaptive flood forecasting in the River Eden at Sheepmount Carlisle. The forecasts of water level are made with a 6-h lead time. Left, shows the calibration event, right, predictions for the January 2005 flood event. Note large difference in the maximum stage for the two events. Shaded area represents 95 per cent prediction limits for the forecasts.

system for the River Severn in the UK, they included rainfall–runoff models based on the DBM methodology of Peter Young described above. The inputs to the forecasting model were measured rainfalls, but the outputs were water levels not discharge. There was therefore no attempt to retain mass balance, even in the calibrated model. Their argument was that the water level is the actual observable and can be measured to good accuracy, whereas conversion to discharge requires the use of a rating curve that introduces error, especially when extrapolation is required to high flood levels. In addition, it is very often the water level that is the required predicted variable in flood warning since it is level that controls when a flood defence will be overtopped or houses will be flooded. The DBM model is based on a non-linear transformation of the rainfall input and a linear transfer function. The adaptive forecasting model in this case uses a combination of a Kalman filter on the states of the transfer function and a final gain updating step to improve the forecasts. Fig. 12.14 shows an application of this methodology to the January 2005 flood in Carlisle. By cascading rainfall to water level and level-to-level components in the main stream river, up to 6-h ahead forecasts were possible in this case (Fig. 12.14) which is adequate for issuing flood warnings to the public. A recent detailed discussion of techniques for adaptive forecasting in hydrology is provided by Young (2009).

Notes

1 The accepted use of the term *effective rainfall* is perhaps unfortunate since, as discussed in Sections 1.2 and 11.3, the discharge from a catchment measured in an event hydrograph may not be all rainfall from that event, but may be, at least in part, stored water from past inputs that is displaced by that storm rainfall. Thus the effective rainfall in an event may not actually be the rain water falling in that event. This may not be important if we only need to estimate the volume of runoff or the peak discharge in the channel. It might be important if trying to predict water quality since the stored water might have a quite different chemistry from the rainwater (see also Section 11.3).
2 See http://www.epa.gov/ceampubl/swater/hspf/index.htm
3 See http://www.es.lancs.ac.uk/cres/captain/

4 See http://www.es.lancs.ac.uk/hfdg/freeware/hfdg_freeware_tfm.htm
5 A demonstration freeware of the basic version of Topmodel can be found at http://www.es.
   lancs.ac.uk/hfdg/freeware/hfdg_freeware_top.htm

References

Abbott, M. B., Bathurst, J. C., Cunge, J. A., O'Connell, P. E. and Rasmussen, J. (1986) An
   introduction to the European hydrological system – Système hydrologique européen 'SHE'
   2. Structure of the physically based, distributed modelling system. *Journal of Hydrology* 87,
   61–77.

Abbott, M. B. and Refsgaard, J.-C. (1996) *Distributed Hydrological Modelling*. Kluwer,
   Dordrecht.

Ambroise, B., Beven, K. J. and Freer, J. (1996) Towards a generalisation of the TOPMODEL
   concepts: topographic indices of hydrological similarity, *Water Resources Research* 32,
   2135–2145.

Bathurst, J. C. (1986) Physically-based distributed modeling of an upland catchment using the
   Système Hydrologique Européen, *Journal of Hydrology* 87, 103–123.

Bathurst, J. C., Ewen, J., Parkin, G., O'Connell, P. E. and Cooper, J. D. (2004) Validation of
   catchment models for predicting land-use and climate change impacts. 3. Blind validation for
   internal and outlet responses. *Journal of Hydrology* 287, 74–94.

Bell, V. A. and Moore, R. J. (1998) A grid-based distributed flood forecasting model for use
   with weather radar data 1: formulation. *Hydrology and Earth System Sciences* 2, 265–281.

Beran, M. (1999) Hydrograph prediction – how much skill? *Hydrology and Earth System
   Sciences* 3, 305–307.

Beven, K. J. (1979) On the generalised kinematic routing method. *Water Resources Research*
   15: 1238–1242.

Beven, K. J. (1987) Towards the use of catchment geomorphology in flood frequency predictions.
   Earth Surface Processes and Landforms V12, 69–82.

Beven, K. J. (1989) Changing ideas in hydrology: the case of physically based models. *Journal
   of Hydrology* 105, 157–172.

Beven, K. J. (1993) Prophecy, reality and uncertainty in distributed hydrological modelling,
   Advances in Water Resources 16, 41–51.

Beven, K. J. (1997) TOPMODEL: a critique, *Hydrological Processes* 11, 1069–1086.

Beven, K. J. (2001) *Rainfall–Runoff Modelling – the Primer*, Wiley, Chichester.

Beven, K. J. (2009) *Environmental Modelling – an Uncertain Future?* Routledge, London.

Beven, K. J. and Freer, J. (2001a) Equifinality, data assimilation, and uncertainty estimation
   in mechanistic modelling of complex environmental systems. *Journal of Hydrology* 249,
   11–29.

Beven, K. J. and Freer, J. (2001b) A dynamic TOPMODEL. *Hydrological Processes* 15,
   1993–2011.

Beven, K. J. and Kirkby, M. J. (1979) A physically based, variable contributing area model of
   basin hydrology. *Hydrological Sciences Bulletin* 24, 43–69.

Beven, K. J., Kirkby, M. J., Schofield, N. and Tagg, A. (1984) Testing a physically-based
   flood forecasting model (TOPMODEL) for three UK catchments. *Journal of Hydrology* 69,
   119–143.

Blazkova, S., Beven, K., Tacheci, P. and Kulasova, A. (2002) Testing the distributed water table
   predictions of TOPMODEL (allowing for uncertainty in model calibration): the death of
   TOPMODEL? *Water Resources Research* 38, W01257, 10.1029/2001WR000912.

Boorman, D. B., Hollis, J. M. and Lilly, A. (1995) *Hydrology of Soil Types: A Hydrologically Based Classification of the Soils of the United Kingdom*. Institute of Hydrology Report
   No. 124, Institute of Hydrology, Wallingford.

Botter, G., Bertuzzo, E. and Rinaldo, A. (2010) Transport in the hydrologic response: travel time distributions, soil moisture dynamics and the old water paradox. *Water Resources Research* (in press).

Cameron, D., Beven, K. J, Tawn, J. and Naden, P. (2000) Flood frequency estimation by continuous simulation (with likelihood based uncertainty estimation). *Hydrology and Earth System Science* 4, 23–34.

Chutha, P. and Dooge, J. C. I. (1990) The shape parameters of the geomorphologic unit hydrograph. *Journal of Hydrology* 117, 81–97.

Cole, S. J. and Moore, R. J. (2008) Hydrological modelling using raingauge and radar-based estimators of areal rainfall. *Journal of Hydrology* 358, 159–181.

Cudenec, C., Fouad, Y., Sumarjo Gatot, I. and Duchesne, J. (2004) A geomorphological explanation of the unit hydrograph concept. *Hydrological Processes* 18, 603–621.

Dawdy, D. R. and O'Donnell, T. (1965) Mathematical models of catchment behaviour. *Proceedings of the ASCE* HY4, 91, 123–137.

De Roo, A. P. J., Wesseling, C. G. and van Deursen, W. P. A. (2000) Physically based river basin modelling within a GIS: the LISFLOOD model. *Hydrological Processes* 14, 1981–1992.

Dooge, J. C. I. (1959) A general theory of the unit hydrograph. *Journal Geophysical Research* 64, 241–256.

Dooge, J. C. I. and O'Kane, J. P. (2003) *Deterministic Methods in Systems Hydrology*, A. A. Balkema, Lisse.

Ebel, B. A. and Loague, K. (2006) Rapid simulated hydrologic response within the variably saturated near surface. *Hydrological Processes* 22, 464–471.

Franchini, M. and O'Connell, P. E. (1996) An analysis of the dynamic component of the geomorphologic instantaneous unit hydrograph. *Journal of Hydrology* 175, 407–428.

Freer, J., McMillan, H., McDonnell, J. J. and Beven, K. J. (2004) Constraining dynamic TOPMODEL responses for imprecise water table information using fuzzy rule based performance measures. *Journal of Hydrology* 291, 254–277.

Golding, B., Clark, O. and May, B. (2005) The Boscastle flood: meteorological analysis of the conditions leading to flooding on 16 August 2004. *Weather* 60, 230–235.

Güntner, A., Uhlenbrook, S., Seibert, J. and Leibendgut, C. (1999) Multi-criteria validation of TOPMODEL in a mountainous catchment. *Hydrological Processes* 13, 1603–1620.

Gupta, V. K., Waymire, E. and Wang, C. T. (1980) A representation of an instantaneous unit hydrogrtaph from geomorphology. *Water Resources Research* 16, 855–82.

Hornberger, G. M., Beven, K. J., Cosby, B. J. and Sappington, D. E. (1985) 'Shenandoah watershed study: calibration of a topography-based, variable contributing area hydrological model to a small forested catchment. *Water Resources Research* 21, 1841–1850.

Horton, R. E. (1945) Erosional development of streams and their drainage basins: hydrophysical approach to quantitative morphology. *Geological Society America Bulletin* 56, 275–370.

Imbeaux, E. (1892) *La Durance: Régime, Crues et Inondations, Annales des Ponts et Chausées, Mémoires et Documents*, 7e séries, Tome III, 5–200.

Institution of Civil Engineers (1996) *Floods and Reservoir Safety: An Engineering Guide*, 3rd edn. Thomas Telford, London.

Iorgulescu, I. and Musy, A. (1997) Generalisation of TOPMODEL for a power law transmissivity profile. *Hydrological Processes* 11, 1353–1355.

Ivanov, V. Y., Vivoni, E. R., Bras, R. L. and Entekhabi, D. (2004) Catchment hydrologic response with a fully distributed triangulated irregular network model. *Water Resources Research* 40, W11102, doi:10.1029/2004WR003218.

Jain, S. K., Storm, B., Bathurst, J.-C., Refsgaard, J.-C. and Singh, R. D. (1992) Application of the SHE to catchments in India. Part 2. Field experiments and simulation studies with the SHE on the Kolar subcatchment of the Narmada River. *Journal of Hydrology* 140, 25–47.

Jakeman, A. J., Littlewood, I. G. and Whitehead, P. G. (1990) Computation of the instantaneous unit hydrograph and identifiable component flows with application to two small upland catchments. *Journal of Hydrology* 117, 275–300.

Kirkby, M. J. (1975) Hydrograph modelling strategies. In Peel, R., Chisholm, M. and Haggett, P. (eds) *Processes in Physical and Human Geography*, Heinemann, London, pp 69–90.

Kirkby, M. J. (1976) Tests of the random network model and its application to basin hydrology. *Earth Surface Processes* 1, 197–212.

Kite, G. and Kouwen, N. (1992) Watershed modelling using land classification. *Water Resources Research* 28, 3193–3200.

Lamb, R., Beven, K. J. and Myrabø, S. (1998) Use of spatially distributed water table observations to constrain uncertainty in a rainfall–runoff model. *Advances in Water Resources* 22, 305–317.

Loague, K., Heppner, C. S., Abrams, R. H., Carr, A. E., VanderKwaak, J. E. and Ebel, B. A. (2005) Further testing of the integrated hydrology model (InHM): event-based simulations for a small rangeland catchment located near Chickasha, Oklahoma. *Hydrological Processes* 19, 1373–1398.

Ministry of Agriculture, Fisheries and Food (MAFF); (1981) *The Design of Field Drainage Pipe Systems*. Technical Booklet 345, MAFF, London.

Maidment, D. R. (ed.) (2002) *ArcHydro: GIS for Water Resources*. ESRI Press, Redlands, CA.

McCuen, R. H., Knight, Z. and Cutter, A. G. (2006) Evaluation of the Nash–Sutcliffe efficiency index. *Journal of Hydrologic Engineering* 11, 597–602.

Minshall, N. E. (1960) Predicting storm runoff on small experimental watersheds. *Journal of Hydraulic Division of the ASCE* 86(HY8), 17–38.

Moore, R. J. (1985) The probability-distributed principle and runoff production at point and basin scales. *Hydrological Sciences Journal* 30, 273–297.

Moore, R. J. (1999) Real-time flood forecasting systems: perspectives and prospects. In: Casale, R. and Margottini, C. (eds) *Floods and Landslides: Integrated Risk Assessment*. Springer, Berlin, pp. 147–189.

Moore, R. J. (2007) The PDM rainfall–runoff model. *Hydrology and Earth System Sciences* 11, 483–499.

Moore, R. J. and Clarke, R. T. (1981) A distribution function approach to rainfall–runoff modeling. *Water Resources Research* 17, 1367–1382.

Naden, P. S., Blyth, E. M., Broadhurst, P., Watts, C. D. and Wright, I. R. (1999) River routing at the continental scale: use of globally-available data and an a priori method of parameter estimation source. *Hydrology and Earth System Sciences* 3, 109–124.

Nash, J. E. (1957) *The Form of the Instantaneous Unit Hydrograph*. IASH Pub. No. 45, 3, 114–121.

Nash, J. E. (1960) A unit hydrograph study, with particular reference to British catchments. *Proceedings of the Institution of Civil Engineers* 17, 249–282.

Nash, J. E. and Sutcliffe, J. V. (1970) River flow forecasting through conceptual models.1. A discussion of principles. *Journal of Hydrology* 10, 282–290.

Nourani, C., Singh, V. P. and Delafrouz, H. (2009) Three geomorphological rainfall–runoff models based on the linear reservoir concept. *Catena* 76, 206–214.

Parkin, G., O'Donnell, G., Ewen, J., Bathurst, J. C., O'Connell, P. E. and Lavabre, J. (1996) Validation of catchment models for predicting land-use and climate change impacts. 1. Case study for a Mediterranean catchment. *Journal of Hydrology* 175, 595–613.

Piñol, J., Beven, K. J. and Freer, J. (1997) Modelling the hydrological response of mediterranean catchments, Prades, Catalonia – the use of distributed models as aids to hypothesis formulation. *Hydrological Processes* 11, 1287–1306.

Post, D. A. and Jakeman, A. J. (1996) Relationship between physical attributes and hydrologic response characteristics in small Australian mountain ash catchments. *Hydrological Processes* 10, 877–892.

Quinn, P. F., Beven, K. J. Chevallier P. and Planchon, O. (1991) The prediction of hillslope flow paths for distributed hydrological modelling using digital terrain models. *Hydrological Processes* 5, 59–79.

Reed, D. W., Johnson, P. and Firth, J. M. (1975) A non-linear rainfall–runoff model, providing for variable lag time. *Journal of Hydrology* 25, 295–305.

Refsgaard, J. C. and Storm, B. (1995) MIKE SHE In: Singh, V. P. (ed.) *Computer Models of Watershed Hydrology*. Water Resources Publications, Englewood, CO, pp. 809–846.

Rodriquez-Iturbe, I. and Rinaldo, A. (1997) *Fractal River Basins; Chance and Self-organization*. Cambridge University Press, Cambridge.

Rodriguez-Iturbe, I. and Valdes, J. (1979) The geomorphic structure of hydrologic response. *Water Resources Research* 15, 1409–1420.

Romanowicz, R. J., Young, P. C., Beven, K. J. and Pappenberger, F. (2008) A data based mechanistic approach to nonlinear flood routing and adaptive flood level forecasting. *Advances in Water Resources* 31, 1048–1056.

Rosso, R. (1984) Nash model in relation to Horton order ratios. *Water Resources Research* 20, 914–920.

Schaefli, B. and Gupta, H. V. (2007) Do Nash values have value? *Hydrological Processes* 21, 2075–2080 (DOi: 10.1002/hyp.6825).

Sefton, C. E. M. and Howarth, S. M. (1998) Relationships between dynamic response characteristics and physical descriptors of catchments in England and Wales. *Journal of Hydrology* 211, 1–16.

Seibert, J., Bishop, K. H. and Nyberg, L. (1997) A test of TOPMODEL's ability to predict spatially distributed groundwater levels. *Hydrological Processes* 11, 1131–1144.

Semenova, O. M. and Vinogradova, T. A. (2009) A universal approach to runoff processes modelling: coping with hydrological predictions in data scarce regions. In: *New Approaches to Hydrological Prediction in Data Sparse Regions. IAHS Publication* 333, 11–19.

Shamseldin, A. Y. and Nash, J. E. (1998) The geomorphpological unit hydrograph – a critical review. *Hydrological Earth System Science* 2, 1–8.

Sherman, L. K. (1932) Streamflow from rainfall by the unit-graph method. *Engineering News Record* 108, 501–505.

Sherman, L. K. (1942) The unit hydrograph method. In: Meinzer, O. E. (ed.) *Hydrology*. Dover Publications, New York, pp. 514–525.

Singh, V. P. (ed.) (1995) *Computer Models of Watershed Hydrology*. Water Resources Publications, Englewood, CO.

Singh, V. P. and Frevert, D. K. (eds) (2002a) *Mathematical Models of Larger Watershed Hydrology*. Water Resource Publications, Highlands Ranch, CO.

Singh, V. P. and Frevert, D. K. (eds) (2002b) *Mathematical Models of Small Watershed Hydrology and Applications*. Water Resource Publications, Highlands Ranch, CO.

Strahler, A. N. (1957) Quantitative analysis of watershed geomorphology. *American Geophysical Soceity Transactions* 38, 913–920.

Thompson, J. R., Sørenson, H. R., Gavin, H. and Refsgaard, A. (2004) Application of the coupled MIKE SHE/MIKE 11 modelling system to a lowland wet grassland in southeast England. *Journal of Hydrology* 293, 151–179.

VanderKwaak, J. E. and Loague, K. M. (2001) Hydrologic response simulations for the R5 catchment with a comprehensive physics-based model. *Water Resources Research* 37, 999–1013.

Wagener, T., Wheater, H. S. and Gupta, H. V. (2004) *Rainfall-Runoff Modelling in Gauged and Ungauged Catchments*. Imperical College Press, London.

Young, P. C. (2001) Data-based mechanistic modelling and validation of rainfall-flow processes. In Anderson, M. G. and Bates, P. D. (eds) *Model Validation: Perspectives in Hydrological Science*. Wiley, Chichester, pp. 117–161.

Young, P. C. (2002) Advances in real time forecasting. *Philosophicial Transactions of the Royal Society of London* A360, 1430–1450.

Young, P. C. (2003) Top-down and data-based mechanistic modelling of rainfall-flow dynamics at the catchment scale. *Hydrological Processes* 17, 2195–2217.

Young, P. C. (2009) Real time updating in flood forecasting and warning. In Pender, G. (ed.) *Flood Risk Management Handbook*. Wiley, Chichester.

Young, P. C. and Beven, K. J. (1994) Data-based mechanistic modelling and the rainfall-flow non-linearity. *Environmetrics* V.5, 335–363.

Young, P. C. and Wallis, S. G. (1993) Solute transport and dispersion in channels. In Beven, K. J. and Kirkby, M. J. (eds) *Channel Network Hydrology*. Wiley, Chichester, pp. 129–175.

Chapter 13

Estimating floods and low flows in the UK

As was seen in Chapter 11, one of the issues that hydrologists and engineers most frequently have to deal with is the estimation of the frequency of different discharges at a site, particularly flood discharges and low flows. In that chapter methods were presented for the analysis of the observations of discharges available at gauging sites. There are, however, only a limited number of sites where gauging station data are available, so the problem that arises in many studies is how to estimate the flood and low flow characteristics at ungauged sites. This is a global problem. The UK has a relatively large number of gauging stations for its area (some 1400 in total; Marsh, 2002); in many developing countries the network is much sparser and often being reduced further for reasons of cost, organisation or war at a time when it is more and more important to monitor the impacts of ongoing catchment change and potential climate change. The International Association of Hydrological Sciences initiated a decadal project in 2003 to address this problem, called 'Prediction in ungauged basins' (PUB;[1] e.g. Sivapalan *et al.*, 2006).

Estimates of flood discharges at ungauged sites are most often required for the design of flood defences and for flood-risk mapping. Estimates of low flows at ungauged sites are most often required for water resources management, particularly setting of minimum acceptable flows for licensing effluents and abstraction in protecting the water quality and ecological status of rivers. In the UK there has been a history of providing practical methods for estimating flows at both floods and low flow stages. Most recently the Centre for Ecology and Hydrology (formerly the Institute of Hydrology; IoH) at Wallingford has developed the *Flood Estimation Handbook* (Institute of Hydrology, 1999) and *LowFlows* study (Holmes *et al.*, 2002a,b) to meet this need. These are good examples of methods of what is often called *regionalisation* in hydrology in which information available at gauged sites is regionalised to the wider set of ungauged sites.

13.1 Background to the *Flood Estimation Handbook*

The term 'flood estimation' usually refers to calculations that seek to relate flood flow to probability, frequency or return period (see Sections 11.6 and 11.7). This is often done to compute a 'design flow' that may be used to define a scenario for testing the hydraulic performance of proposed works in the floodplain or for flood mapping. Hydrologists refer to 'flood estimation' to emphasise that we can never know the precise value of a design flow of a specified probability of exceedance. Rather, we have to make the best use of the available data, as well as understanding the physical processes that give rise to flooding, to obtain an estimate that can be used with confidence.

In the UK, the standard approaches for flood estimation are those set out in the *Flood Estimation Handbook* or *FEH* (Institute of Hydrology, 1999), along with supplementary research reports and guidance. The *FEH* was a development of the earlier *Flood Studies Report* (Natural Environment Research Council; NERC, 1975), which was included in earlier editions of *Hydrology in Practice*. The *FEH* advanced flood estimation procedures by using methods that can be updated continuously to include the latest gauged flow data and by the use of gridded digital catchment descriptors (CDs) distributed as a spatial database known as the *FEH* CD-ROM. The *FEH* procedures are supported by a software package called WINFAP-FEH. Details of the range of *FEH* publications, data and software can be found via the Centre for Ecology and Hydrology web site.[2] After the publication of the *FEH*, a national programme of work was undertaken to assess the quality of flood flow data and to publish gauged peak flows and catchment metadata via a web site known as a HiFlows-UK.[3] The HiFlows-UK data set is periodically updated and is the primary data source for the *FEH* statistical method.

The *FEH* presents two alternative routes to flood estimation. One is based on fitting probability distributions to river flow data (the 'statistical method'), the other is based on rainfall–runoff modelling driven by statistical estimates of rainfall for 'design events'. Both methods can be used either for catchments with flow data or for ungauged locations, which is often the requirement in practice since it is relatively rare that a site for which flood estimates are required already has a long-term stream gauge. This is the *regionalisation* problem introduced above. In the *FEH* a variety of methods are used in regionalisation including trying to relate parameters describing the catchment response and flood frequency curves to the physical or climatic characteristics of a catchment.

13.2 The index flood and growth curve concepts in the *Flood Estimation Handbook*

13.2.1 The index flood method and estimation of QMED, the median annual maximum flow

In the *FEH*, the statistical analysis of peak flows is based on an *index flood* procedure. The index flood can be thought of as a medium-sized flood for a particular catchment. In the *FEH*, the index flood is the median annual maximum flood flow, QMED. This is the flood that is exceeded on average in exactly half of all years, thus it has a return period of 2 years and has an annual exceedance probability (AEP) of 0.5 (see Section 11.6 for discussion of the definition of exceedance probability). The median, QMED, is a robust index in that it is not affected by one or two exceptional floods in the data series, unlike the mean annual maximum flood, QBAR, which was used in the earlier *Flood Studies Report* (Natural Environment Research Council, 1975).

A flood *growth curve* is a probability distribution fitted to annual maximum flows that have been standardised by dividing each value in the series by QMED. Growth curves from different catchments can then easily be compared because this means that they are all scaled to have a value of one at a return period of 2 years (the return period of the index flood). Growth curves can therefore be combined to allow effective use of flood data from more than one gauging station.

The flood frequency curve at the site of interest is simply the product of the flood growth curve and QMED at the site. Ideally, QMED is estimated directly from the median of a set of flood data. This is a straightforward procedure for annual maximum flood data. If no gauged flow data are available, then QMED can be estimated from an empirical formula based on catchment descriptors, although this gives much more uncertain estimates compared with even a short gauged record, and should therefore be regarded as a last resort. It is usual in this case to transfer estimates of QMED from a nearby gauged catchment to enhance confidence in estimates based on catchment descriptors.

The original *FEH* formula for QMED on rural catchments was:

$$QMED = 1.172 \, AREA^{\left[1-0.0150\ln\left(\frac{AREA}{0.5}\right)\right]} \left(\frac{SAAR}{1000}\right)^{1.560}$$

$$\times FARL^{2.642} \left(\frac{SPRHOST}{100}\right)^{1.211} 0.0198^{RESHOST} \qquad (13.1)$$

where

$$RESHOST = BFIHOST + 1.30\frac{SPRHOST}{100} - 0.987$$

The catchment descriptors are defined in Table 13.1. The descriptor FARL attempts to represent the attenuation of flood flows caused by on-line reservoirs and lakes. For a catchment of total area A containing $n(i = 1,\ldots, n)$ on-line water bodies, the attenuation effect is indexed by

$$FARL = \prod_{i=1,\ldots n} \left(1 - \sqrt{\tfrac{a_i}{A_i}}\right)^{A_i/A} \qquad (13.2)$$

Table 13.1 Some catchment descriptors used in the *Flood Estimation Handbook*

AREA	Catchment drainage area derived from a digital terrain model (km^2)
FARL	Index of flood attenuation due to reservoirs and lakes (1.0 indicates no attenuation)
SPRHOST	Standard percentage runoff derived using the HOST (hydrology of soil types) classification of Boorman et al. (1995).
BFIHOST	Baseflow index derived using the HOST classification
SAAR	Standard-period (1961–90) average annual rainfall (mm)
DPSBAR	Mean drainage path slope ($m\,km^{-1}$) to the point in the channel network of interest
DPLBAR	Mean drainage path length (km)
PROPWET	Proportion of time when calculated soil moisture deficit for the catchment area was $\leq 6\,mm$ during the standard period 1961–90
URBEXT$_{1990}$	Extent of urban and suburban land cover in 1990

where a_i is the surface area of the ith reservoir or lake, which drains a sub-catchment of area A_i. The FARL index value is therefore equal to one where there are no reservoirs or lakes and decreases as attenuation effects are expected to become more important because of the influence of on-line water bodies.

The descriptors SPRHOST and BFIHOST capture information about the expected hydrological response of the catchment related to a classification of UK soil types according to their expected runoff generation characteristics. This is the *hydrology of soil types* (HOST) classification, with 29 different soil classes, that was developed in IH Report 126 (Boorman *et al.*, 1995). The descriptor SPRHOST, which is discussed further in Section 13.5.3, provides an estimate of the standard percentage runoff (SPR), which is the percentage of rainfall that causes the short-term increase in stream flow in response to a storm event. The descriptor BFIHOST estimates the proportion of flow that occurs as baseflow, defined by the baseflow index (BFI) developed in the IoH *low flow studies* (Institute of Hydrology, 1980).

The variables used in the empirical model (13.1) suggest that QMED increases with catchment size and wetness, is higher on catchments with poorly drained soils (where SPRHOST is larger) and is moderated by attenuation due to reservoirs and lakes. This model was based on the empirical analysis of gauged data from rural catchments largely free of artificial influences such as pumped drainage. For urban catchments, there is an urban adjustment factor in the *FEH* that attempts to represent the physical influence of urban growth on flood response, in particular the decreasing influence of soil properties on runoff (see Chapter 18). The urban adjustment factor represents the net effect of urbanisation, that is, it includes the consequences of flood mitigation works. For this reason, the use of a QMED model to project the effect of future urbanisation is highly uncertain and not recommended (one of the design event rainfall–runoff models can be used instead, see later).

A revised equation for QMED was derived by Kjeldsen *et al.* (2008) using longer and higher quality flood peak records than in the original *FEH* research. The revised formula,

$$\text{QMED} = 8.3062 \, \text{AREA}^{0.8510} \, 0.1536^{\frac{1000}{\text{SAAR}}} \, \text{FARL}^{3.4451} \, 0.0460^{\text{BFIHOST}^2} \qquad (13.3)$$

no longer includes SPRHOST. This equation performs significantly better than the original *FEH* equation and is therefore to be recommended. Both equations are recommended for application only to catchments of greater than $0.5 \, \text{km}^2$.

At an ungauged site, estimates of QMED can be improved by using a data transfer from a gauge on a similar catchment. A data transfer is essentially a local correction to the estimate of QMED obtained from (13.1) or (13.3). The procedure follows from the premise that the ratio between an estimate of QMED derived from flow data and a (more uncertain) estimate derived from catchment descriptors, QMED_{CD}, should be approximately constant between catchments of similar characteristics. This can be written as

$$\left(\frac{\text{QMED}}{\text{QMED}_{\text{CD}}} \right)_{ungauged} = \left(\frac{\text{QMED}}{\text{QMED}_{\text{CD}}} \right)_{gauged} \qquad (13.4)$$

where the subscripts refer to two 'similar' catchments, one gauged and one ungauged. A simple rearrangement shows that the right-hand side of (13.4) is effectively an adjustment factor that can be used to re-scale the initial estimate of QMED for the ungauged site derived from the catchment descriptor model. The gauged catchment used to provide the correction factor is known as a '*donor*' catchment.

The *FEH* provided guidance on selecting sites that could be regarded as 'similar enough' to be used in for data transfer, although it was a somewhat subjective process with no precise rules. A revised data transfer method was introduced by Kjeldsen *et al.* (2008) based on more rigorous analysis of errors in the QMED catchment descriptor model. The revised procedure uses a single local donor, selected purely on the basis of distance between catchment centroids and with no requirement for the donor to be on the same water-course as the subject site, although in practice this is likely if the catchment centroids are close. The adjustment ratio is moderated by a power term, *a*, that is derived from a statistical analysis of the errors surrounding the QMED model (13.3) and is applied such that (13.4) becomes:

$$\left(\frac{QMED}{QMED_{CD}} \right)_{ungauged} = \left[\left(\frac{QMED}{QMED_{CD}} \right)^{a} \right]_{gauged} \tag{13.5}$$

where *a* reduces with distance between the catchment centroids, so that the adjustment has its full effect when the donor gauge site is very close to the ungauged location and the effect declines to quite small once the centroids are more than 10 km apart.

13.2.2 Pooled growth curves

Where there is a reliable and long record at the site of interest and when the target return period, *T*, is much shorter than the record length, it is acceptable to fit a probability distribution to the gauged flow data. This is an ideal case, however, and rather unusual in practice. Most flood estimates are required at ungauged locations and for return periods much longer than the typical gauging station record length (a return period of 100 years or 1 per cent AEP is a typical choice for flood risk assessments or design calculations). In such cases a pooled analysis allows data from several sites to be combined to improve confidence in the estimate at the gauged or ungauged 'subject' site. The *FEH* forms pooling groups using a region-of-influence approach, a flexible method in which the group is specifically adapted to the site of interest. Unlike earlier methods such as the *Flood Studies Report* (Natural Environment Research Council, 1975), which used fixed geographical regions, the pooling group is chosen to be *hydrologically similar* to the subject site and a different pooling group can be used for every subject site.

Ideally, the pooling group contains catchments that have similar hydrological characteristics as the subject site so that the additional data represents an unbiased sample that can reduce the variance about the estimated growth curve. In the original *FEH* methods, this was achieved by grouping together catchments of similar size (as measured by catchment descriptor AREA), soils (BFIHOST) and wetness (SAAR). In the revised statistical method of Kjeldsen *et al.* (2008), a slightly different procedure has been used, based on assessing similarity in terms of AREA, SAAR, FARL and a measure of floodplain extent.

13.3 Probability distributions used in the Flood Estimation Handbook

The analysis of discharge data to assess flood frequencies and return periods has been explained earlier in Section 11.7. In that section an example was given of fitting the Gumbel extreme value distribution to a set of annual maximum peaks. The Gumbel distribution is a special case of the generalised extreme value (GEV) distribution, which was the basis of annual maximum flood analysis in the *Flood Studies Report* (Natural Environment Research Council, 1975). With the longer series available to the *FEH*, the generalised logistic (GL) distribution was selected as better representing annual maximum series. The advantage of the GL distribution over the GEV is that fitting the GEV more often results in a distribution with an upper bound (an EV3 distribution; see Section 11.8). It is perhaps debatable whether we should expect floods to have some upper bound on the basis of meteorological and hydrological conditions in any region. Reaching such a bound would normally occur at such a long return period that for practical purposes the assumption of an unbounded distribution can be taken as reasonable; hence the choice of the GL distribution in the *FEH*.

13.3.1 The generalised logistic distribution

Like the GEV distribution, the GL distribution is a three-parameter distribution with a special form, the logistic distribution, for the case where a parameter is equal to zero as:

$$X(F) = a + \frac{b}{k}\left\{1 - \left[\frac{1-F(X)}{F(X)}\right]^{k}\right\} \quad \text{for} \quad k \neq 0$$

$$X(F) = a + b\ln\left\{\frac{F(X)}{1-F(X)}\right\} \qquad \text{for} \quad k = 0$$

(13.6)

For the GL distribution, the median, $QMED$ is found by substituting $F(X) = 0.5$ into (13.6) yielding simply

$$QMED = a \tag{13.7}$$

The GL growth curve is then defined by

$$x(F) = 1 + \frac{\beta}{k}\left\{1 - \left[\frac{1-F(x)}{F(x)}\right]^{k}\right\} \tag{13.8}$$

where $\beta = b/a$; or in terms of return period, T,

$$x_T = 1 + \frac{\beta}{k}\left[1 - (T-1)^{-k}\right] \tag{13.9}$$

13.3.2 The generalized Pareto distribution

A sample of peaks-over-threshold (POT) series should approximate the form of the generalised Pareto (GP) distribution and this was chosen as the basis for the analysis of POT series in the *FEH*. It is not normally used for the analysis of annual maximum series, when the GEV and GL distributions are preferred. The GP distribution is defined by

$$X(F) = a + \frac{b}{k}\left\{1 - [1 - F(X)]^k\right\} \qquad \text{for} \quad k \neq 0$$

$$X(F) = a + b\ln\{F(X)\} \qquad\qquad \text{for} \quad k = 0$$

(13.10)

When $k = 0$, this has the form of an exponential distribution with mean shift a. Another special case of the GP distribution occurs when $k = 1$, when it has the form of the uniform distribution on the interval $a \leq X \leq a + b$. For the GP distribution, substituting $F(X) = 0.5$ into (13.10) gives the median as:

$$QMED = a + \frac{b}{k}\left(1 - 2^{-k}\right)$$

(13.11)

The GP growth curve is then defined by

$$x(F) = 1 + \frac{\beta}{k}\left\{2^{-k} - [1 - F(x)]^k\right\}$$

$$\beta = \frac{b}{a + \frac{b}{k}\left(1 - 2^{-k}\right)}$$

(13.12)

or in terms of return period, T,

$$x_T = 1 + \frac{\beta}{k}\left[2^k - T^{-k}\right]$$

(13.13)

13.3.3 Fitting frequency distributions

The *FEH* recommends fitting the GL and GP distributions by the method of L-moments (Hosking and Wallis, 1997). This is the method used in the WINFAP software. WINFAP also provides estimates of the uncertainty in the parameter values for the fitted distributions and confidence limits for the estimated extremes similar to those shown in Table 11.9 and Fig. 11.16. Where a three-parameter distribution is fitted and the k parameter (as defined above for the GEV, GL and GP distributions) is close to zero relative to the estimated standard error, then it is recommended that the equivalent two-parameter distribution (Gumbel, logistic or Pareto) should be used. Pooled growth curves are fitted using pooled L-moment statistics, which are calculated by taking a weighted average of the L-moment ratios of the sites in the pooling group. Longer records are given more weight, as are sites that are most similar to the subject site.

13.4 Example: *Flood Estimation Handbook* peak flow estimates for the River Brett, Suffolk

The Brett is a tributary of the River Stour in the county of Suffolk in south-east England. The Brett flows though a rural landscape made up of arable land, grassland, and some areas of ancient woodland. The underlying geology is chalk overlain by semi-pervious boulder clay. Soils in the catchment are mixed between freely draining acid loamy soils, and lime-rich loamy clayey soils with slightly impeded drainage. Average annual rainfall for the area is 586 mm.

13.4.1 Requirements for flood estimation

Peak flow estimates were required for a location on the River Brett just downstream of the village of Monks Eleigh. The catchment draining to this point has an area of 85 km² and is as shown in Fig. 13.1. *FEH* catchment descriptors show that the catchment is moderately impermeable (SPRHOST = 45.7, which is consistent with the soils and land cover) with almost no flow attenuation from lakes or reservoirs expected (FARL = 0.99). *FEH* methods are considered appropriate because the catchment is larger than the recommended limit of 0.5 km² and is not heavily urbanised. The proximity of a gauging station at Hadleigh just downstream of the study reach favours the use of the *FEH* statistical method. The revised (2008) *FEH* statistical methods are adopted.

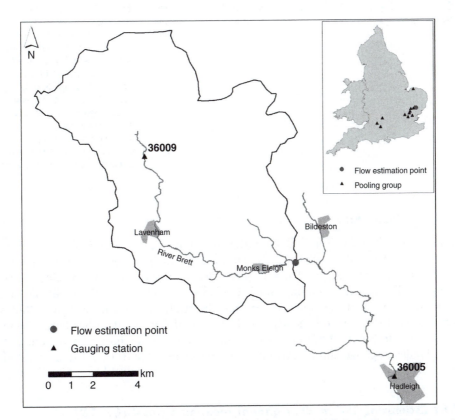

Fig. 13.1 FEH study catchment at Monks Eleigh on the River Brett, Suffolk showing nearby gauging stations. Inset map shows location of the study site and gauging stations selected for pooled growth curve analysis.

Table 13.2 Gauging stations near to the study site on the River Brett

Station name	Drained area (km²)	Period of record	Suitable for QMED estimation?	Suitable for pooling?	Other comments on station and flow-data quality
36005 (Hadleigh)	156	1962–2006	Yes – just acceptable	Yes – good fit at high flows	'Essex' profile (modified Flat V Crump) weir with low-flow side weir and high-flow rated spillway. Downstream water level recorder to allow for drowning. Naturalised flows from 1962 to 1976. Since 1976, adjustments for artificial influences are no longer made to the gauged daily mean flows. High flows gauged by bridge gauging downstream of the gauging station
36009 (Cockfield)	26	1968–2006	Yes – good for QMED	Yes – just acceptable for pooling	'Essex' profile (modified Flat V Crump weir). No spillway. Modular limit of 0.66 m theoretically derived. No telemetry but planned for future. Naturalised flows from 1969 to 1976, only minimal adjustments needed since. High flows measured by cableway gauging upstream of the gauging structure. High flows were calculated or estimated. Also some float-runs

13.4.2 Available data

The latest river flow data set available at the time of the study was HiFlows-UK Version 2.2.1 issue of December 2008. Catchment descriptors were obtained from the *FEH* CD-ROM v3.0 dated 2009. There are two river flow-gauging stations close to the study site (Table 13.2). Station 36005 at Hadleigh is closest to the study site and has a catchment area 1.8 times larger than the study catchment. Station 36009 at Cockfield is also close to the study site. However, this gauge data has been rejected as a donor site for the *FEH* analysis because it has a poor rating curve compared to the Hadleigh gauge and its catchment is smaller by a factor of 3.3 than the study catchment.

13.4.3 Estimating the median annual maximum flood (QMED)

The *FEH* empirical formula for estimating QMED from catchment descriptors (13.3) predicts QMED = $8.9 \, \mathrm{m^3 \, s^{-1}}$. This figure was adjusted using (13.5) and an estimate of

QMED from gauged data at the nearby Hadleigh gauging station to give a final estimate of $8.1 \, \text{m}^3 \, \text{s}^{-1}$. The close proximity of the catchment gauged by the Hadleigh station to the study catchment (a separation of 3.9 km in terms of catchment centroids) and the very similar physical and climatic characteristics mean that the QMED adjustment increases confidence in the analysis.

13.4.4 Pooled frequency analysis

The pooling group for the study site is shown in the inset map in Fig. 13.1. Stations for pooling are selected based on the similarity of their catchments to the study area. It can be seen that most of the stations selected for pooling are clustered in the same geographic area as the study site. There are four pooling group catchments that are much further away, but which have been selected based on their similar physical characteristics. Analysis of the pooling group includes checks on the suitability of each gauging station. In this case, one station was removed because of significant attenuation from lakes and reservoirs within its catchment area. Another station was highlighted as discordant owing to a large annual maximum (AMAX) flow value that appeared as a statistical outlier. On inspection, this station was retained, as the 'outlying' high flow was believed to be related to accurate gauging of real flood flows.

The final pooling group contains 528 station years of AMAX data. The GL distribution (13.1) fits the data well, with parameter values as follows: 'location' $a = 1.00$, 'scale' $b = 0.264$ and 'shape' $k = -0.057$. The growth curve for the pooling group is shown in figure along with the individual growth curves for each catchment within the group.

Peak-flow estimates for the site follow easily from multiplication of the growth factor Q/QMED for a given return period with the estimated value of QMED. The estimated peak flow for a 100-year return period (1 per cent AEP) is $19.3 \, \text{m}^3 \, \text{s}^{-1}$, or 2.4 times QMED. The pooling group gauges have record lengths of between 36 and 67 years. Assuming the pooling group is unbiased, the scatter between the background curves in Fig. 13.2 is indicative of the sampling uncertainty associated with the gauged records. For further discussion of the theory and analysis of uncertainty in pooled flood frequency analysis, see Rosbjerg and Madsen (1995), Kjeldsen and Jones (2006) and Kjeldsen et al. (2010).

13.4.4.1 Methodology checks

Although the study site has no gauging station, it is good practice to compare the estimated flood frequency curve derived from the pooling group with gauged AMAX flows in the record at Hadleigh. Applying the growth curve to the Hadleigh record would imply that in the 44 years of record, ten gauged annual maxima had return periods of between 2 and 5 years, two had return periods of between 5 and 10 years, three between 10 and 50, and the largest recorded event, in October 1987, would have had an approximate return period of 100 years. These estimates are considered plausible. Specific runoff for the catchment area at the estimated 100-year peak flow is approximately $0.8 \, \text{mm} \, \text{h}^{-1}$, which is a plausible value for a relatively permeable catchment with mild slopes and rural land cover.

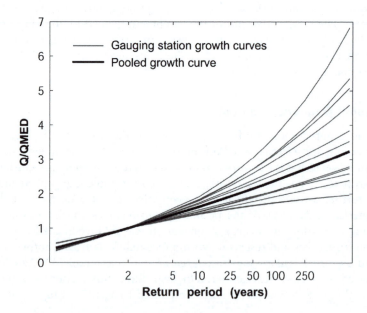

Fig. 13.2 Pooled dimensionless growth curve for the River Brett study site superimposed on growth curves for individual gauging stations within the pooling group.

13.5 Design event methods in the *Flood Estimation Handbook*

13.5.1 The FEH event models

The original *FEH* published in 1999 contained an event-based rainfall–runoff model, a modified version of that used in the previous *Flood Studies Report* of 1975 (Natural Environment Research Council, 1975). It is an example of a model that is based on the concepts of effective rainfall and storm runoff discussed in Section 12.3 and 12.4. It is primarily intended for use at ungauged sites where there is information about rainfalls, but no discharge data available for calibration, particularly in predicting the flood hydrograph that might arise in a catchment for a particular extreme rainstorm event or design storm. The purpose of such a prediction might be in designing a flood runoff detention basin for a new commercial development, or in a dam safety assessment, including the estimation of a probable maximum flood, or for providing the inputs to a flood inundation model for flood risk mapping and planning. The *FEH* event model essentially combines the calculation of effective rainfall described below in Section 13.5.2, the triangular unit hydrograph described in Section 13.5.4, and a baseflow component.

13.5.2 Estimation of a design storm event

The *FEH* flood hydrograph method is a single event-based rainfall–runoff model. Thus, an important part of the method is the choice of design rain event, as defined in terms of storm duration, total depth and profile, together with the antecedent conditions for the catchment.

Following the *FEH* recommendations, the storm duration for a particular catchment is defined as:

$$D = T_p \left(1 + \frac{SAAR}{1000}\right) \tag{13.14}$$

where T_p is the estimated time to peak for the catchment, and *SAAR* is the standard annual average rainfall for the period 1961–90. The total storm depth is estimated using the *FEH* depth–duration–frequency curves (see Section 9.6.1). Depth–duration–frequency analysis provides standard tools for estimating the total storm volume and duration for different probabilities of exceedance. A complete description of a design storm for use in the event model requires more, however. In particular we will be interested in the average rainfall over a catchment area rather than rainfall at a point, so that the estimated storm depth should be adjusted by an areal reduction factor (see Section 9.4).

We will also require a storm profile, although in the analyses leading to the *FEH*, the estimated peak discharges were found to be relatively insensitive to the choice of storm profile. Thus, two profiles were suggested for use in design, the 75 per cent winter profile for use in rural catchments and the 50 per cent summer profile for use in urbanised catchments. The percentages refer to the quantiles over all storms analysed, when normalised by storm depth and duration (see Fig. 13.3). These simple profiles have been criticised as too simple, particularly for large catchments with reservoirs where the critical event may be several days long and made up of multiple events; and the Institution of Civil Engineers (1996) has recommended using the profile of the severest sequence of storms of the required duration in the local observed rainfall records.

Finally, the original *FEH* event model method makes an allowance for the fact that we do not expect a rainfall event of a given probability of exceedance to produce a

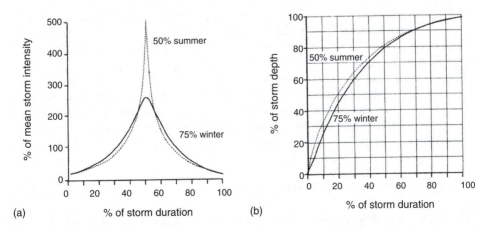

Fig. 13.3 FEH design storm profiles for rural and urban catchments in form of: (a) profiles; and (b) as cumulative percentages of storm depth (from *FEH* Vol. 4, 3.5, Copyright NERC (CEH)).

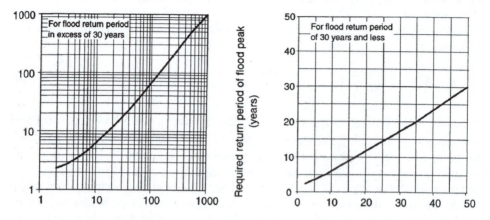

Fig. 13.4 FEH recommended storm return periods to estimate a flood peak of a required return period (from *FEH* Vol. 4, 3.2, IoH, 1999, Copyright NERC (CEH)).

flood peak with the *same* probability of exceedance. This is because of catchment characteristics and antecedent wetness effects. An extreme rainfall on dry ground in a rural catchment in summer might produce a smaller flood peak than a less extreme rainfall falling on a wet catchment in winter. Thus the *FEH* method makes recommendations about the choice of rainfall return period that will best estimate a flood peak of the required return period (Fig. 13.4).

13.5.3 Estimation of percentage runoff

Given a design rainfall, the next stage in the design event method is to estimate the percentage runoff. To do this, *FEH* provides a method to estimate the standard percentage runoff, which is then modified to reflect particular conditions. The estimation of SPR is based on empirical equations for percentage runoff derived from the analysis of the records from UK catchments up to 500 km^2 in area. Once a percentage runoff is defined for a storm, it is applied as a constant proportional multiplier at each time increment of rainfall (see Fig. 12.2b). Note that this will result in a different time distribution of effective rainfall to the Φ index approach shown in Fig. 12.2a, even if both can match a required volume of effective rainfall.

The calculation of effective runoff for a rural catchment takes account of the soil characteristics of the catchment, the event rainfall volume and a catchment wetness index effect. It consists of two parts, a SPR and dynamic components (dynamic percentage runoff; DPR) that depend on the antecedent catchment wetness prior to an event (DPR_{CWI}) and the event magnitude (DPR_{RAIN}). Percentage runoff (PR) for a rural catchment is then calculated as:

$$PR_{rural} = SPR + DPR_{CWI} + DPR_{RAIN} \qquad (13.15)$$

The *FEH* gives a number of ways of determining SPR. When a site is gauged, then SPR can be determined as an average value from an analysis of observed events. A quicker method is provided by a relationship to the BFI, for a catchment that is tabulated for all gauged catchments in the UK. This is:

$$SPR = 72 - 66.5\,BFI \qquad\qquad (13.16)$$

If an ungauged site is being studied, however, then SPR must be estimated from catchment characteristics. The method for doing so depends on the HOST classification introduced earlier in Section 13.2. Each HOST class is associated with a standard percentage runoff (Table 13.3). The SPR is given by a simple weighted sum over all HOST classes.

$$SPR = \sum_{i=1}^{29} SPR_i HOST_i \qquad\qquad (13.17)$$

where $HOST_i$ is the proportion of the catchment mapped to HOST class i. For the other components of (13.3)

$$DPR_{CWI} = 0.25\,(CWI - 125)$$

$$DPR_{RAIN} = 0.45\,(P - 40)^{0.7} \qquad \text{for } P > 40\,\text{mm} \qquad\qquad (13.18)$$

$$DPR_{RAIN} = 0 \qquad\qquad\qquad\qquad\; \text{for } P \leq 40\,\text{mm}$$

This dynamic adjustment requires a value of the catchment wetness index (CWI). This changes over time as the catchment wets and dries, although for design purposes the *FEH* also gives a relationship between a design CWI and the standard annual average rainfall in a catchment (Fig. 13.5).

For particular events, a more complex calculation of the initial CWI is carried out, depending on the antecedent rainfall and soil moisture deficit for that event (see *FEH*, Vol. 4, Appendix A).

Table 13.3 Recommended standard percentage runoff (SPR) values for different HOST classes

HOST class	SPR (%)	HOST class	SPR (%)
1	2.0	16	29.2
2	2.0	17	29.2
3	14.5	18	47.2
4	2.0	19	60.0
5	14.5	20	60.0
6	33.8	21	47.2
7	44.3	22	60.0
8	44.3	23	60.0
9	25.3	24	39.7
10	25.3	25	49.6
11	2.0	26	58.7
12	60.0	27	60.0
13	2.0	28	60.0
14	25.3	29	60.0
15	48.4		

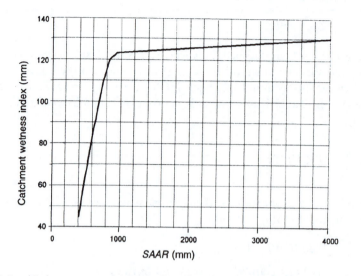

Fig. 13.5 Relationship between standardised value of catchment wetness index and standard annual average Rainfall, SAAR (*FEH* Vol. 4, 3.7, Copyright NERC (CEH)).

A final modification can take account of the effects of urban development in a catchment as:

$$PR = PR_{RURAL}\,(1 - 0.615\ URBEXT) + 70\,(0.615\ URBEXT) \qquad (13.19)$$

where *URBEXT* is the fraction of urban area in the catchment.

The calculations required for the *FEH* estimation of effective rainfall are greatly facilitated by the availability of the *FEH* software that allows the calculation of the relevant catchment characteristics for any point on the river networks of the UK. The method has provided a standard for estimating flood runoff in the UK and continues to do so through the evolution of the 'revitalised' flood hydrograph method (see later). What is currently missing from the *FEH* methodology, however, is any consideration of uncertainty in the estimation methods outlined above.

13.5.4 The FEH unit hydrograph

The UK *Flood Estimation Handbook* (Institute of Hydrology, 1999) continues the practice of the earlier *Flood Studies Report* in using a standardised triangular form of the unit hydrograph (Fig. 13.6). It is defined by only two parameters which can be related to catchment characteristics for application to ungauged catchments.

The *FEH* suggests a number of methods of estimating the peak of the instantaneous unit hydrograph, T_p, including the analysis of several flood events at gauging sites or by scaling from a nearby donor gauging site. At ungauged sites, where no events are

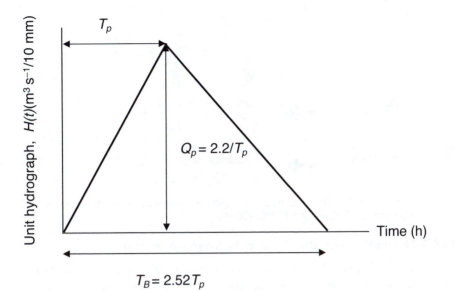

Fig. 13.6 The FEH unit hydrograph.

available for analysis, the regression relationship

$$T_P = 4.270\, DPSBAR^{-0.35}\, PROPWET^{-0.80}\, DPLBAR^{0.54}$$
$$\times (1 + URBEXT_{1990})^{-5.77} \quad \text{(h)} \tag{13.20}$$

where the variables describing the catchment characteristics are defined in Table 13.1.

The base time of the unit hydrograph (Fig. 13.6) is then given by

$$T_B = 2.52 T_P \quad \text{(h)} \tag{13.21a}$$

and the peak by

$$Q_P = 2.2/T_P \quad (\text{m}^3\,\text{s}^{-1} \text{ per } 10\,\text{mm over } 1\,\text{km}^2) \tag{13.21b}$$

where the constant 2.2 depends on the units and on the constraint that the area of the triangle should represent a certain volume of effective rainfall. The standard value of 2.2 is for 10 mm of effective rainfall and for a discharge defined in m^3 s^{-1} over a catchment area of 100 km^2. Thus, more generally, for an effective rainfall of R_{eff} mm and a catchment area of A km^2, q_p in m^3 s^{-1} would be given by

$$Q_P = 2.2 \frac{R_{eff}}{10} A T_P \quad \text{m}^3\,\text{s}^{-1} \tag{13.22}$$

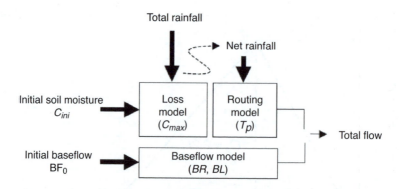

Fig. 13.7 Structure of the *FEH* 'revitalised' flood hydrograph method (*ReFH*) design event model (after Kjeldsen, 2007, Copyright NERC (CEH)).

13.5.5 The FEH 'revitalised' flood hydrograph method

The *FEH* 'revitalised' flood hydrograph method (*ReFH*) method was developed to make improvements to the main components of the *FEH* design event approach in order to address problems identified with the original method in large-scale comparison studies (Webster and Ashfaq, 2003), in particular a widely held view that the design flows it produced were too high and that it produced overly symmetrical hydrographs. The *ReFH* method is described in *FEH* Supplementary Report No. 1 (Kjeldsen, 2007). The basic structure of *ReFH* (Fig. 13.7) retains the fundamental form of the *FEH* design event model, that is, a loss module combined with a unit hydrograph 'routing' module, superimposed on a baseflow hydrograph.

The loss model in the *ReFH* method uses the concept of the probability distributed soil moisture model or PDM (which is described in Section 12.8.1). Soil moisture is continually updated throughout the event, which in effect allows for an increase in percentage runoff throughout the storm, answering one criticism of earlier event-based rainfall models. The parameter C_{ini} is the soil moisture content at the start of the storm. The parameter C_{max} represents the maximum localised soil moisture storage capacity in the catchment. For gauged catchments it is best estimated from flow and rainfall data using an optimisation method that simultaneously estimates the time to peak. Otherwise, it can be estimated from catchment descriptors:

$$C_{max} = 596.7 \, BFIHOST^{0.95} \, PROPWET^{-0.24} \tag{13.23}$$

The output from the loss model is net rainfall, which is routed to the catchment outfall using a 'kinked triangle' unit hydrograph, as shown in Fig. 13.8. This avoids the excessively symmetrical hydrographs that could be derived from the *FEH* rainfall–runoff method and also results in a reduced peak flow. It must be borne in mind that the time to peak for *ReFH* is not the same as that used in the *FEH* rainfall–runoff method. It can be estimated by calibration against event data or from catchment descriptors as

$$T_p = 1.563 \, PROPWET^{-1.09} \, DPLBAR^{0.60} \, (1 + URBEXT_{1990})^{-3.34}$$
$$\times \, DSPBAR^{-0.28} \tag{13.24}$$

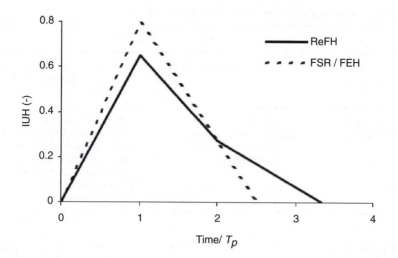

Fig. 13.8 Instantaneous Unit Hydrographs (IUH) used in the *ReFH*, Flood Studies Report (FSR) and *Flood Estimation Handbook* (FEH) rainfall-runoff methods (after Kjeldsen, 2007, Copyright NERC (CEH)).

The constant baseflow value used in the *FEH* rainfall–runoff model has been replaced in the *ReFH* baseflow model, which allows variation through the event. The model is based on a linear reservoir concept, with inflow to the reservoir being recharge which is assumed to come from saturated land that is also producing runoff; elsewhere it is assumed that rainfall inputs are retained as soil moisture that will become evapotranspiration and longer term recharge. The ratio of runoff to recharge is controlled by the baseflow recharge parameter, *BR*. Baseflow lag (BL) controls the decay of baseflow from the linear reservoir, which is an exponential function in time (see equations 12.6). Both the baseflow parameters are best calculated by fitting recession curves to flow data. Otherwise they can be estimated from catchment descriptors, as follows:

$$BR = 3.75\ BFIHOST^{1.08}\ PROPWET^{0.36} \tag{13.25}$$

$$BL = 25.5\ BFIHOST^{0.47}\ DPLBAR^{0.21}\ PROPWET^{-0.53}$$

$$\times (1 + URBEXT_{1990})^{-3.01} \tag{13.26}$$

The *ReFH* rainfall–runoff model and the design event inputs have been calibrated so as to match, on average, flood frequency curves derived from pooled statistical analysis at 100 gauging stations. The model was calibrated up to return periods of 150 years, relatively long compared with the *FEH* rainfall–runoff model, which was only calibrated to the 10-year return period. The number of catchments used to calibrate the two methods is similar; however, calibration of the *ReFH* method used more data at each site, and included larger events. The *ReFH* method differs from the original *FEH* design event model in that it has been calibrated such that the T-year design rainfall

Table 13.4 Comparison of *FEH* rainfall–runoff and *ReFH* methods

	FEH	ReFH
Losses calculation	Constant percentage runoff	Varies through event, calculated by continuous accounting of soil moisture
Unit hydrograph shape	Triangle	Kinked triangle
Baseflow calculation	Constant	Varies through event
Specification of return period	Rainfall return period different from flow (winter event)	Calibrated such that design rainfall return period is the same as the required flow return period. Losses adjusted for return period
Seasonality	Winter/summer rainfall profiles	Rainfall depth and initial soil moisture vary with season

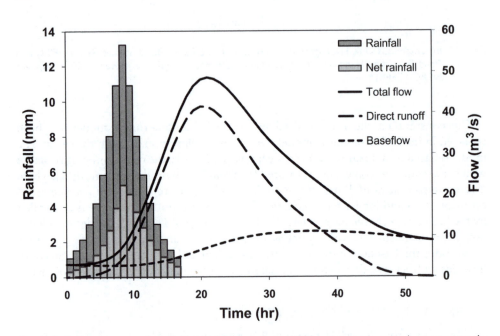

Fig. 13.9 Design inputs and modelled hydrograph for a 1000-year return period event over the River Brett.

is used to generate the T-year design flow. The calculation of *ReFH* model parameters and design inputs also accounts for seasonality. Table 13.4 summarises the main differences between the original *FEH* and the *ReFH* design event methods.

Fig. 13.9 shows the design rainfall, net rainfall and runoff components estimated using the *ReFH* method for a 1000-year return period event on the River Brett, as described earlier. The change in effective percentage runoff during the event can be seen by comparison of the design rainfall and net rainfall hyetographs. The *ReFH* model was used in this study to obtain a hydrograph estimate for the long return period of 1000 years. For such a rare event, the pooled *FEH* statistical method, which used 500 station-years of record, is not considered adequate for confident estimation. The design event method is chosen instead so as to make use of information in rainfall frequency estimates, which are regarded as more reliable for very rare events, because many rainfall records are significantly longer than flow records.

13.6 Low-flow estimation in the UK

A key element in water resources planning is to assess the low-flow resource available from water-courses. This provides information on whether there is a surplus or deficit of water available to meet current licensed abstractions. Scenarios are usually evaluated with reference to the flow rate that can be expected for a specified number of days in the year, i.e. the flow–duration curve. Hydrological analysis is therefore required to estimate mean flow and flow–duration curve statistics. In 1998 a survey found that there were 4400 requests for low-flow statistics made each year in the Environment Agency (Gustard *et al.*, 2004).

The approach used in the UK for assessing low flow for abstraction management planning is to produce flow–duration curves for a range of situations including the natural, current and future demand scenarios. Flow–duration curves can be derived directly from the many continuous recording flow gauges in England and Wales. However, despite this dense gauging network, by international standards, over 95 per cent of river reaches in England and Wales are distant from a flow-measuring station. At these sites a model to regionalise the low-flow statistics is required.

There are many techniques that have been applied in different parts of the world for estimating statistics describing the low-flow regime at ungauged sites. In Europe, studies include Martin and Cunnane (1976) in Ireland, Lundquist and Krokli (1985) in Norway, and Gustard *et al.* (1989) in northern and western Europe. Tallaksen and van Lanen (2004) give an overview of the low-flow hydrology methods.

The 1980 *Low Flow Studies Report* (Institute of Hydrology, 1980) was the first major study of the relationships between low-flow regimes and physiographic and climatic catchment characteristics in the UK. Subsequently there have been many regional low-flow estimation procedures developed for application within the UK including Pirt and Douglas (1982), Gustard *et al.* (1987) and Young *et al.* (2000). The basis for the methods now used in the UK was presented in detail by Holmes *et al.* (2002a, b) and are incorporated into the *LowFlows* software (see Section 13.6.4).

13.6.1 Naturalised river flows

Natural river flow regimes are dependent on rainfall, temperature and evaporation. On a local scale, the flow regime is controlled by the physical properties of a catchment, including geology, land use and the presence of surface water bodies. River flow regimes are also affected directly and indirectly by human activities. A review of over 1600 gauging stations has identified that <20 per cent of gauged catchments within the UK can be regarded as being natural (Gustard *et al.*, 1992). The impacts of these activities vary considerably and are dependent to a certain extent on the characteristics of the catchment.

For water management purposes, it is essential to differentiate between the natural and artificial components of stream-flow data. Fig. 13.10, for example, shows gauged and naturalised discharges for the River Thames at Kingston. Flow naturalisation is the process of adjusting an observed flow hydrograph to remove the effects of artificial influences. Artificial influences include surface and

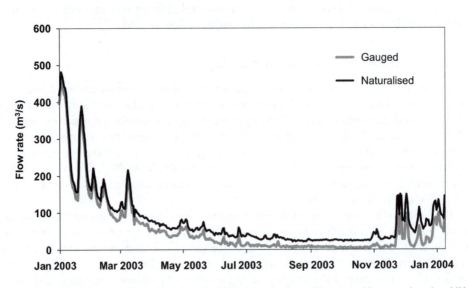

Fig. 13.10 Gauged and naturalised flow data for 2003 form the River Thames at Kingston, London, UK.

groundwater abstractions, discharges from sewage treatment plants and industrial sources, impounding reservoirs, canal transfers and inter-basin transfer schemes. The difficulties in obtaining reliable data on these influences and the need for assumptions means that naturalisation is notoriously difficult to carry out and may be associated with errors up to something on the order of 40 per cent. Despite the uncertainty, separation of the components enables assessment of the natural reliable yield of the catchment, based upon the climatically driven variability of the natural stream flow. The impacts of actual and planned water resource management scenarios can then be compared with the natural flow regime to assess yield and environmental impact.

13.6.2 Estimation of low-flow statistics in the UK

Constraints on exploitation of water resources are concentrated around the low flows, and so estimation of the natural flow regime in water resources is often synony-mous with assessing the low-flow statistics. Four methods for estimating low flows are commonly applied in the UK. These are:

(1) calculation of low-flow statistics from continuous gauged flow data series;
(2) direct measurement of flows by an occasional programme of flow measurement using current meters or temporary gauges;
(3) estimation of time series of river flow using rainfall–runoff models;
(4) estimation of flow statistics by using generalised models which relate low flows to the physical and climatic characteristics of the catchment.

Where continuous flow data are available for the catchment of interest, method 1 is the most accurate and preferred technique. However, flow estimates are often required for ungauged catchments and therefore method 4 is commonly used.

Table 13.5 Methodology for estimating natural and artificially influenced low-flow statistics for an ungauged location, as used in the UK in the '*LowFlows*' system

Step		Data, model or output
1	Estimation of key natural low-flow statistics for the ungauged catchment	Mean flow, monthly mean flow, monthly flow–duration curves and mean monthly minima
2	Identification of all upstream artificial influences	Abstractions from surface and groundwater sources, discharges to surface water, impacts of impounding reservoirs
3	Quantification of all individual upstream artificial influences	Actual values of monthly abstraction rates, discharge returns and reservoir compensation flows (represented as release profiles)
4	Simulation of the reduction in stream flow associated with abstractions from groundwater sources	Analytical model based on source and aquifer properties
5	Construction of a monthly artificial influence profile for upstream abstractions. Construction of release duration profiles for each upstream impounding reservoir	Represents the net impact of all upstream abstractions and discharges (excluding those above any impounding reservoirs) within the catchment
6	Combination of the estimated natural monthly low-flow statistics with the artificial influence profiles	Generates estimated artificially influenced monthly low-flow statistics
7	Aggregation of monthly artificially influenced low-flow statistics	Generates annual artificially influenced flow statistics, mean flow and flow–duration curves
8	Estimation of natural and artificially influenced low-flow statistics for discrete river stretches	Generates residual flow diagrams

The impact of artificial influences is most severe during periods of low flows when absolute volumes of water transfers represent a significantly higher proportion of the natural flow regime. Natural low-flow statistics and artificial influence data are therefore analysed on a monthly basis and allows the seasonal variations to be taken into account. The overall methodology for estimating both natural and artificially influenced low-flow statistics in ungauged catchments is summarised within eight steps listed in Table 13.5.

13.6.3 Models for estimation of natural low-flow statistics

For regionalised analysis of flow duration statistics, it is usual to express the individual catchment daily flows as a percentage of the long-term mean flow for the catchment to remove the majority of the influence of hydrological scale. Low-flow estimation methods in the UK use the HOST hydrological response classification (Boorman *et al.*, 1995) of soils as an indication of the hydrogeological and soil characteristics within a catchment, alongside data about average annual runoff. The procedure for estimating the annual flow–duration curve and mean flow from catchment characteristics is shown in Fig. 13.11.

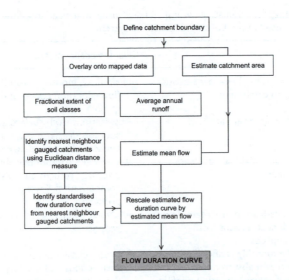

Fig. 13.11 Estimation of naturalised low-flow statistics for ungauged catchment.

The methods are based around using a digital terrain model to generate a catchment boundary for any location on a watercourse. Using this boundary, the physical catchment characteristics are extracted from digital grids.

A region-of-influence (ROI) regionalisation approach is used as described by Holmes *et al.* (2002a). The ROI approach develops an estimate of a flow statistic at an ungauged 'target' catchment from observed values of that flow statistic made at a set of gauged catchments, which are considered to be hydrologically similar to the target. This set is considered a 'region' within the space defined by the parameters used to assess similarity (note that this is not necessarily geographical space). In application to a catchment, the methods can be summarised as the following steps:

(1) catchment similarity is assessed by calculating a weighted Euclidean distance, in HOST parameter space, between the target catchment and the catchments within a similar climatic pool;
(2) a 'region' is formed around the target catchment by ranking all of the catchments in the data pool by their weighted Euclidean distance in HOST space and selecting the five catchments that are closest to the target catchment;
(3) a standardised annual flow duration curve is estimated for the ungauged site by taking a weighted combination of the standardised flow duration curves for the gauged catchments within the region.

In step 1, the Euclidean between the target catchment, indexed by t, and the ith catchment from the climatically similar data pool is calculated as

$$d_{it} = \sum_{m=1}^{M} W_m \left(X_{mi} - X_{mt} \right)^2 \tag{13.27}$$

where W_m is the weight applied to the mth of M catchment characteristics and X_{mi} is the standardised value of that catchment characteristic for catchment i. The catchment characteristics used are the fractional extents of the HOST classes within a catchment, which will vary between zero and unity. Differing weights are applied for each HOST class to reflect the fact that relatively small proportions of certain HOST classes strongly influence the variability of the flows within a catchment.

The process for estimating the standardised annual flow–duration curve for the target site from those within the region involves weighting each of the donor catchment flow–duration curves by the inverse absolute distance, in weighted HOST space, of the catchment from the target catchment. Thus, greater weight was given to catchments that are more similar in HOST characteristics to the target catchment.

The standardised flow–duration curve has to be re-scaled by multiplication with an estimate of the mean flow to compute the required flow–duration statistics for resource assessment. An estimate of the long-term natural mean flow is obtained by re-scaling an estimated value of annual runoff by catchment area. For the UK low-flows procedures, an annual average runoff grid was derived from the output of a daily time-step, regionalised soil moisture accounting model based on the Penman drying curve and calibrated against stream flow data (Holmes *et al.*, 2002b).

Long-term natural mean monthly flows are calculated using a ROI approach where similarity is measured with respect to both HOST classes and average annual rainfall. Long-term natural average flow–duration curves for specific calendar months are generated in an identical manner to the long-term average flow–duration curves. These standardised curves are re-scaled by the long-term natural mean monthly flows.

13.6.4 Software tools

The UK-standard methods described above are implemented in the *LowFlows* software[4] (Young *et al.*, 2003; Holmes *et al.*, 2005), which is widely used by the Environment Agency in the development of *Catchment Abstraction Management Strategies* (CAMS; see Chapter 17) in England and Wales, and by SEPA in the implementation of the Controlled Activity Regulations in Scotland.

13.6.5 Estimation of artificially influenced flow statistics

Versions of the *LowFlows* software used operationally include a geo-referenced database of influence features, such as surface and groundwater abstractions, impounding reservoirs and discharges. The information held for these features may be very complex; e.g. a licence to abstract may relate to tens of distinct sites, which in turn may be licensed for abstraction relating to multiple purposes. The licensed quantities for the whole licence, its constituent sites and individual purposes may also be highly interdependent.

Abstraction and discharge points are quantified in terms of a typical monthly volume for each calendar month within the year; this is termed a monthly profile. In the case of abstraction licences and discharge consents, the monthly volumes relate to water that is either abstracted or discharged. The software will use actual recorded data, if loaded, to represent the monthly profile for a site, or if no actual data are available, the software will estimate a profile based on authorised volumes and patterns observed in historical data.

For abstractions from groundwater, an estimate of the impact of the monthly abstraction profile on the nearest river reach is derived. This is derived using an algorithm based upon the Jenkins superposition method (Jenkins, 1970), applied to the Theis analytical solution for predicting the impact of a groundwater abstraction from an unconfined aquifer. This algorithm requires the user to define values for aquifer transmissivity and storativity. The distance of the abstraction site from the nearest stream is calculated automatically using the grid reference of the site in conjunction with a stored digital river network.

The method for adjusting natural flows for the impact of impounding reservoirs is equivalent to replacing natural river flows and artificial influences upstream of the dam site by 12 monthly reservoir release duration curves, which combine mean monthly compensation flows, reservoir spill and augmentation releases, or freshets if appropriate.

Based on the natural and artificially influenced estimation methods, the *LowFlows* system can be used to estimate the flow–duration curve for current or future abstractions scenarios.

Fig. 13.12 illustrates an example where one such scenario is compared with the (estimated) naturalised and artificially influenced flow regime, showing large differences in resource availability at low flows. This figure also illustrates the typical complexity

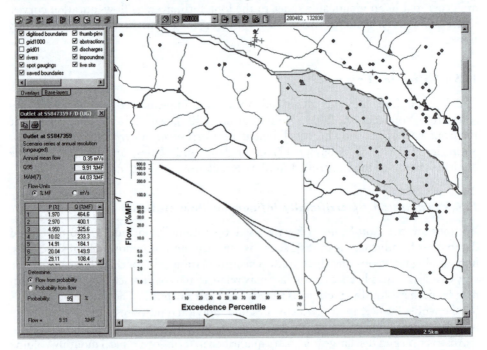

Fig. 13.12 Flow duration curves for natural and influenced flow regimes obtained from LowFlows. Reprinted from Environmental Modelling & Software, Volume 20 Issue 2, M. G. R. Holmes, A. R. Young, T. H. Goodwin and R. Grew, A catchment-based water resource decision-support tool for the United Kingdom, Pages 197–202, Copyright (2005), with permission from Elsevier.

of artificial influence features within a catchment where water resources are being exploited.

13.7 Some final comments on methods of regionalisation

The *FEH* and *LowFlows* methods are, as noted in the introduction to this chapter, a response to the requirement in practical applications of estimating the nature of the hydrological response for a catchment anywhere in the UK, with and without a local flow gauge. They depend, in estimating the response of ungauged basins, very heavily on statistical methods of regionalisation of the discharge frequency characteristics against catchment descriptor variables. Other countries have addressed the regionalisation problem using related statistical methods (e.g. Jennings *et al.*, 1994; Pilgrim 1999; Turnipseed and Ries, 2007). It is known that such methods will produce rather uncertain estimates, but the methods are often used deterministically as current 'best estimates'.

It has been necessary to resort to these statistical relationships because our representations of hydrological processes, as embodied in the hydrological models of Chapter 12 are not yet good enough to be easily applied to ungauged catchments purely on the basis of knowledge of soil, geology and land management information. An international programme, the prediction of ungauged basins (PUB)[1] initiative of the International Association of Hydrological Sciences, is currently under way to try to improve this situation and find ways of reducing the uncertainty in flood estimation more generally.

In the UK there have also been attempts to develop alternative methods for design flow estimation in ungauged based on *continuous simulation*, where gauged or simulated rainfall data are applied as inputs to a rainfall–runoff model such as PDM or TOPMODEL (see Chapter 12) to generate a long synthetic river flow series, which can be analysed using statistical techniques such as the flow–duration curve, *FEH* methods or plotting position formulae (Chapter 11). For a review of the development and application of the continuous simulation approach for flood estimation, see Lamb (2005). The approach is particularly useful for including temporal changes in climate input data, model parameters or artificial influences (e.g. abstractions or flood control structures). Regionalisation of the PDM and TATE models for flood frequency estimation in the UK has been described by Calver *et al.* (2005), whilst uncertainty about regionalised estimates was studied by Lamb and Kay (2004). Low flows have been modelled using a version of the PDM parameterised for the UK by Young (2002). The regionalisation of conceptual rainfall–runoff models in general has been discussed by Wagener *et al.* (2004) in the UK, Parajka *et al.* (2007) in Austria and Zhang *et al.* (2008) in the USA.

Notes

1 See http://pub.iwmi.org/UI/Content/Default.aspx?PGID=0
2 http://www.ceh.ac.uk/feh2/FEHintro.html
3 http://www.environment-agency.gov.uk/hiflows-uk
4 http://www.hydrosolutions.co.uk/lowflows.html

References

Acreman, M. C. and Horrocks, R. J. (1990) Flood frequency analysis for the 1988 Truro floods. *Journal of the Institution Water and Environmental Management* 4, 62–69.

Archer, D. R., Leesch, F. and Harwood, K. (2007) Assessment of severity of the extreme River Tyne flood in January 2005 using gauged and historical information. *Hydrological Sciences Journal* 52, 1–12.

Bayliss, A. C. and Reed, D. W. (2001) *The Use of Historical Data in Flood Frequency Estimation.* Report to Ministry of Agriculture, Fisheries and Food, Centre for Ecology and Hydrology, Wallingford.

Black, A. R. and Law, F. (2004) Development and utilisation of a national web-based chronology of hydrological events. *Hydrological Sciences Journal* 49, 237–246.

Boorman, D. B., Hollis, J. M. and Lilly, A. (1995) *Hydrology of Soil Types: a Hydrologically Based Classification of the Soils of the United Kingdom.* Institute of Hydrology, Report No. 126, Wallingford.

Calver, A., Crooks, S., Jones, D., Kay, A., Kjeldsen, T. and Reynard, N. (2005) *National River Catchment Flood Frequency Method using Continuous Simulation Modelling.* R&D Technical Report FD2106/TR, Department for Environment, Food and Rural Affaris, London.

Gustard, A., Bullock, A. and Dixon, J. M. (1992) *Low Flow Estimation in the United Kingdom,* Institute of Hydrology, Report No. 108, Wallingford.

Gustard, A., Marshall, D. C. W. and Sutcliffe, M. F. (1987) *Low Flow Estimation in Scotland.* Institute of Hydrology, Report No. 101, Wallingford.

Gustard, A., Roald, L., Demuth, S., Lumadjeng, H. and Gross, R. (1989) *Flow Regimes from Experimental and Network Data (FREND)* [in 2 vols]. Institute of Hydrology, Wallingford.

Gustard, A., Young, A. R., Rees, H. G. and Holmes, M. G. R. (2004) Operational hydrology. In: Tallaksen, L. M. and van Lanen, H. A. J. (eds) *Hydrological Drought.* Developments in Water Science, Vol. 48. Elsevier, Amsterdam, pp. 455–485.

Holmes, M. G. R., Young, A. R., Gustard, A. and Grew, R. A. (2002a) A new approach to estimating mean flow in the UK. *Hydrology and Earth System Sciences* 6, 709–720.

Holmes, M. G. R., Young, A. R., Gustard, A. and Grew, R. A. (2002b) A region of influence approach to predicting flow duration curves within ungauged catchments. *Hydrology and Earth System Sciences* 6, 721–731.

Holmes, M. G. R., Young, A. R., Goodwin, T. H. and Grew, R. (2005) A catchment-based water resource decision-support tool for the United Kingdom. *Environmental Modelling and Software* 20, 197–202.

Hosking, J. R. M. and Wallis, J. R. (1997) *Regional Frequency Analysis: an Approach Based on L-Moments.* Cambridge University Press, Cambridge.

Institution of Civil Engineers (1996) *Floods and Reservoir Safety*, 3rd edn. Thomas Telford, London.

Institute of Hydrology (IoH) (1980) *Low Flow Studies.* IoH, Wallingford.

Institute of Hydrology (1999) *Flood Estimation Handbook.* IoH, Wallingford.

Interagency Advisory Committee on Water Data (1982) *Guidelines for Determining Flood Flow Frequency.* Bulletin 17-B of the Hydrology Subcommittee: US. Geological Survey, Office of Water Data Coordination, Reston, VA, 183 pp. [Available from National Technical Information Service, Springfield, VA 22161 as report no. PB 86157278 or see http://www.fema.gov/mit/tsd/dl_flow.htm]

Jenkins, C. T. (1970) Computation of rate and volume of stream depletion by wells. In: *Techniques of Water Resources Investigations of the USGS. Book 4. Hydrologist Analysis and Interpretation.* US Government Printing Officer, Washington, DC.

Jennings, M. E., Thomas, W. O. Jr. and Riggs, H. C. (1994) *Nationwide Summary of U.S. Geological Survey Regional Regression Equations for Estimating Magnitude and Frequency of Floods for Ungaged Sites.* US Geological Survey Water-Resources Investigations Report 94–4002, 196 pp.

Kjeldsen, T. R. (2007) *The Revitalised FSR/FEH Rainfall–Runoff Method.* Flood Estimation Handbook Supplementary Report No. 1. Centre for Ecology and Hydrology (CEH), Wallingford. Also available at: www.ceh.ac.uk/*FEH2/FEH*ReFH.html

Kjeldsen, T. R. and Jones, D. A. (2006) Prediction uncertainty in a median-based index flood method using L moments. *Water Resources Research* 42, W07414, doi:10.1029/2005WR004069.

Kjeldsen, T. R., Jones, D. A. and Bayliss, A. C. (2008) *Improving the FEH Statistical Procedures for Flood Frequency Estimation.* Environment Agency/Defra Science Report SC050050.

Kjeldsen, T. R., Lamb, R. and Blazkova, S. (2010) Uncertainty in flood frequency analysis. In: Beven, K. and Hall, J. (eds) *Applied Uncertainty Analysis for Flood Risk Management Book.* Imperial College Press, London, 500 pp.

Lamb, R. (2005) Rainfall–runoff modelling for flood frequency estimation. In: Anderson, M. G. and McDonnell, J. J. (eds) *Encyclopedia of Hydrological Sciences*, Vol. 3, Chapter 125 Wiley, Chichester.

Lamb, R. and Kay, A. L. (2004) Confidence intervals for a spatially generalized, continuous simulation flood frequency model for Great Britain. *Water Resources Research* 40, W07501, doi:10.1029/2003WR002428.

Lundquist, D. and Krokli, B. (1985) *Low Flow Analysis.* Norwegian Water Resources and Energy Administration, Oslo.

Marsh, T. J. (2002) Capitalising on river flow data to meet changing national needs – a UK perspective. *Flow Measurement and Instrumentation* 13, 291–298.

Martin, J. V. and Cunnane, C. (1976) *Analysis and Prediction of Low-flow and Drought Volumes for Selected Irish Rivers.* The Institution of Engineers of Ireland, Dublin.

Natural Environment Research Council (NERC) (1975) *Flood Studies Report.* NERC, London.

Parajka, J., Blöschl, G. and Merz, R. (2007) Regional calibration of catchment models: potential for ungauged catchments. *Water Resources Research* 43, W06406, doi:10.1029/2006WR005271.

Pilgrim, D. H. (ed.) (1999) *Australian Rainfall and Runoff – a Guide to Flood Estimation.* Institution of Engineers, Australia, Engineers Australia, 426 pp.

Pirt, J. and Douglas, R. (1982) A study of low flows using data from the Severn and Trent catchments. *Journal of the Institution of Water Engineers and Scientists* 36, 299–309.

Rosbjerg, D. and Madsen, H. (1995) Uncertainty measures of regional flood frequency estimators. *Journal of Hydrology* 167, 209–224.

Sivapalan, M., Wagener, T., Uhlenbrook, S., Zehe, E., Lakshmi, V., Liang, X., Tachikawara, Y. and Kumar, P. (eds) (2006) *Predictions in Ungauged Basins (PUB): Promises and progress.* IAHS Redbook Publ. no. 303, 520pp. ISBN 1-901502-48-1

Stedinger, J. R. and Cohn, T. A. (1986) Flood frequency analysis with historical and paleoflood information. *Water Resources Research* 22, 785–793.

Tallaksen, L. M. and van Lanen, H. A. J. (eds) (2004) *Hydrological Drought, Vol. 48: Processes and Estimation Methods for Streamflow and Groundwater.* Elsevier Science, Amsterdam, 579 pp.

Turnipseed, D. P. and Ries, K. G. III (2007) *The National Streamflow Statistics Program: Estimating High and Low Streamflow Statistics for Ungaged Sites.* US Geological Survey Fact Sheet XXX-07, 4 pp.

Wagener, T., Wheater, H. S. and Gupta, H. V. (2004) *Rainfall–Runoff Modelling in Gauged and Ungauged Catchments.* Imperial College Press, London, 306 pp.

Webster, P. and Ashfaq, A. (2003) Comparison of UK flood event characteristics with design guidelines. *Water and Maritime Engineering* 156, 33–40.

Young, A. R. (2002) River flow simulation within ungauged catchments using a daily rainfall-runoff model. In: Littlewood, I. (ed.) *British Hydrological Society Occasional Paper No. 13: Continuous River Flow Simulation: Methods, Applications and Uncertainties*, 80 pp.

Young, A. R., Gustard, A., Bullock, A., Sekulin, A. E. and Croker, K. M. (2000) A river network based hydrological model for predicting natural and influenced flow statistics at ungauged sites, LOIS Special Volume. *Science of the Total Environment* 251/252, 293–304.

Young, A. R., Grew, R. and Holmes, M. G. R. (2003) Low Flows 2000: a national water resources assessment and decision support tool. *Water Science and Technology* 48, 119–126.

Zhang, Z., Wagener, T., Reed, P. and Bhushan, R. (2008) Reducing uncertainty in predictions in ungauged basins by combining hydrologic indices regionalization and multiobjective optimization. *Water Resources Research* 44, W00B04, doi:10.1029/2008WR006833.

Chapter 14

Flood routing

One of the most common problems facing a practising hydrologist or hydraulic engineer is the estimation of the hydrograph of the rise and fall of a river at any given point on the river during the course of a flood event. The problem is solved by the technique of *flood routing*, which is the process of following the behaviour of a flood water upstream or downstream along the river and over the flood plain. There are two primary uses of flood routing models. The first is to provide maps of areas at risk of inundation for design events with a chosen probability of exceedance. In the UK, as required under Section 105 of the Water Resources Act, 1995 and by the EU Floods Directive (see Chapter 16), there are national maps of the area at risk of inundation. It is possible to examine such maps on the web site of the Environment Agency for any postcode in England and Wales.[1] The Scottish Environmental Protection Agency (SEPA) has a similar site for flood risk areas in Scotland[2] and, in the USA, the Federal Emergency Management Agency (FEMA) is developing similar tools.[3] These are 'indicative' flood maps, produced by approximate estimates of the 1 in 100 year annual peak discharge for each reach of river using hydraulic models based on a relatively fine topographic model of the flood plain with a resolution of 5–10 m. In recent flood-plain inundation studies, the Environment Agency has required additional model predictions that make an allowance for potential climate change. There are estimates of the area at risk of inundation from the 0.001 probability (1 in 1000 year) event, although it should be remembered from Section 11.7 that the uncertainty in estimating the discharge for such a low probability of exceedance can be rather high.

The second use is to provide flood forecasts at a downstream site in real time, given some estimates of flows or water levels at an upstream site. The upstream levels might come directly from observations of water levels or might come from predictions from a rainfall–runoff model, but the routing of the flood wave downstream will affect both the magnitude and timing of the peak. Accurate predictions of the arrival of the flood peak ahead of time can then be important in issuing flood warnings to the public or deploying temporary flood defences.

A flood hydrograph is modified in two ways as the storm water flows downstream. Firstly, and obviously, the time of the peak rate of flow occurs later at downstream points. This is known as *translation*. Secondly, if there are no major inputs to the channel, the magnitude of the peak discharge is diminished at downstream points, the shape of the hydrograph flattens out, and the volume of flood water takes longer to pass a lower section. This modification to the hydrograph is called *attenuation* (Fig. 14.1).

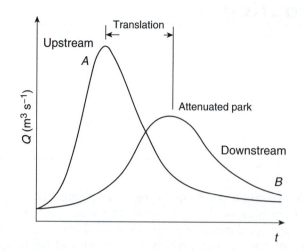

Fig. 14.1 Modification of a flood wave showing translation and attenuation of the hydrograph from an upstream to a downstream site.

A further consideration that can be important in determining the magnitude and shape of the downstream hydrograph is the volume of *lateral inflows* to the channel between the two sites, particularly if there are significant inputs from tributary streams.

The derivation of downstream hydrographs like B in Fig. 14.1 from an upstream known flood pattern A is essential for river managers concerned with forecasting floods in the lower parts of a river basin. The design engineer also needs to be able to route flood hydrographs in assessing the capacity of reservoir spillways, in designing flood protection schemes or in evaluating the span and height of bridges or other river structures. In any situation where it is planned to modify the channel of a river, it is necessary to know the likely effect on the shape of the flood hydrograph in addition to that on the peak stage, i.e. the whole hydrograph of water passing through a section, not just the peak instantaneous rate.

Flood routing methods may be divided into two main categories differing in their fundamental approaches to the problem. One category of methods uses the principle of continuity, and a relationship between discharge and the temporary storage of excess volumes of water during the flood period. The calculations are relatively simple and reasonably accurate and often give satisfactory results. The second category of methods, favoured by hydraulic engineers, adopts the more rigorous equations of motion for unsteady flow in open channels, but in the complex calculations, assumptions and approximations are often necessary, and some of the terms of the dynamic equation might be omitted in certain circumstances to obtain solutions.

The choice of method depends very much on the nature of the problem and the data available. Flood routing computations are more easily carried out for a single reach of river that has no tributaries joining it between the two ends of the reach. According to the length of reach and the magnitude of the flood event being considered, it may be necessary to assess contributions to the river from lateral inflow, i.e. seepage or overland flow draining from, and distributed along, the banks. Further, river networks can be very complex systems, and in routing a flood down a main channel, the calculations must be done for separate reaches with additional hydrographs being introduced for major tributaries. In order to develop an operational flood routing procedure for a major river system, detailed knowledge of the main stream and the various feeder channels is necessary. In addition, the experience of several major flood events with discharge measurements made at strategic points on the drainage network will be useful to both calibrate and evaluate model predictions, especially if information on the extent of inundation is available from surveys after past events.

In this chapter, a selection of flood routing methods will be presented, beginning with the simplest using the minimum of information and progressing through to the more complex methods requiring significant computer power for their application.

14.1 Simple non-storage routing

It has been said that 'engineering is the solution of practical problems with insufficient data'. If, in an application to a particular river, there are no gauging station data available and therefore no measurements of discharge, the engineer may have to make do with *stage* measurements. In such circumstances, it is usually the flood peaks that have been recorded, and indeed it is common to find the people living alongside a river have marked on a wall or bridge pier the heights reached by notable floods. Hence the derivation of a relationship between peak stages at upstream and downstream points on a single river reach may be made (Fig. 14.2) when it is known that the floods are caused by similar notable conditions.

This is a very approximate method, and should not be used if there are major tributaries or significant lateral inflows between the points with the stage measurements, which would cause the relationship to change between events. However, with enough stage records it may be possible to fit a curve to the relationship to give satisfactory forecasts of the downstream peak stage from an upstream peak stage measurement.

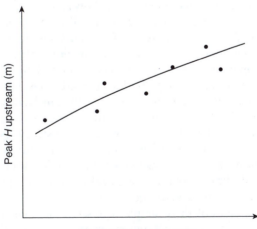

Fig. 14.2 Peak stage relationship between upstream and downstream sites.

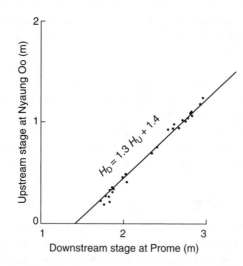

Fig. 14.3 Peak stage relationship for two sites on the River Irrawaddy, Burma.

For example, on the River Irrawaddy in Burma a linear relationship exists between the peak stages of an upstream gauging station at Nyaung Oo and a station at Prome, 345 km downstream. Thirty-five comparable stages (m) for irregular flood events over 5 years (1965–69) are shown in Fig. 14.3. An equation $H_D = 1.3H_U + 1.4$ relating H_D, the downstream stage to H_U, the upstream stage, can then give forecast values of H_D from H_U.

The time of travel of the hydrograph crest (peak flow) also needs to be determined; curves of upstream stage plotted against time of travel to the required downstream point can be compiled from the experience of several flood events. (The time of travel of the flood peaks between Nyaung Oo and Prome on the Irrawaddy ranged from 1 to 4 days.)

A typical stage-time of travel plot in Fig. 14.4 shows the time of travel at a minimum within the stage range; this occurs when the bankfull capacity of the river channel is reached. After reaching a minimum at this bankfull stage, the time of travel tends to increase again as the flood peak spreads over the flood plain and its downstream progress is retarded owing to storage effects on the flood plain.

The complexities of rainfall–runoff relationships are such that these simple methods allow only for average conditions. Flood events can have very many different causes, and spatial distributions of runoff production, that will produce flood hydrographs of different shapes. Flood hydrographs at an upstream point, with peaks of the same magnitude but containing different flood volumes, in travelling downstream will produce different peaks at a downstream point. Modifications to the flow by the channel conditions will differ between steep, peaky flood hydrographs and gentle fat hydrographs with the same peak discharges.

The principal advantages of these simple regression methods are that they can be developed for stations with only stage measurements and no rating curve, and they are quick and easy to apply, especially for warning of impending flood inundations when the required answers are immediately given in stage heights. The advantages of

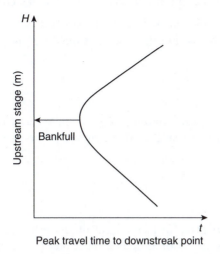

Fig. 14.4 Expected relationship between water levels (stage) and travel time for in-bank and out of bank flows.

speed and simplicity are less important now that fast computers are available, and more accurate and comprehensive real-time techniques can be used.

14.2 Storage routing

When a storm event occurs, an increased amount of water flows down the river channel and, in any one short reach of the channel, there is a greater volume of water than usual contained in temporary storage. If, at the beginning of the reach, the flood hydrograph (above a normal flow) is given as I, the inflow (Fig. 14.5), then during the period of the flood, T_1, the channel reach has received the flood volume given by the area under the I hydrograph. Similarly, at the lower end of the reach, with an outflow hydrograph O, the flood volume is again given by the area under the curve. In a flood situation, relative quantities may be such that lateral and tributary inflows can be neglected, and thus, by the principle of continuity, the volume of inflow equals the volume of outflow,

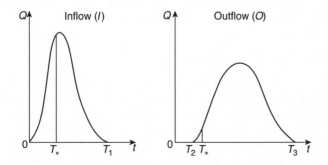

Fig. 14.5 Flood hydrographs for a river reach.

$V = \int_0^{T_1} I\,dt = \int_{T_2}^{T_3} O\,dt$, i.e. the flood volume $\int_0^{T_1} I\,dt$ has entered the reach and an amount $\int_{T_2}^{T_3} O\,dt$ has left the reach. The difference must be stored within the reach, so the amount of storage, S_*, within the reach at time $t = T_*$ is given by:

$$S_* = \int_0^{T_*} (I - O)\,dt \tag{14.1}$$

where I and O are the corresponding rates of inflow and outflow. An alternative statement of this equation is that the rate of change of storage within the reach at any instant is given by:

$$\frac{dS}{dt} = I - O \tag{14.2}$$

This, *the continuity equation*, forms the basis of all the storage routing methods. The routing problem consists of finding O as a function of time, given I as a function of time, and having information or making assumptions about S. Equation (14.2) cannot be solved directly.

Any procedure for routing a hydrograph generally has to adopt a numerical approximation such as the finite difference technique. Choosing a suitable time interval for the routing period, Δt, the continuity equation can be represented in a finite difference form as:

$$\frac{(I_1 + I_2)\,\Delta t}{2} - \frac{(O_1 + O_2)\,\Delta t}{2} = S_2 - S_1 \tag{14.3}$$

The subscripts 1 and 2 refer to the start and end of any Δt time step. The routing time step has to be chosen small enough such that the assumption of a linear change of flow rates within the time step is acceptable (as a working guide, Δt should be less than one-sixth of the time of rise of the inflow hydrograph).

At the beginning of a time step, all the variables in (14.3) are known except O_2 and S_2. Thus with two unknowns, a second equation is needed to solve for O_2 at the end of a time step. A second equation is obtained by relating S to O alone, or to I and O together. The two equations are then used recursively to find sequential values of O through the necessary number of Δt intervals until the outflow hydrograph can be fully defined. It is the nature of the second equation for the storage relationship that distinguishes two methods of storage routing.

14.2.1 Reservoir or level-pool routing

For a reservoir or a river reach with a determinate control, such as a weir or overspill crest, upstream of which a nearly level pool is formed, the temporary storage can be evaluated from the topographical dimensions of the 'reservoir', assuming a horizontal water surface (Fig. 14.6). For a level pool, the temporary storage, S, is directly and uniquely related to the head, H, of water over the crest of the control. The discharge from the 'pool' is also directly and uniquely related to H. Hence S is indirectly but uniquely a function of O.

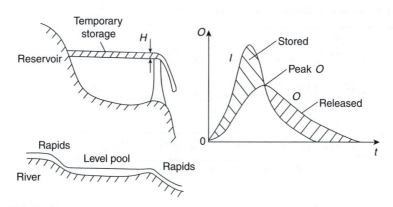

Fig. 14.6 Level-pool routing.

It is convenient to rearrange (14.3) to move the unknowns S_2 and O_2 to one side of the equation and to adjust the O_1 term to produce:

$$\left(\frac{S_2}{\Delta t} + \frac{O_2}{2}\right) = \left(\frac{S_1}{\Delta t} + \frac{O_1}{2}\right) + \frac{(I_1 + I_2)}{2} - O_1 \tag{14.4}$$

Since S is a function of O, $[(S/\Delta t) + (O/2)]$ is also a specific function of O (for a given Δt), as in Fig. 14.7. Replacing $[(S/\Delta t) + (O/2)]$ by G, for simplification, (14.5) can be written:

$$G_2 = G_1 + I_m - O_1 \tag{14.5}$$

where $I_m = (I_1 + I_2)/2$. Fig 14.7 defines the relationship between O and $G = (S/\Delta t + O/2)$, and this curve needs to be determined by using the common variable H to fix values of S and O and then G, for a specific Δt.

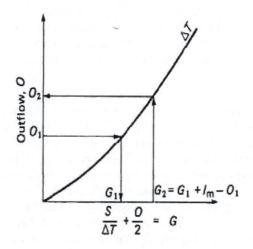

Fig. 14.7 Outflow from a level pool as a function of $(S/\Delta t + O/2)$.

Table 14.1 Order of calculations for level-pool routing

T	A I	B I_m	C O	D $(I_m - O)$	E G
1	I_1		O_1		G_1
		$(I_1 + I_2)/2$		$\{(I_1 + I_2)/2\} - O_1$	
2	I_2		O_2		G_2
⋮	⋮				
i	I_i		O_i		G_1
		$(I_i + I_j)/2$		$\{(I_i + I_j)/2\} - O_i$	
j	I_j		O_j		G_3

Equation (14.5) and the auxiliary curve of Fig. 14.7 now provide an elegant and rapid step-by-step solution. At the beginning of a step, G_1 and O_1 are known from the previous step (or from conditions prior to the flood for the first step). I_m is also known from the given inflow hydrograph. Thus all three known terms in (14.5) immediately lead to G_2 at the end of the time step, and then to O_2 from Fig. 14.7.

The recursive calculation of values for the outflow hydrograph is easily carried out in a table or, if the relationship of Fig. 14.7 is known in a functional form (such as for flow over a weir), in a spreadsheet. The calculation starts with all the inflows known (column A in Table 14.1). The mean inflows for a time step Δt given by $I_m = (I_1 + I_2)/2$ are then entered in column B. The initial outflow O_1 is known from the discharge before the flood (column C). Thus the first value in column E, G_1, may be determined from Fig. 14.7. The mean inflow during the first time interval less the initial outflow gives the first entry in column D and this is added to the first column E value to give the second entry in column E, G_2, (14.5). Then the second value of the outflow hydrograph O_2 can be read off from Fig. 14.7 (or calculated from a known functional relationship). This now becomes the O_1 value for the next interval. The sequence of calculations is continued until the outflow hydrograph is completed or until the required outflow discharges are known.

A useful check on the validity of any level pool routing calculation is that the peak of the outflow hydrograph should occur at the intersection of the inflow and outflow hydrographs on the same plot (Fig. 14.6). At that point, $I = 0$, so $dS/dt = 0$, i.e. storage is a maximum and therefore O is a maximum. Thereafter, the temporary storage is depleted.

Example: Discharge from a reservoir is over a spillway with discharge characteristic $Q(\text{m}^3\,\text{s}^{-1}) = 110\,H^{1.5}\,\text{m}^3\,\text{s}^{-1}$, where H m is the head over the spillway crest. The reservoir surface area is $7.5\,\text{km}^2$ at spillway crest level and increases linearly by $1.5\,\text{km}^2$ per metre rise of water level above crest level. The design storm inflow, assumed to start with the reservoir just full, is given by a triangular hydrograph, base length 36 h and a peak flow of $360\,\text{m}^3\,\text{s}^{-1}$ occurring 12 h after start of inflow. Estimate the peak outflow over the spillway and its time of occurrence relative to the start of the inflow. *Solution.* A level water surface in the reservoir is assumed. Temporary storage above crest level is given by integrating the change in surface area with increasing depth in

Table 14.2 Derivation of O and G for the auxiliary curve of O versus G

H	O	104H	$0.53\sqrt{H}$	$10+0.53\sqrt{H}+H$	G
0.2	10	20.8	0.24	10.26	213
0.4	28	41.6	0.34	10.38	432
0.6	51	62.4	0.41	10.47	653
0.8	79	83.2	0.47	10.55	878
1.0	110	104.0	0.53	11.53	1199
1.2	144	124.8	0.58	11.78	1470
1.4	182	145.6	0.63	12.03	1752
1.6	223	136.4	0.67	12.27	2042
1.8	265	187.0	0.71	12.51	2339
2.0	312	208.0	0.75	12.75	2652

the reservoir as:

$$S = \int_0^H A dh = 10^6 \int_0^H (7.5+1.5h)dh = 10^6 \left(7.5H+0.75H^2\right) \text{ m}^3$$

Outflows over the crest are given by $O = 110H^{1.5} \text{ m}^3 \text{ s}^{-1}$. Taking a value of $\Delta t = 2$ h $= 7200$ s. Then:

$$G = \left(\frac{S}{\Delta t}+\frac{O}{2}\right) = \left[\frac{10^6}{7200}\left(7.5H+0.75H^2\right) + 55H^{1.5}\right]$$

$$= 104H\left(10+0.53\sqrt{H}+H\right) \text{ m}^3\text{s}^{-1}$$

The derivation of O and G values are shown in Table 14.2. O is plotted against G and a curve drawn through the points (Fig. 14.8). (Note that the relationship is specifically for $\Delta t = 2$ h.)

The inflows into the reservoir are evaluated for each 2h period from the beginning of the storm inflow and the outflows are computed recursively using the storage equation (equation 14.5 and the auxiliary curve). The results are given in Table 14.3 and plotted on Fig. 14.9. The peak outflow over the spillway is $180 \text{ m}^3 \text{ s}^{-1}$ and it occurs 24h after the commencement of the inflow. The approach to the peak outflow becomes obvious during the computations as the net inflow rate $(I_m - O)$ declines, and after the maximum O is reached, $(I_m - O)$ becomes negative.

14.2.2 River routing

For a river channel reach where the water surface cannot be assumed horizontal, the stored volume becomes a function of the stages at both ends of the reach, and not just at the downstream (outflow) end only.

In a typical reach, the different components of storage may be defined for a given instant in time as in Fig. 14.10. Again, the continuity equation (14.2) holds at any

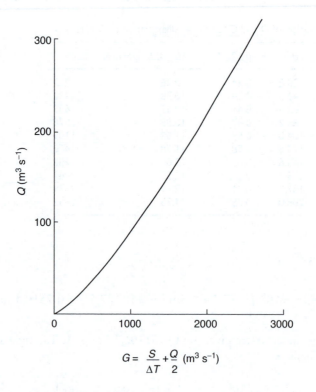

$$G = \frac{S}{\Delta T} + \frac{Q}{2} \ (m^3 \ s^{-1})$$

Fig. 14.8 Auxiliary curve for outflow O v. G for $\Delta t = 2\,h$.

Table 14.3 Calculations for level-pool routing

t	I	I_m	O	$I_m - O$	G
0	0				0
2	60	30	0	30	30
4	120	90	1	89	119
6	180	150	6	144	133
8	240	210	14	196	459
10	300	270	31	239	698
12	360	330	55	275	973
14	330	345	85	260	1233
13	300	315	114	201	1434
18	270	285	138	147	1581
20	240	255	158	97	1378
22	210	225	171	54	1732
24	180	195	178	17	1749
26	150	135	180	−15	1734
28	120	135	178	−43	1391
30	90	105	173	−68	1323
32	60	75	133	−88	1535
34	30	45	152	−107	1428
36	0	15	139	−124	1304
38		0	123	−123	1181

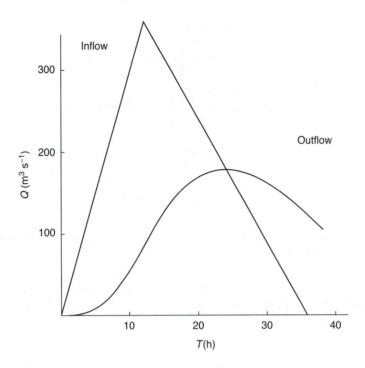

Fig. 14.9 Results of level pool routing over a reservoir spillway.

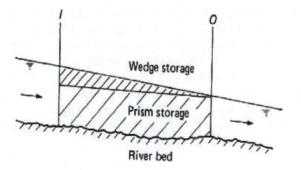

Fig. 14.10 Storage for flow routing in a river reach.

given time but the total storage, S, is now the sum of *prism storage* and *wedge storage*. The prism storage is taken to be a direct function of the stage at the downstream end of the reach; the simple assumption ignores the effects of the slope of the water surface and takes the downstream stage and the outflow to be uniquely related, and thus the prism storage to be a function of the outflow, O. The wedge storage exists because the inflow, I, differs from O and so may be assumed to be a function of the difference between inflow and outflow (I − O).

Three possible conditions for wedge storage are shown in Fig. 14.11: during the rising stage of a flood in the reach, I > O, and the wedge storage must be added to

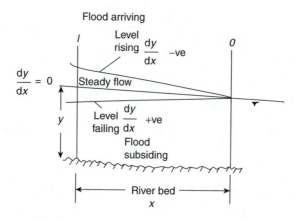

Fig. 14.11 Storage in a river reach showing three possible wedge storage profiles for rising levels, steady flow and falling levels.

the prism storage; during the falling stage, $I < O$, and the wedge storage is negative to be subtracted from the prism storage to obtain the total storage. The total storage, S, may then be represented by:

$$S = f_1(O) + f_2(I - O) \tag{14.6}$$

with due regard being paid to the sign of the f_2 term.

In the level-pool method, S was a function only of O. Here it is a more complex function involving I as well as O at the ends of a river reach. However, there are again two equations, (14.2) and (14.6) and again, it should be possible using a finite-difference method to solve for the unknown, S_2 (14.3), at the end of a routing interval, Δt. One such method of solution is the Muskingum method (named after the Muskingum river in Ohio where it was first applied), which will be described in the next section. The Muskingum method is still widely used as a simple, computationally cheap, flow-routing method, even though it has some intrinsic disadvantages, as discussed later.

14.2.3 The Muskingum method

McCarthy (1938) made the bold assumption that in (14.6), $f_1(O)$ and $f_2(I - O)$ could both be simple straight-line functions, i.e. $f_1(O) = K.O$ and $f_2(I - O) = b(I - O)$. Thus:

$$S = bI + (K - b)O = K\left[\frac{b}{K}I + \left(1 - \frac{b}{K}\right)O\right] \tag{14.7}$$

and writing $x = b/K$, (14.7) becomes:

$$S = K[xI + (1 - x)O] \tag{14.8}$$

Thus x is a dimensionless weighting factor indicating the relative importance of I and O in determining the storage in the reach. The value of x has limits of zero and 0.5, with typical values in the range 0.2 to 0.4. K has the dimension of time.

Substituting for S_2 and S_1 in the finite difference form of the continuity equation (14.3):

$$\frac{(I_1+I_2)\Delta t}{2}-\frac{(O_1+O_2)\Delta t}{2}=K\left[xI_2+(1-x)O_2\right]-K\left[xI_1+(1-x)O_1\right] \quad (14.9)$$

Collecting terms in O_2, the unknown outflow, on to the left-hand side:

$$O_2(-0.5\Delta t-K+Kx)=I_1(-Kx-0.5\Delta t)+I_2(Kx-0.5\Delta t)+O_1(-K+Kx+0.5\Delta t) \quad (14.10)$$

which can also be expressed in the form

$$O_2 = c_1 I_1 + c_2 I_2 + c_3 O_1 \quad (14.11)$$

where

$$c_1 = \frac{\Delta t + 2Kx}{\Delta t + 2K - 2Kx}$$

$$c_2 = \frac{\Delta t - 2Kx}{\Delta t + 2K - 2Kx} \quad (14.12)$$

$$c_1 = \frac{-\Delta t + 2K - 2Kx}{\Delta t + 2K - 2Kx}$$

To maintain mass balance, the sum $\sum_c = 1$ so that when c_1 and c_2 have been found $c_3 = 1 - c_1 - c_2$. Thus the outflow at the end of a time step is the weighted sum of the starting inflow and outflow and the ending inflow, as per (14.11). Earlier editions of *Hydrology in Practice* provide a practical method of determining the parameters K and x. Here we will note that (14.11) has the form of a linear transfer function as in (12.18) with c_1 and c_2 as b parameters, c_3 as a single a parameter and no time delay. Methods for fitting the general linear transfer function can also therefore be used for fitting a flood routing model of this type. The lack of a time delay can give rise to problems with the Muskingum method (see Section 14.2.5).

14.2.4 The Muskingum–Cunge method

One of the modified methods that has worked well is the Muskingum–Cunge method. As shown previously, the Muskingum method is based on the storage equation with the coefficients K and x derived by trial and error. Jean Cunge showed that K and x could be determined by considering the hydraulics of the flow (Cunge, 1969).

From above, with $S = K[xI + (1 - x)O]$ and $dS/dt = I - O$ the differential with respect to t gives

$$K\frac{d}{dt}[xI+(1-x)O]=I-O \quad (14.13)$$

If this is expressed in finite-difference form with subscripts 1 and 2 representing the beginning and end of a time increment, ΔT, then

$$\frac{K}{\Delta t} x \left(I_2 - I_1 \right) + \frac{K}{\Delta t} \left(1 - x \right) \left(O_2 - O_1 \right) = \frac{I_2 + I_1}{2} - \frac{O_2 + O_1}{2} \qquad (14.14)$$

K may be shown to be approximately equal to the time of travel of a flood wave through the reach and this assumption was used by Cunge (1969), i.e. $K = \Delta L / c$, where c is the average speed of the flood peak and ΔL is the length of the river reach. Substituting for K in (14.14) gives:

$$\frac{\Delta L}{c \Delta t} x \left(I_2 - I_1 \right) + \frac{\Delta L}{c \Delta t} \left(1 - x \right) \left(O_2 - O_1 \right) + \frac{1}{2} \left(O_2 - I_2 + O_1 - I_1 \right) = 0 \qquad (14.15)$$

Multiplying by $c / \Delta L$ and rearranging gives

$$\frac{x \left(I_2 - I_1 \right)}{\Delta t} + \frac{\left(1 - x \right) \left(O_2 - O_1 \right)}{\Delta t} + \frac{c}{2 \Delta L} \left(O_2 - I_2 + O_1 - I_1 \right) = 0 \qquad (14.16)$$

With $K = \Delta L / c$ as an acceptable approximation, a means is required of obtaining x, the coefficient governing the discharge weighting. Cunge derived the following expression for x from channel properties to match the numerical dispersion of the method with the expected dispersion of the real flood wave. Thus:

$$x = \frac{1}{2} - \frac{\overline{Q}}{2 S_o \overline{B} c \Delta L} \qquad (14.17)$$

where $\overline{Q}$ is a reference discharge, S_o is average bed slope, $\overline{B}$ is mean channel width and with the other variables as defined previously. In practice, the Muskingum coefficients are evaluated according to each reach forming the subdivisions of the total length of the river reach being considered. Then the routing of the inflow hydrograph can proceed as before to obtain the outflow hydrograph by recurrent application of the Muskingum equation (14.11):

$$O_2 = c_1 I_1 + c_2 I_2 + c_3 O_1$$

with the coefficients c_1, c_2 and c_3 being evaluated from K and x for each reach as before. The success and accuracy of the Muskingum–Cunge flood routing method depend on the choice of ΔL and Δt.

14.2.5 A three-parameter Muskingum model

Both the original Muskingum method of Section 14.2.2 and the modified Muskingum–Cunge method of Section 14.2.3 apply to situations where there is no lateral inflow to the river reach between the upstream and downstream gauging stations. In most rivers, this constrains the routing reaches to be rather short, generally terminating at tributaries, and requires gauged or estimated tributary inflows to be added to the

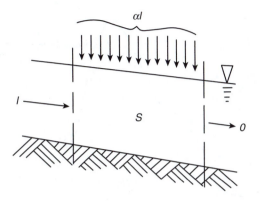

Fig. 14.12 Inflow storage and lateral inflow for a reach of river. (Reproduced from O'Donnell, 1985.)

main channel inflow term. In turn, this means using many reaches in the total routing procedure. A second modified Muskingum method has been developed by Terrence O'Donnell that incorporates a simple lateral inflow model. This modified method has two further advantages: (a) it provides a numerical and direct best-fit solution technique to obtain the parameters K and x; and (b) by treating the whole river as one reach, it avoids the need for multiple routings (and multiple parameter determinations) over many sub-reaches.

The lateral inflow model used is shown in Fig. 14.12 (O'Donnell, 1985) and assumes that the total rate of lateral inflow over the whole reach is directly proportional to the upstream inflow rate. The proportionality constant, α, is taken to be fixed for any one event but takes different values for different events. The original two-parameter Muskingum model (K, x) is thus extended to a three-parameter model (K, x, α).

The three coefficients, c_i of the routing equation (14.11) can be related to K, x and α (O'Donnell, 1985) and vice versa. A direct least-squares solution by a matrix inversion technique yields a set of best-fit values for the c_i coefficients from the set of equations formed via equation (14.11) applied to all the Δt intervals in an observed event. The three coefficients no longer sum to unity as in the original two-parameter Muskingum method since it is expected that the mass outflow will be greater than upstream inflow because of the lateral inputs.

O'Donnell *et al.* (1987) applied this extended Muskingum procedure in a split sample test to a number of flood events over a single 50 km reach on the Grey River, New Zealand. Half the events were used for calibration, i.e. to establish average values for K and x for the reach. (The latter parameters are postulated to be fixed properties of the reach, whereas each event has its own α value, a property of the causative storm.) The average K and x values were then applied to reconstruct the outflow hydrographs for the events not used in the calibration from their individual inflow hydrographs and α values. Fig. 14.13 shows such a reconstruction for an event in which the value of α was 6.92. The volume of outflow at Dobson for this event was nearly seven times the volume of inflow at Waipuna due to the very substantial lateral inflow between the two stations.

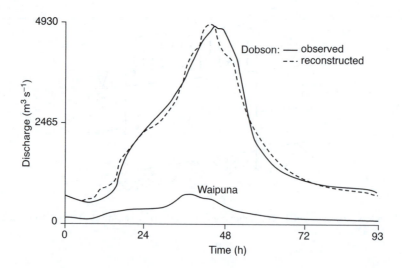

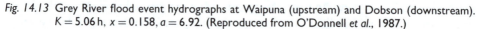

Fig. 14.13 Grey River flood event hydrographs at Waipuna (upstream) and Dobson (downstream). $K = 5.06$ h, $x = 0.158$, $a = 6.92$. (Reproduced from O'Donnell *et al.*, 1987.)

14.2.6 The variable parameter Muskingum–Cunge–Todini model (VPMCT)

The simplicity of the Muskingum–Cunge method has made it popular as a technique for flood routing (particularly where detailed information about channel cross-sections is not available) but the relationships for x and K provided by Cunge (1969) show that we should expect the values to vary with discharge because of their dependence on the wave celerity c and the reference discharge Q. It has been shown, however, that applying the method with variable parameter values leads to mass balance errors that become greater with decreasing slope. Todini (2007) shows that the origins of this mass balance error lie in the original derivation of the Muskingum method. He proposes an alternative derivation that is mass conservative that takes the form:

$$c_1 = \frac{\Delta t + 2\,[Kx]_{t+\Delta t}}{\Delta t + [2K - 2Kx]_t}$$

$$c_2 = \frac{\Delta t - [Kx]_t}{\Delta t + [2K - 2Kx]_{t+\Delta t}}$$

$$c_1 = \frac{-\Delta t + [2K - 2Kx]_t}{\Delta t + [2K - 2Kx]_{t+\Delta t}}$$

(14.18)

where t and $t + \Delta t$ indicate that the values of x and K are evaluated at the start and end of a time step. Since at the start of the time step, the values of c and Q and therefore x and K will not be known, this method requires an iterative solution at each time step. Ezio Todini shows that the method gives a very good approximation to a full solution of the 1D dynamic equations (see Section 14.3) below, at least for simple channel shapes.

14.2.7 Practical application of the Muskingum routing method

The Muskingum routing method has been very widely applied since it was first developed in the 1930s. It is still an option in a number of river-modelling software packages currently in use, such as in the Infoworks software package from Wallingford Software. It must be used, however, with some care. It was noted earlier that, mathematically, the Muskingum equations have the form of a linear transfer function (see Section 12.8.3) with one a parameter, two b parameters and no time delay. The values of K, x and α will determine the equivalent values of a and b coefficients in the general linear transfer function equation. While the values of K and x can determine the expected attenuation of a flood wave in moving downstream, they can only indirectly account for any pure time delay between the start of a rise in discharge at the upstream site and the start of the rise at a downstream site. If this time delay is long enough, then fitting the Muskingum parameters will often give rise to parameter values that correspond to a transfer function that initially has a negative response to an upstream rise, before the expected positive impulse response (see Fig. 14.14). This provides the delay necessary, but clearly only in a non-physical way.

There are two ways around this in practice. The first is to subdivide the reach into smaller reaches, all with the same fitted values of K, x and α. Output at a time step from the first reach then becomes the input at the next time step to the next reach. If there are at least as many reaches as time steps required to represent a pure delay, then the Muskingum transfer function for each reach should remain positive. Routing the input through multiple Muskingum models in this way will, however, also have an effect on the attenuation of the flood wave (as with the Nash cascade model of the unit hydrograph discussed in Section 12.5.1, when the higher the number of stores in series, the greater the smoothing of the impulse–response).

The second is to recognise that the Muskingum model is only a specific example of the more general family of linear transfer functions and take advantage of the routines available to fit transfer functions with time delays (such as in the DBM models discussed in Section 12.8.2). This is discussed in the next section.

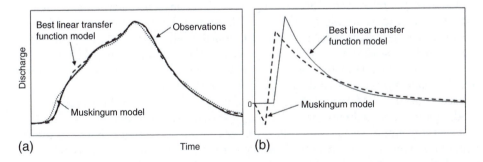

Fig. 14.14 Comparison of extended Muskingum and DBM Flood routing results, River Wyre, Lancashire: (a) predicted flows against time for Muskingum and best-fit DBM model; (b) fitted transfer functions (after Young, 1986).

14.2.8 Data-based mechanistic modelling of flood waves

The data-based mechanistic (DBM) modelling methodology of Young (2001, 2003) has been introduced in Section 12.8.2. It is an approach that is based in the theory of linear transfer functions (see also Dooge and O'Kane, 2003), but which recognises that, in hydrology and hydraulics, it may be necessary to apply a non-linear transformation to the inputs, producing an effective input that is more linearly related to the outputs. Section 12.8.2 dealt with the application to this methodology to rainfall–runoff modelling problems, but the methods have also been applied to flood routing (e.g. Romanowicz *et al.*, 2008). Thus, given an upstream input to a reach, and a downstream output from a reach, then a DBM model can be fitted that can then be used to predict other flood events, including for forecasting in real time (see Section 12.10). One of the tenets of the DBM approach is to let the data define an appropriate model structure. Here this means the form of the non-linear transformation, the number of *a* parameters, the number of *b* parameters and the time delay. Unlike the Muskingum model, the time delay is included explicitly in the model (if needed) so that a DBM model can be applied to any length of reach. In addition, any lateral inflows in the reach will be implicitly included in the magnitude of the fitted *b* coefficients. Fig. 14.14, shows a comparison of the DBM and Muskingum 3 parameter transfer functions and the predictions for a flood wave in the River Wyre, south of Lancaster (the example used in O'Donnell, 1985), that demonstrates the advantage of including an explicit time delay quite nicely.

Experience with the application of the DBM approach to flow routing has suggested that it can provide very good fits to observed flood waves (and, indeed, even better fits to the outputs of non-linear hydraulic routing models that are discussed in the next section). A point of particular interest is that it can be applied not only to upstream and downstream discharge data but also to water level (stage) measurements. This is important because, remember, it is water level that is generally measured at gauging sites not discharge. Thus, we need to apply a rating curve to obtain an estimate of discharge. At flood levels, the rating curve may not be well defined so that discharges at higher stages may be quite uncertain. Water level continues to be measured directly and rather accurately, however (or at least until the gauge goes under water or gets washed away). Thus a level to level transfer function model might be very useful, particularly in flood-forecasting situations.

However, the relationship between upstream level and downstream level might be expected to be more non-linear than the relationship between upstream and downstream discharges. This is because the relationship will reflect arbitrary changes in the channel cross-sections as well as the effects of the routing process. The non-linear transform in the DBM methodology, however, seems to be able to handle this rather well in applications to actual data (e.g. Leedal *et al.*, 2008).

Success in the DBM approach to flood routing will depend on how well the fitted model parameters including the time delay stay constant across events. Hydraulic theory, for example, suggests that the effective routing delay might change as the flow goes out of bank. Flood wave speeds will tend to increase as the river approaches bank full, slow down during the initial stages of overbank flow and then speed up again as storage builds up on the flood plain (see Fig. 14.4 where this is expressed in terms of peak travel time). This picture is also complicated by the full three-dimensional geometry of the flood plain and flood defences between upstream and downstream gauging sites (the flow will not go out of bank everywhere at the same time, for example). In practice, it has been found that the use of a constant time delay can give quite good results, and that in forecasting a new event, any differences from calibration events can be largely accounted for by an updating algorithm (see Section 12.10 and the results of Romanowicz *et al.*, 2008 and Leedal *et al.*, 2008). This might be necessary in any flow routing algorithm, since the next event will always be different from those seen before (in time distribution of the inputs, lateral inflows, magnitude of the peak, etc.).

14.3 Hydraulic routing

The methods of flood routing that have been considered so far have been rather simplistic, treating the river as a storage responding to upstream inputs. Full hydraulic theory for open channel flow has not been considered but is, in fact, well developed and computer models based on hydraulic theory are widely available and used. The hydraulic methods of flood routing are based on the solution of the two basic differential equations governing gradually varying non-steady flow in open channels, known as the *Saint Venant equations* after the Barré de St Venant, who first published a derivation of the equations in 1871. These are the continuity (mass balance) equation and energy or momentum balance equations.

14.3.1 The continuity equation

For a small length of channel, Δx, (Fig. 14.15) and considering a small time interval, Δt, the mass balance or continuity equation can be written in discrete form as:

$$\Delta S = Q\Delta t - (Q + \Delta Q)\Delta t \tag{14.19}$$

where ΔS is the increment of storage during Δt, Q is the inflow and $(Q + \Delta Q)$ is the outflow.

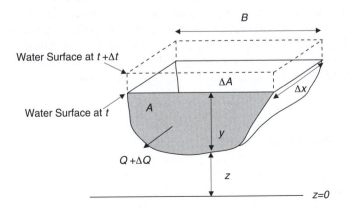

Fig. 14.15 Changes in flow over a short increment of time, Δt, in a short river reach of length Δx.

If A is the average cross-section in the reach then $S = A\Delta x$, so that

$$\Delta S = \Delta A \Delta x = Q \Delta t - (Q + \Delta Q) \Delta t \tag{14.20}$$

and dividing by $\Delta x \Delta t$

$$\frac{\Delta A}{\Delta t} = -\frac{\Delta Q}{\Delta x} \tag{14.21}$$

As the increments become very small, they can be written as differentials (the partial differential symbol is used because the equation has differentials with respect to more than one variable, here both t and x) so that

$$\frac{\partial A}{\partial t} = -\frac{\partial Q}{\partial x} \tag{14.22}$$

or

$$\frac{\partial Q}{\partial x} + \frac{\partial A}{\partial t} = 0 \tag{14.23}$$

Now $Q = A.v$, i.e. cross-sectional area times mean velocity, so differentiating with respect to x and substituting in (14.23) gives

$$A \frac{\partial v}{\partial x} + v \frac{\partial A}{\partial x} + \frac{\partial A}{\partial t} = 0 \tag{14.24}$$

If $dA = B dy$, where B is the width of the water surface, then

$$A \frac{\partial v}{\partial x} + B v \frac{\partial y}{\partial x} + B \frac{\partial y}{\partial t} = 0 \tag{14.25}$$

14.3.2 The energy balance equation

Equation (14.25) has two unknowns of velocity and depth at each cross-section. Thus another equation is required to effect a solution. The second fundamental equation can be derived from considering either the energy or momentum equation for the short length of channel, Δx. In Fig. 14.16, v is a mean velocity averaged over the cross-section at distance x in the reach, and g is the gravity acceleration. The loss in head over the length of the reach, Δx, has two main components:

$$h_f = S_f \Delta x \tag{14.26}$$

the head loss due to friction, and:

$$h_a = \frac{1}{g} \frac{dv}{dt} \Delta x = S_a \Delta x \tag{14.27}$$

the head loss due to acceleration.

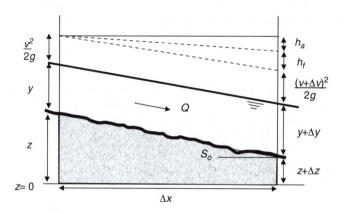

Fig. 14.16 Definition diagram for the energy equation expressed in terms of energy per unit weight or head (which has units of length); $(z+y)$ represents the potential head, $v^2/2g$ represents the velocity head at a point; while S_o is the bed slope, h_f represents the friction head loss, and h_a represents the acceleration head loss over a short reach of length Δx.

If it is assumed that the channel bed slope is small and the vertical component of the acceleration force is negligible, then the combined loss of head is $(h_f + h_a)$. Using the Bernoulli expression for total head, H:

$$H = z + y + \frac{v^2}{2g} \tag{14.28}$$

then the change in H over Δx is $-\Delta H = h_f + h_a$. Thus:

$$\frac{\Delta H}{\Delta x} = -S_f + S_a = \frac{d}{dx}\left(z + y + \frac{v^2}{2g}\right) \tag{14.29}$$

from which:

$$S_f = -\frac{\partial z}{\partial x} - \frac{\partial y}{\partial x} - \frac{v}{g}\frac{\partial v}{\partial x} - S_a$$

or

$$S_f = S_o - \frac{\partial y}{\partial x} - \frac{v}{g}\frac{\partial v}{\partial x} - \frac{1}{g}\frac{\partial v}{\partial t} \tag{14.30}$$

Equations (14.25) and (14.30) provide two equations in the unknowns v and y. The *friction slope*, S_f, however also depends on depth and velocity. To complete the system of equations it is common to assume that the rate of head loss due to friction under dynamic flow conditions is the same as if the flow was steady and uniform with a water surface slope equal to the friction slope. Under this assumption, one of the uniform flow equations can be used to derive the friction slope. For example, the Manning equation is widely used in hydraulic routing models in the form

$$S_f = v^2 n^2 R_h^{-4/3} \tag{14.31}$$

where R_h is the hydraulic mean radius given by cross-sectional area divided by wetted perimeter ($R_h = A/P$) and n is the Manning roughness coefficient that depends on the nature of the channel. For more rigorous derivations of the equations describing unsteady open channel flow, the reader is referred to the text on open channel flow and the solution of the St Venant equations (e.g. Chow, 1959; Abbott and Minns, 1998). The specification of channel and flood plain roughness values is considered in Section 14.4.2 below.

14.3.3 Two-dimensional depth-averaged flow equations

The development of the St Venant equations above has looked at only one downstream dimension. In flood inundation problems, however, it would be useful to have predictions in two dimensions, where mean depth and mean velocity are allowed to vary across the flood plain. The equivalent equations in two dimensions (2D), known as the *shallow water equations*, now have three unknowns: depth y and two orthogonal velocity components v_1 and v_2 in directions x_1 and x_2. Thus three equations are required and are given by:

Continuity equation

$$y\left(\frac{\partial v_1}{\partial x_1} + \frac{\partial v_2}{\partial x_2}\right) + v_1\frac{\partial y}{\partial x_1} + v_2\frac{\partial y}{\partial x_2} + \frac{\partial y}{\partial t} = 0 \tag{14.32}$$

Energy balance equation

$$S_{f1} = -\frac{\partial z}{\partial x_1} - \frac{\partial y}{\partial x_1} - \frac{v_1}{g}\frac{\partial v_1}{\partial x_1} - \frac{1}{g}\frac{\partial v_1}{\partial t}$$

and $\hspace{10cm}$ (14.33)

$$S_{f2} = -\frac{\partial z}{\partial x_2} - \frac{\partial y}{\partial x_2} - \frac{v_2}{g}\frac{\partial v_2}{\partial x_2} - \frac{1}{g}\frac{\partial v_2}{\partial t}$$

Simplified forms of the St Venant equations are also used for flood routing. These correspond to assuming that different terms in the energy balance equations (14.30 and 14.33) are sufficiently small to be neglected. Thus, in one dimension, if in (14.30) the last two terms on the right-hand side are neglected, then the resulting equations are called the *diffusion approximation*. If the last three terms on the right-hand side of (14.30) are neglected, such that if it assumed simply that $S_f = S_o$ (i.e. assuming that the flow is uniform everywhere) then the resulting equations are called the *kinematic approximation*. Both are simpler to solve than the full dynamic equations, but are good approximations only under certain conditions (see e.g. Daluz Vieira, 1983).

14.4 Flood routing in practice

The St Venant hydraulic routing equations (equations 14.25 and 14.30 or 14.32 and 14.33) are non-linear in v and y. The use of the St Venant equations in flood routing by their integration down the length of a channel depends on the use of approximate solutions. Three sorts of approximate solutions have been used in the past. Approximate analytical solutions depend on linearising the equations around some reference flow state and relying on changes around that state being small. This results in a form of linear transfer function. More accurate solutions can be obtained by the *method of characteristics*, which uses a mathematical property of partial differential equations in that the solution can be separated into ordinary differential equations (the *characteristic equations*) that represent how changes in the upstream and boundary conditions propagate in the space–time domain (e.g. Amein, 1966). It is simpler to integrate these ordinary differential equations. Where the characteristic lines cross, solutions for v and y can be obtained, but to obtain values at specific points in space and time may then require interpolation. The method of characteristics is interesting because the characteristic equations are the expression of how the *wave velocity* or *celerity* propagates any change in the flow through space and time. For sub-critical flow, the celerity, c, is given, in one dimension, by:

$$c = v \pm (gy)^{0.5} \tag{14.34}$$

The two celerities given by (14.34) represent propagation in the upstream and downstream directions (think of the waves resulting from the perturbation caused by throwing in a stone into a deep river). Both are different from the mean flow velocity. This explains both why a flood wave propagates downstream at a rate faster than the flow, and why changes downstream can have a *backwater effect* in the upstream direction (such as upstream of a confining bridge during a flood event). For super-critical flow, $v > (gy)^{0.5}$, and disturbances can have an effect only in the downstream direction (try throwing a stone into the fast, shallow, super-critical flow downstream of a weir crest). The kinematic approximation to the full St Venant equations, referred to in the previous section, has only a downstream celerity and therefore cannot represent the backwater effects that will affect flood wave propagation in sub-critical flows.

Approximate numerical solutions are more frequently used now to solve the St Venant equations, particularly finite difference solutions for the one-dimensional (1D) case (Abbott and Minns, 1998). Finite element and finite volume solutions are also used in solving the two-dimensional (2D) equations, with the advantage that the element size can be more easily made smaller where more detail in the solution is needed. There are many different computer packages now available, both commercially and freeware, that solve the 1D and 2D St Venant equations (or simpler versions of the equations that neglect one or more terms in the energy equation).

Examples of 1D models that are widely used are MIKE 11 from the Danish Hydraulics Institute (DHI),[4] SOBEK from Deltares in the Netherlands,[5] ISIS from Halcrow in the UK[6] and HEC-RAS from the US Army Corps of Engineers.[7] Examples of 2D models are MIKE FLOOD from DHI,[8] TUFLOW from BMT WBM Pty. Ltd,[9] RMA-2 from the US Army Corps of Engineers,[10] TELEMAC from EdF/SOGREAH in France,[11] and DIVAST-TVD from the University of Cardiff (Liang *et al.*, 2007). Examples of simplified 2D models are LISFLOOD FP[12] (e.g. Hunter *et al.*, 2006;

Bates *et al.*, 2006) and JFLOW (Bradbrook, 2006). Both of these last models use the diffusion approximation to the full St Venant equations for routing flow across the flood plain, either coupled to a 1D kinematic solution for the channel or, for large-scale mapping purposes under steady flow assumptions, with an allowance made for the capacitance of the channel in each reach (see Chapter 16).

As an example of the use of this type of model, the 2D LISFLOOD-FP has been used to simulate the inundation in Carlisle as a result of the January 2005 flood. LISFLOOD is a grid-based 2D model, with simplified dynamics. For this application, the simulation made use of a LIDAR survey of the flood plain topography, and a grid resolution of 25 m. After the Carlisle event, the maximum extent of the flooded area was surveyed by differential GPS positioning of wrack marks and water marks on buildings. Fig. 14.17 shows a comparison of the depths and maximum extent of inundation predicted by the model to the flooded area estimated from the survey.

The JFLOW model has been used by JBA Consulting to produce national maps of areas at risk of flooding, also making use of remotely sensed topographic survey data. Chapter 16 describes a regional application of this technology for flood risk modelling to support catchment flood management planning in north-east England.

14.4.1 Stability and uncertainty in hydraulic models

The full St Venant equations are hyperbolic partial different equations. The celerity puts some constraints on the time step that can be used in an approximate numerical

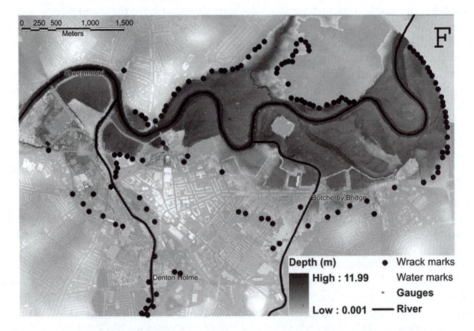

Fig. 14.17 Application of LISFLOOD FP to predict flooding at Carlisle during the 2005 flood event. Colours show predicted maximum depths during the event, wrack marks and water marks show maximum depths from water marks and wrack marks surveyed after the event (after Neal *et al.*, 2009).

solution since (on a gridded model) the approximate solution will deteriorate or go *unstable* if the local celerity is sufficient to propagate the peak (or other disturbance) more than one grid length in one time step. Most of the numerical codes described above have automatic time step controls to avoid this, but this can add significantly to computer run times because the celerity has to be calculated everywhere in the flow domain to check the maximum time step allowable, especially if in a finite element or finite volume solution the element size is very different in different parts of the flow domain. It is also the case that particular solutions can generate instabilities at time steps less than the theoretical limit imposed by the local celerity (this may be the case at the edge of the inundated area when part of the domain may oscillate between being wet and dry; or where there is inadequate description of the channel transition between a tributary and the main stream). Thus other conditions on the time step are often imposed and the actual time steps used in the solution are often very short.

A further consideration in the application of flood routing models based on the St Venant equations is uncertainty in the upstream and downstream boundary conditions and parameters that need to be specified to run the model. The St Venant equations involve two unknown variables, mean velocity and depth, at every boundary point. It is normal to have measurements of stage at both upstream and downstream boundaries in a forecasting problem, or an estimate of a design discharge in a flood risk mapping problem. In both cases, therefore, one condition on depth and velocity can be specified but, because there are two unknowns, another is needed. Very often this is imposed in a model by assuming that, at the boundary, the friction slope can be specified (generally by assuming it is equal to the bed slope so that flow at the boundary is uniform). One of the uniform flow equations, Manning or Chezy, can then be used to provide the second condition relating velocity and depth. This is, however, an approximation that is somewhat inconsistent with using the fully dynamic St Venant equations to route the flood wave (which will predict water surface slopes that are different on the rising and falling limbs of the hydrograph such that there is a *hysteresis* in the stage-discharge relationship).

14.4.2 The specification of channel and flood plain roughness

The uniform flow equations require the specification of a roughness parameter. Roughness parameters, in fact, need to be specified everywhere in the flow domain because, as explained earlier, they are used in the calculation of the local friction slope. Classically, roughness is thought of as a way of representing the loss of energy due to friction at the local boundary. It is often estimated by measuring a velocity profile (or the mean velocity in a channel cross-section) and back-calculating a local roughness value. Tables of such values can be found in hydraulics text books and even now on a USGS web site[13] where photographs of different types of channels are labelled with their measured roughness values. Estimates of roughness for different types of flood plain surface can also be found in the literature. In 1D flood routing models, most of the available packages require the form of the channel cross-section to be specified, with roughness values for both channel and flood plain parts of the cross-section. The model will then calculate a combined roughness for the cross-section for different depth of flow. Two-dimensional models generally require roughness values to be specified for channel and flood plain elements. These can be different in different parts of the domain but are usually assumed constant over all depths of flow.

While flood routing models are often applied with roughness values specified in this way, this is a rather simplistic view of roughness because there are other sources of energy loss that need to be accounted for that arise because of the three-dimensional nature of the flow. In particular, to obtain good results from a model, we need an estimate, not of local roughness, but of the *effective* roughness that represents all the energy losses in a calculation element (or between two cross-sections in a 1D model). This must include the effects of form roughness as the channel geometry changes, the interactions with any obstructions (including e.g. hedges and walls on flood plains that might otherwise be neglected), and the characteristic eddying that occurs at the edge of a channel in overbank flow because of the strong shear between fast flow in the channel and relatively slow flow on the flood plain. Such effects may also be depth dependent, such as when in deep floods, high-velocity flows spill out of the channel and start to by-pass meanders.

Some of these types of effects have been studied experimentally in laboratory channels, such as the flood channel facility at HR Wallingford (e.g. Knight and Selling, 2007). The effects on effective roughness values have been included into the conveyance estimation system (CES Knight *et al.*, 2010).[14] To get good model simulations, however, the effective roughness may also need to reflect how the channel and flood plain geometry has been represented (resolution of the topographic model, representation of infrastructure on the flood plain, etc.). Thus, there may be some uncertainty with which effective roughness values (and the way in which they change with depth of inundation) can be specified (see, e.g. Pappenberger *et al.*, 2007). It is clear that this type of flood routing model can be valuable in producing distributed predictions (as in the case of the Carlisle flood shown in Fig. 14.17) but that the predictions should, where possible, be checked against field information and the uncertainty in the predictions assessed.

14.5 Pollutant transport in rivers

As well as the routing of flood waves in rivers, it is also important to understand the advection and dispersion of pollutants in the flow so as to be able to make predictions about pollution incidents in rivers. The processes of transport and mixing in rivers are rather different to the transport and dispersion of pollutants in groundwater (which are described in Section 15.7.1) but the mathematical description that is generally used, called the *advection–dispersion equation*, is very similar. This is because, in both cases, while the pollutant is being advected by the mean velocity of the flow, the dispersive mixing is predominantly a result of the wide distribution of flow velocities in any cross-section of the flow. There would also be some mixing, even if the water was still as a result of molecular diffusion gradually reducing concentration differences, but this is generally a small effect relative to the macroscopic dispersion caused by velocities differences in both the vertical and across the stream.

Consider a pollution incident occurring in a major river. An example is the accidental release of ammonium sulphate that occurred on the River Eden, Cumbria, in March 1993. The liquid was intended for use as a pasture fertiliser and was being pumped from a tanker lorry to a farm bowser. As the bowser came close to being full it toppled over.

The ammonium sulphate ran into a ditch which led to a small tributary of the Eden, the Ploughlands Beck. Here it was retained for a while by damming the channel. The dam was removed once the water appeared to be clear of pollutant, but there was sufficient pollutant held back in the soil by the dam that once it reached the Eden it caused the largest recorded fish kill in one of the most pristine rivers in England.

In fact, one of the uses of predictive models of pollutant transport in rivers is to try and trace back from a recorded pollution incident to determine where the origin of the pollution might have been (e.g. Whitehead et al., 1986). To do so, we need to know both the mean velocity characteristics of the river and the way in which any pollutant is likely to disperse as a result of differences in velocity in the river.

14.5.1 The advective–dispersion model

We can take a similar control volume approach to looking at dispersion to that used in Section 14.3 above to derive the flow routing equations (see Fig. 14.15), but here applied for a steady flow rate Q. In doing so, we need to distinguish between two types of downstream flux of the pollutant or tracer: one is the advective flux due to the mean velocity of the flow ($v = Q/A$) and one is the dispersive flux associated with the net effect of deviations in velocity within the cross-section, A. For non-conservative pollutants, we might also need to consider a loss within the reach. For simplicity, we will assume that the loss is linearly proportional to the mass in the reach.

Thus, for a small length of channel, Δx, and considering a small time interval, Δt, the mass balance or continuity equation for the pollutant can be written in discrete form as:

$$\Delta M = QC\Delta t - Q(C + \Delta C)\Delta t + AJ\Delta t - A(J + \Delta J)\Delta t - \alpha M \Delta t \qquad (14.35)$$

where ΔM is the increment of pollutant mass change in the reach during Δt, C is the average inflow concentration, J is the net dispersive influx per unit cross-sectional area, and α is a loss coefficient.

If A is assumed constant if Q is not changing, then $M = AC\Delta x$. We can also make use of the relationship $Q = vA$ so that

$$\Delta M = A\Delta C\Delta x = vAC\Delta t - vA(C + \Delta C)\Delta t - \alpha AC\Delta x \Delta t \qquad (14.36)$$

And dividing by $A\Delta x\Delta t$

$$\frac{\Delta C}{\Delta t} = -v\frac{\Delta C}{\Delta x} - \frac{\Delta J}{\Delta x} - \alpha C \qquad (14.37)$$

As the increments become very small, they can be written as differentials (the partial differential symbol is used because the equation has differentials with respect to more than one variable, here both t and x) so that

$$\frac{\partial C}{\partial t} = -v\frac{\partial Q}{\partial x} - \frac{\partial J}{\partial x} - \alpha C \qquad (14.38)$$

As noted above, we expect the net dispersive flux J to be dominated by the velocity distribution in the reach. In general, this is poorly known, and so resort is often made to a general relationship between the dispersive flux and the gradient in the mean concentration, called Fick's law, as originally used to described random molecular diffusion:

$$J = -D\frac{\partial C}{\partial x} \qquad (14.39)$$

Thus we would expect that, if there was a steep concentration gradient (from high to low) in the downstream direction, then as a result of the turbulent mixing in the river flow, there would be a net dispersive flux in the downstream direction in addition to the net advection downstream. This leads to strong concentration differences being gradually reduced over time as the pollutant moves downstream (Fig. 14.18).

Combining these last two equations gives:

$$\frac{\partial C}{\partial t} = D\frac{\partial^2 C}{\partial x^2} - v\frac{\partial C}{\partial x} - \alpha C \qquad (14.40)$$

This is the advection–dispersion equation (ADE). Note that, in developing the ADE, we have used average concentrations in the cross-section of the flow. This means that the ADE applies strictly only after the pollutant is well mixed across the cross-section. This is normally some way downstream of the point of entry of the pollutant, beyond a distance called the *mixing length*. The mixing length will vary with the nature of the entry point and the flow characteristics but can be expected to be of the order of 10–30 times the channel width (e.g. Rutherford, 1994).

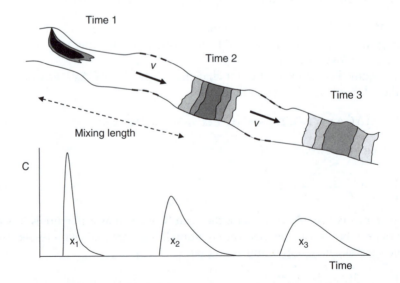

Fig. 14.18 Snapshots of advection and dispersion of a discrete mass of pollutant or tracer in a river reach originating from a source at the river bank in a flow of mean velocity *v* at three different times, and concentration curves at three different distances (x_1, x_2, x_3) downstream from the input point. Darker shading represents higher concentrations.

The ADE has a simple analytical solution for flows in which the velocity and dispersion characteristics of the flow stay constant downstream of the mixing length. This is clearly rare in real rivers but can sometimes be a useful approximation. For more complex cases, it is necessary to use an approximate numerical solution. For the simplest case of an instantaneous input of mass M and no losses, in a flow of cross-section A and mean velocity v, the analytical solution is given by

$$C(x, t) = \frac{M}{2A (\pi Dt)^{0.5}} \exp\left[-\frac{(x - vt)^2}{4Dt} \right] \tag{14.41}$$

This is exactly the same equation as the Gaussian distribution that is commonly used in statistical problems. It predicts that, as the input of pollutant disperses in the flow, it does so as a symmetric Gaussian curve in space, i.e. with distance downstream (Fig. 14.19a). Normally, however, we do not take measurements of concentration at the same time along a river. It is much more convenient to follow a concentration curve in time at a particular point in space. In this case, as the pollutant passes that point, the tail of the concentration curve will have been subject to more dispersion than the rising limb. Thus, the concentration curve in time will not be symmetric but slightly skewed (Fig. 14.19b).

Application of (14.41) or (14.40) requires estimates of the two characteristics of the flow, the mean velocity, v, and the dispersion coefficient, D. These are not constants. We expect the mean velocity to increase with discharge and the dispersion coefficient increases approximately with the square of discharge. A variety of equations have been suggested to allow the estimation of the dispersion coefficient in relation to the characteristics of the flow (e.g. Rutherford, 1994; Young and Wallis, 1993). When a pollutant is expected to be non-conservative, losing mass as the cloud moves downstream, application of (14.40) also requires a value for the loss coefficient, α. The loss coefficient will vary with the nature of the pollutant and the flow conditions and is difficult to generalise.

When concentration curves are examined from pollution incidents or tracer experiments in real rivers, however, it is generally found that the curves are highly skewed, with much longer tails than predicted by the ADE (Fig. 14.20). This is generally interpreted in terms of 'dead zones' in the river, caused by backwater eddies, storage within

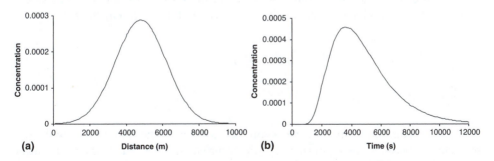

(a)   Distance (m)   (b)   Time (s)

Fig. 14.19 Solution of (14.41) with $v = 0.5\,\text{m}\,\text{s}^{-1}$ and $D = 100\,\text{m}^2\,\text{s}^{-1}$ for: (a) the pattern of concentration with distance downstream for a fixed time at 9600 s; and (b) the pattern of concentration with time for a fixed distance of 2000 m.

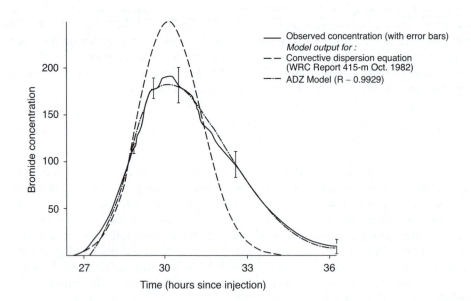

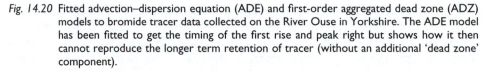

Fig. 14.20 Fitted advection–dispersion equation (ADE) and first-order aggregated dead zone (ADZ) models to bromide tracer data collected on the River Ouse in Yorkshire. The ADE model has been fitted to get the timing of the first rise and peak right but shows how it then cannot reproduce the longer term retention of tracer (without an additional 'dead zone' component).

vegetation growth in the river, or other effects that lead to part of the volume of water having much lower downstream velocities than the main stream. Then, when pollutant gets into a dead zone, it takes much longer to return to the main stream than the timescale of transport in the faster flowing part of the flow. This is what leads to the long tails of concentration commonly observed.

There have been two main approaches to making better predictions of pollutant transport. The first has been to modify the ADE by adding a transient storage component in which it is assumed that transfers to and from the storage are proportional to the concentration difference between main stream and transient storage concentration (Bencala and Walters, 1983). Thus (14.41) can be modified by the addition of a simple representation of exchange between the main river and the transient storage:

$$
\frac{\partial C}{\partial t} = D\frac{\partial^2 C}{\partial x^2} - v\frac{\partial C}{\partial x} - \alpha C - \alpha_D\left(C - C_D\right)
$$

$$
\frac{dC_D}{dt} = \alpha_D\left(C - C_D\right)
$$

(14.42)

where C_D is concentration in the dead zone and α_D is a coefficient controlling the rate of exchange between the main river and the transient storage. Thus, as concentration rises, pollutant will be transferred into the transient storage so that C_D will increase, while after the peak concentration has passed and the main stream concentration falls below that of the transient storage, then the concentration difference is reversed and pollutant will move back into the main flow. There is an implicit assumption in this approach that the pollutant is instantaneously mixed with the transient storage volume. In a numerical solution, the transient storage can be distributed along each increment of river to represent 'dead zones' that are not well mixed with the main flow in the river. Briggs et al. (2009) have also introduced the use of two dead zone components: one to represent dead zones in the water body itself, and one to represent mixing with waters in the hyporheic zone at the edge of the channel. This is a good example of where the wish to incorporate more process understanding into modelling reality also introduces additional parameters that may be difficult to estimate or to identify from data.

14.5.2 The aggregated dead zone model

There is, however, an alternative approach that is based on the direct analysis of concentration curves. This makes use of the type of DBM transfer function methodology that has already been described in Section 14.2.6 above). The resulting model has been called the *aggregated dead zone* (ADZ) model (Wallis et al., 1989; Young and Wallis, 1993). It is easiest to apply when the flow is approximately steady and the pollutant or tracer is already well mixed, but does not require that discharge be constant downstream. It then treats the advection of pollution as a pure time delay and the dispersion as the result of mixing in one or more linear stores representing the aggregated effect of dead zone dispersion. If only a single store is needed, then there are only two parameters, the advective time delay, τ, and the mean residence time, T. The sum of these two times is the mean residence time in the reach. The ratio

$$DF = \frac{T}{T + \tau} \qquad (14.43)$$

is called the *dispersive fraction*. The dispersive fraction can also be interpreted under the steady flow assumption as the ratio of an effective mixing volume to the total volume of water participating in the flow in the reach. It is worth noting that this may not be the total volume in the reach if there are real 'dead zones' in the reach that do not interact with either the flow or the pollutant.

These parameters are easily determined for a reach of river where well-mixed concentration curves are available at the entry to and exit from a river reach. Fig. 14.20 shows how well the ADZ model can fit a concentration curve observed during a tracer experiment. For a number of reaches, such curves have been available for a variety of discharges so that relationships between the advective time delay and the dispersive fraction can be developed (Fig. 14.21). Interestingly, in many (but not all) reaches studied with this model, the dispersive fraction has proven to be a near constant (Fig. 14.21b), while the slope of the log discharge v. log advective travel time relationship appears to be rather similar for many (but not all) reaches. This implies a scaling relationship between the effective mixing volume and the total volume in the reach as the total volume increases with discharge. There is no physical reason why this should be the case, it is a purely empirical result. It means, however, that it is relatively simple to set up a model of pollutant transport in a reach.

Thus, in studying the 1994 pollution incident in the River Eden, for example, there was some uncertainty about the timing of the release of the ammonium sulphate that caused the fish kill, so a series of tracer experiments was carried out in the river at a

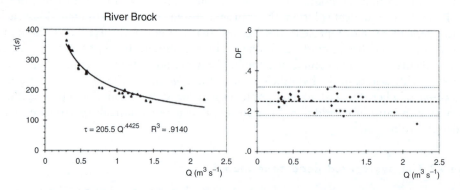

Fig. 14.21 Example of change in the aggregated dead zone model parameters with discharge for a reach of the River Brock, south of Lancaster, UK derived from tracer experiments: (a) advective time delay; and (b) dispersive fraction.

similar discharge ($4.1 \, \mathrm{m^3 \, s^{-1}}$) to that on the day of the pollution incident ($3.1 \, \mathrm{m^3 \, s^{-1}}$). The difference in discharge would, however, make a significant difference to the travel times and dispersion. Thus, relationships of the form of those shown in Fig. 14.21 were used to estimate the parameters of the ADZ model at the lower discharge, which could then be used to predict the transport of the pollutant and resolve the timing of the input (Fig. 14.22).

The ADZ model can also be applied to the initial mixing problem from an estimate of the point input of tracer, but higher order DBM models are then generally required. Lees *et al.* (2000), by matching the solutions of the ADE and ADZ models under steady flow conditions have provided relationships that better match the long tails of real concentration curves.

14.6 Some final comments

This chapter has considered a variety of ways of predicting the routing of flood waves and the transport of pollutants in river reaches and networks. The methods have varied in complexity from those based on mass and energy balance equations, including the prediction of maps of inundation, to those based simply on an analysis of input and output data for a reach. In all cases, the predictive accuracy of the different models is going to be dependent on the quality of the input data. In particular, those models that require information about discharges are going to be dependent on the accuracy of the rating curves at the input and output sites (and for flood routing, their accuracy in the over-bank range of water levels). The impact of rating curve accuracy has been the subject of some recent studies (e.g. Pappenberger *et al.*, 2006). Those models that require information about flood plain topography and channel cross-sections are going to be dependent on the accuracy of the available survey information. Even models that depend only on the analysis of input and output curves, will be dependent on how well the observations represent the real variables (e.g. whether the observations of concentration used to fit an ADZ model are really well-mixed in the river cross-section). Examples of post-flood surveys of depths and extent of flooding, such

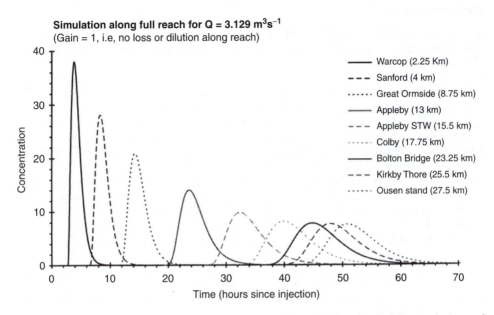

Fig. 14.22 Predicted concentration curves at sites on the River Eden at the discharge at the time of the pollution incident (simulations produced by Hannah Green).

as those shown in Fig. 14.17 that can be used to calibrate and test flow models are relatively rare. Similarly, while tracer experiment and pollutant cloud observations that can be used to calibrate and test transport models have been carried out on many rivers, there are relatively few cases where observations are available for multiple discharges or for non-conservative pollutants. As stressed elsewhere in this book, quality assessment of the available data should therefore be an important part of any study.

Notes

1 http://www.environment-agency.gov.uk/homeandleisure/floods/31656.aspx
2 http://www.multimap.com/clients/places.cgi?client=sepa
3 http://msc.fema.gov/webapp/wcs/stores/servlet/FemaWelcomeView?storeId=10001&catalogId=10001&langId=-1
4 http://www.dhigroup.com/Software/WaterResources.aspx
5 http://delftsoftware.wldelft.nl/
6 A limited freeware version of ISIS can be downloaded from http://www.halcrow.com/isis/default.asp. ISIS is also built into the InfoWorks set of modelling packages from Wallingford Software (see http://www.wallingfordsoftware.com/products/infoworks/)
7 http://www.hec.usace.army.mil/software/hec-ras/
8 http://www.dhigroup.com/Software/Release2008/Flooding.aspx
9 http://www.tuflow.com/
10 http://chl.erdc.usace.army.mil/chl.aspx?p=s&a=ARTICLES;369&g=75
11 http://www.telemacsystem.com/
12 http://www.ggy.bris.ac.uk/research/hydrology/models/lisflood
13 http://wwwrcamnl.wr.usgs.gov/sws/fieldmethods/Indirects/nvalues/
14 http://www.river-conveyance.net/

References

Abbott, M. B. and Minns, A. W. (1998) *Computational Hydraulics*. Ashgate, Aldershot.

Amein, M. (1966) Stream flow routing on computer by characteristics. *Water Resources Research* 2, 123–130.

Bates, P. D., Wilson, M. D., Horritt, M. S., Mason, D., Holden, N. and Currie, A. (2006) Reach scale floodplain inundation dynamics observed using airborne Synthetic Aperture Radar imagery: data analysis and modelling. *Journal of Hydrology* 328, 306–318.

Bencala, K. E. and Walters, R. A. (1983) Simulation of solute transport in a mountain pool and riffle stream: a transient storage model. *Water Resources Research* 19, 718–724.

Bradbrook, K. (2006) JFLOW: a multiscale two-dimensional dynamic flood model. *Water and Environment Journal* 20, 79–86.

Briggs, M. A., Gooseff, M. N., Arp, C. D. and Baker, M. A. (2009) A method for estimating surface transient storage parameters for streams with concurrent hyporheic storage. *Water Resources Research* 45, W00D27, doi:10.1029/2008WR006959.

Chow, V. T. (1959) *Open Channel Hydraulics*. McGraw-Hill, New York, p. 540.

Cunge, J. A. (1969) On the subject of a flood propagation method. *Journal Hydraulic Research IAHR*, 7, 205–230.

Daluz Vieira, J. (1983) Conditions governing the use of approximations for the St Venant equations for shallow water flow. *Journal of Hydrology* 60, 43–58.

Dooge, J. C. I. and O'Kane, J. P. (2003) *Deterministic Methods in Systems Hydrology*, A. A. Balkema, Lisse.

Green, H. M., Beven, K. J., Buckley, K. and Young, P. C. (1994) Pollution incident prediction with uncertainty. In Beven, K. J., Chatwin, P. C. and Millbank, J. H. (eds) *Mixing and Transport in the Environment* Wiley, Chichester, pp. 113–140.

Hunter, N. M., Bates, P. D., Horritt, M. S. and Wilson, M. D. (2006) Improved simulation of flood flows using storage cell models. *Proceedings of the Institution of Civil Engineers Water Management* 159, 9–18.

Knight, D. W. and Selling, R. H. J. (2007) The SERC Flood channel facility. *Water and Environment Journal* 1, 198–204.

Knight, D. W., McGahey, C., Lamb, R. and Samuels, P. (2010) *Practical Channel Hydraulics: Roughness, Conveyance and Afflux*. CRC Press, Boca Raton, 460 pp.

Lees, M. J., Camacho, L. A. and Chapra, S. C. (2000) On the relationship of transient storage and aggregated dead zone models of longitudinal solute transport in streams. *Water Resources Research* 36, 213–224.

Liang, D., Lin, B. L. and Falconer, R. A. (2007) Simulation of rapidly varying flow using an efficient TVD-MacCormack scheme. *International Journal of Numerical Methods in Fluids* 53, 811–826.

Leedal, D. T., Beven, K. J., Young, P. C. and Romanowicz, R. (2008) Data assimilation and adaptive real-time forecasting of water levels in the river Eden catchment, UK. In: Samuels, P., Huntingdon, S., Allsop, W. and Harrop, J. (eds) *Flood Risk Management: Research and Practice*. CRC Press/Balkema, Leiden.

McCarthy, G. T. (1938) *The Unit Hydrograph and Flood Routing*. Unpublished mss, Conference of US Army Corps of Engineers.

Neal, J. C., Bates, P. D., Fewtrell, T. J., Hunter, N. M., Wilson, M. D. and Horritt, M. S. (2009) Distributed whole city water level measurements from the Carlisle 2005 urban flood event and comparison with hydraulic model simulations. *Journal of Hydrology* DOI: 10.1016/j.jhydrol.2009.01.026.

O'Donnell, T. (1985) A direct three-parameter Muskingum procedure incorporating lateral inflow. *Hydrological Sciences Journal* 30, 479–496.

O'Donnell, T., Pearson, C. P. and Woods, R. A. (1987) An improved three-parameter Muskingum routing procedure. *Journal of Hydraulic Engineering, ASCE* 114, 5.

Pappenberger, F., Frodsham, K., Beven, K. J., Romanovicz, R. and Matgen, P. (2007) Fuzzy set approach to calibrating distributed flood inundation models using remote sensing observations. *Hydrology and Earth System Sciences* 11, 739–752.

Romanowicz, R. J., Young, P. C., Beven, K. J. and Pappenberger, F. (2008) A data based mechanistic approach to nonlinear flood routing and adaptive flood level forecasting. *Advances in Water Resources* 31, 1048–1056.

Rutherford, J. C. (1994) *River Mixing.* Wiley, Chichester.

Todini, E. (2007) A mass conservative and water storage consistent variable parameter Muskingum-Cunge approach. *Hydrology and Earth System Sciences* 11, 1645–1659.

Wallis, S. G., Young, P. C. and Beven, K. J. (1989) Experimental investigation of the aggregated dead zone model for longitudinal solute transport in stream channels. *Proceedings of the Institution of Civil Engineers Part II* 87, 1–22.

Whitehead, P. G., Williams, R. J. and Hornberger, G. M. (1986) On the identification of pollutant or tracer sources using dispersion theory. *Journal of Hydrology* 84, 273–286.

Young, P. C. (1986) Time-series methods and recursive estimation in hydrological systems analysis. In: Kraijenhoff, D. A. and Moll, J. R. (eds) *River Flow Modelling and Forecasting* D. Reidel, Dordrecht. pp. 129–180.

Young, P. C. (2001) Data-based mechanistic modelling and validation of rainfall-flow processes. In: Anderson, M. G. and Bates, P. D. (eds) *Model Validation: Perspectives in Hydrological Science* Wiley, Chichester, pp. 117–161.

Young, P. C. (2003) Top-down and data-based mechanistic modelling of rainfall-flow dynamics at the catchment scale. *Hydrological Processes* 17, 2195–2217.

Young, P. C. and Wallis, S. G. (1993) Solute transport and dispersion in channels. In: Beven, K. J. and Kirkby, M. J. (eds) *Channel Network Hydrology* Wiley, Chichester.

Chapter 15

Groundwater

15.1 The importance of groundwater

In many catchments, there are groundwater bodies that are important for water supply and the maintenance of low flows in rivers. The proper management of groundwater is therefore an important topic in hydrology and all hydrologists need to understand the basic principles that govern the analysis and prediction of groundwater. Some of the necessary theory has already been covered in Chapter 5 where the basic principles, parameters and terminology were introduced in considering the monitoring of soils and ground waters. Here we will look at how the way in which groundwater bodies might react to changes in recharge or pumping conditions can be predicted. In doing so, we will recognise that such predictions are now commonly made using modelling programs such as MODFLOW and ASM (see Section 15.5). In applying such models, however, it is important to have a proper understanding of the concepts and assumptions on which they are based.

Groundwater is an important source of water supply in the UK. The principal aquifers of the UK are shown in Fig. 15.1a, with water table levels in the major aquifer that underlies East Anglia shown in Fig. 15.1b.

We may distinguish between confined and unconfined aquifers. The former are fully saturated permeable layers beneath a confining impermeable layer. They are particularly important for water supply purposes in synclinal basins where the groundwater system is recharged at the edges of the basin but where there is a useful storage of water in the lower part of the syncline. There is such a basin under London, where a confined aquifer in the chalk is overlain by impermeable London clay. Wells drilled into the chalk provide water of good quality and, before the aquifer was extensively developed for water supply purposes, wells penetrating into the confined part of the chalk aquifer used to flow without pumping in some places. This was because the pressure in the groundwater (the *piezometric head*) was greater than that needed to raise the water to the surface. Such wells are called *artesian wells* (Fig. 15.2).

Abstraction for water supply purposes from the confined chalk under London in the late nineteenth century and early twentieth century caused the piezometric head to decline steadily (Fig. 15.2). In the second half of the twentieth century, however, abstraction started to decrease until a balance of recharge and abstraction was reached in about the 1960s, with abstractions at a level of about 480 ML day^{-1}. By this time, groundwater levels in the centre of the London Basin had fallen by about 65 m.

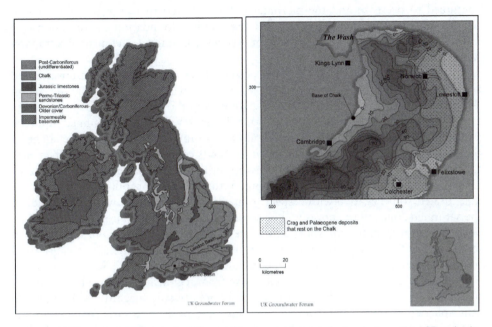

Fig. 15.1 (a) The major aquifers in the UK; and (b) water table levels in the chalk aquifer of East Anglia. (Figures courtesy of the UK Groundwater Forum.)

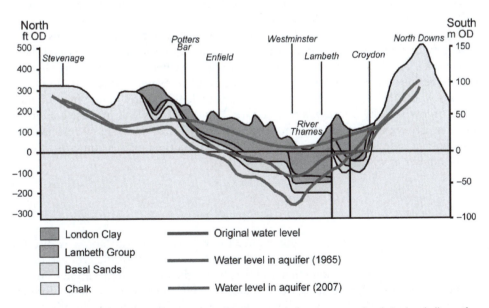

Fig. 15.2 Schematic representation of the London Basin and changing water levels in the chalk aquifer. (Figure courtesy of UK Groundwater Forum.)

With declining demand for groundwater for industrial uses in London, abstractions continued to decrease and the piezometric head started to rise again, with potential consequences for flooding of tunnels and building foundations. Abstractions are now managed through the Environment Agency's London Catchment Abstraction Management Scheme (see Chapter 17).

An unconfined aquifer is a saturated permeable layer overlain by an unsaturated zone. Unconfined aquifers are found in the recharge areas for the chalk in the London Basin (Fig. 15.2), but also in the shallow aquifers in the fluvioglacial deposits of the London gravels nearer to the surface in the Thames Valley. In unconfined aquifers, patterns of recharge from the surface will have an important effect on the shape of the saturated zone, the upper boundary of which is called the water table. The water table represents a surface where the pressure in the water in the pore space is zero relative to atmospheric pressure. In Chapter 5, the concept of capillary potential in an unsaturated soil (negative with respect to atmospheric pressure) was introduced. In the saturated zone, all the pores will be saturated and the local pore pressures will be positive with respect to atmospheric pressure. Pressure is also commonly expressed in terms of 'head' or energy per unit weight of water. This is convenient because head has units of length (see also use of head in describing river flows in Section 14.3.2).

In all groundwater bodies, water will move from regions of high total head to low total head. Total head will be sum of the local pore pressure (negative with respect to atmospheric pressure in unsaturated conditions and positive in saturated conditions) and elevation above some datum level, where elevation is defined to be zero. It does not really matter what elevation datum is used but it is important that it is the same for all points in the analysis. Thus, mean sea level is a suitable datum, but the ground elevation is generally not (only if all the points considered are in the vertical profile below a particular point on the surface).

Expressing total pressure in terms of head makes it easy to draw diagrams, called *flow nets*, to show how the groundwater should move, particularly for vertical slices. In confined aquifers, water will move from regions where the piezometric head is high to regions where it is low. In unconfined aquifers, water will tend to flow from regions where the water table is high to regions where it is low. Flow in aquifers is often of low velocity and *laminar* in nature. Flow velocities will in these conditions vary linearly with the head gradient, as described by Darcy's law. Flow velocities are usually expressed in this context as Darcian velocities, the flux per unit cross-section of the aquifer, q which can then be expressed as:

$$q = -K_s \frac{dh}{dx} \tag{15.1}$$

where K_s is the saturated hydraulic conductivity, h is total head and x is distance in the direction of flow. In a porous medium, of course, not all the cross-section is flowing water; part of it is solid material that forms the matrix of the aquifer. In fact, in the distribution of pore spaces in the cross-section, there may be many different flow velocities. Any solute or pollutant flowing with the water will locally follow this distribution of velocities and, as a result, be subject to significant dispersion (see Section 15.7 on transport processes in groundwater below). Thus, another descriptive velocity that is used is the mean pore water velocity. This is related to the Darcy velocity as

$$q = v_p n \tag{15.2}$$

where q is the Darcy velocity, v_p is the *mean pore water velocity* and n is the water-filled porosity of the porous medium.

Models based on Darcy's law have been the basis for nearly all models used in groundwater management when we need to understand questions like what will be the effects of adding a new water supply well on the yield of other wells or low river flows; what is the capture zone for a particular well and are there potential sources of pollution at the surface; what will be the travel pathways and times of a pollutant; and what will be the impacts of climate change on the groundwater resource if natural recharge rates are changed?

It is important, however, to recognise the limitations of the simple groundwater theory that will be presented here. In deep aquifers, the pressures may result in deformation of the pore space and changes in the density and viscosity of the water. It may then be necessary to take account of the *compressibility* of the aquifer materials. There may also be circulations developing, not only as a result of head gradients but also because of temperature and density gradients. Density-driven flows are also important where a freshwater aquifer interacts with sea water in a coastal or island aquifer with direct connections to the sea. Finally, there are *karstic aquifers* and *fractured aquifers*, where the flows may not be laminar and will then not be well described by Darcy's law. Karstic aquifers are typically found in limestones, where solution of the carbonate rock by percolating rainwater results in channels and caves that can provide complex networks of rapid pathways through the relatively impermeable solid rock. Similarly, fractures in otherwise impermeable rock can also provide storage and permeability. The chalk aquifers that are so important for water supply purposes in many parts of the world are again a particularly interesting case. They tend to exhibit a form that consists of high porosity but relatively low permeability blocks of matrix (originally formed from the carbonate remains of diatoms). Between the blocks there are joints and fractures, that provide little additional storage but add greatly to the permeability, and, occasionally, the chalk also exhibits karstic features that give rise to very rapid flow pathways.

15.2 Groundwater flow equations

Here we will only consider groundwater flows that can be assumed to be well described by Darcy's law (see Chapter 5 for the relevant definitions). Flow will therefore take place from a point of high potential to a point of low potential. The constant of proportionality, which we call the hydraulic conductivity, is normally assumed to be a constant for saturated aquifers (whereas we saw earlier that it can vary over orders of magnitude for unsaturated conditions as a soil or rock dries out). Methods of estimating hydraulic conductivity and a table of typical values for different types of aquifers can be found in Chapter 5.

To address groundwater management issues we need to be able to model the patterns of potential and velocities in an aquifer in response to recharge and pumping. This requires a dynamic model that can be defined by combining Darcy's law with the

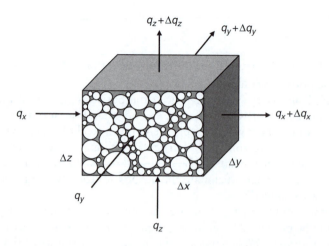

Fig. 15.3 Definition diagram for flow in and out of a block of aquifer. Darcian fluxes indicated by q_x, q_y and q_z. Block dimensions are Δx, Δy and Δz.

continuity (mass balance) principle. Fig. 15.3 shows an element of saturated earth or rock with sides of length Δx, Δy and Δz. Considering water movement in the x, y and z directions, and using the principle of continuity, the following mass balance equality can be written for the difference between inflow and outflow:

Inflow − outflow = Change in storage

$$q_x \Delta y \Delta z + q_y \Delta x \Delta z + q_z \Delta x \Delta y - q_{x+\Delta x} \Delta y \Delta z - q_{y+\Delta y} \Delta x \Delta z - q_{z+\Delta z} \Delta x \Delta y = S_s \frac{\Delta h}{\Delta t} \Delta x \Delta y \Delta z$$

$$\left(q_x - q_{x+\Delta x}\right) \Delta y \Delta z + \left(q_y - q_{y+\Delta y}\right) \Delta x \Delta z + \left(q_z - q_{z+\Delta z}\right) \Delta x \Delta y = S_s \frac{\Delta h}{\Delta t} \Delta x \Delta y \Delta z$$

$$- \Delta q_x \Delta y \Delta z - \Delta q_y \Delta x \Delta z - \Delta q_z \Delta x \Delta y = S_s \frac{\Delta h}{\Delta t} \Delta x \Delta y \Delta z$$

$$- \frac{\Delta q_x}{\Delta x} - \frac{\Delta q_y}{\Delta y} - \frac{\Delta q_z}{\Delta z} = S_s \frac{\Delta h}{\Delta t} \tag{15.3}$$

Here q_x is the Darcian velocity (flux per unit area of aquifer), S_s is called the specific storage, which relates a change in storage to a change in head and $\Delta h/\Delta t$ is the change in potential during a short time step Δt. Note that if, in any direction (here, e.g. the x direction), the input flux is greater than the output flux, such that storage should be increasing, then the gradient term $\Delta q/\Delta x$ is given a negative sign because the Darcian velocity is *decreasing* in the direction of increasing x, i.e. the gradient is negative. In confined aquifers, the specific storage, S_s, is small and is related to the compressibility of the porous medium and the density of water with changes in head. In unconfined aquifers it will be larger as a result of changes in storage in the unsaturated zone as the water table rises and falls with changes in head.

If the increments in space and time are reduced to infinitely small values, we can write the mass balance equation as a continuous partial differential equation in three dimensions:

$$-\frac{\partial q_x}{\partial x} - \frac{\partial q_y}{\partial y} - \frac{\partial q_z}{\partial z} = S_s \frac{\partial h}{\partial t} \qquad (15.4)$$

Applying Darcy's law for q_x, q_y and q_z in the form:

$$q_x = -K_x \frac{dh}{dx} \qquad (15.5)$$

and allowing that the aquifer might be *anistropic* (i.e. it has different hydraulic conductivities in the different directions) we get:

$$\frac{\partial}{\partial x}\left(K_x \frac{\partial h}{\partial x}\right) + \frac{\partial}{\partial y}\left(K_y \frac{\partial h}{\partial y}\right) + \frac{\partial}{\partial z}\left(K_z \frac{\partial h}{\partial z}\right) = S_s \frac{\partial h}{\partial t} \qquad (15.6)$$

In large deep aquifers it may be necessary to take account of the changes in hydraulic conductivity as a function of the temperature and viscosity of water (see the discussion of the *intrinsic permeability* of the medium in Section 5.3). For a homogeneous, *isotropic* material, the hydraulic conductivities (K) are equal and constant and (15.6) reduces to:

$$K\left(\frac{\partial^2 h}{\partial x^2} + \frac{\partial^2 h}{\partial y^2} + \frac{\partial^2 h}{\partial z^2}\right) = S_s \frac{\partial h}{\partial t} \qquad (15.7)$$

This simplifies further for *steady* flow when $\partial h/\partial t = 0$ and, dividing through by K, becomes:

$$\frac{\partial^2 h}{\partial x^2} + \frac{\partial^2 h}{\partial y^2} + \frac{\partial^2 h}{\partial z^2} = 0 \qquad (15.8)$$

This is called the *Laplace equation*. Note that it does not include the hydraulic conductivity. Thus solutions for *steady* flow will not depend on the hydraulic conductivity. This means that, in this special case of steady state flow, identical patterns of head could be predicted regardless of the permeability of the aquifer. In fact, this strictly applies only where the boundary conditions for the aquifer can be fixed at constant potential values (these are called *Dirichlet boundary conditions*) or assumed impermeable (*Neumann boundary conditions*). If there are recharge boundary conditions or other specified flux boundary conditions such as pumped wells (*Cauchy boundary conditions*) then the right-hand side of (15.8) is not everwhere zero and (15.7) must be used with an additional source-sink term in the form:

$$K\left(\frac{\partial^2 h}{\partial x^2} + \frac{\partial^2 h}{\partial y^2} + \frac{\partial^2 h}{\partial z^2}\right) = S_s \frac{\partial h}{\partial t} + R(x, y, z, t) \qquad (15.9)$$

where $R(x, y, z, t)$ is a specified flux as a function of space and time. For fixed head boundary conditions, however, the solution of (15.8) gives the hydraulic head h in terms of x, y and z. The solution of the full equation for transient flow in an anisotropic medium gives h in terms of t as well as x, y and z. Both (15.6) and (15.7) are difficult to solve in all but very simple situations. That is why, in real applications, approximate numerical solutions using computer models are often used (see Section 15.5). However, it can still be useful to obtain insights into a groundwater problem to be able to interpret a *flow net*. The construction of flow nets is easiest where flow is steady and conditions allow a problem to be reduced to two dimensions, e.g. water movements in a vertical $x - z$ plane through layered aquifers or in the horizontal $x - y$ plane relative to some convenient baselevel of z. Such simplification often enables satisfactory answers to be obtained rather simply to problems of well yields or in the assessments of areal groundwater resources.

15.3 Flow nets

Flow nets comprise the intersection of two sets of lines: lines of *equipotential* (contours of potential head) and *streamlines* or flowlines. Equipotentials and streamlines always intersect at right angles in an isotropic aquifer (they are *orthogonal*). Flow nets can represent the velocities and directions of groundwater movement in two dimensions. The closer the equipotentials, the faster will be the Darcy velocity, while the streamlines indicate the direction of flow everwhere. One of the features of a flow net is that there can be no net flow across a streamline. Streamlines can therefore be drawn so that the flow between two adjacent streamlines is the same everywhere. In the case of a confined aquifer of constant thickness, this means that, if the hydraulic gradient increases, the equipotentials will get closer together and, since the velocity is also increasing, to keep a constant flow rate between streamlines the streamlines must also get closer together (see Fig. 15.4). This will happen e.g. where there is flow towards a well in a confined aquifer (see later).

15.3.1 Graphical solutions

The use of flow nets is most easily demonstrated for the case of a confined aquifer where the totally saturated groundwater body is overlain by an impermeable aquiclude. A confined aquifer with several vertical piezometers installed is portrayed diagrammatically in Fig. 15.4a. Piezometer A from ground level 450 m penetrates to a depth of 150 m above datum and the water level rests at 375 m. The total potential or hydraulic head h is then equal to the sum of the pressure head and the elevation head z (height of pressure measuring point above a fixed datum).

Thus, knowing the height of the land surface and the length of the piezometer, from a measure of the water depth d, the value of h can be obtained. For A, the piezometer length is 300 m, d is 75 m and z is 150 m so that:

$$h_A = 300 - 75 + 150 = 375 \text{ m} \tag{15.10}$$

At piezometer B, where the land surface is also 450 m, d is at 150 m and therefore $h = 300$ m. If the distance Δx between A and B is 300 m, then the mean hydraulic

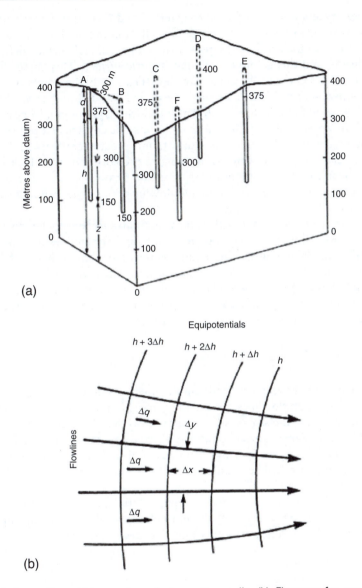

(a)

(b)

Fig. 15.4 (a) An aquifer with a number of observation wells. (b) Flow net for two-dimensional groundwater flow in plan.

gradient is:

$$-\frac{\Delta h}{\Delta x} = \frac{375 - 300}{300} = 0.25 \tag{15.11}$$

Thus at the elevation head, $z = 150\,\mathrm{m}$, there is a difference in potential from A to B and therefore there will be a component of Darcian velocity q_{AB} from A to B of $0.25\,K_s\,\mathrm{m\,s^{-1}}$, where K_s is the saturated hydraulic conductivity of the medium. In the 'field' of piezometers in Fig. 15.4a, piezometers C and E from different surface levels have h values of 375 m, at piezometer D, h is 400 m, and at B and F, h values are 300 m. Within the block of land there is a three-dimensional surface defined by the hydraulic heads, and this is known as a *piezometric surface*. This passes through all the water rest levels in the wells. On a plan view (Fig. 15.4b) the points A, C and E lie on an *equipotential line* (375 m). Through the points B and F on the two-dimensional plan runs the equipotential line of 300 m. Once the equipotential lines have been determined for an isotropic aquifer, *streamlines* may be constructed perpendicular to the equipotential lines in the direction of maximum potential gradient. In the example, it is obvious that the groundwater is draining to the corner of the block between B and F, and hence, three streamlines with direction arrows have been drawn on the plan. The pattern of equipotential lines and flowlines constitutes a *flow net*, of which Fig. 15.4b is a very simple example.

Flow nets drawn under certain rules allow flow rates to be calculated very simply. Fig. 15.4b shows a flow net of equipotential lines and streamlines drawn for a two-dimensional groundwater flow. The equipotentials have equal drops of head, Δh, between any adjacent pair. Taking a typical cell in which the distance between the equipotential lines is Δx, then the Darcian velocity of flow through the cell is $q_v = K\Delta h/\Delta x$. For unit thickness of aquifer (perpendicular to the flow net), the flow rate through the cell bounded by flow lines Δy apart, is:

$$\Delta q = q_v \Delta y\, 1 = K\frac{\Delta h}{\Delta x}\Delta y = K\Delta h \frac{\Delta y}{\Delta x} \tag{15.12}$$

Since Δq is constant between two adjacent flowlines (no flow can cross them), all the cells between two such flowlines having the same Δh must have the same width to length ratio, $\Delta y/\Delta x$. If the streamlines are drawn so that Δq is the same between all pairs of adjacent flowlines, then the ratio $\Delta y/\Delta x$ will be the same for *all* the cells in the flow net. In addition, the spacings can be chosen such that $\Delta y = \Delta x$ and all cells then become part of a curvilinear square grid. Following such rules, then $\Delta q = K\Delta h$, per unit thickness of aquifer.

If there are N drops of Δh between equipotential boundaries whose potential difference is H, then $\Delta h = H/N$. If there are M 'flowtubes' between impermeable boundaries, then the total flow rate (per unit thickness of aquifer), is:

$$Q = M \quad \Delta q = MKH/N \tag{15.13}$$

Summarizing the properties and requirements of flow nets in homogeneous, isotropic media:

(a) equipotential lines and flowlines must all intersect at right angles;
(b) constant-head boundaries are equipotential lines;
(c) equipotential lines meet impermeable boundaries at right angles;
(d) if a square grid is used, it should be applied throughout the flow net (although difficulties will arise near sharp corners and towards remote or infinite boundaries).

The steady-state equations that underlie the description of groundwater flows are linear equations. This has a real advantage in the construction of more complex flow nets because it allows the use of the principle of superposition. This means e.g. that, where a pumped well is affecting a regional groundwater flow, the flow nets for each can be constructed separately and added together.

For example, we can consider the effect of adding a pumped well to the flow net of Fig. 15.4b. The flow net for a pumped well in the absence of other disturbances consists of a radial system of M flow tubes converging towards the well. M can be chosen so that the flow rate between each streamline is equal to the flow between streamlines of the original flow net. As the flow converges towards the well, the distances between the streamlines gets smaller. To preserve the correct proportionality, the spacing between equipotentials must also get smaller. This results in a correct reflection of the *cone of depression* in the piezometric surface or water table normally found around a pumped well (Fig. 15.5b).

Then, having created the two flow nets with the same flow rates and equipotential spacing, they are easily superimposed to give the final joint flow net (e.g. Fig. 15.5d). The results will be a good approximation to the effect of a well on the actual system, providing that the effect of the well does not have a significant effect on the boundaries of the flow domain (e.g. if the cone of depression intersects

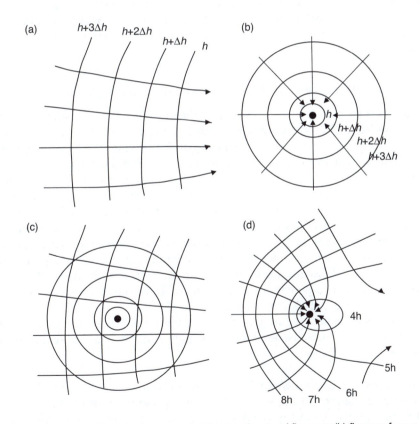

Fig. 15.5 Linear superposition of equipotentials: (a) original regional flow net; (b) flow net for a pumped well showing cone of depression in equipotentials; (c) superposition of equipotentials from (a) and (b) before summation; and (d) resulting flow net showing final equipotentials and flow lines.

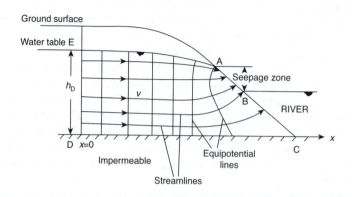

Fig. 15.6 Schematic of a vertical flow net for steady unconfined flow from an aquifer towards a river.

and impermeable boundary). There are techniques for dealing with these additional complications (see, e.g. Cedergren, 1997) but it is now more normal to construct a numerical model of an aquifer to deal with complex situations. Once a model has been constructed, exploring the effects of different boundary conditions, complex patterns of hydraulic conductivity, well positions or different pumping rates are easily implemented.

15.4 Unconfined flow and groundwater recharge

The lateral seepage from a river bank into a river produced by unconfined flow from a porous aquifer with a well-defined water table is portrayed in Fig. 15.6. The aquifer is assumed to be isotropic and homogeneous, and it is underlain by an impermeable stratum. The flow pattern in the combined unsaturated and saturated layers of such an aquifer is not so easily identified as in a completely saturated confined aquifer. The free surface of the water table has an increasingly downward slope to meet the river bank at point *A*, the top of a seepage surface. The water-table slope reflects a hydraulic gradient and the line of the water table describes a stream line if no recharge occurs from the ground surface. In effect, the water table is the upper boundary of a flow net in the saturated part of the aquifer. The river bed/bank interface is an equipotential line and hence all streamlines meeting it must turn to meet *BC* at right angles. The seepage surface *AB* is not an equipotential and the water table flowline is tangential to the river bank at *A* (Cedergren, 1997).

To be able to calculate the flow to the river, several approximations are generally made. These approximations, developed by Dupuit and Forchheimer, make the major assumptions:

(a) the hydraulic gradient dh/dx is equal to the slope of the water table;
(b) the related specific discharge is constant throughout the depth of flow; and
(c) the line *ED* may be taken to act as a constant head boundary with water table height h_D.

Thus, the streamlines are taken to be horizontal and the equipotential lines vertical.

Making these assumptions for steady-flow conditions, the discharge per unit width (Q) over the depth of saturation (h), with hydraulic conductivity (K), is given at all sections by:

$$Q = -Kh\frac{dh}{dx} \tag{15.13}$$

The integral of hydraulic conductivity over the depth of saturation (here Kh) is also known as the *transmissivity*, T, of the aquifer. In a real aquifer, there would be non-horizontal flow components near the water table and near ABC. However, making the Dupuit–Forchheimer assumption of a horizontal flow pattern, satisfactory results are obtained for the discharge when the water table slope is small and the variation in the unconfined aquifer depth relative to that depth is small.

An extension of the seepage flow in an unconfined aquifer can include recharge from water percolating downwards through the unsaturated zone. This is a problem encountered by drainage engineers designing channels for leading off surplus water. In Fig. 15.7, drainage from the surface is assumed to give a uniform recharge rate R at the ground surface of an isotropic, homogeneous aquifer of hydraulic conductivity K. It is often required to know the depth of the water table relative to the water level in the river or channel. With steady-state conditions and adopting the Dupuit–Forchheimer assumptions, the Darcian velocity or specific discharge q at distance x, from Darcy's law, is $q = -K\,dh/dx$. Then the flow Q at x per unit width for the aquifer depth h is $Q = -Kh\,dh/dx$. However, from mass balance at steady state, the total recharge at any point is also given by $Q = Rx$. If all the flow in the aquifer is assumed to come from the recharge and none across the boundary ED at $x = 0$, then:

$$Rx = -Kh\frac{dh}{dx}$$

or

$$Kh\,dh = Rx\,dx \tag{15.14}$$

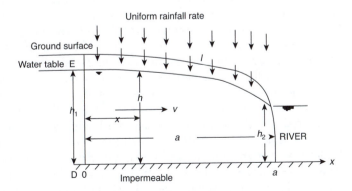

Fig. 15.7 Recharging unconfined flow contributing lateral inflow to a river.

Integrating between h_1 and h_2 with x going from 0 to a (see Fig. 15.7) then:

$$K\left(h_1^2 - h_2^2\right) = Ra^2 \tag{15.15}$$

whence:

$$h_1 = \left[\frac{R}{K}a^2 + h_2^2\right]^{0.5} \tag{15.16}$$

Thus, h_1, the water table depth at distance a from the river can be found, knowing the water level in the channel, the infiltration rate and K (ignoring the seepage surface). Under the steady recharge assumption, baseflow to the river per unit length of channel, Q, can be expressed in terms of recharge as:

$$Q = Ra \tag{15.17}$$

Example: **Calculation of water table heights in an unconfined aquifer**
In Fig. 15.7, it is required to know the height of the water table at 150 m from the river when irrigation water is applied at a constant rate of 0.8 mm day^{-1} assuming a steady state is attained with the uniform irrigation rate, $K = 4.3$ m day^{-1} and a drain level at 12 m above the impermeable stratum. Substituting into (15.16):

$$h_1 = \left[\frac{0.008}{4.3}150^2 + 12^2\right]^{0.5}$$

$$= 12.173 \, \text{m}$$

15.4.1 Groundwater flooding

One of the consequences of periods of high rainfall on unconfined aquifers is to enhance the recharge to the water table. Recharge rates will generally be much higher in winter than in summer, because the near-surface soil remains wetter and losses to evapotranspiration are smaller. During extreme wet periods, the water table may rise to the surface and cause flooding of normally dry valley bottoms. The *bournes* of the chalk in south-east England (such as at Sittingbourne, Nailbourne and Petham Bourne in Kent) are examples of dry valleys subject to occasional flooding of this type. This type of local flooding occurred quite widely in this area in the winter of 2000/2001 (Marsh and Dale, 2007) when the highest recharge rates were recorded since 1920. Groundwater flooding can cause flooding of basements and lower floors of buildings; flooding of buried services and other assets below ground level; flooding of farmland, roads, commercial, residential and amenity areas; overflowing of sewers and drains; and reduction in the strength of foundations. One of the characteristics of groundwater floods is the delay between the rainfall and a widespread water table response. This means that groundwater flooding can occur with a significant delay after a period of high rainfall inputs and can also last for long periods. These floods are sometimes called 'clear water' floods since the return flow from an aquifer will often be of good quality.

In the past, however, there has been little modelling of groundwater flooding and no warning system in operation. Following the 2007 floods, this is now being remedied as a result of recommendations in the Pitt Report.

15.5 Numerical groundwater models

15.5.1 Assumptions and boundary conditions

Complex and transient (non-steady) flow conditions require an approximate numerical solution to the groundwater flow equations. There are a variety of numerical solution methods available, including finite difference, finite element, finite volume and boundary element methods. All are general methods for solving partial differential equation, such as that based on Darcy's law used to describe groundwater flows. Similar solvers are now available in many general mathematics packages such as Matlab, Mathcad and Mathematica. *The finite-difference method* is the most straightforward and easily understood numerical method and is the basis for groundwater packages such as the widely used MODFLOW package, originally developed by the US Geological Survey but now also available with graphical interfaces from a variety of different vendors (such as Visual MODFLOW). MODFLOW, and other groundwater packages can be used to solve problems in one, two and three dimensions with a variety of boundary conditions and for both steady-state and transient flow conditions.

In the finite difference method, the groundwater flow domain is subdivided into blocks in a similar way to that used to derive the flow equations in Fig. 15.2 (e.g. Fig. 15.8). In solving the equations, however, it is more convenient to use a grid of nodes at the centre of each block. The nodes of the grid are the points at which the values of head are required or can be specified as boundary conditions for the flow domain. At each node, the h value is assumed to be representative of the block centred on that node. If the value of h at A is unknown, it can be calculated from considerations of the flow pattern around the block.

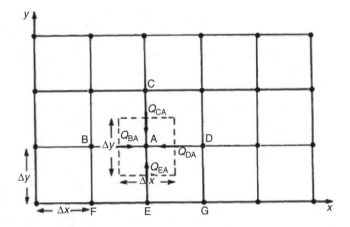

Fig. 15.8 Definition diagram for a block-centred finite difference scheme. The block around the point A is indicated by dashed lines. Flows Q are to be calculated given heads at the centres of the neighbouring blocks, B, C, D and E.

Then the finite difference estimate of the flux from A to B is

$$q_{AB} = -K_x \left(\frac{h_A - h_B}{\Delta x} \right) \tag{15.18}$$

To obtain a complete finite difference form for the full transient flow equation (15.9), we need to define finite difference approximations for differential terms of the form

$$\frac{\partial}{\partial x} \left(K_x \frac{\partial h}{\partial x} \right) \tag{15.19}$$

centred around the point A. This can be achieved by using a Taylor series expansion of h around the point A. By including higher order terms in the expansion, we can get a closer approximation to the values of the differential terms. Most finite difference models use a second-order approximation for the spatial derivatives of the form (assuming that K_x is constant everywhere):

$$\frac{\partial}{\partial x} \left(K_x \frac{\partial h}{\partial x} \right) = \frac{K_x}{\Delta x} \left[\left(\frac{h_D - h_A}{\Delta x} \right) - \left(\frac{h_A - h_B}{\Delta x} \right) \right]$$

$$= K_x \left[\frac{h_D - 2h_A + h_B}{(\Delta x)^2} \right] \tag{15.20}$$

Similar expressions can be used for the differentials in the y and z directions. If K_x is not isotropic, then the change in the conductivities in the different directions can be included in formulating the flux terms (this is clearly easiest if the nodes are on the boundaries between regions of the flow domain of different conductivity). If the conductivities are not homogeneous because of changes in aquifer properties, lithology, occurrence of faults or other features, then different conductivities can be used in different elements of the discretisation of the flow domain.

For steady-state flows the spatial difference approximations of (15.20) will be sufficient to formulate a finite difference solution for all the nodes in the grid. Effectively, since the solution at A will depend on the solution at the surrounding points B, C, D, E, which are also initially unknown except at fixed head boundary points, the result is a set of linear simultaneous equations, including equations for the boundary conditions. Solution techniques for such cases are well known, and involve iterating the solution from an initial estimate until a final solution is reached.

For transient problems, we also need to define a finite difference estimate for the time differential $\partial h/\partial t$ in (15.9).

We also need to now index the nodal head values by a time step index, $t = 1, 2, 3.....T$ for time steps of length Δt. There are two approaches to solving such time-stepping problems. The easiest to implement is an *explicit* solution. This makes the solution at time t a function only of the known values at the previous time step at $t - \Delta t$ (Fig. 15.9a). In this case, given a pattern of heads for the first time step (the *initial conditions*) at time $t = 0$, the spatial differential terms are evaluated at time $t = 0$, and the solution at time $t = \Delta t$ is obtained using the forward difference approximation to

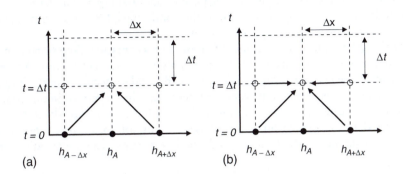

Fig. 15.9 Schematic diagram for: (a) explicit; and (b) implicit time-stepping schemes for finite difference representation of the time derivative. Filled circles represent values that are known at time step t_1. Open circles represent values to be calculated. Arrows indicate dependencies in the calculation of $h_{A,t=\Delta t}$.

the time derivative

$$\frac{\partial h_A}{\partial t} = \frac{h_{A,t=\Delta t} - h_{A,t=0}}{\Delta t} \qquad (15.21)$$

Then having solved the set of equations for time $t = \Delta t$, the process can be repeated for successive time steps. The calculations are relatively simple because, at each time step, the only unknown in each nodal equation is now $h_{A,t}$, all the other terms use the values of h at time $t - \Delta t$, which are known. This formulation of the time stepping is only first order correct and can be subject to stability problems if there are strong gradients in time or space, or rapid changes in boundary conditions, unless rather short time steps are used.

An *implicit* solution can also be used (Fig. 15.9b). In this case, at each time step the time differential is treated in a similar way but the spatial differentials are treated as a weighted sum of the values at the times t and $t + \Delta t$. Thus, in an implicit formulation for a typical node at position A:

$$\frac{\partial}{\partial x}\left(K_x \frac{\partial h}{\partial x}\right) = \Omega K_x \left[\frac{h_{D,t+\Delta t} - 2h_{A,t+\Delta t} + h_{B,t+\Delta t}}{(\Delta x)^2}\right]$$

$$+ (1 - \Omega)K_x \left[\frac{h_{D,t} - 2h_{A,t} + h_{B,t}}{(\Delta x)^2}\right] \qquad (15.22)$$

where Ω is a time-weighting coefficient. The values of h at time t are known at the start of the time step, but those at time $t + \Delta t$ are not. This leads to a system of linear simultaneous equations in the unknown values of h at time $t + \Delta t$. Efficient techniques for solving such equations are well developed, but involve a computational cost greater than for an explicit solution for each time step. The advantage of using an implicit solution, however, is that it is much less likely to suffer from stability problems so that much longer time steps can be used. The overall computational cost may then be less. The USGS groundwater modelling package, MODFLOW, for example, uses an implicit finite difference approximation of the continuous differential flow equations for both confined and unconfined aquifer cases.

For both explicit and implicit finite difference approximations, the starting point at time $t = 0$ does depend on being able to specify a set of initial conditions. Accuracy in reproducing the behaviour of the real system will depend on accuracy in the initial conditions. In some cases it will be possible to interpolate the pattern of initial conditions from measured heads from observation wells. In other cases, a steady-state solution might be used as the starting point for a transient solution, or a 'run-in' period might be used to obtain a consistent initial condition before simulating a period of interest.

15.5.2 Groundwater modelling packages

There are now a variety of groundwater modelling packages around from both commercial and public domain sources. MODFLOW 2005 can be freely downloaded from the United States Geological Survey (USGS) site.[1] It has also been packaged by several companies to give a more user friendly graphic interface to help in setting up models and visualising the results of model runs. MODFLOW is widely used (see, e.g. Todd and Mays, 2004). It is based on a finite difference solution to the groundwater flow equations, can handle two- and three-dimensional (3D) problems, includes model calibration routines to fit aquifer parameters by matching observations, and can be linked to pollutant transport models (see next section). The International Groundwater Modelling Centre at the Colorado School of Mines[2] also maintains a list of freeware groundwater modelling software, including MODFLOW and the United States Department of Agriculture (USDA)'s Hydrus 2D/3D,[3] which is based on finite element solutions of the flow equations, which give more flexibility in the representation of irregular aquifers. An alternative freeware program for Windows is the Aquifer Simulation Model (ASMWin),[4] which can be downloaded from the Institute of Environmental Engineering (IfU) at ETH Zurich, Switzerland. This is also based on finite difference approximations to the flow equations. ASMWin is limited to two-dimensional problems and relatively small grids, but includes both finite difference and random particle tracking transport components, a model parameter calibration routine and visualisation routines. Fig. 15.10 shows a typical finite difference grid for an application of MODFLOW. Where more detail is required in some parts of the simulation, e.g. around an abstraction well, 'nested' grids can be used, where a finer grid is embedded locally in the main grid, as in the ZOOM 3D modelling package developed by the British Geological Survey (BGS[5]).

Many sophisticated commercial packages for groundwater flow (and pollutant transport) modelling are also available, such as the finite element based FEFLOW (see a review by Trefry and Muffels, 2007). Fig. 15.11 shows the finite element mesh that might be used to simulate the flow field in an aquifer with element refinement close to a number of wells.

These groundwater modelling packages often involve large numbers of options in discretising the flow domain, setting up initial and boundary conditions, setting up the time stepping, specifying and optimising the model parameters, and so on. There is a danger that, with poor choices of time step, for example, that the solution will be affected by numerical artefacts such as numerical dispersion or, in some cases,

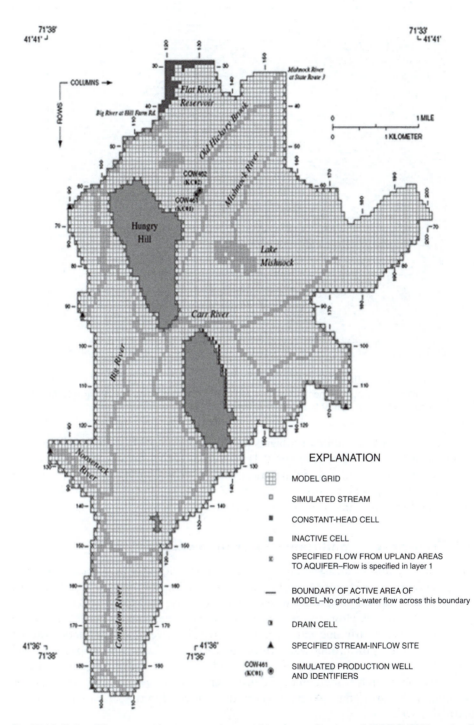

Fig. 15.10 Finite difference grid representation used in the application of MODFLOW to the Big River aquifer, Rhode Island, USA. (Reproduced from Granato and Barlow, 2004, courtesy of United States Geological Survey.)

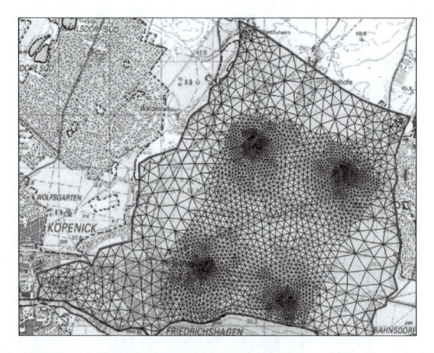

Fig. 15.11 Plan view of a finite element discretisation of an aquifer containing wells to simulate an aquifer remediation scheme. (From a FEFLOW test problem, Trefry and Muffels, 2007, with kind permission of John Wiley & Sons.)

mass balance errors or stability problems (see also Section 15.7). Thus it is important to consider any outputs from such models with care. Techniques for the visualisation of model outputs have improved dramatically in recent years and can be very useful in checking and comparing results, but might also make incorrect solutions which look convincing if the programs are not used carefully.

15.6 Transport of pollutants in groundwater

There is a very important issue about the use of groundwater as a source of water supply. Potentially, because of the filtering effect of the soil and unsaturated zone, the geochemical quality of groundwater is normally expected to be better than most surface water sources. However, if a groundwater body becomes polluted, then it can become unusable for drinking water. Many countries have very strict regulations about land uses in important groundwater recharge areas. Sweden, for example, has road signs indicating when a recharge area is being entered, while in the UK maps of source protection zones can be found on the Environment Agency web site.[6] In the UK, three protection zones are defined with respect to time of travel in the saturated zone to an abstraction well:

(1) a travel time of less than 50 days representing an expected safe decay time for any bacteriological contamination;

(2) a travel time of less than 400 days to protect against more persistent pollutants that degrade and attenuate more slowly;

(3) the complete catchment area *capture zone* draining to the well.

Defining the areas of these zones requires making assumptions about the characteristics of the aquifer and the prevailing hydraulic conditions (or constructing a numerical model of the aquifer). In the European Union, more generally, groundwater vulnerability maps are required to be developed under the Water Framework Directive legislation.

In the past, however, there has been much careless and thoughtless pollution of groundwater. On Long Island, New York, for example, over 50 per cent of the water supply wells have had to be closed, primarily because of pollution by salt applied to de-ice roads in winter. Close to the ocean, over-pumping of wells can lead to salt-water intrusion into fresh water aquifers. The application of agricultural fertilisers to fields in recharge areas can also lead to groundwater exceeding limits for drinking water supply of nitrates. There are also sites polluted by chemical spills and industrial wastes. In the USA, a program of clean-up at over 1000 'Superfund' sites was launched in 1980, following the Love Canal, Niagara, disaster when a school and new houses were built on top of a contaminated infill site. Many of these sites involved soil and groundwater contamination. Some $9bn has been spent, without a great deal of success, though some sites have been delisted. The lessons learned from this program and clean-up efforts elsewhere, is that groundwater, once contaminated can be very difficult to clean up and can remain contaminated for long periods of time. It is known that in some places in the Chalk aquifer in the south of England, much of the nitrate from the application of fertilisers is still in the unsaturated zone. Rates of vertical movement down towards the water table can be of the order of $0.5 - 1$ m year^{-1} and the water table can be more than 50 m below the surface. Certain areas in the UK have been declared nitrate vulnerable zones to try to reduce nitrate additions and the potential for pollution of drinking water sources in the future.[7] It can, however, be difficult to predict the rates of transport in groundwater systems, partly because of heterogeneities in soil and aquifer properties.

15.6.1 Advection and dispersion in groundwater systems

Transport of solutes through porous media is generally described by the advection-dispersion equation (ADE). The ADE can be formulated by looking at the mass fluxes of a solute into and out of a control volume in the same way as the subsurface flow equation (see also the derivation of the ADE for transport in a river in Section 14.5). For the transport equation, we have to consider two types of flux, advection and dispersion. The joint effects of advection and dispersion on a point source of solute are shown, in plan, for different times in Fig. 15.12. Real patterns of concentration tend to be much more complicated due to both complex aquifer structures, and complex geochemistry and microbiological effects. There have been a number of large-scale tracer experiments involving the transport of multiple substances in groundwater that have greatly improved process understanding and helped to improve transport models (e.g. the Borden aquifer experiments: Sutton and Barker, 1985; and Cape Cod experiments: Fig. 15.13; Le Blanc *et al.*, 1991).

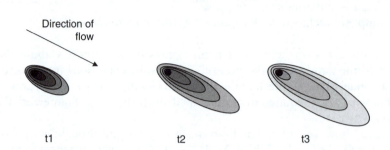

Fig. 15.12 Effects of advection and longitudinal and transverse dispersion on concentrations of a solute in an aquifer at three different times, t1 < t2 < t3. Black circle represents origin, darkness of shading represents concentration.

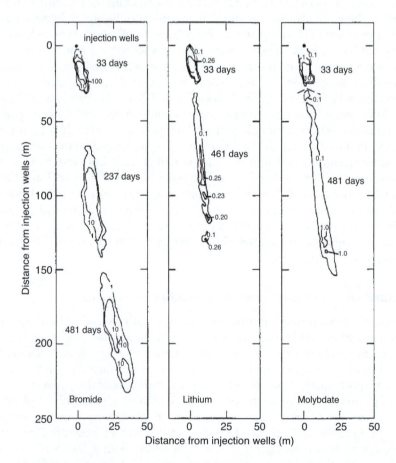

Fig. 15.13 Actual groundwater concentration maps for bromide, lithium and molybdate at different times from a tracing experiment at Otis Air Force Base, Cape Cod, MA (after LeBlanc et al., 1991, reproduced by permission of American Geographical Union).

Advection is the flux due to the mean pore water velocity of the flow (remembering that the mean pore water velocity will be faster than the Darcian velocity or specific discharge of the flow). As already noted, within the pores, solute moving with the water may be subject to a whole range of velocities. This means that some solute will move faster than the mean pore velocity and some more slowly. This gives rise to what is called a dispersive flux of solute. This is generally represented as a form of Fick's law, which describes the expected flux due to simple diffusion of solute molecules as a linear function of a concentration gradient. In dispersion, the process involves difference in flow velocity as well as molecular diffusion but the dispersive flux is also treated as a linear function of a concentration gradient. Thus, the dispersive flux, J, is represented as:

$$J_D = -D\frac{dC}{dx} \tag{15.23}$$

where C is solute concentration, D is called the dispersion coefficient (with units of L^2T^{-1}) and x is distance as before. This relationship includes the effect of random molecular diffusion, but this is usually has an effect on mixing that is orders of magnitude smaller than dispersion due to the pore water velocity distribution.

Making use of (15.23), total longitudinal flux of solute per unit area of flow in the direction of flow by both advection and dispersion is then given by

$$J_x = q_x C + nD_L\frac{dC}{dx} \tag{15.24}$$

where n is the porosity as before and D_L is the longitudinal dispersion coefficient. In groundwater systems, the source of a solute will often be quite localised. Thus there will usually also be a concentration gradient transverse to the direction of the flow. In two dimensions that means that we also need to take account of the transverse spreading of the solute plume (as in Fig. 15.14). This can be represented by a similar flux law but, in general, transverse dispersion is much less efficient than longitudinal mixing in subsurface flow, so that the transverse dispersion coefficient will be smaller.

Going straight to the partial differential form of the ADE we can then write, in two dimensions, for a conservative pollutant or tracer:

$$n\frac{\partial C}{\partial t} = \frac{\partial}{\partial x}\left(nD_L\frac{\partial C}{\partial x}\right) + \frac{\partial}{\partial y}\left(nD_T\frac{\partial C}{\partial y}\right) - \frac{\partial}{\partial x}\left(Cq_x\right) \tag{15.25}$$

where D_T is the transverse dispersion coefficient, x is the direction of flow and y is orthogonal to the direction of flow.

Experimental evidence suggests that the dispersion coefficients in groundwater systems are a function of the mean pore water velocity. This dependence is generally represented in the form:

$$D_L = \alpha_L\frac{q_x}{n}$$

$$D_T = \alpha_T\frac{q_x}{n} \tag{15.26}$$

The coefficients α_L and α_T are called the longitudinal and transverse *dispersivities* (with units of length) and vary with the properties of the porous medium. A special case of (15.25) occurs when the transport is purely one-dimensional. This requires, however, that the source of the solute is spread equally throughout the flow where it is introduced. There is then no transverse concentration gradient, and no transverse dispersion so that (15.25) reduces to (assuming that D_L and n are both constant in space):

$$\frac{\partial C}{\partial t} = D_L \frac{\partial^2 C}{\partial x^2} - \frac{\partial}{\partial x}(Cv_x) \tag{15.27}$$

where v_x is the mean pore water velocity in the direction of flow $= q_x/n$. This simplification is not very realistic in real groundwater systems (where there may be concentration gradients in all three dimensions) but can be useful for the analysis of dispersion experiments in the laboratory, for example, in cores where a fully mixed solute tracer is introduced at one end. For constant v_x and D_L, (15.12) has an analytical solution for certain simple boundary conditions (with an identical form to that of equation 14.41). These conditions are often used to fit a value of the dispersion coefficient to a tracer experiment (e.g. van Genuchten, 1986).

More commonly, the equations for solute transport are solved by numerical approximations. The ADE requires, however, more care in the selection of a solution than the flow equation. That is because the numerical approximation (e.g. the finite difference approximation used for the flow equation earlier) introduces numerical dispersion into the solution, which, if not handled carefully, can result in small errors in the solution growing exponentially and leading to oscillations and instabilities in the solution, particularly when dealing with sharp concentration fronts. There are numerical techniques to avoid such numerical artefacts. The MT3DMS package builds on MODFLOW flow stimulations to predict transport (Zheng and Bennett, 2002).

An alternative approach is to track representative particles of solute through the flow field numerically. Advection is then handled by moving a particle at the mean pore water velocity; dispersion by adding a random velocity to the particle in a way consistent with the dispersion expected. Thus, in one dimension, at a time step of length Δt, the position of the pth particle ($p = 1, 2, \ldots\ldots N_p$) that starts at position x_p at time t can be updated as:

$$x_p^{t+\Delta t} = x_p^t + v\Delta t + z\sqrt{2D_L\Delta t} \tag{15.28}$$

where z is a random variable selected by a random number generator from a normal distribution with mean of zero and standard deviation of unity (N[0,1]), v is the mean pore water velocity and D_L is the longitudinal dispersion coefficient. By moving a very large number of particles in this way the overall dispersion of a pollutant can be predicted. These *random particle tracking* techniques have become more widely used in groundwater modelling as computers and graphical visualisation tools have become more powerful and are built into many groundwater modelling packages, including MODFLOW and ZOOM mentioned earlier.

15.6.2 Immobile pore space and structural porosity

So far, we have treated flow in groundwater systems as if it was always through a porous medium. This is a good representation for many aquifers, e.g. the chalk and sandstone aquifers in the UK. There are also, however, other rocks where the greater part of the volume is rather impermeable or non-porous and any groundwater flow takes place through the structural pore space of fissures, joints and fault lines. Such rocks are more properly described as aquitards than aquifers, though they can sometimes be used for small-scale water supply. The description of flow through fractured systems has assumed greater importance in recent years because of the planning of subsurface waste repositories, in particular for the storage of radioactive wastes. Both Yucca Mountain in the USA and Sellafield in the UK, for example, have been the subject of feasibility studies for deep repositories. A public inquiry for a planning proposal for a deep experimental facility at Sellafield hinged critically on the potential role of fractures in the Borrowdale Volcanic rock series (which would normally be considered as an aquifer) in bringing radionuclides back to the surface (see Hetherington, 1998). While the experimental facility had been proposed to study this in more detail, the difference between different model predictions produced to support or oppose the proposals and based on different interpretations of the information about the site obtained from 15 deep boreholes was so great that the planning inspector recommended that the experimental facility should not be allowed.

15.6.3 Sorption and mass loss

So far in this section on the transport of solutes in groundwater systems we have assumed that the solutes involved can be considered conservative (that is that the total mass in solution does not change). This is, however, rarely the case in real systems. Solutes in transport can be subject to biogeochemical processes such as chemical reactions, volatilisation, sorption on to the solids making up the aquifer and microbial processes. It may also be that not all the pores contain flowing water even under saturated conditions. There can also be diffusion into and out of 'immobile' pore space leading to retardation of the transport. All of these processes can result in changes to the total mass in solution. In prediction transport of a particular solute, these processes are generally treated as additional source/sink terms in the flow equation (15.25). Finally, not all groundwater pollutants are solutes. Some pollutants may be sorbed on to fine colloidal particles that can move through larger and structural pores; others may be transported fluids that do not mix well with water. Oils, for example, can pollute groundwater systems but will tend to travel as droplets separate from the water. This class of pollutants are known as *non-aqueous phase liquids* (NAPLs) and are distinguished by their density with respect to water. Denser liquids (DNAPLs) will tend to settle to the bottom of an aquifer under the influence of gravity; lighter liquids (LNAPLs) will tend to collect near the water table, or upper surface of a confined aquifer. NAPLs may also be broken down by microbial action, particularly under anoxic conditions and one methodology formulated to deal with groundwater pollution is to allow for natural attenuation (National Academy, 2000). As noted earlier, however, any remediation of polluted groundwater may take extended periods of time. More detail on complex groundwater pollution problems can be found in texts such as Younger (2007). A number of groundwater modelling packages now include geo-chemical calculations in addition to predicting transport, including MT3DMS, which allows for reactions amongst multiple chemical species.

15.7 Calibration of groundwater flow and transport models

One of the most problematic issues in the application of groundwater flow and transport models in practice is the calibration of model parameter values. Since they are distributed models with large numbers of calculation elements, they need large numbers of parameter values of transmissivity and specific storage to be defined. Large-scale problems may need other parameters such as compressibility of the different materials to be defined. Transport problems will require dispersivities and reaction coefficients to be defined. In some cases we can estimate parameter values from experiments, such as the derivation of transmissivity and specific storage parameters from well-pumping tests described in Section 15.5 above. We also know something of the different ranges of parameters that might be expected for different types of aquifer materials (see Section 5.1). However, in practical applications we may have only limited information on the geological structures and boundary conditions involved, especially in systems subject to faulting. Thus it is common to try to fit parameter values in a model to match whatever observations are available on pressures, water table elevations and, in transport problems, concentrations at different locations of a pollutant of interest (see, e.g. Hill and Tiedeman, 2007). This generally requires some simplification of the parameter space to be identified, such as by assuming that hydraulic conductivities and dispersivities are constant for a different rock type. The model identification problem may remain challenging, however.

Notes

1 See http://water.usgs.gov/software/lists/ground_water/
2 See http://typhoon.mines.edu/
3 See http://www.ars.usda.gov/services/docs.htm?docid=8939
4 See http://www.ifu.ethz.ch/publications/software/asmwin/index_EN
5 See http://www.bgs.ac.uk/science/3Dmodelling/zoom.html
6 http://www.environment-agency.gov.uk/homeandleisure/37833.aspx
7 http://www.defra.gov.uk/environment/water/quality/nitrate/nvz2008.htm

References

Cedergren, H. R. (1997) *Seepage, Drainage and Flow Nets*. Wiley, Chichester.

Fetter, C. W. (2001) *Applied Hydrogeology*. Prentice-Hall, New York.

Granato, G. E. and Barlow, P. M. (2004) Effects of alternative instream-flow criteria and water-supply demands on ground-water developement options in the Big River area, Rhode Island. *USGS Scientific Investigations Report 2004–5301*. United States Geological Survey, Departmet of the Interior, Washington, DC.

Hetherington, J. (1998) Nirex and deep disposal: the Cumbrian experience. In Barker, F. (ed.) *Management of Radioactive Wastes: the Issues for Local Authorities*. Thomas Telford, London.

Hill, M. C. and Tiedeman, C. R. (2007) *Effective Groundwater Model Calibration, with Analysis of Sensitivities, Predictions and Uncertainty*. Wiley, New York.

LeBlanc, D. R., Garabedian, S. P., Hess, K. M., Gelhar, L. W., Quadri, R. D., Stollenwork, K. G. and Wood, W. W. (1991) Large-scale natural-gradient tracer test in sand and gravel, Cape Cod, Massachusetts: 1. Experimental design and observed tracer movement. *Water Resources Research* 27, 895–910.

Marsh, T. J. and Dale, M. (2007) The UK Floods of 2000–2001: A hydrometeorological appraisal. *Water and Environmental Journal* 16, 180–188.

National Academy (2000) *Natural Attenuation for Groundwater Remediation*. Committee on Intrinsic Remediation, National Academy Press, Washington, DC.

Schartz, F. W. and Zhang, H. (2002) *Fundamentals of Ground Water*. Wiley, Chichester.

Sutton, P. and Barker, A. J. F. (1985) Migration and attenuation of selected organics in a sandy aquifer – A natural gradient experiment. *Ground Water* 23, 10–16.

Todd, D. K. and Mays, L.W. (2004) *Groundwater Hydrology*. Wiley, New York.

Trefry, M. and Muffels, C. (2007) FEFLOW: A finite-element ground water flow and transport modeling tool. *Ground Water* 45, 525–528.

van Genuchten, M. Th. (1986) Solute dispersion coefficients and retardation factors. In: Klute, A. (ed.) *Methods of Soil Analysis, Part 1 Physical and Mineralogical Methods*, 2nd edn., American Society of Agronomy, Madison, pp. 1025–1054.

Young, P. C. and Wallis, S. G. (1993) Solute transport and dispersion in channels. In Beven, K. J. and Kirkby, M.J. (eds) *Channel Network Hydrology*. Wiley, Chichester, pp. 129–175.

Younger, P. L. (2007) *Groundwater in the Environment: an Introduction*. Blackwell, Oxford.

Zheng, C. and Bennett, G. D. (2002) *Applied Contaminant Transport Modelling*, 2nd edn., Wiley, New York.

Chapter 16

Flood risk management

16.1 Drivers for flood risk management

Fortunately in the UK loss of life from flooding is relatively rare, compared to the scale of events in some other parts of the world. Even so, the deaths of 13 people were linked to flooding during the summer of 2007 in north-east, central and south-west England. During these floods, there was damage to approximately 48 000 homes and 7000 businesses (Cabinet Office, 2008) and the Association of British Insurers[1] (ABI) reported around 165 000 insurance claims totalling £3 billion. Flooding of a home or business is a difficult experience with considerable follow-on effects on health, security and financial viability. Table 16.1 shows the largest insured losses from natural hazards in the UK. Whilst some of the losses shown were caused by a combination of flooding and other perils, it is clear how important flooding is (and how significant was the flooding in the summer of 2007). Note that the figures shown are insured losses, and hence not the same as estimates of the wider economic consequences of such events.

Flooding is by no means a new phenomenon. There are many historic references to major floods. In the UK, a detailed database of historical flood events has been compiled by the British Hydrological Society in the Chronology of British Hydrological Events[2] (Black and Law, 2004). What has changed over time is the social and economic impact of flooding. The consequences of water entering homes and businesses in terms of financial loss are now much greater than they once were. Earlier buildings with solid stone walls, stone flagged or seasoned timber floors, no electricity and relatively few possessions would not have suffered from the same losses as a modern timber frame house with chipboard floors filled with high-value consumer items. Similarly, flooding of modern industrial buildings fitted with expensive electronic equipment results in greater losses than flooding of more resilient heavy industrial plant. An ABI study in November 2007 showed an average insured loss of a flood-damaged house of the order of £20 000.

Over the last century, there has also been extensive development of land which is subject to periodic flooding. There was little need to use this land hundreds of years ago other than for summer grazing. Today floodplains can be occupied by houses, factories, transport systems, supermarkets and other infrastructure. Flooding is not a problem that is likely to be eliminated. The challenge is to manage the risk and do what is possible to reduce it. Hopefully risk to life will be contained and the rate of increase of economic loss reduced, but complete flood prevention is now considered an unrealistic aspiration.

Table 16.1 Insurance losses (2007 adjusted figures) for the ten largest natural catastrophe events in the UK. (Guy Carpenter, 2007, incorporating figures from the Association of British Insurers.)

Date	Event	Adjusted loss 2007 (£m)
January/February 1990	Storms and flooding across UK	3370
June/July 2007	Floods	3000
October 1987	Storm	2089
October/November 2000	Floods	914
December 1981/January 1982	Arctic weather/rapid thaw	630
January 1987	Severe snowstorm	551
December 1995/January 1996	Severe snowstorm	427
December 1997/January 1998	Storms and flooding	342
January 2007	Windstorm kyrill	342
February 1991	Severe snowstorm and flood	281

16.2 Flood management and land drainage responsibilities in the UK

16.2.1 Flood management responsibilities

The lead government department for flood risk in England is Defra (Department for Environment, Food and Rural Affairs). In Scotland, it is the Scottish Government, in Wales the National Assembly for Wales and in Northern Ireland the Department of Agriculture and Rural Development (DARD). A summary of flood risk management responsibilities is given in Fleming (2002). In England and Wales, the day-to-day responsibility is largely devolved to the Environment Agency. Flood risk management accounts for over half of the Agency's £1bn budget (House of Commons, Environment, Food and Rural Affairs Committee, Flooding, Fifth Report of Session 2007–8, Vol. 1).

16.2.1.1 The Environment Agency

The Environment Agency was established by the Environment Act 1995 and is empowered (although not legally obliged) under the Water Resources Act 1991 to manage flood risk from designated 'main rivers' and the sea. The Agency also provides flood warning and forecasting services, and has to carry out flood defence work through the mechanism of the Regional Flood Defence Committees (RFDCs). There are 12 such committees, whose membership structure is set out by the Environment Act 1995. They make decisions about improvement and maintenance work to be carried out by the Agency and financial allocations, as well as other issues related to flood management and land drainage within their regions.

16.2.1.2 Internal drainage boards

Internal Drainage Boards (IDBs) are independent bodies responsible for land drainage in 1.2 million hectares of lowland England. They are long established bodies operating predominantly under the Land Drainage Act 1991 and have permissive powers to undertake drainage works and water level management. Much of their work involves the improvement and maintenance of rivers, drainage channels and pumping stations. There are now around 170 in England and Wales, concentrated in East Anglia, Yorkshire, Somerset and Lincolnshire. Internal drainage board membership includes elected members representing the occupiers of the land in the district and members nominated by local authorities to represent other interests.

Internal drainage boards have flood management powers for watercourses not designated as 'main river'. These are known as 'ordinary watercourses'.

16.2.1.3 Local authorities and highways authorities

Local authorities may undertake flood defence works under the Land Drainage Act 1991 on watercourses not designated as 'main river' and not managed by an IDB. There are also 88 maritime district councils that have powers to protect the land against coastal erosion under the Coast Protection Act 1949.

16.2.1.4 Water companies

Water and wastewater companies are required under the Water Industry Act 1991 to 'effectively drain' areas for which they are responsible. Recent government thinking is leading towards transfer of responsibility for private sewers and lateral drains to the water and wastewater companies, as considered in the Water Strategy (Defra, 2008a).

16.2.2 Land drainage

Land drainage is closely related to flood management and may indeed be a part of flood management in a legal sense. Here land drainage is considered in terms of legal rights and obligations. Land drainage responsibilities in the UK vary significantly. European Union (EU) legislation is having a major impact on UK legislation, particularly concerning the environment. In England, the law is not all set out in legislation, but has developed over many years from either common customs (common law) or out of court decisions (case law). While case law provides most of the law in England and Wales, parliaments are becoming increasingly active in making new laws and statute law (primary and secondary legislation) is coming to dominate common law.

Land drainage responsibilities generally relate to riparian rights and drainage authority powers. Riparian rights are dependent on ownership of land abutting a 'watercourse' and the rights have usually come about from common and civil case law. Drainage authority powers have usually come about from statute law, i.e. legislation.

16.2.3 Riparian rights

The common law rights of a riparian owner come from the ownership of property adjoining a natural watercourse (or a lake). Riparian rights are normally subject to reasonable use of the watercourse by all who have access to it. A riparian owner has the right to use the water (from a watercourse or lake) for domestic purposes, drinking water for cattle and for irrigation purposes (provided that any diversion for irrigation is not continuous and it is returned without diminution, other than by evaporation and absorption).

A riparian landowner has a right at common law to raise the banks of a watercourse on the owner's land as necessary to prevent flooding on to the land behind the banks. In doing so the owner must not damage the property of others, nor increase the flow in the watercourse to the detriment of downstream owners.

Riparian owners do have liabilities as well as rights. On diversion of a watercourse into a new channel it is the duty of the person diverting to ensure that the new channel will carry off the water as before and will be liable for any deficiency of damage irrespective of the capacity of the old natural channel.

16.2.4 Critiques of the historical arrangements for flood management

16.2.4.1 Institutional arrangements

The institutional arrangements in the UK mean that numerous organisations have responsibility for different aspects of flood management. There are, broadly speaking, four types of flood event that can be distinguished. They are river (fluvial) flooding, coastal flooding, groundwater flooding and surface water (sometimes called 'pluvial') flooding. To date most flood management organisations have concentrated on river and coastal flood. In England, the lead government bodies (Defra and the Environment Agency) have set in place a programme of catchment flood management plans (CFMPs) and shoreline management plans (SMPs) to provide strategic 'overview' plans for river and coastal flooding.

The House of Commons Environment Committee report on flooding (House of Commons, 2008) noted that the main responsibilities of the Environment Agency, including defences, maps and warning systems, are geared to river and coastal flooding only. The Committee reported that this focus in river and coastal flooding led to apparent limitations during the extensive flooding in England in 2007 in which significant damage and disruption was caused by surface water flooding. No organisation had overall responsibility either nationally or locally and yet, as the data in Table 16.1 show, this was one of the most damaging floods in recent history. The Committee remarked that local authorities did not have information to help in predicting which streets in a city would be most affected by heavy rainfall, nobody was responsible for issuing warnings to residents and it was difficult even to determine who was responsible for drainage assets placed under strain by the exceptional rainfall.

16.2.4.2 Lessons learned from the floods of summer 2007

The experience of 2007 has raised questions about the organisation of responsibilities for flood risk management in England. There are clear reasons for continuing to maintain a high profile for river and coastal flooding. Although 'surface water' flooding was an important component of the 2007 flooding (especially in Hull and South Yorkshire), it is over-simplifying matters to neglect rivers. The independent review of the summer 2007 floods led by Sir Michael Pitt (Cabinet Office, 2008) included estimates of the numbers of properties affected by flooding ranging from 48 000 to 55 357, but with 20 238 properties affected by flooding from main rivers.

Of course the real hydrological and drainage system is an integrated whole, and so there are many interactions between 'surface water drainage' and 'river flood'. For example, high river levels can reduce the discharge capacity of surface drains because of reduced hydraulic gradient, prevent them from discharging entirely because of flapped outfalls, or even back up into drains thus contributing to surcharging of water on to streets. Conversely, a surface water drainage system that conveys water very efficiently may discharge storm water quickly into rivers, contributing to flood risk downstream. It is therefore not surprising that several reports and enquiries have recommended a national body having a strategic overview for inland flooding, which must include rivers and surface water drainage. In England, the Environment Agency has welcomed the roles of strategic overview, coordination and advice.

At the local level, the complex arrangements for land drainage summarised above can hinder effective planning for flood risk and operational responses to flooding. The summer 2007 floods highlighted that, in some cases, it can be difficult even to determine which organisation owns certain drainage assets. An example cited was the ambiguity about when a highway drain under local authority responsibility might become a public sewer, under water company responsibility. The government's water strategy advocates local authorities taking the lead role on surface water drainage. Such a role could involve changes in law to empower authorities to carry out works on assets that they do not own and recover costs from the owner, or to sub-contract responsibility to the owners of parts of the drainage system (water companies or IDBs, e.g.) and require their cooperation.

16.2.4.3 A national 'flood agency'?

The concept of a dedicated, national flood agency appears to be a simple approach to unify the responsibilities for flood risk management. However, it has been opposed by many on the grounds that it would require too radical a restructuring of current powers and funding mechanisms. There is also an objection that flood management should not be isolated from other aspects of water and environmental management. The Commons Efra Committee considered that, even until recently, flooding has been dealt with in isolation at a policy level, although government strategy now recognises benefits in an integrated approach.

Integration of water management across river catchments has developed with national bodies for England and Wales (the Environment Agency) and Scotland (Scottish Environment Protection Agency), although from 1974 to 1989, the then Regional Water Authorities led to a more geographically fragmented structure. However, the Efra Committee also noted that functional integration of water management (floods, water supply and demand, drought, pollution and ecology) has not been strong historically, and nor has integration of land and water management.

16.2.4.4 A technocratic 'floods business'?

In the UK, there has been an ancient principle of individual responsibility for land drainage and flood avoidance amongst riparian land owners (Parker, 2000). However, over time there has been centralisation of flood management responsibilities and some have argued that this could disempower individuals, leading to a sense that people might expect 'someone to do something about it'. In England and Wales, the government's vision for flood management has been set out in a programme called *Making Space for Water*, which includes the aspiration to have '... local participation in decision-making, ... within a context of national standards and nationwide information on flood risks and prioritisation' (Defra, 2005). The 'context of national standards and nationwide information and prioritisation' implies a systematic approach to assessing flood risk. This assessment relies heavily on the creation of knowledge, e.g. about the estimated probabilities of rare, damaging events. The creation of this knowledge usually involves technically complex and expensive analysis (e.g. the statistical analysis of river flows, see Chapters 11 and 13) as well as infrastructure (especially the river flow and rain-gauge networks). As a consequence, flood management has become a highly specialised, technologically advanced business, which can potentially create barriers between institutions and individuals, especially members of the public in affected communities.

Some researchers have expressed concern that the technocratic 'floods business' alienates the populations actually at risk and have begun to investigate alternative strategies to include those people in the creation of scientific knowledge about flooding. One example is the local 'competency groups', set up in Ryedale, North Yorkshire and the River Uck catchment, East Sussex, to capture local knowledge of flooding. The Ryedale Flood Research Group reported in a document called *Making Space for People* (Ryedale Flood Research Group, 2008) in a deliberate contrast to the government vision *Making Space for Water*.

This approach aims to demonstrate that, by dissolving the divide between 'experts' and 'local people', a collective competence can be formed that will in fact offer a cost-effective alternative to the conventional, 'industrialised' way of managing flooding. This claim will be based on comparing the costs of establishing flood management preferences through the competency group mechanism with conventional project planning and appraisal, which takes place within a standardised national framework.

The idea of competency groups stems from a desire to democratise the creation of knowledge about flooding. This aim may not, in principle, be incompatible with the emergence of the technological flood management business because this has also created a market for specialists such that consultants (and sometimes academics, for that matter) are available for hire to institutions or individuals. However, if the scope of what is considered to be admissible knowledge is not broadened, then individuals may be denied full access to the decision-making process.

Other conventional means by which individuals can take responsibility for flood risk are through adaptation measures to reduce the consequences of a flood (rather than to reduce the chance of being flooded) and through insurance to offset the damages caused by flooding. Both aspects expose tensions between individual and collective responsibility that hinge on questions about social equity and solidarity. There may be difficult decisions about the extent to which people living in low-risk areas should contribute to preventing losses in higher risk areas, or to what extent those in high-risk areas should have to bear the costs, or whether incentives should be provided to encourage people and businesses to relocate or take loss prevention measures. Similar questions emerge both in deciding policy on state investment in flood defences and in the creation of insurance pools.

16.2.4.5 Insurance

In the UK, flood cover has been a standard feature of household insurance policies since the early 1960s. According to the ABI, the insurance industry was able to make this commitment on the understanding that the government would provide effective flood defences. In the wake of widespread flooding in 2000, the insurance industry identified three priorities for government action, namely greater investment in flood defences, radical curtailment of development in flood risk areas and faster, more consistent decisions on flood defences. In a follow-up report in 2005, the ABI noted progress in all three areas but also called for substantial further investment by the state, claiming that over 280 000 homes and 100 000 commercial properties were not 'protected to a standard that allows the insurance market to operate freely'.

One of the advances over the period 2000–5 that the ABI cited was improved national information about flood risk. It is hard to escape the fact that much of this type of information will be created by the technically specialised flood management industry. Important concepts and techniques are outlined in the next section below. However, some argue that experts alone cannot make the fundamental decisions about what is a fair or effective distribution of risk because these decisions are value judgements. It is interesting to look at emerging flood insurance markets like those in Hungary and Turkey, where uptake has been low. A pilot study (Vári et al., 2003) carried out in the Upper Tisza river in Hungary involved stakeholders who, despite initial strong differences, eventually reached a consensus view of what an effective insurance/state flood management system should look like. Their model involved restricting government post-disaster recovery assistance to households carrying partial insurance, but opposed mandatory insurance. The findings were contrasted by Linnerooth-Bayer et al. (2005) with the difficulties faced in establishing a Turkish flood insurance pool, where government, World Bank and insurers were reported to have designed a scheme from the top down. The process of arriving at a consensus view in the Tisza basin revealed that building an effective private and public policy for managing flood risk requires place-specific knowledge, similar to the competency groups advocated in the Ryedale study discussed earlier.

16.3 Flood risk concepts

16.3.1 Standard of protection

For many years the paradigm in flood management was to 'protect' or 'defend' against flooding, typically through a combination of engineered structures like flood banks or storage reservoirs and channel maintenance to improve conveyance of flood waters downstream. In this approach, the performance of the flood defence is typically expressed as a 'standard of protection' (SoP; see Fig. 16.1). The SoP represents what size of flood the defence should protect against, usually written in terms of the return period. In design calculations, the SoP would be used to determine the crest level of a flood bank or size of a storage pond.

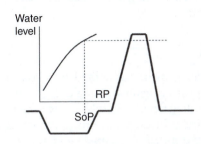

Fig. 16.1 Channel cross-section with flood bank showing the definition of standard of protection (SoP) in terms of return period (RP). The difference between the SoP level and the actual crest level of the defence is a freeboard allowance.

16.3.2 Fragility curves

Even if a fixed standard of protection is used as a design condition, in reality flood defences may not always perform to the standard, as a result of deteriorating structural condition caused by repeated loading, settlement, wave action or scour.

Detailed geotechnical analysis is needed to understand fully the structural condition of the defence and predict its performance under flood conditions. This level of analysis is not always practical because data collection and setting up a suitable geotechnical model may be difficult or expensive. Detailed modelling is not usually feasible for large-scale risk models.

A defence is more likely to fail when it is subjected to extreme hydraulic loading. The most obvious definition of 'loading' is the water level on the exterior of the defence (although other variables such as velocity could be considered). Different failure modes can be defined, which might include simple over-topping of the defence when the water level is higher then the defence crest level. Uncertainty about the performance of a particular flood defence asset can then be expressed as a conditional probability distribution for the failure of the defence, given the hydraulic loading (Fig. 16.2). This conditional distribution is called a fragility curve. Different fragility

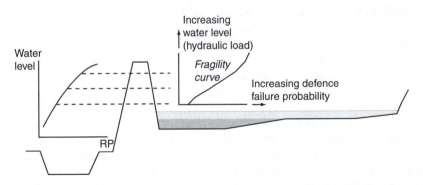

Fig. 16.2 Channel cross-section with flood bank showing a relationship between the return period (RP) of the hydraulic load on the defence, represented by the water level, and the probability of the defence failing. This relationship is called a 'fragility curve'.

curves can be defined for different types of defence, classes of structural condition, ages or maintenance regimes.

16.3.3 Risk assessment

In recent years, flood management has shifted to a concept based on risk analysis. Rather than seeking to protect against a single design threshold, the premise is to understand the risk of flooding from all floods, and to accept that it may not be possible to eliminate the risk entirely. In England, this approach is aligned with the general *Guidelines for Environmental Risk Assessment and Management* (DETR, 2000), that established risk assessment approaches for structured decision-making across different areas of government.

Risk is generally thought of as having two components, which are the probability of a harmful event occurring and consequences of the event. A simple and commonly used definition of risk is 'risk = probability × consequence'. This equation tends to imply that both probability and consequence can be expressed in numerical terms, whereas both factors may be difficult to summarise in this way. For example, the probability of climate change projections may be very difficult to determine confidently, and the consequences of projected changes may be complicated and hard to represent on a common numerical scale.

A general framework for assessing risk comprises the following five steps (DETR, 2000).

Stage 1: Identify the hazard (e.g. 'this town could flood').

Stage 2: Identify the consequences (e.g. 'property damage, disruption of critical infrastructure, long-term health effects').

Stage 3: Estimate the magnitude of consequences (e.g. '£X million of damage could occur, Y hospitals could be closed, Z number of people could suffer health problems'). Ideally, this step should consider the spatial scale of the consequences, the duration, and also, where relevant, how quickly the harmful effects of a hazard would be felt.

Stage 4: Estimate the probability of the consequences. This usually requires estimates for the probability of the hazard occurring, the degree of exposure to the hazard and the degree of harm resulting from the exposure. For flood risk, this part of the assessment has come to be approached using the risk models described in the following sections.

Stage 5: Evaluate the significance of a risk (e.g. 'the current level of flood risk here is not severe compared to some places, but will be mitigated by taking actions A, B and C').

In Stage 5 of the risk assessment, there is clearly some value judgement about what constitutes acceptable risk, or about where the priorities of society (or simply of the person carrying out the risk assessment) lie. However, there may be less obvious value judgements made throughout the process, e.g. in the weights applied to different types of consequence, or even the choice of using, say, an economic measure of consequence rather than a social one. Ideally, the strength of the structured risk assessment process should be to make these judgements and assumptions explicit, rather than hidden.

Table 16.2 Future 'stories' considered by the UK Foresight study

Scenario	Governance	Values	Greenhouse gas emissions	Projected change in flood risk
National enterprise	Autonomy	Consumerism	Medium–high	
Local stewardship	Autonomy	Community	Medium–low	
World markets	Interdependent	Consumerism	High	
Global sustainability	Interdependent	Community	Low	

16.3.4 Scenario-based risk assessment

Where the uncertainties are very significant, it is still possible to examine risk by considering alternative 'story lines' that may have different characteristics and allow differing opinions about their chances of happening. The UK Foresight[3] study into future flood risk identified four such future scenarios (Table 16.2). Each of these 'stories' corresponded to a different set of assumptions about future flood risk incorporating changes in climate, social and political attitudes, investment and development. The 'stories' were expressed in terms of the balance between 'interdependence' and 'autonomy', on one axis representing types of governance, and 'consumerism' and 'community' on the other axis, representing societal values.

Foresight presented results in terms of the following metrics:

(1) probability of flooding;
(2) number of people living within the floodplain;
(3) number of people at 'high' risk of flooding (greater than 1/75 annual probability);
(4) expected annual damage – residential and commercial;
(5) expected annual damage – agricultural;
(6) social flood vulnerability;
(7) expected annual damage (residential and commercial) adjusted by gross domestic product.

Expected annual damage (EAD) is a commonly used summary measure of flood risk, and will be discussed in more detail below. In the Foresight study, the headline results for expected annual damages for the 2080s were a rise from the present level of approximately £1 billion to about £25 billion in the worst case, 'national enterprise' scenario. A slight reduction in EAD was shown for 'local stewardship' and 'global sustainability', whilst the 'world markets' scenario led to an increase in EAD by the 2080s, but less than half as dramatic as for the worst case.

The 'story line' analysis carried out within Foresight is in part a response to the many, often unquantified, uncertainties about flood risk, climate change, economic value and social and political futures. This is acknowledged in the report and the numerical values within the analysis should therefore be treated very cautiously, although it remains the case that information about uncertainties is often not communicated fully when data are quoted and re-used, even if the original sources do make the uncertainties clear.

16.3.5 The 'source–pathway–receptor' concept

One general conceptual approach that has been adopted for environmental risk is the 'source–pathway–receptor' (S–P–R) model. For a risk to arise there must be a hazard, i.e. something that could potentially lead to damage. It can be useful to think of the hazard as having a 'source', which is the physical condition or event that creates the risk, and one or more 'receptors', which suffer the consequences of the risk. The 'pathway' is the link between the source and the receptor. Examples are flood routes including overtopping of defences, breached defences or overland flow. More generally, the 'pathway' transfers and modifies the risk between source and receptor.

The idea of 'a receptor' seems rather abstract but what this usually means in practice is a measure of the consequences of flooding for people and property, usually an economic loss.

In reality there are often multiple sources, pathways and receptors within a flood risk management system. A method for estimating the risk of a consequence occurring must then be able to integrate the various sources, pathways and receptors. This can add significant complexity when the sources and pathways interact, e.g. if the performance of a defence system (pathway) depends on the water level loading (source).

16.3.6 Probabilistic risk models

In the traditional approach based on standard of protection, the probability of the source received attention, e.g. through calculation of flood frequency curves and water levels for a specified design standard such as a '1/100 years' flow. When there are multiple possible sources and pathways, this simple approach becomes difficult to justify, especially if we consider the uncertainties in each of the S–P–R components and the interactions between them. Probabilistic risk-based approaches to flood management have been developed as one response to these complexities. Initially, the approaches were largely geared towards providing a more sophisticated analysis of the risk of flooding based on reliability analysis of flood defences. In 1992, the US Army Corps of Engineers (USACE) issued a draft engineering circular on risk-based analysis for evaluation of flood damage (USACE, 1992). The methodologies subsequently became Corps policy, USACE (1996). In England and Wales, similar approaches have been adopted (Sayers *et al.*, 2002) for national planning or for asset management.

An objective in risk-based methods is usually to determine the EAD from flood hazard along a section of river or coastline. By using a model of the risk, it is then possible to compare changes in the EAD according to different management scenarios.

The calculation method starts with the flood discharge, Q, which is equalled or exceeded on average once in T years. The value T is therefore the return period of the discharge Q. The annual exceedance probability (AEP) is the corresponding probability

$$p = 1/T \tag{16.1}$$

that the discharge Q is equalled or exceeded in any given year (see Chapters 11 and 13).

For a flood of annual probability p, assume that a corresponding value of flood damage $D(p)$ can be estimated. In the definition of methods used by the USACE, this was based on the depth of inundation of the floodplain and on the value of the inundated structures. Regardless of exactly how the damages are calculated, the EAD is the notional long-run average value of such damages taken over floods of all different AEPs. Mathematically this is the integration of the economic damages $D(p)$ over all exceedance probabilities,

$$EAD = \int_0^1 D(p)dp \tag{16.2}$$

The calculation of EAD can be conveniently split into parts. For the original USACE procedures, there were three parts, as follows:

(1) calculate flood frequencies, which describe the probability of floods equal to or greater than some discharge Q occurring within a given period of time;
(2) calculate stage discharge rating curves to 'convert' the discharge into water level;
(3) calculate relationships between economic damage and water level.

The calculations are a mix of stochastic (probability distributions for flood frequency) and deterministic (rating and damage curves) relationships. A Monte Carlo procedure can be used to generate a large number, N, of random pseudo 'observations' of flood flow, each one leading to a damage value. The EAD is then calculated from the simulated damage values by approximating (16.2) as

$$EAD \approx \frac{1}{N} \sum_{i=1}^{N} D(p) \tag{16.3}$$

where p is randomly sampled from a uniform distribution between 0 and 1.

Fig. 16.3 illustrates the procedure, tracing the calculation of damage for one sample value (dashed line).

The risk calculation described above had only one probabilistic component, the flood frequency distribution representing the 'source' of the risk. However, it is generally unrealistic to regard all other links in the chain as being completely deterministic, if only because of the uncertainties in the specified relationships. Instead, there may be uncertainties about each component of the analysis, as illustrated in Fig. 16.4.

The uncertainties can be incorporated in the risk analysis by generating Monte Carlo samples from each of conditional distributions The outcome is a value for EAD that inherently includes the specified uncertainties in the component relationships. This capability to build the uncertainty into the overall measure of risk is an advantage of the risk-based methods, although it also means that there is a danger of regarding the whole, complex, calculation as something of a closed box.

A similar probability integration can be structured from the outset using the 'S–P–R' concept, as illustrated in Fig. 16.5.

The method contains the same essential elements as the USACE approach. The probability integration for EAD is now

$$EAD = \int_0^{y_{max}} D(y)f_Y(y)dy \tag{16.4}$$

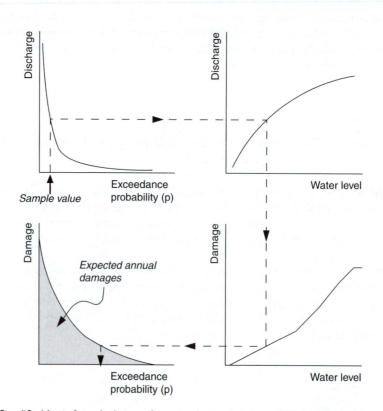

Fig. 16.3 Simplified basis for calculation of expected annual damage (EAD). (Adapted from National Research Council, 2000.)

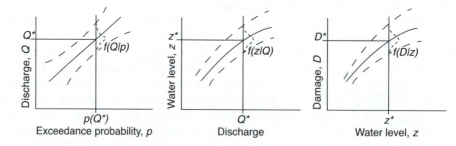

Fig. 16.4 Uncertainties in the discharge (Q), water level (z) and damage (D) relationships within a risk analysis calculation. Left panel shows the relationship between discharge and exceedance probability (p). The solid curve is the best estimate of this relationship, the dotted lines show confidence intervals of the function $f(Q|p)$, which is the probability density function describing uncertainty in discharge for a given exceedance probability. The second panel shows the relationship between discharge and water height (i.e. the rating curve). The solid line is the best estimate and $f(z|Q)$ represents uncertainty in the water level estimate for a given discharge. The third panel shows the relationship between damage and water height, again with confidence intervals and an illustration of the conditional probability density function for a given water level. Adapted from National Research Council (2000).

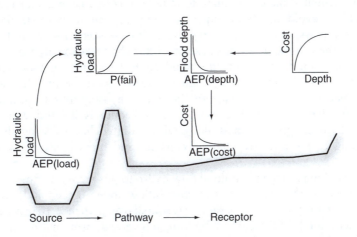

Fig. 16.5 'Source–pathway–receptor' concept for flood risk analysis for an idealised system made up of a river channel ('source'), defence system ('pathway') and floodplain containing capital assets, property, people or other 'receptors' of the risk. The same concept is applicable for a coastal flood risk system. AEP(X) denotes the exceedance probability of X (the probability of observing a value of X greater then some specified threshold value), P(fail) denotes the probability of a food defence failure. The 'pathway' is typically regarded as being the flood defence assets, such as embankments as shown here, but may also include flow routing over the floodplain or other processes that can modify the risk. 'Receptors' may be represented in terms of an economic cost function, as shown here, but could also include other generic 'costs' such as measures of social consequence.

where $D(y)$ is the flood damage for a given flood depth y and $f_Y(y)$ is the probability density function of flood depths. The integration may be bounded in practice by a maximum flood depth y_{max} corresponding to an extremely rare event. The calculation of $f_Y(\bullet)$ can include treatment of defence system performance using the fragility curve concept.

The methods discussed above have been implemented in national scale risk assessments and for analysis of specific flood defence systems (see, e.g. Hall *et al.*, 2003; Gouldby *et al.*, 2008).

Large-scale flood risk models now rely on two-dimensional (2D) modelling of flow routing on the floodplain. This can take advantage of readily available digital terrain model (DTM) data, typically a grid of elevation values, combined with automated flood mapping methods that allow many kilometres of floodplain length to be modelled efficiently (Bradbrook, 2006). An example of this type of regional flood risk analysis is a model for the north-east of England including 19 scenarios representative of different flood risk management measures and a wide range of flood flow probabilities. This strategic investigation of flood risk was primarily aimed at providing a consistent evidence base for catchment flood management plans, such as the Tyne CFMP, which is discussed later.

The hydrological analysis involved generating hydrographs at hydrological 'inflow nodes' located approximately every 200 m along each watercourse. This step made use of *Flood Estimation Handbook* (*FEH*) methods for river flow frequency analysis (see Chapter 13). The natural conveyance capacity of all channels was assumed equal to the estimated median annual flood flow (QMED, i.e. a 2-year return period or 1/2 AEP) to avoid the (very expensive) requirement to have detailed channel cross-section surveys everywhere. To represent different scenarios about flood defence systems, only the volume of water in excess of this channel capacity was then considered as an input for routing over the flood plain. Using this method, the nominal design performance of the defences can be treated as a pseudo-channel capacity and a volumetric adjustment made to the inflow hydrographs to account for the increased effective capacity. The modelling procedure is illustrated in Fig. 16.6, where each hydrograph is plotted along with a dotted line showing the flow rate corresponding to the effective channel capacity and the volume in excess of this capacity.

Over 1.25 million independent flood flow hydrographs were generated in this study to represent each of the 19 scenarios at every inflow node. At each node, and for each scenario, the relevant flood hydrograph was routed over a 10 m × 10 m horizontal resolution DTM using JFLOW-GPU, a 2D diffusion wave approximation to the shallow water flow equations adapted to run on fast parallel processing hardware (Lamb *et al.*, 2009). Inflows are added to the DTM grid over a cross-section set up perpendicular to main flood flow direction (see Fig. 16.6) and routed downstream for at least 1 km until a grid of maximum depths is attained. For each scenario, the maximum depth grids were merged together to form a single regional depth map, using geographical information system (GIS) tools.

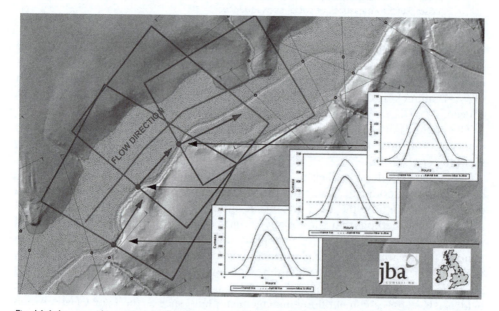

Fig. 16.6 Large-scale automated flood mapping procedure. Overlapping boxes highlight three individual two-dimensional hydraulic model runs where flows enter along a cross-section aligned perpendicular to predominant flow direction; dots indicate the centre of the cross-section at the river centre line. Feint lines and dots show other 'inflow' cross-sections. Each inflow comprises a flood hydrograph, displayed as inset plots, which show the total hydrograph, the assumed channel capacity (dotted line) and the resulting inflow hydrograph.

Table 16.3 Simulation matrix for scenario construction, north-east Region. 'ST' represents an assumed defence system design standard of protection against the T year flood, 'QT' the design flow having a return period of T years. In both cases the corresponding annual exceedance probability (AEP) is 1/T. Shadow indicates cells left blank because they cannot logically contain a simulation

	5% AEP Q20	1.3% AEP Q75	1% AEP Q100	0.5% AEP Q200	0.1% AEP Q1000
S2	S2Q20	S2Q75	S2Q100	S2Q200	S2Q1000
S5	S5Q20	S5Q75	S5Q100	S5Q200	S5Q1000
S20		S20Q75	S20Q100	S20Q200	S20Q1000
S75			S75Q100	S75Q200	S75Q1000
S100				S100Q200	S100Q1000

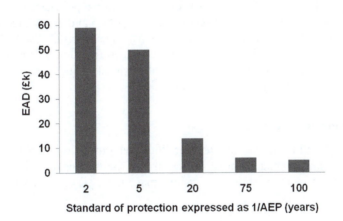

Fig. 16.7 Expected annual damage (EAD) estimates for one administrative area as a function of the assumed standard of protection of flood defences (SoP). Example from the River Wear catchment.

Each entry in Table 16.3 represents one regional grid of modelled flood depths corresponding to a given standard of protection ('S' prefix) and severity of hydrological event ('Q' prefix). Economic damages were estimated based on (deterministic) depth–damage relationships (Penning-Rowsell *et al.*, 2005) and spatial information about properties and agricultural land use. For each of the defence performance scenarios, the simulation matrix therefore provides estimates of quantiles of the probability distribution of depth or damages, allowing the EAD to be computed. The example calculations in Fig. 16.7 show the EAD estimated for each defence performance scenario in a single administrative area. The data show little difference in EAD between defence design standards of 1/75 and 1/100 AEP, suggesting that, in this area at least, the additional damage caused by very infrequent larger events (up to 1/1000 AEP) does not add greatly to the long-term average damages brought about by more frequent, but less severe flooding.

16.4 Flood management policy and delivery

There is a complex network of policies and mechanisms for their delivery that determines how flood risk is managed in the UK. The precise details vary between England, Wales, Scotland and Northern Ireland, although there are roughly equivalent policies, activities and management frameworks in all cases. Here the focus is on England and Wales, but some of the equivalents are noted.

16.4.1 High level policies and legislation

In England, Defra sets policy for flood management. In devolved regions, the Scottish Government, Northern Ireland Assembly and Welsh Assembly take on this role. Government policy on flood management is encapsulated within *Making Space for Water* (http://www.defra.gov.uk/environ/fcd/policy/strategy.htm). This is linked to the overall water strategy *Future Water* (Defra, 2008a), which considers flood management as one part of an overall environmental and social strategy for floods, drainage, water supply and water quality. Sitting above these national policies are two important European Directives, the Floods Directive (2007/60/EC) and the Water Framework Directive (2000/60/EC). The Flood and Water Management Bill (2009–10) includes provisions that will transpose requirements set out in the Floods Directive into law in England and Wales. The Bill seeks to establish the strategic overview role for the Environment Agency for all sources of flood risk.

16.4.2 EU Floods Directive

Directive 2007/60/EC of the 23 October 2007 set out a common framework for the assessment and management of flood risks in the European Union. The Floods Directive requires member states to put into place three types of provision for flood management:

- preliminary flood risk assessments (to be completed in 2011);
- flood hazard and risk maps (to be completed in 2013);
- flood risk management plans (to be completed in 2015).

16.4.2.1 Preliminary flood risk assessments and flood maps

Preliminary assessments include basic, readily derived information such as river basin boundary maps, topography, land used and descriptions of historical flooding. The Directive requires maps for three broad ranges of probability, described as 'extreme events' (low probability), 'medium probability', with a suggested return period of at least 100 years and 'high probability' (unspecified return period). Flood extents are required for each probability range, and also water depths and flow velocities 'where appropriate'. These parameters constitute the hazard component. Risk is interpreted as meaning the combination of probability and adverse consequence. For places where coastlines have been protected to a high standard by flood defences or where groundwater is thought to be the cause of flooding, then only the 'extreme' flood event maps are required.

16.4.2.2 Flood risk management plans

Flood risk management plans set out objectives for reducing potential adverse consequences of flooding for human health, the environment, cultural heritage and economic activity. Note that this aspect of risk management is stated in terms of reducing consequence and not merely of improving 'protection' from flooding. The plans should identify measures that are to be taken to achieve those objectives, including analysis of costs and benefits, flood extent, flood conveyance routes and areas which have the potential to retain flood water, such as natural floodplains.

The scope of the plans includes all aspects of flood risk management focusing on prevention, protection and preparedness. This may therefore include provision of flood forecasting and warning services, and also promotion of sustainable land use practices, improvement of water retention and controlled flooding of areas that provide flood storage.

16.4.2.3 Adoption in the UK and Ireland

The Floods Directive leaves many details of implementation to member states, allowing for the differences across Europe in existing flood management practice, data and also different priorities for reducing adverse consequences. In England and Wales the existing programmes of national flood mapping, catchment flood management plans, shoreline management plans and strategic flood risk assessments (SFRAs) clearly correspond to many of the requirements of the EU Directive.

In Scotland the Flood Risk Management (Scotland) Bill transposes the Directive into law and will place considerable responsibility on the Scottish Environment Protection Agency (SEPA) for the provision of mapping and flood risk management plans. In the Republic of Ireland, the Office of Public Works is the primary agency for flood risk management and has been using a programme of catchment flood risk assessment and management (CFRAM) studies.

16.4.3 Flood risk assessments

Flood risk assessments (FRAs) are important mechanisms for the practical delivery of flood risk management policy via the planning process. There are three tiers of FRA, known as regional flood risk appraisals (RFRAs), SFRAs and, at the smaller scale, site-specific FRAs. Table 16.4 shows the relationships between each tier of FRA and the planning process.

In the light of its strategic overview role for flood risk, the Environment Agency in England and Wales works in partnership with the regional and local planning

Table 16.4 Hierarchy of flood risk assessments

Flood risk management tool	Applicable to	Prepared by
Regional flood risk appraisal	Regional spatial strategy	Regional planning body
Strategic flood risk assessment	Local development document or area action plans	Local planning authority
Flood risk assessment	Specific sites	Developer/land owner

authorities to draw up the RFRAs and SFRAs. For the local FRAs that go with individual development proposals, the assessment may often be carried out by the developer. The Environment Agency may recommend that an FRA is carried out to mitigate flood risk at a site on the floodplain or risk downstream of the site. It is usually private consultants who carry out the technical work for FRAs.

16.5 Catchment flood management plans

A catchment flood management plan is a high-level strategic planning tool to identify broad policies for sustainable flood risk management that make sense in the context of the whole catchment and for the next 50–100 years. Whilst it is not possible to understand in detail what will occur over this time horizon, general trends can be projected to test the sustainability of plans. The first round of about 80 CFMPs is now in place and these plans will be implemented by use in guiding more detailed strategies and through the planning system. The CFMPs will be reviewed and updated periodically.

16.5.1 Example: the River Tyne CFMP

The Tyne is a catchment in north-east England containing two main tributaries, the Rivers North Tyne and South Tyne, and a number of small streams (Fig. 16.8). The lower reach of the main River Tyne and short sections of its smaller tributaries, the Ouseburn, Team and Don, are tidal. Upstream of the tidal reach the topography and river channels are steep. The catchment is made up of sedimentary carboniferous rock overlain by relatively impermeable glacial sediments. Soils in upland areas are generally peaty with good water retention properties but their overall effect on runoff is complex.

Over three-quarters of the Tyne catchment is rural. Arable farming is limited to small areas of available floodplain. Most upland areas are managed grasslands. Urban and commercial areas are concentrated in the lower reaches of the catchment with Newcastle and Gateshead having a population in excess of 900 000. The catchment relief is predominantly steep slopes. With its shallow soils and short contributing streams, this leads to high runoff rates and rapid flood travel times, although there is storage and attenuation within the lower floodplains.

16.5.1.1 Current flood risks and management

The catchment has experienced a number of extreme flood events, with most parts of the River Tyne system experiencing flooding at some time. In recent times there have been few instances of significant flooding to residential property. The larger events have generally happened in winter. Snowmelt has contributed to some of the most extreme flood events on record, although it is not a common source of flood risk.

There is a significant number of people and properties in areas mapped as having a 1 per cent or greater annual probability of flooding in the lower reaches of the catchment, especially in Haydon Bridge, Hardaugh near Warden, Hexham, Corbridge, Low Prudhoe/Ovingham and areas of the Tyneside conurbation along the tidal Tyne, Ouseburn, Team and Don rivers. The potential agricultural damages are much less

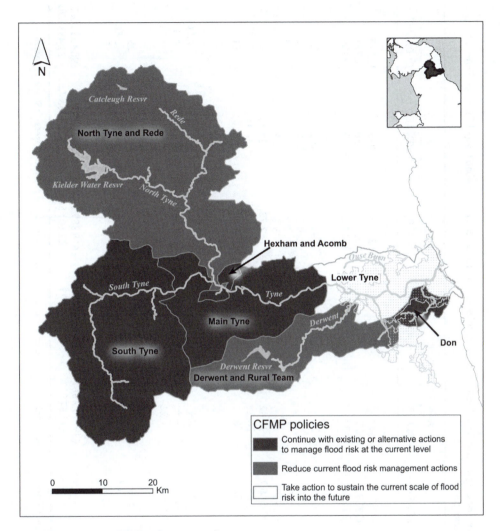

Fig. 16.8 River Tyne CFMP policy appraisal.

than property damages throughout the CFMP area. Important communication routes cross significant sections of the floodplain within the River South Tyne and River Tyne sub-catchments and are therefore vulnerable to the risk of disruption through flooding. Table 16.5 is a summary of some current indicators of flood risk for the CFMP area. This is expressed in terms of the potential consequences of flooding within outlines that correspond everywhere to the specified annual exceedance probability (AEP). It is important to be aware that these represent notional catchment-wide outlines and not an individual flood event of the given AEP.

Flood risk management measures discussed in the CFMP comprise an extensive flood warning system and lengths of engineered, structural flood defences. Most measures are implemented for the key areas at risk concentrated along the lower reaches of the River South Tyne and reaches of the non-tidal River Tyne. There is about 50 km of flood

Table 16.5 A summary of the consequences of flooding for notional flood outlines of different annual exceedance probability (AEP), River Tyne catchment flood management plan (CFMP) (June 2009)

	CFMP area	Modelled flood outlines					
		5% AEP	1.30% AEP	1% AEP	0.50% AEP	0.10% AEP
Social						
Population	942 747	6613	10 341	11 341	12 995	15 573
Residential properties	409 890	2875	4496	4931	5650	6771
Commercial properties	25 202	1499	1929	2014	2170	2484
Infrastructure						
Gas and electricity substations	1299	71	86	93	103	127
Hospital, surgery and health care centre	256	4	6	6	6	6
Fire, ambulance and police stations	110	2	2	2	2	3
Nursery, school, college and university	752	2	5	5	8	8
Main roads (km)	546.7	25.42	31.79	32.73	34.74	40.1
Railway (km)	189	26.2	33.87	34.18	37.41	45.12
Sewage treatment works	35	6	7	8	8	11
COMAH sites (top tier)	8	1	1	1	1	1
Environmental						
SSSI (km^2)	524.6	5.4	6.15	6.65	6.55	7.06
SAC (km^2)	373	2.7	2.85	3.05	3.57	4.01
SPA (km^2)	194	1.22	1.22	1.21	1.21	1.21
Cultural						
SAMs	524	19	21	21	22	25
Registered parks and gardens	6.7	0.58	0.58	0.58	0.58	0.58
Campsites	59	4	5	5	7	7
Economic						
Property damages		£397 677 561	£585 593 231	£618 389 500	£729 702 915	£988 358 087
Agricultural damages		£1 757 912	£2 059 529	£2 101 568	£2 222 063	£2 453 690

COMAH, control of major accident hazards; SAC, special area of conservation; SAM, scheduled ancient monument; SPA, special protection areas; SSSI, site of special scientific interest.

banks or walls, offering varying standards of protection, predominantly around the 1/5 to 1/30 AEP level. There are also reservoirs in the upper reaches of the North Tyne and Derwent rivers that provide some flood attenuation, along with the natural flood plains along the River Rede. Routine maintenance is carried out annually along over 500 km of river channel, including grass cutting, trimming of trees and shrubs and other work to maintain channel conveyance. Flood warning areas included 1169 properties considered to be at risk of flooding. Of those at risk, 48 per cent requested to be included on the Environment Agency's flood warning messaging system, a level of uptake less than the national target at the time (77.5 per cent) for those covered by a warning opting to receive it.

16.5.1.2 Future changes

The Tyne CFMP discusses possible future changes in urbanisation, land use, land management and climate change to explore the robustness of alternative flood risk management policies.

Future plans for urban development are thought likely to be concentrated in existing developed areas within the Tyneside conurbation and established towns to the west. The CFMP discusses how government policies are likely to help limit the geographical extent of urbanisation. However, this may place more pressure on existing drainage structures. The CFMP flood risk management policies (see later) include recommendations to assess the risk of flooding from surface water through more localised surface water management plans and to develop system asset management plans (SAMPs) to determine the most sustainable approach to managing flood defence assets.

Depending on their location, forests may have the potential for reducing flood flows (through improved soil infiltration, slowing snowmelt, tree canopy evapotranspiration and holding back runoff). About 12 per cent of the CFMP area is occupied by Kielder Forest, the largest and most productive forest area in England. Most of the forest is at a mature stage of growth and felling is carried out on a sustainable basis, hence the overall impact on flooding is likely to be relatively unchanged into the future. Additionally, a large part of the forest is upstream of Kielder Water Reservoir and so changes in flood runoff would be expected to have a reduced impact further downstream.

There has been research to examine whether the development of forest plantations, and also deforestation, has a discernible impact on river flows (e.g. Robinson and Newson, 1986, Robinson and Dupeyrat, 2004). Whilst changes have been found at small scales, in general the findings are of complex changes that are difficult to discriminate from other possible sources of variability, especially in larger catchments (Defra, 2008b). Working with data from the nearby River Irthing catchment, Archer (2003) found evidence of impacts on the smaller peaks in flow, but the evidence for larger floods, which could threaten large populations or critical infrastructure, seems highly uncertain. It remains an open question about whether, or under what conditions, changes in land management might be able to make a recognisable contribution to flood risk reduction at a large catchment scale, although other environmental benefits can be derived from land-use and management changes as well as a reduction in runoff. These issues are considered in more detail in Chapter 19.

Owing to uncertainties in how to represent the impacts of changes in land management, only climate change impacts were modelled in the Tyne CFMP to make a quantitative assessment of future risk. Methods for assessing climate change impacts are discussed in Chapter 19. In the CFMP, a single future scenario was tested based on guidance available at the time (see Table 19.2). Given that this involved applying increases to peak river flows, it is not surprising that the CFMP indicates increased future flood risk; e.g. the population considered to be at risk within the notional 1 per cent AEP flood outline increased to 13 202 from the current estimate of 11 341 given in Table 16.5.

16.5.1.3 Policies and actions

The CFMP policy appraisal process considers the key effects and potential significance of a set of generic policies that accept, eliminate, reduce, transfer or share flood risk. The generic policies can be summarised as follows:

(1) no active intervention (including flood warning and maintenance), continue to monitor and advise;
(2) reduce existing flood risk management actions (accepting that flood risk will increase with time);
(3) continue with existing or alternative actions to manage flood risk at the current level;
(4) take further action to sustain the current scale of flood risk into the future (responding to the potential increases in flood risk from urban development, land use change, and climate change);
(5) take further action to reduce flood risk (now and/or in the future);
(6) take action with others to store water or manage runoff in locations that provide overall flood risk reduction or environmental benefits (locally or elsewhere in the catchment).

For its CFMP, the River Tyne catchment was split into geographical policy units and the standard generic policies tested for each unit.

Fig. 16.8 shows the policy options proposed for each unit of the Tyne CFMP, while Table 16.6 summarises the current flood risk in each policy unit. As it is a strategic document, the CFMP stops short of detailing the actual capital works, maintenance, regulation or other concrete actions that will deliver the chosen policies. However, it does include action plans that set out specific next steps for implementing the selected policies.

For example, the selected policy for the Main Tyne policy unit is to continue with existing or alternate actions to manage flood risk at the current level. Actions to implement the policy include the following:

• Produce and implement a system asset management plan to determine the most sustainable approach to managing assets to ensure that the current standard of protection is maintained at the current level.
• Work in partnership to develop an assessment into the risk of flooding from surface water through undertaking a surface water management plan. Where locations of surface water flood risk are identified, ensure that cross-boundary issues are taken into account and fed into the management of surrounding policy units.

Table 16.6 Flood risk metrics for River Tyne catchment flood management plan policy units

	Policy unit						
	Hexham and Acomb	Don	North Tyne and Rede	Main Tyne	Derwent and rural team	Lower Tyne	South Tyne
Risk metrics for 1% AEP flood outline							
Properties within 1% AEP outline							
Residential	155	83	150	462	128	3 115	141
Commercial	125	17	42	60	52	1 966	37
Population within 1% AEP outline	356	191	345	1 063	294	7 164	324
Expected annual damages (EADs)							
EAD without defences (£)							
Property	2 600 000	125 000	460 000	943 000	327 000	23 000 000	287 000
Agriculture	2 600	<1 000	25 000	6 800	6 800	6 000	20 000
Flood defences							
Number of raised defences	5	4	78	58	2	0	13
Length of raised defences	0.7	1.9	30.2	14.7	0.7	0	1.4
Reduction in EAD attributed to defences (£)	>2 000 000	22 000	83 000–300 000	748 000	50 500	N/A	60 000–100 000
Flood warning							
Properties covered by flood warning areas	253	0	147	241	34	478	59
Properties receiving flood warning direct service	106	0	116	126	18	336	25

- Work with the owners of the caravan parks to ensure that the residual risks of flooding are fully understood. Ensure public awareness is raised through the installation of information boards. Ensure emergency evacuation plans are in place and registration for flood warnings, where available, is carried out.
- Investigate the potential for improving the lead time given by flood warnings in the policy unit, e.g. as a result of new rainfall radar installation.
- Identify the implications of changing the flood regime on environmentally designated sites. This work should identify the potential positive, neutral and negative effects of flooding, taking into account climate change. These findings should be fed into future plans to change the flood regime of the policy unit and ensure that the condition of sites is maintained, and where possible improved.
- Determine in greater detail the risk of flooding to transport and the consequences of road and rail closures during flooding. Where practically possible, ensure that key routes remain operational during a flood event. Following the identification of flood risk to these facilities, ensure alternative routes and emergency plans are developed and reviewed periodically.
- Develop a role for a Sustainable Land Management Officer to promote sustainable land management practices in order to adapt to the implications of climate change and where possible reduce surface water runoff and soil erosion. Examples of land management techniques that should be explored include moorland grip management, afforestation, buffer strips and managed grazing stock densities.
- Further define the fluvial flood risk to sewage treatment works and establish the consequences of these sites flooding. Where practically possible, ensure that sites remain operational during flood events.

The policy selected for the Main Tyne is justified on the basis of the level of risk to people and property within the policy unit and the potential for this to increase in the future. The CFMP considers the current flood defence system to be cost effective, with around £130 000 spent per annum providing over £700 000 of benefits per annum. Were the defences not maintained, the potential for damage in a large flood of 1 per cent AEP is significant at over £1.7 million. Channel maintenance is also considered justified to continue to manage the risk of blockage at structures. Existing defences and flood risk measures are considered adequate for present and future flood risk and so the CFMP does not consider the cost of extra protection to be warranted.

Notes

1 http://www.abi.org.uk/
2 http://www.dundee.ac.uk/geography/cbhe/
3 http://www.foresight.gov.uk/

References

Archer, D. (2003) Scale effects in the hydrological impact of upland afforestation and drainage using indices of flow variability: the River Irthing, England. *Hydrology and Earth System Sciences* 7, 325–338.

Black, A. R. and Law, F. M. (2004) Development and utilization of a national web-based chronology of hydrological events. *Hydrological Sciences Journal* 49, 237–246.

Bradbrook, K. F., Waller, S. G. and Morris, D. (2005) National floodplain mapping: datasets and methods – 160000 km in 12 months, *Natural Hazards* 36, 103–123.

Cabinet Office (2008) *The Pitt Review – Learning Lessons from the 2007 floods*. Cabinet Office, 22 Whitehall, London SW1A 2WH. Available at http://www.cabinetoffice.gov.uk/

Defra (2005) *Making Space for Water, First Government Response to the Autumn 2004 Making Space for Water Consultation Exercise*. Department for Environment, Food and Rural Affairs, London. http://www.defra.gov.uk/environ/fcd/policy/strategy/firstresponse.pdf (accessed December 2008).

Defra (2008a) *Future Water*. Cm 7319, February 2008. The Stationery Office, London.

Defra (2008b) *Analysis of Historical Data Sets to Look for Impacts of Land Use and Management Change on Flood Generation*, FD2120 Final Report, February 2008. Department for Environment, Food and Rural Affairs, London.

DETR (Department of the Environment, Transport and the Regions) (2000) *Guidelines for Environmental Risk Assessment*. The Stationery Office, London. http://www.defra.gov.uk/environment/risk/eramguide/index.htm

Evans, E., Ashley, R., Hall, J., Penning-Rowsell, E., Saul, A., Sayers, P., Thorne, C. and Watkinson, A. (2004) *Foresight. Future Flooding. Scientific Summary*: Vol. I. *Future Risks and their Drivers*. Office of Science and Technology, London.

Fleming, G. (ed.) (2002) *Flood Risk Management: Learning to Live with Rivers*. Thomas Telford Publishing, London, 250 pp.

Gouldby, B., Sayers, P., Mulet-Marti, J., Hassan, M. and Benwell, D. (2008) A methodlogy for regional-scale flood risk assessment. *Proceedings of the ICE, Water Management* 161, 169–182.

Guy Carpenter (2007) *Wrong Type of Rain: Impact and Implications of 2007 UK Floods*. Guy Carpenter & Company, LLC, Report K.UK.12.07, December 2007, London.

Hall, J. H., Dawson, R. J., Sayers, P. B., Rosu, C., Chatterton, J. B. and Deakin, R. (2003) A methodology for national-scale flood risk assessment. *Proceedings of the ICE, Water and Maritime Engineering* 156, 235–247.

House of Commons (2008) Environment, Food and Rural Affairs Committee, *Flooding, Fifth Report of Session 2007–8*, Vol. 1, HC 49-I. The Stationery Office, London. Available at www.parliament.uk/ (accessed 14 July 2008).

Lamb, R., Crossley, A. and Waller, S. (2009) A fast 2D floodplain inundation model. *Proceedings of the Institution of Civil Engineers, Water Management* 162, 1–9, doi: 10.1680/wama.2009.162.1.1

Linnerooth-Bayer, J., Vári, A. and Mechler, R. (2005) Designing a disaster insurance pool: participatory and expert approaches in Hungary and Turkey. In: *Catastrophic Risks and Insurance, Policy Issues in Insurance*. Organisation for Economic Co-operation and Development (OECD) Proceedings No. 8. OECD Publishing, Paris, Chapter 14, pp 267–290.

National Research Council (2000) *Risk Analysis and Uncertainty in Flood Damage Reduction Studies*. Committee on Risk-Based Analysis for Flood Damage Reduction, Water Science and Technology Board. National Academy Press, Washington, DC, 216 pages.

Parker, D. J. (2000) Introduction to floods and flood management. In: Parker, D. J. (ed.) *Floods*, (Vol. I), pp. 3–39. Routledge (Taylor & Francis), London.

Penning-Rowsell, E. C. E., Johnson, C., Tunstall, S., Tapsell, S., Morris, J., Chatterton, J., Coker, A. and Green, C. (2005) *The Benefits of Flood and Coastal Risk Management: A Manual of Assessment Techniques*. Middlesex University Press, Enfield (the Multi-Coloured Manual or MCM).

Robinson, M. and Dupeyrat, A. (2004) Effects of commercial timber harvesting on streamflow regimes in the Plynlimon catchments, mid-Wales. *Hydrological Processes* 19, 1213–1226.

Robinson, M. and Newson, M. D. (1986) Comparison of forest and moorland hydrology in an upland area with peat soils. *International Peat Journal* 1, 49–68.

Ryedale Flood Research Group (2008) *Making Space for People: Involving Local Knowledge in Flood Risk Research and Management in Ryedale, Yorkshire*, 28 October 2008. http://knowledge-controversies.ouce.ox.ac.uk/ryedaleexhibition/Making_Space_for_People.pdf (accessed December 2008).

Sayers, P. B., Hall, J. W. and Meadowcroft, I. C. (2002) Towards risk-based flood hazard management in the UK. *Proceedings of the Institution of Civil Engineers, Civil Engineering* 150, 36–42.

USACE (1992) *Risk-based Analysis for Evaluation of Hydrology/Hydraulics and Economics in Flood Damage Reduction Studies* (EC 1105-2-205). US Army Corps of Engineers, Washington, DC.

USACE (1996) *Risk-based Analysis for Evaluation of Hydrology/Hydraulics, Geotechnical Stability, and Economics in Flood Damage Reduction Studies*. ER 1105-2-101. US Army Corps of Engineers, Washington, DC.

Vári, A., Linnerooth-Bayer, J. and Ferenz, Z. (2003) Stakeholder views on flood risk management in Hungary's Upper Tisza Basin. *Risk Analysis* 23, 537–627.

Chapter 17

Water resources management

17.1 Drivers for water resources management

The subject of water resources provided the greatest motivation for the growth in hydrological studies in the twentieth century. An expansion of hydrological measurements, especially in the second half of the twentieth century, was driven largely by needs for the evaluation of water resources, which initiated widespread and comprehensive study of the hydrological environment. In the UK, legislation for water dated from the Victorian era, when Acts of Parliament were passed to grant powers at a local level to provide water supply and sewerage for the rapidly expanding population centres. It was the Water Resources Act of 1963 that consolidated water resources management at a regional and national level, creating River Authorities responsible for enforcing the law in relation to water resources. The 1963 Act set the legal basis for regulation of abstractions and impoundment of water.

The establishment of hydrological measurement networks resulted from the legal requirement laid on the former river authorities to evaluate the water resources of their areas. Before the new gauging stations had time enough to produce the long-term information required by the planners, hydrologists had to devise techniques for making assessments from the limited data, which stimulated development of statistical methods and modelling techniques in hydrology, see Chapters 11–14. It will be seen that demand for reliable and safe public water supply is critical to the management of water resources, albeit placed in an environmental regulatory framework. Hydrological studies for water resources in the UK are often linked to balancing requirements for water supply with environmental considerations through mechanisms such as water resources management plans and abstraction strategies. It is only possible to cover some of the relevant techniques here. Detailed guidance on the subject of water supply, including hydrological analysis and water quality standards can be found in Twort *et al.* (2000) and other specialist texts.

The British Isles are often characterised as having a temperate, maritime climate, with annual average rainfall over England and Wales of 890 mm, reaching more than 2500 mm in hilly regions in the north and west. However, parts of the south and east of England have high population densities and relatively low annual rainfall of less then 200 mm. The 'water exploitation index' is the amount of freshwater resources abstracted as a proportion of effective rainfall. It is a useful summary measure of whether a region can be considered to be under 'water stress'. In the south-east of England the water exploitation index is greater than 20 per cent. Fig. 17.1 shows

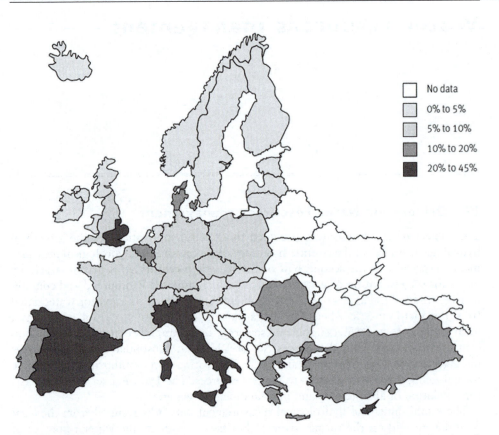

Fig. 17.1 Water exploitation index for Europe (Environment Agency, 2009, reproduced with permission).

how this compares to other parts of Europe. It can be seen that south-east England is anomalous compared with most other parts of northern and western Europe. Only in much drier countries is there greater water stress.

17.2 Water resources management policy in the UK

17.2.1 European legislative framework

The Water Resources Act of 1963 can be seen as the beginning of river basin management in the UK. Now, in the early twenty-first century, water resources management falls under the over-arching framework of European legislation that takes this concept further. The European Water Framework Directive 2000/60/EC (WFD) established an integrated approach to the protection, improvement and sustainable use of rivers, lakes, estuaries, coastal waters and groundwater. It introduced two significant advances in the environmental framework for management of water resources. One of these was a move from regulation based on specific Environmental Quality Standards (EQS) and emissions control of dangerous chemical pollutants to a more ecosystem-based approach. The second was the formalisation of large-scale, integrated river basin management planning across Europe.

The WFD introduced broad ecological objectives to promote sustainable use of water resources. This 'ecological health' is assessed by a system of classification and assessment that considers both the ecological potential of water bodies and the present and planned ecological status. There are five classes of ecological status: high, good, moderate, poor and bad. Surface water bodies are sub-divided in rivers, lakes, transitional (e.g. estuaries) and coastal. The concept of ecological potential allows for the fact that there are also some water bodies, such as harbours and canals, which are heavily modified for social or economic use or entirely artificial. Ecological status is assessed through a combination of biological, chemical and morphological indicators (Table 17.1).

The WFD introduced a river basin management planning system as the mechanism for ensuring the integrated management of water resources and the water needs of terrestrial ecosystems. The river basin planning system provides a decision-making

Table 17.1 Indicators for assessment of the ecological status of surface water bodies in the Water Framework Directive 2000/60/EC classification scheme recommended for UK waters by UKTAG (2009)

Quality elements	Surface water category			
	Rivers	Lakes	Transitional	Coastal
Biological	Benthic invertebrates			
	Phytoplankton			
		Fish		
	Macrophytes and phytobenthos		Macroalgae and angiosperms	
Chemical	Oxygenation (dissolved oxygen concentration)			
	Thermal conditions			
	Nutrients (Soluble reactive phosphorus concentration)	Nutrients (total phosphorus concentration)	Nutrients (dissolved inorganic nitrogen)	
			Transparency	
	Salinity			
	Acidification status:			
	pH	acid-neutralising capacity		
Morphological	Depth variation			
	Bed structure and substrate			
	Quantity and dynamics of water flow		Freshwater flow	Direction of dominant currents
				Wave exposure
	Connection to ground water			
	Continuity	Residence time		
	Riparian zone structure	Shore structure	Structure of the inter-tidal zone	
	Width variation			

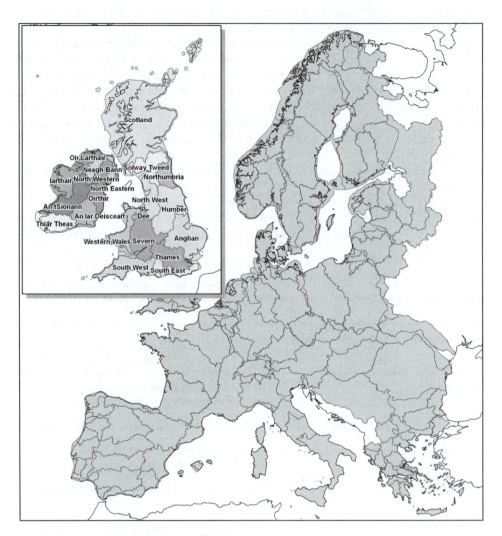

Fig. 17.2 European river basin districts (RBDs).

framework for setting environmental objectives and identifying cost-effective combinations of measures to achieve the objectives. River-basin planning is based geographically around river basin districts (RBDs), which are catchments or groups of
catchments, together with associated coastal waters, as shown in Fig. 17.2. The WFD
requires a River Basin Management Plan to be published and reviewed every 6 years for
each RBD, setting out the environmental objectives for water bodies within the RBD
and how they will be achieved. In England and Wales plans are due to be published
in 2009.

Other European directives that affect water resources management are the Drinking Water Directive (98/83/EC) and the Habitats Directive (92/43/EEC). The Drinking
Water Directive sets quality standards for drinking water and obliges member states

to regular monitoring of drinking water quality. The Habitats Directive (29/43/EEC) requires the identification of special areas of conservation (SACs). The Birds Directive (79/409/EEC) designates Special Protection Areas (SPAs). There are over 240 SPAs in the UK as a whole. Together with sites identified as SACs, these form the 'Natura 2000' network of protected conservation sites. Many of these sites are within rivers, or the whole length of a river, or are affected by river management practices. Where river reaches fall within an SAC or SPA, they are also likely to receive designation as protected areas under the WFD. This can result in restrictions being placed on water resources management practices that could impact on the sites, such as abstractions, discharges or the deployment of physical structures.

17.2.2 UK legislation

After the 1963 Water Resources Act, the next major legislation in the UK was the 1973 Water Act which created ten regional water authorities with responsibility for water supply, sewage and river basin management, working with public water companies and local authorities. In the late 1980s the water supply and sewage functions were privatised with the Water Act 1989, which saw the creation of the National Rivers Authority (NRA). The Water Industry Acts (1991, 1999) also set the law for regulation of the water industry, including such matters as regulation of pricing and consumer protection, which established the duties of Ofwat, the Water Services Regulation Authority, as the economic regulator of the water and sewerage industry in England and Wales.

The Environment Act 1995 established the Environment Agency as the non-departmental public body responsible for protecting and improving the environment of England and Wales, including managing water resources. The Environment Agency combined the functions of the NRA with the activities of several other organisations, including former local authority water regulation functions.

Together with the Water Resources Act 1991, the Environment Act 1995 set out administrative and legislative basis for the water abstraction licensing system in England and Wales. The Water Act 2003 introduced time limits for new abstraction licences and the facility to revoke abstraction licences causing serious environmental damage without compensation, along with greater flexibility to adjust licensing thresholds. It also made water company drought plans and water resource management plans become statutory requirements.

In Scotland, the Environment Act 1995 set up the Scottish Environmental Protection Agency (SEPA) as the regulatory authority with a similar role to the Environment Agency in England and Wales. Since 1998, legislation for water resource management and sewage has been the responsibility of the Scottish Government. The Water Industry (Scotland) Act 2002 established Scottish Water as body responsible for water supply and sewage. Unlike the industry in England and Wales, which consists of private regional monopoly water companies, Scottish Water is a publicly owned company. However, The Water Services etc (Scotland) Act 2005 introduced competition in Scotland by creating a market for retail sales of water and sewage services, allowing private companies to compete for business, using the infrastructure operated by Scottish Water.

In Northern Ireland, responsibility for water supply and sewage were transferred to a publicly owned company, Northern Ireland Water, in 2007 under the Water and Sewerage Services (Northern Ireland) Order 2006. Northern Ireland Water operates within a regulatory framework similar to the England and Wales system, with environmental regulation including abstraction licensing through the Northern Ireland Environment Agency.

17.2.3 Key organisations

There is a complex web of organisations that have a stake in water resources management, either through statutory responsibilities or as representatives of specific interest groups. Fig. 17.3 shows some of the most important stakeholders identified by

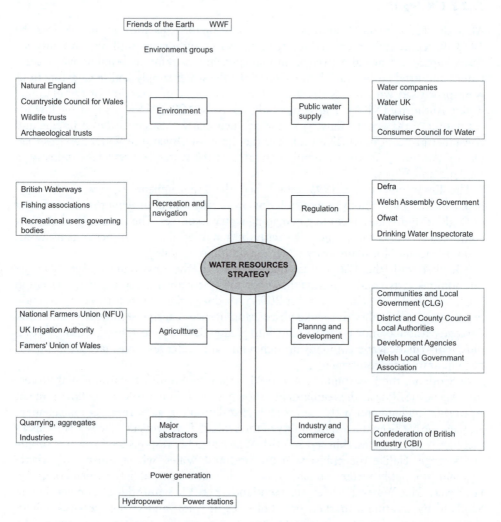

Fig. 17.3 Major stakeholders involved with the management of water resources in England and Wales (after Environment Agency, 2009).

Table 17.2 Key organisations with responsibilities for water resources management in the UK

Organisation	Scope
Defra	All aspects of water policy in England, including water supply and resources, environmental and water industry regulation, water quality sewage treatment and reservoir safety
Scottish Government	Sets policies on drinking water quality, reservoir safety, marine and freshwater quality, sewage treatment, conservation and use of water resources and provision of adequate water and sewerage services in Scotland
Welsh Assembly	The Environment Strategy for Wales sets out the Welsh Assembly Government's strategies for sustainable water resources management sustainably and drinking water quality
Environment Agency	England and Wales. Manages water resources and enforces water quality standards
SEPA (Scottish Environment Protection Agency)	Regulates activities which impact on the water environment in Scotland, including discharges to groundwater and surface water, abstractions, river basin planning
Northern Ireland Environment Agency	Responsible for protection of the aquatic environment. Abstraction and Impoundment Licensing function monitors and controls water body resources in Northern Ireland
Ofwat	Responsible for economic regulation of the water industry in England and Wales. Regulates the 21 regional monopoly water companies
WICS (Water Industry Commission for Scotland)	Manages regulatory framework of the Scottish water industry. Licenses private retail water supply companies
Drinking Water Inspectorate (DWI)	Responsible for assessing the quality of drinking water in England and Wales, taking enforcement action if standards are not being met, and appropriate action when water is unfit for human consumption
DWQR (Drinking Water Quality Regulator)	DWQR ensures quality of public drinking water supplies in Scotland. Role created under Water Industry (Scotland) Act 2002
Water companies	There are 21 private regional monopoly water companies in England and Wales, ten providing both water and sewerage services and 11 are water only. Scottish Water is the publicly owned company that operates the network of water and sewerage in Scotland, as well as selling wholesale water services. Since April 2008, business customers in Scotland can choose from competing retail supply companies, licensed by WICS. Northern Ireland Water is a publicly owned company set up in April 2007 to provide water and sewerage services in Northern Ireland

the Environment Agency in respect of their Water Resources Strategy (Environment Agency, 2009).

Table 17.2 summarises key organisations with responsibilities for water resources management in the UK. In England and Wales, Ofwat and the Environment Agency play central roles for economic and environmental regulation respectively (with similar roles performed by the Water Industry Commission for Scotland and SEPA in Scotland).

17.3 Water resources strategies and assessment processes

17.3.1 Policies, strategies and processes

High-level policies for water resources management are set by government and inform the long-term strategies of the environmental and economic regulators. Key policies and strategies in England and Wales are summarised in Table 17.3. Abstraction licensing takes place within the context of catchment abstraction management strategies (CAMS), which are produced by the Environment Agency to summarise information about water resource availability and licensing within a catchment and so provide a consistent approach to local water resource management.

When the Environment Agency considers an application for a licence they have to ensure that it would not cause flows or water levels to fall below the minimum ecological requirement (except under circumstances when they would otherwise do so). A proposed abstraction should also not impinge upon existing lawful uses of the water. Through CAMS the Agency can form a balanced assessment of these requirements and the needs of abstractors. The strategies indicate available water resources within catchments and so help to identify areas where future resource development may take place or where current licensed abstractions exceed the resources available.

Table 17.3 Summary of policy, strategies and processes for water resources management in England and Wales

Policies, strategies or processes	Scope
Government water strategy	*Future Water* is Defra's 2008 water strategy for England, covering all aspects of the water cycle including water resources, supply and regulation, as well as drainage and flooding
Environment Agency water resources strategy	Published in March 2009, the strategy sets out the Environment Agency's view of how water resources should be managed in England and Wales to 2050 and beyond
Water companies periodic review process	Ofwat sets price limits for water and sewerage companies for each of the following 5 years, after assessing companies' water resource management plans to ensure that they maintain the balance between supply and demand and provide value for money, taking account of environmental and social costs
Water company water resources management plans	Companies have a duty to maintain the security of their supplies and produce water resources plans that forecast supply and demand over a 25-year horizon
Water company drought plans	Under the Water Act, water companies in England and Wales agree detailed drought plans every 3 years with the Environment Agency. The plans examine a range of drought situations that may occur and the actions a company would expect to take in response
Catchment abstraction management strategies (CAMS)	CAMS are used by the Environment Agency to manage the water resources of a catchment and contribute to implementation of the Water Framework Directive
Abstraction licensing	The process of setting and enforcing individual licence permits

Table 17.4 Summary of water industry periodic review (PR) and asset management plan (AMP) cycles since privatisation. (Source: Environment Agency.[1])

Period	PR	AMP	Approximate value of AMP investment plans
1989–95	—	AMP1	£6bn
1995–2000	PR94	AMP2	£5bn
2000–5	PR99	AMP3	£5bn
2005–10	PR04	AMP4	£7.5bn
2010–15	PR09	AMP5	£17.4bn

17.3.2 Periodic price review and water resources planning processes

Every 5 years Ofwat sets limits on the prices that water companies can charge to customers. This review process is referred to as the Periodic Review or Price Review, usually abbreviated as 'PR'. The process requires water companies to set out their 5-year plans for investment and pricing, also referred to as the Asset Management Plan (AMP) cycle. The first AMP followed privatisation in 1989 and set out a £6 billion investment programme for 1989–95. Subsequent AMPs followed price reviews as shown in Table 17.4.

17.3.3 Water resource management plans

Ofwat sets price limits so as to allow water companies to finance their functions, whilst protecting the interests of their customers by ensuring they receive value for money and reliable services. One of the sources of information in the review of price limits is an assessment of the balance between supply and demand. Water companies must show how they propose to manage the supply–demand balance over a 25-year period by producing a water resources management plan (WRMP). Since the Water Act 2003 this has been a statutory requirement in England and Wales, with the WRMPs reviewed by the Environment Agency. The development of WRMPs is aligned where possible with the Periodic Review of price limits, but the two are separate processes.

17.3.4 Catchment abstraction management plans

The programme of CAMS in England and Wales was begun by the Environment Agency in 2001. CAMS are non-statutory plans, but they are used to inform the development of water company WRMPs and in assessing abstraction licensing, hence they provide an important source of information in the statutory process of water resource management.

The key objective of CAMS implementation in England and Wales is to provide a consistent and structured approach to local water resources management. It has made information on water resources publicly availability and provides an opportunity for greater public involvement in the process of managing abstraction at a catchment level. The balance between the available resource and the environmental needs of the

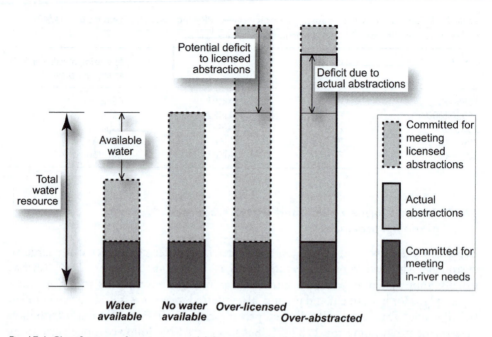

Fig. 17.4 Classification of resource availability status used in catchment abstraction management strategies (CAMS).

river, current actual abstractions and licensed abstractions determines the status of a water-course at specific assessment points. The CAMS approach is to produce flow–duration curves (see Sections 11.4 and 13.6) for a range of situations, including the natural, current and future demand scenarios. This enables the identification of river reaches that have the potential for further development, are over-abstracted, over-licensed or have no water available for further development, as illustrated in Fig. 17.4. The CAMS approach enables time-limited (normally 12-year) abstraction licenses to be issued in contrast to the historical practice of issuing licenses for perpetuity.

The CAMS divide catchments geographically into water resource management units, which can be classified using the scheme above. These may be based on surface water catchments, river reaches for larger river corridors or groundwater units.

17.4 Supply and demand

The supply of water to a resident population is dependent on population size and rate of consumption. Additionally, water resources are consumed by industry and by agricultural users for irrigation, which is seasonal and weather dependent. Public water supply is the largest category of water use, accounting for an average of between 15 000 and 20 000 megalitres per day (ML day^{-1}) over the period since 1971 (Fig. 17.5). Consumption by industry has declined in recent decades, reflecting changes in the industrial base in the UK. It can be seen that the largest water user after public supplies is the electricity-generating industry. Under normal circumstances much of this is non-consumptive abstraction that is returned to water-courses with limited

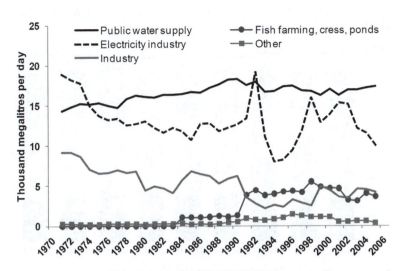

Fig. 17.5 Abstractions from freshwater in England and Wales (data source, Environment Agency).

environmental impact. The proportion of the total abstraction from non-tidal surface waters decreased from 2000 to 2007, whereas abstraction from tidal waters increased, mostly for electricity generation.

Total annual freshwater abstractions in England and Wales are around 35 000 ML, against a licensed total of over 70 000 ML. However, about 70 per cent of the water taken for public water supply is ultimately returned to the environment as treated effluent. Consumption from public water supplies is approximately 52 per cent for domestic use, with around 23 per cent non-domestic users and a similar figure lost to leakage (Defra, 2008). Agricultural abstraction amounts to only a relatively small fraction of the total water use. Spray irrigation in particular accounts for only about 1 per cent of abstractions. However, it is most needed during hot, dry summers when it can then amount to 20 per cent of total water use, with almost all of that amount being lost as evapotranspiration.

In recent history, the general trend in demand for water has been a continual increase due to the growth in populations and to higher standards of living. This applies in most countries. In the UK, the population grew from 38.3 million in 1901 to 60.8 million in 2007, with projections of further increases to 67.2 million in 2021, 74.3 million in 2041, 79.8 million in 2061 and 85.3 million in 2081 (Hicks and Allen, 1999; Office for National Statistics, 2008, 2009).

The amount of water consumed is used as an indicator of demand. Estimates of per capita water consumption vary according to the assumptions made in the calculations. Both *Water UK*, which represents the UK water industry and *Water-wise*, a non-governmental organisation promoting water efficiency, estimate average domestic water use in the UK of around 150 L per person per day (Water UK, 2008; Waterwise, 2009). This figure is somewhere in the centre of a range estimated for other European countries, with Estonia and the Czech Republic estimated to consume only around 100 L per person per day, Germany 127, France 150, Italy and Ireland 190 and Spain 265 (Defra, 2008).

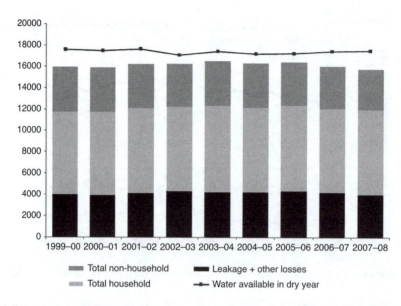

Fig. 17.6 Dry year 'supply demand balance' for water resources in England and Wales (Environment Agency, 2008, reproduced with permission). Vertical axis is in units of megalitres per day. The difference between the estimated demand total and available supply is known as the 'headroom'.

The water industry follows a 'twin-track' approach to ensuring security of supply. One track is demand-side water efficiency, which is promoted through educational and publicity campaigns, investment in infrastructure improvements, reducing leakage and use of water meters. The second track is enhancing supply where necessary, which may include planning new storage facilities or reservoirs, improving connections between different parts of supply areas and desalination plants.

One of the important considerations in managing water resources is the balance between supply and demand. This is assessed in water company plans making use of analysis of available resources contained in CAMS. Fig. 17.6 shows the balance between supply and demand for England and Wales for a 'dry year' (Environment Agency, 2008). The difference between the estimated available resource and the estimated demand is known as the 'headroom'. Whilst the headroom appears to offer security of supply for England and Wales as a whole, the total figures mask some important geographical differences. Fig. 17.7 shows water supply areas in England and Wales shaded according to whether or not the estimated supply meets the target headroom.

17.5 Water resources yield assessment

17.5.1 Definitions of yield

The *yield* is a generic term for the reliable supply of water from various sources or groups of sources. For example, the yield of a reservoir can be defined in terms of the

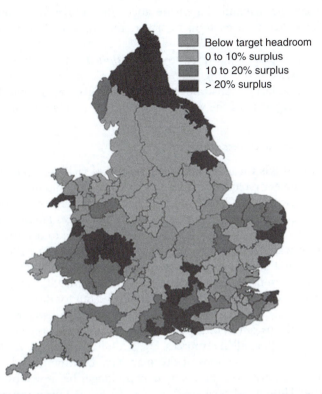

Below target headroom
0 to 10% surplus
10 to 20% surplus
> 20% surplus

Fig. 17.7 Geographical variation in relative security of public water supplies in England and Wales expressed in terms of supply as a percentage of demand (Environment Agency, 2008, reproduced with permission).

volume of water regularly available over a unit period of time; it is often expressed in units of millions of litres (megalitres) per day (ML day^{-1}). Techniques for reservoir yield analysis were mentioned in Section 11.12 and are covered in detail by McMahon and Adeloye (2005).

In studying the yield of a reservoired catchment area, it is important to appreciate more fully the implications of the term with respect to the variabilities in the catchment runoff. A definition of yield suggested by the Institution of Water Engineers and modified by Law (1955) is as follows:

> The uniform rate at which water can be drawn from the reservoir *throughout a dry period of specified severity* without depleting the contents to such an extent that withdrawal at that rate is no longer feasible.

Thus given a fixed storage capacity in the reservoir, over a drought period of greater severity than that of the design, the regular amount of water available for supply, the yield, would have to be reduced. The occurrence of any such drought governs the yield for a given storage. Alternatively, if a constant yield is required beyond the capabilities of a reservoir designed to fulfil requirements over a drought of specified severity and duration, the capacity of the storage would need to be increased to meet the demand. To ensure adequate storage for a constant yield or to evaluate the yield of an existing reservoir, the study of low flows is therefore fundamental.

In water resources planning, the *deployable output* is a more specific assessment of yield that can be sustained during a dry year from a commissioned water source, or group of sources, subject to constraint by abstraction licences, required water quality standards and the capacity of the water treatment and supply systems.

17.5.2 Regional yield assessment

Recommended methods for assessing water resources yield and the deployable output of water resources systems in the UK are discussed in guidance published by UK Water Industry Research (UKWIR[2]) and the Environment Agency (2000) and in the Environment Agency's periodically updated *Water Resources Planning Guideline* and related documents. The guidance requires water companies to determine the deployable output from water resource systems that could be sustained through the worst historic drought conditions. Typically the analysis involves complex modelling of the water resources system using software packages such as *MOSPA* and *Aquator*.

The assessment of deployable output is often carried out for discrete *water resources zones*. A resource zone is the largest possible area in which all water resources can be shared and hence all customers or a water company could experience the same risk of supply failure from a shortfall in resources. Fig. 17.8 shows resource zones in the north-west of England (previous versions of this map have been published showing major water transfer routes, however these may may no longer be shown owing to security concerns). United Utilities Water Plc supplies water to the four water resource zones, with the integrated resource zone serving a population of 6.5 million, the Carlisle resource zone, serving 106 000, the North Eden resource zone, serving 14 000 and the West Cumbria resource zone, serving 152 000. The supply network within the integrated zone has a high degree of inter-connection, and serves 95 per cent of the region's population. The other three zones are relatively small, and are remote from the regional network.

The yield from each resource zone is shown in Table 17.5, which is data published in the United Utilities WRMP of September 2009.[3] The figures give a projection of future yield incorporating the effects of reductions caused by changes in abstractions licence conditions ('sustainability reductions') and by climate change (see Chapter 19). Whilst the great majority of the abstractions are not thought to impact adversely on the environment, it is expected that the amount of water that can be abstracted from certain water bodies during dry-weather conditions will be reduced to help protect salmon and other aquatic species from potential harmful effects due to low-flow conditions.

The models used to assess yield in each resource zone are based on complex networks of components used to represent physical features such as abstractions, links (pipelines, aqueducts or channels), reservoirs and lakes, flow regulators, diversions, pumping stations and water treatment works. In addition, limits and constraints can be applied to represent abstraction licences or behavioural features such as reservoir release control curves.

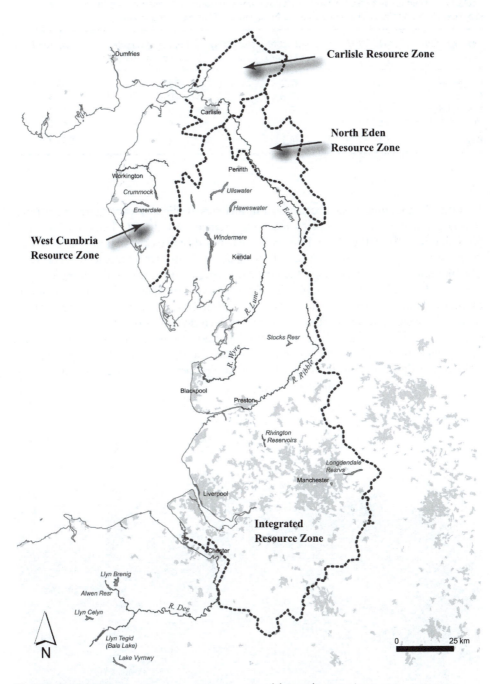

Fig. 17.8 United Utilities water resource zones, major lakes and reservoirs.

Table 17.5 Yield assessment for United Utilities water resources zones (in megalitres per day). (Source: United Utilities Water Resources Management Plan, September 2009)

Resource zone	Yield 2007–08	Impact of sustainability reductions	Impact of climate change at 2034–35	Yield 2034–35
Integrated	1908	−32.9	−28.1	1863.6
Carlisle	36.5	−3.8	−0.3	32.4
North Eden	10.3	0	0	10.3
West Cumbria	57.9	−9.4	−0.2	48.2
Region	2012.6	−46.1	−28.6	1954.7

In the West Cumbria resource zone, a model was applied to investigate operational changes that may be required to meet requirements of the EU Habitats Directive (see Chapter 19) in the River Eden SAC. This is downstream of Ennerdale Water, where outflows are controlled via a weir, fish pass and valve structures seen in Fig. 17.9. The diagram in Fig. 17.10 is a schematic of the model, illustrating the complexity of events in this comparatively small water resources system. The model was particularly useful to test alternative scenarios for the operational use of various links and controls within the system, allowing these scenarios to include conditions that had not previously been experienced and hence could not be assessed using past observations. Based on assessments of deployable output and water demand, the WRMP identified a likely 8 ML day^{-1} deficit in resources by 2014–15. It also set out measures to address this,

Fig. 17.9 Ennerdale Water (image courtesy of United Utilities).

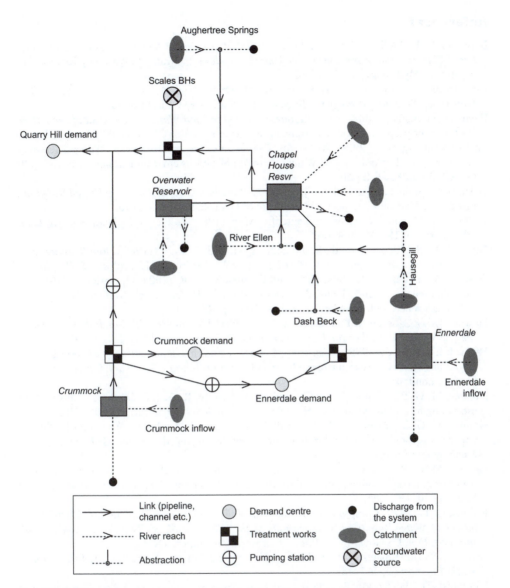

Fig. 17.10 Water resource system model for the West Cumbria resource zone (data used courtesy of United Utilities).

including reductions in leakage, helping customers to use water more efficiently and develop a new groundwater source in the region.

Notes

1 http://www.environment-agency.gov.uk/business/sectors/33065.aspx
2 www.ukwir.org
3 http://www.unitedutilities.co.uk/WaterResourcesPlan.htm

References

Boorman, D. B., Hollis, J. M. and Lilly, A. (1995) *Hydrology of Soil Types: A hydrologically-based Classification of the Soils of the United Kingdom.* Institute of Hydrology Report 126. Institute of Hydrology, Walling ford.

Department for Environment Food and Rural Affairs (Defra) (2008) *Future Water: The Government's Water Strategy for England.* The Stationery Office, London.

Environment Agency (2008) *Water Resources in England and Wales – Current State and Future Pressures.* http://publications.environment-agency.gov.uk/pdf/GEHO1208BPAS-e-e.pdf.

Environment Agency (2009) *Water for People and the Environment: Water Resources Strategy for England and Wales.* http://publications.environment-agency.gov.uk/pdf/GEHO0309BPKX-E-E.pdf

Gustard, A., Bullock, A. and Dixon, J. M. (1992) *Low Flow Estimation in the United Kingdom.* Institute of Hydrology Report 108. Institute of Hydrology, Wallingford.

Gustard, A., Marshall, D. C. W. and Sutcliffe, M. F. (1987) *Low Flow Estimation in Scotland.* IH Report No. 101, Wallingford.

Gustard, A., Roald, L., Demuth, S., Lumadjeng, H. and Gross, R. (1989) *Flow Regimes from Experimental and Network Data (FREND).* (2 vols). Institute of Hydrology, Wallingford.

Gustard, A., Young, A. R., Rees, H. G. and Holmes, M. G. R. (2004) Operational hydrology. In: Tallaksen, L. M. and Henny, A. J. van Lanen (eds) *Hydrological Drought.* Devlopments in Water Science, Vol. 48. Elsevier, Amscerdam. pp. 455–485.

Hawnt, R. J. E., Joseph, J. B. and Flavin, R. J. (1981) *Experience with Borehole Recharge in the Lee Valley. Journal of the Institution of Water Engineers and Scientists* 35, 437–351.

Hicks, J. and Allen, G. (1999) *A Century of Change: Trends in UK Statistics since 1900.* House of Commons Library Research Paper. House of Commons Library, Social and General Statistics Section, London.

Holmes, M. G. R., Young, A. R., Gustard, A. and Grew, R. A. (2002a) A new approach to estimating mean flow in the UK. *Hydrology and Earth System Sciences* 6, 709–720.

Holmes, M. G. R., Young, A. R., Gustard, A. and Grew, R. A. (2002b) A region of influence approach to predicting flow duration curves within ungauged catchments. *Hydrology and Earth System Sciences* 6, 721–731.

Holmes, M. G. R., Young, A. R., Goodwin, T. H. and Grew, R. (2005) A catchment-based water resource decision-support tool for the United Kingdom. *Environmental Modelling and Software* 20, 197–202.

Jenkins, C. T. (1970) *Computation of Rate and Volume of Stream Depletion by Wells. Techniques of Water Resources Investigations of the USGS.* Book 4. Hydrologist Analysis and Interpretation. US Government Printing Officer, Washington, DC.

Law, F. (1955) Estimates of the yield of reservoired catchments. *Journal of the Institution of Water Engineers* 9, 467–487.

Lundquist, D. and Krokli, B. (1985) *Low Flow Analysis.* Norwegian Water Resources and Energy Administration, Oslo.

Martin, J. V. and Cunnane, C. (1976) *Analysis and Prediction of Low-flow and Drought Volumes for Selected Irish Rivers.* The Institution of Engineers of Ireland, Dublin.

McMahon, T. A. and Adeloye, A. (2005) *Water Resources Yield.* Water Resources Publications, Colorado, 200 pp.

Natural Environment Research Council (NERC) (1980) *Low Flow Studies Report.* NERC, Wallingford.

Office for National Statistics (2008) *National Population Projections 2006-based.* PP2 26. Palgrave Macmillan, Basingstoke.

Office for National Statistics (2009) *Population Trends. 136.* Palgrave Macmillan, London. http://www.statistics.gov.uk/downloads/theme_population/Popular-Trends136.pdf.

Pirt, J. and Douglas, R. (1982) A study of low flows using data from the Severn and Trent catchments. *Journal of the Institution of Water Engineers and Scientists* 36, 299–309.

Shaw, E. M. (1989) *Engineering Hydrology Techniques in Practice*. Ellis Horwood, Chichester, 349 pp.

Stedinger, J. R., Vogel, R. M. and Foufoula-Georgiou, E. (1993) Frequency analysis of extreme events. In: Maidment, D. R. (ed.) *Handbook of Hydrology*. McGraw Hill, New York, pp. 18.1–18.66.

Tallaksen, L. M. and van Lanen, H. A. J. (eds) (2004) *Hydrological Drought*, Vol. 48: *Processes and Estimation Methods for Streamflow and Groundwater*. Elsevier Science, Amsterdam, 579 pp.

Twort, A. C., Ratnayaka, D. D. and Brandt, M. J. (2000) *Water Supply*, 5th edn. IWA Publishing, London, 676 pp.

UKTAG (2009) *Recommendations on Surface Water Classification Schemes for the Purposes of the Water Framework Directive*. UK Technical Advisory Group on the Water Framework Directive. Published 2007, revised June 2009. http://www.wfduk.org/UKCLASSPUB/LibraryPublicDocs/class%20report

UKWIR and the Environment Agency (2000) *A Unified Methodology for the Determination of Output from Water Sources*. UK Water Industry Research Report 00/WR/18/1.

Water UK (2008) *Sustainable Water: State of the Water Sector Report*. Water UK, London. www.water.org.uk

Waterwise (2009) *Reducing Water Wastage in the UK – the Facts*. http://www.waterwise.org.uk/reducing_water_wastage_in_the_uk/the_facts/the_facts_about_saving_water.html (accessed August 2009).

Young, A. R., Grew, R. and Holmes, M. G. R. (2003) Low Flows 2000: a national water resources assessment and decision support tool. *Water Science and Technology* 48, 119–126.

Young, A. R., Gustard, A., Bullock, A., Sekulin, A. E. and Croker, K. M. (2000) A river network based hydrological model for predicting natural and influenced flow statistics at ungauged sites. LOIS special volume. *Science of the Total Environment* 251/252, 293–304.

Zaidman, M. D., Keller, V. and Young, A. R. (2002) *Low Flow Frequency Analysis: Guidelines for Best Practice*. Environment Agency R&D Technical Report W6-64/TR1.

Zaidman, M. D., Keller, V., Young, A. R. and Wall, A. (2003) Adapting low flow frequency analysis for used with short-period records. *CIWEM Journal* 17, 74–79.

Chapter 18

Urban hydrology

Throughout history, there have been periods when populations have tended to congregate together in towns and cities. The reasons for town development are numerous; in the early empires of the Middle East, water supply was a prime factor and in the city states of the Mediterranean lands, safety from raiders and pirates dictated collective defence. Sources of fresh water also played a part in the establishment of nucleated settlements in the lowlands of Great Britain, and the advantages of safety in numbers led to the growth of the medieval walled towns. With increases in populations in more peaceful times, people were encouraged to spread out to make their homes where they could grow more food, and, often spurred on by religious persecution, many travelled across the world to establish themselves in new lands. The Industrial Revolution and the growth of manufacturing industries brought people together again. The establishment of factories meant that livelihoods became dependent on employment rather than on subsistence farming through self-endeavour. This process of urbanization, the congregation of people together to live in towns, escalated through the twentieth century. In the older developed countries, like the UK, the well-established towns and cities have continued to expand; all the benefits of a high standard of living are much more economically provided in a centralised community. In the developing countries, the cities are an attraction to expanding rural populations seeking factory employment; throughout the world, urban centres are growing and, in some countries, the expansions are not planned or controlled. Thus, with much of the world's population living in urban environments, the effect of such developments on elements of the hydrological cycle assumes a significant importance.

A high proportion of the activities of water engineers is directed towards urban construction, including the provision of a potable water supply and the removal and treatment of surplus and waste water. In England and Wales most of the population receives a piped water supply from a public authority or a private water company, and the vast majority of households are connected to a public sewerage system. In addition to these major modifications to the natural water balance of an area, there are marked effects of extensive built-up areas on different components of the hydrological cycle.

18.1 Hydrological regime modifications

Studies in the UK have failed to relate annual and seasonal rainfall differences to urban development (Tabony, 1980). In the London area, rainfall differences are influenced mainly by altitudinal differences. On a short timescale, the proportionally greater incidence of severe thunderstorms in built-up areas compared with rural areas is well noted and can be ascribed to greater concentrations of condensation nuclei in the air, increased turbulence and urban overheating. Information on such extreme events is naturally more readily available (and perhaps more likely to be recorded) in centres of population, and the hydrological consequences are usually of immediate concern. Storm magnitudes and their frequency of occurrence are of greater importance than annual rainfall totals in urban hydrology.

Urbanisation can have a dramatic effect on all aspects of the terrestrial hydrological cycle. Urban development changes the relative proportions of water that infiltrates the ground, evaporates, travels as subsurface drainage or is carried as overland flow and surface drainage. Fig. 18.1 shows generalised estimates of the partition of water into different flow pathways given by Endreny (2005), based on assumed typical land cover characteristics. Of course, in any particular catchment, the actual impacts of these types of changes may vary greatly depending on the actual proportions of pervious and impervious surfaces, their spatial arrangement, the design of the drainage system and the connectivity of flow pathways.

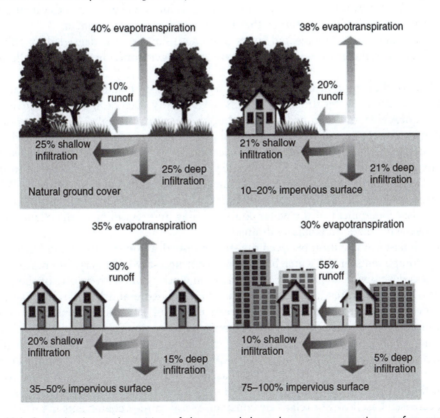

Fig. 18.1 Generalised typical partition of the water balance between evaporation, surface runoff, shallow subsurface infiltration and deeper infiltration for four different land cover classes defined in terms of the percentage impervious surface area. Reprinted from The Encyclopedia of Hydrological Sciences, Chapter 117, Endreny, T. A., Land Use and Land Cover Effects on Runoff Processes: Urban and Suburban Development, Pages 1775–1804, Copyright (2005) with permission from John Wiley & Sons.

The covering of the land surface by a large proportion of impervious materials means that a much larger proportion of any rainfall forms immediate runoff. In addition to extensive ground coverage by the buildings in a city, the paved streets and car parks contribute large areas to the impervious surfaces. Any slope of the land also greatly enhances the runoff response of a paved area. In a defined catchment area, the effect on the stream discharge is dependent on the extent of the impervious area. Contributions to groundwater are limited to rainfall on the remaining pervious surfaces, where normal infiltration into the soil and percolation into the underlying strata can take place.

After major urban developments in a catchment, the following differences in the river flow from that of an equivalent rural catchment can be identified:

(1) for a specific rainfall event, the response of the catchment is accelerated, with a steeper rising limb of the flow hydrograph; the lag time and time to peak (see Chapter 13) are reduced;
(2) flood peak magnitudes are increased;
(3) in times of low flows, discharges are decreased since there is a reduced contribution from the groundwater storage that has received less replenishment;
(4) water quality in streams and rivers draining urban areas is degraded by effluent discharges, increased water temperature and danger from other forms of pollution.

Many of these modifications are promoted by structural changes made to drainage channels. Surface water drainage systems have usually been designed to remove rain water quickly from developed areas, although measures may also be taken to delay and attenuate the runoff response, particularly in modern sustainable drainage design (see sustainable urban drainage systems; Section 18.7). In many old established settlements, storm water runs into the domestic waste water sewers, but in some countries, e.g. Australia, the cities have separate storm water and sewerage systems. When an area is newly developed, it is sometimes expedient to modify the natural stream channels; re-alignment of the water courses, lining and regrading of the channels are improvements made to facilitate drainage.

The long-term partition between the stream flow draining an urban catchment and other components of the water balance may not necessarily be very different to that of a less developed catchment. For example, in catchments with a scale of the order of $100 \, \text{km}^2$ in the area around Atlanta in the USA, Rose and Peters (2001) studied stream flow and precipitation data between 1958 and 1996, and found that annual runoff coefficients did not differ significantly when comparing catchments with different percentage urban areas. However, the same study showed that in the more urban catchments peak discharges were 30–100 per cent greater for the 25 largest storm events with 25–35 per cent less baseflow and higher flows during the recession limb of the storm hydrographs. Brun and Band (2000) found a similar pattern of decreased baseflow but relatively constant overall runoff coefficients in an 18-year study of suburban growth in a $170 \, \text{km}^2$ watershed in Maryland, USA.

Similar patterns are seen in the response of The Cut at Binfield, a catchment in the Bracknell area west of London (Fig. 18.2). This catchment has rural headwaters and around 30 per cent urban area in total. Flows have been monitored in the channel upstream and downstream of outfalls from the urban drainage system. Whilst the

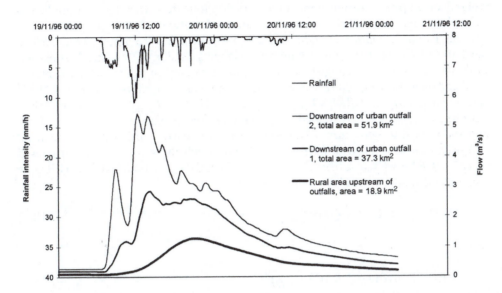

Fig. 18.2 Storm hydrographs for The Cut at Binfield showing rapid urban runoff response. Reprinted from The hydrology of the UK: a study of change, Chapter 2, Robinson, M., Boardman, J., Evans, R., Heppell, K., Packman, J. and Leeks, G., Land Use Change, Pages 30–54, Copyright (2000) with permission from Routledge.

total runoff scales approximately with drained area, the influence of the urban system is clearly seen in the hydrographs with much higher peak flow rates and more rapid responses.

The interaction of the artificial nature of urban catchments and the need to accommodate the changed hydrological characteristics is complex. The solving of one drainage problem may easily exacerbate another feature of the catchment runoff, e.g. rain events on the planned surface drainage of a new housing estate could produce higher peaks downstream than formerly, and these might cause flooding at previously safe points along the channel.

18.2 Catchment response modifications

The quantifying of urbanisation effects on the rainfall–runoff relationship has been studied widely. Much of the work has concentrated on the modifications made to the volume and time distribution of surface water runoff hydrographs from single rainfall events. The various hydrograph parameters such as peak discharge, Q_p, time to peak, t_p and lag time (various definitions) are usually related to catchment characteristics, including area of impervious surfaces or proportion of area urbanised, in order to obtain quantitative rainfall–runoff relationships. A thorough description of many of the studies was given by Packman (1980). The various formulae that have resulted from individual studies are only applicable to the areas where they have been derived, and it is not advisable to use them for areas with different climates and topography. Instead, rainfall–runoff modelling is used, although most widely applied rainfall–runoff models still rely on empirical formulae for some of their parameters when they are applied to sites where there is insufficient measured data for calibration.

In the *Flood Estimation Handbook* (*FEH*; Institute of Hydrology, 1999) described in Chapter 13, the net effect of urbanisation is represented in the statistical estimation of design flows at an ungauged site by treating a catchment as if it were essentially rural and then applying adjustment factors. For moderately urbanised catchment (*FEH* catchment descriptor $URBEXT_{1990} < 0.125$ or $URBEXT_{2000} < 0.150$) an urban adjustment factor (UAF) is applied to estimation of the median annual flood (QMED) and the pooled growth curve. The adjustment was derived using flood data from 115 urban catchments (with $URBEXT > 0.05$), of which 34 were heavily urbanised ($0.125 < URBEXT_{1990} < 0.500$ or $0.150 < URBEXT_{2000} < 0.600$). For QMED, the recommended adjustment is currently given as

$$QMED_{urban} = UAF \times QMED_{rural} \tag{18.1}$$

where

$$UAF = PRUAF \left(1 + URBEXT_{2000}\right)^{0.66} \tag{18.2}$$

and

$$PRUAF = 1 + 0.47 \times URBEXT_{2000} \left(\frac{70}{SPRHOST} - 1\right). \tag{18.3}$$

Here $URBEXT_{2000}$ is the *FEH* digital catchment descriptor for urban extent dated to year 2000 and *SPRHOST* is the catchment descriptor for standard percentage runoff as a function of the HOST soil classification; the *FEH* catchment descriptors and HOST were discussed in Chapter 13.

The growth curve applied to derive the flood frequency curve is also adjusted for urbanisation using an empirical formula, which is given by the *FEH* as

$$x_T = 1 + \frac{\left(xrural_T - 1\right)\left(\dfrac{xrural_{1000}}{UAF} - 1\right)}{xrural_{1000} - 1} \tag{18.4}$$

where x_T is the adjusted growth curve factor for the T-year return period, $xrural_T$ is the 'as rural' growth factor for the T-year return period and UAF is as given above. The effect of the adjustment is to modify the underlying 'as rural' growth curve progressively with the greatest modification occurring at low return periods. The rural and urban growth curves are assumed to converge at a return period of 1000 years. This represents the concept that, within a moderately urbanised catchment, the urbanisation will have proportionally the greatest impact on runoff for smaller storms, whereas for a very large flood flow, the soils are likely to be uniformly saturated (reducing the difference in runoff production between permeable and impermeable areas) and the sheer depth of rainfall will tend to dominate over differences in local runoff coefficients (see Fig. 18.3). The basic urban adjustment is relevant for the degree of urbanisation observed in the URBEXT descriptor for the year 2000. For analysis of past or projected

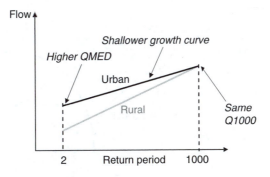

Fig. 18.3 *FEH urban growth curve adjustment.*

future changes, the URBEXT value can be adjusted according to a general national model of urban development that is encoded within the software packages that support the FEH, or by map-based analysis of urban locations.

The *FEH* rainfall–runoff model and *ReFH* methods described in Chapter 13 are generally recommended for catchments with a high proportion of urban land cover, and for extremely heavily urbanised catchments (e.g. $URBEXT_{1990} > 0.5$) the advice is to use sewer design methods such as the *Wallingford Procedure* discussed below. The revised flood hydrograph method *ReFH* was calibrated with only seven urban catchments and so the general guidance is not to use it for heavily urbanised catchments, though further research may change this position in the future.

18.3 The Transport and Road Research Laboratory rational method

Hydrologists are often concerned with evaluating the runoff from the subareas to be drained in order to design the necessary storm water sewers. The peak runoff from the selected design storm determines the size of sewer pipes, which are dependent on the extent of each sub-area to be drained. At the head of a catchment sub-area, the required pipe size may be quite small, but downstream, as the sewer receives water from a growing area through a series of junctions, the pipe size gradually needs to be increased.

The problem of estimating the runoff from the storm rainfall is very much dependent on the character of the catchment surface. The degree of urbanisation (extent of impervious area) greatly affects the volume of runoff obtained from a given rainfall. Retention of rainfall by the initial wetting of surfaces and absorption by vegetation and pervious areas reduces the amount of storm runoff. These surface conditions also affect the time distribution of the runoff. Thus the method used to obtain the runoff from the rainfall should allow for the characteristics of the surface area to be drained.

Calculation of runoff is the most straightforward for small impervious areas, such as roof areas and paved surfaces, in which there is very little or even no part of the ground surface into which rainfall could infiltrate. Over such limited areas, the storm rainfall can be assumed to be uniformly distributed with 100 per cent runoff occurring. The response of the impervious surfaces is rapid, resulting in a short time of concentration

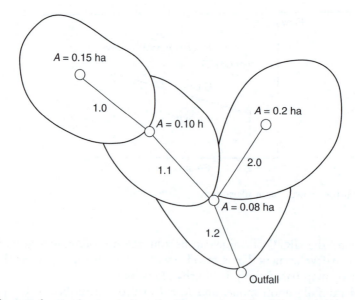

Fig. 18.4 Schematic for simple pipe design using Transport and Road Research Laboratory (TRRL) methods.

of the flow in the drainage system. The Rational Formula introduced in Chapter 12 can then provide a simple estimate for peak flow.

A recommended method for the design of a piped sewer drainage system using the Rational Method was given in the Transport and Road Research Laboratory (TRRL) Road Note 35 (TRRL, 1976). The procedure may be explained by considering the simple pipe design in Fig. 18.4. The sequence of pipes must be numbered according to the convention shown. The first pipe of a branch is always labelled 1.0, 2.0, etc. and the following pipes in a line are labelled sequentially, 1.1, 1.2, etc. Here there is a line of three pipes leading to an outfall and a tributary area (pipe 2.0) drains into the junction at the end of the second pipe in line 1, pipe 1.1. The computations to determine the required pipe sizes are shown in Table 18.1. The first four columns give the surveyed particulars of level differences along each pipeline, the required length and the calculated gradient.

At the outset of the design procedure, the selected return period for a design storm will have been decided. Storm water sewers are usually designed for return periods of much less than 100 years. In the example, the expected annual storm intensities are used (1-year return period). The type of pipe will also have been chosen; the internal roughness governs the flow characteristics, and a roughness coefficient, k_s, must be selected from published tables (Ackers, 1969) in order to use the Colebrook–White equation to determine the flow velocity in the pipe. Velocities and discharges for standard-sized pipes computed from this complex formula are published in tabular form for different pipe sizes, assuming full-bore conditions, a hydraulic gradient equal to the pipe gradient and appropriate roughness (Ackers, 1969). Flows larger than those derived from the tables or charts would require hydraulic gradients greater than the pipe gradient stipulated, and these could only occur by ponding (or surcharging) of

Table 18.1 Rational method drainage design example data (Pipe no. = Pipe number; Level diff. = Level difference; Grad. = Gradient; Trial pipe da. = Trial pipe diameter; V = Velocity; Q = Discharge; Time of conc = Time of concentration; Imp. A cum = Cumulative impervious area; Storm Q = Storm discharge)

Pipe no.	Level diff. (m)	Pipe length (m)	Grad. (1 in)	Trial pipe da. (mm)	Pipe V (ms^{-1})	Pipe Q (Ls^{-1})	Time of flow (min)	Time of conc (min)	Rate of rain (mmh^{-1})	Imp. A cum (ha)	Storm Q (Ls^{-1})	Comment
1.0	1.00	65	65	150	1.26	23.0	0.86	2.86	67.5	0.15	28.1	Surcharge partial flow
				225	1.64	67.5	0.66	2.66	69.2		28.8	
1.1	0.90	70	78	225	1.50	61.7	0.78	3.44	63.2	0.25	43.9	Partial flow
2.0	1.50	60	40	150	1.61	29.4	0.62	2.62	69.5	0.20	38.6	Surcharge partial flow
				225	2.10	86.0	0.48	2.48	70.7		39.3	
1.2	0.90	50	56	225	1.77	72.8	0.47	3.91	60.2	0.53	88.6	Surcharge partial flow
				300	2.13	156.0	0.39	3.83	3.83	60.7	89.4	

water in the manholes at the pipe junctions. The design objective is to avoid such surcharging.

The design procedure begins with the choice of a trial pipe size for pipe 1.0 (150 mm in the example in Table 18.1). From the published tables and for $k_s = 0.6$ for a normal concrete pipe, the velocity and discharge for a gradient of 1 in 65 are noted, 1.26 m s^{-1} and 23.0 Ls^{-1}, respectively. A flow greater than 23.0 Ls^{-1} would result in surcharging.

The time of flow along the pipe is next calculated from the velocity and length of pipe and comes to 0.86 min. The time of concentration at the end of the first pipe is then 0.86 min plus an assumed allowance of 2 min, for the time of entry, which is assumed to cover the lag time between the onset of the storm rainfall and the entry of the overland flow into the leading manhole. With the time of concentration of the drainage to the end of the first pipe known, the design return period rainfall intensity, i over this duration to give the peak flow can be obtained from intensity–duration–frequency data. Ideally, data from a local rain gauge will be analysed to assess the design storm depth. Where there is not adequate data, a generalised model may have to be used. For durations longer than 1 hour, the *FEH* depth–duration–frequency model (see Chapter 9) are available, whilst shorter durations were considered in the older *Flood Studies Report* rainfall analysis. In this example, the rates of rainfall are taken from Table 18.2 for a location in Southern England (TRRL, 1976). The storm peak discharge for this sub-area is then calculated from the rate of rainfall, i, and its cumulated impervious area using the rational method for comparison with the unsurcharged full bore pipe flow. The first trial pipe of 150 mm diameter would clearly be surcharged, so the calculations are repeated with the next size pipe, diameter 225 mm. The calculated storm discharge, 28.8 L s^{-1}, would be easily contained by the larger pipe.

The calculations proceed for each pipe in turn, with the previous time of concentration being added to the new time of flow to give the combined times of concentration at the end of sequential pipes. The drainage areas are also accumulated. It will be noted

Table 18.2 Rainfall intensities (in millimetres per hour) for specified durations and return periods (point location in Southern England)

Duration (min)	Return period (years)		
	1	2	5
2.0	75.6	93.4	120.5
2.5	70.5	87.5	113.4
3.0	66.3	82.3	107.2
3.5	62.8	77.8	101.7
4.0	59.6	73.8	96.8

that the 2.0 min time of entry is also added to the flow time of pipe 2.0, since it is at the start of a branch pipeline. The time of concentration for the last pipe 1.2, is the sum of the time of concentration of pipe 1.1 and the flow time of pipe 1.2. The extra contribution from the greatly increased area drained by the tributary pipe results in a much larger discharge requiring the next size larger pipe, 300 mm diameter. (The pipe diameters are given as rounded metric equivalents to the old 6-, 9- and 12-in diameter pipes.)

Thus in the simple pipe design for the system in Fig. 18.4, pipes 1.0, 1.1. and 2.0 need to be 225 mm in diameter and the last pipe 1.2 must be of 300 mm diameter. These requirements conform to the normal concrete pipes specified ($k_s = 0.6$) and 1-year return period design storm intensities, with an assumed 2 min time of entry.

It will be appreciated that the computations become complicated as more branch pipe lines are incorporated into the system. This method is most satisfactory for small impervious areas, but if more pervious fractions are included within the catchments, results from the Rational Method become less acceptable.

18.4 The Transport and Road Research Laboratory hydrograph method

In designing storm water sewerage systems for towns, city suburbs and new developments of around 200–400 ha with varied surface characteristics, a method is required that also takes into account differences in storm rainfall over the catchment area. Developed from the time–area concept of catchment response (Chapter 12), the TRRL hydrograph method (Watkins, 1962) was applied widely in the UK.

In the time-area method, the total catchment area is deemed to be contributing to the flow after the time of concentration, T_c, the time it takes for the rain on the furthest part of the catchment to reach the outfall. Thus, in Fig. 18.5, for two drains receiving uniform rainfall from areas A_1 and A_2, with drain 2 joining the main channel, drain 1, a relationship of contributing area, A, versus time, T, is constructed. From the beginning of the flow in drain 1 at $T = 0$, there is a steady increase in area contributing until $T = T_1$ which is the value of T_c for area A_1. Drain 2 begins to contribute to the outfall flow at $T = T_3$ before $T = T_1$. After a further period, T_2, area 2 reaches its own T_c at time $T = (T_2 + T_3)$. Between times T_3 and T_1 both drains have been flowing and the

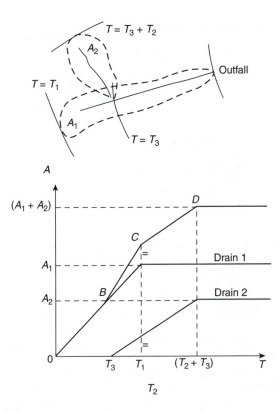

Fig. 18.5 Time–area diagram.

joint contributing area (at C) at $T = T_1$ is given by

$$A_1 + \frac{T_1 - T_3}{T_2} A_2 \tag{18.5}$$

From $T = (T_2 + T_3)$, both areas are contributing fully. The time–area curve for the combined drains is the composite line $OBCD$.

The principle of the TRRL hydrograph method is outlined in Fig. 18.6. In Fig. 18.6(a), a catchment area, divided into four subareas, is drained by a single channel to the outfall where the hydrograph is required. Sub-area 1 begins contributing to the flow first, to be followed sequentially by the other three sub-areas. The individual time–area curves are shown in Fig. 18.6(b) and the composite curve for the whole catchment is drawn by summing the sub-area contributions at regular time intervals. The choice of time unit is dependent on the surface characteristics of the catchment and may range from 1 min for highly impervious areas to about 30 min for nearly natural catchments. The incremental contributing areas after each time interval are then read from the composite curve, a_1, a_2, a_3, etc. In the diagram, the time of concentration for the whole area is time units.

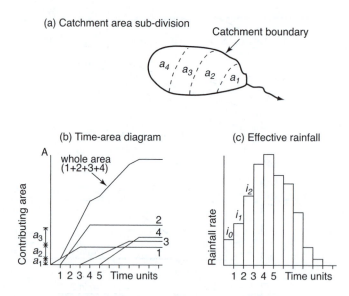

Fig. 18.6 Transport and Road Research Laboratory (TRRL) hydrograph method.

The next stage in the method involves the storm rainfall. The values of the areal rainfall are calculated from the rain-gauge measurements by one of the standard methods (Chapter 10) for each of the chosen time unit intervals throughout the duration of the storm. Since some of the rainfall will infiltrate the pervious areas, not all the storm rainfall will contribute to the direct runoff from the catchment area. An effective rainfall rate must be assessed for each time unit. The effective rainfalls may be obtained by assuming a runoff coefficient and, applying this to each time unit rainfall in turn, or a constant loss rate can be assumed and subtracted from each time unit rainfall rate. Fig. 18.6(c) shows the effective rates of rainfall for each time unit, i_0, i_1, i_2, etc., for the storm duration (ten time units).

The discharge rates after each time unit interval are given by

$$q_1 = i_0 a_1$$
$$q_2 = i_1 a_1 + i_0 a_2 \qquad\qquad (18.6)$$
$$q_3 = i_2 a_1 + i_1 a_2 + i_0 a_3 \text{ etc.}$$

A worked example is shown in Table 18.3. There are four increments of area (ha) resulting in a time of concentration for the catchment equivalent to four time units. The storm duration extends over ten time units. A runoff coefficient of 0.64 has been assumed and thus the total areal rainfalls in column 2 have been multiplied by 0.64 to give the corresponding effective rainfalls (i mm h^{-1}). The values of q for each area increment a and effective rainfall rate i are calculated from the basic form of the rational formula

$$q(\text{Ls}^{-1}) = \frac{i(\text{mm h}^{-1}) \cdot a(\text{ha})}{0.36} \qquad\qquad (18.7)$$

Table 18.3 The Transport and Road Research Laboratory hydrograph method (runoff coefficient 0.64)

Time unit	Areal rate	Rainfall	Area increment (ha)				Discharge
	Total mm h^{-1}	Effective rainfall *i* (mm h^{-1})	a_1 0.25	a_2 0.82	a_3 0.92	a_4 0.34	q (L s^{-1})
I	13.7	8.8	6.1				6.1
2	90	57.6	40	20			60
3	59.4	38	26.4	131.2	22.5		180.1
4	18.3	11.7	8.1	86.6	147.2	8.3	250.2
5	16.8	10.8	7.5	26.7	97.1	54.4	185.7
6	13.7	8.8	6.1	24.6	29.9	35.9	96.5
7	5.3	3.4	2.4	20	27.6	11.1	61.1
8	5.1	3.3	2.3	7.7	22.5	10.2	42.7
9	6.1	3.9	2.7	7.5	8.7	8.3	27.2
10	4.6	2.9	2	8.9	8.4	3.2	22.5
11				6.6	10	3.1	19.7
12					7.4	3.7	11.1
13						2.7	2.7

where 0.36 is a units conversion factor. The summation of the rows across a_1 to a_4 gives the discharge values after each time increment, and thus the required hydrograph. It will be noted that the peak flow occurs after the fourth time interval, the time of concentration of the catchment. This does not always happen, e.g. with late peaking rainfalls.

Two further considerations are necessary. A time of entry from the onset of the storm rainfall to the time of flow into the pipe is usually taken to be 2 min and must be allowed for in the computations. Secondly, experience has shown that there is a certain amount of retention of water in the pipe channel, and amendments to the hydrograph must be made to account for pipe storage (Watkins, 1962).

The procedure of the TRRL method, as presented, applies to one drainage unit, i.e. one pipe, in a system. The calculations have to be carried out for each pipe in a sewerage network as demonstrated earlier in the rational method of design. In practice, the application of the TRRL hydrograph method to even a simple configuration of drainage pipes becomes too complex for manual computation and so it has been applied in practice using software packages.

18.5 The Wallingford procedure

From the 1970s, access to computers meant that simplifications of former methods were no longer necessary and many more relevant factors could be introduced into runoff design calculations using software packages. In 1981 the *Wallingford Procedure* was published. The Wallingford Procedure describes the hydraulic design and analysis of pipe networks for both new schemes and existing systems. It can accommodate both independent storm water sewers and combined sewers, but the waste water flows must be given as inflows at the appropriate junction in a combined sewer. The hydraulic analysis of a range of structures controlling the flow in a pipe system can be made and certain economic factors are also incorporated into the procedure. The whole package provides a range of methods from which a series of calculation techniques can be selected to suit the conditions of any particular design scheme. Basic data required for the design techniques were included for the whole of Great Britain and Northern Ireland so that the procedure could be applied nationwide. Almost three decades later the methods, although updated and revised, remain in widespread use within the water industry.

The Wallingford Procedure comprised several methods, as follows:

(1) a *modified rational method* giving peak flows only, including a routing coefficient in addition to a volumetric runoff coefficient and recommended for initial designs and for use in homogeneous catchments up to 150 ha in area;
(2) a *hydrograph method* to model surface runoff and pipe flow and provide a pattern of discharge in time;
(3) an *optimizing method*, to set the design of pipe depth and gradient as well as diameter;
(4) a *simulation method* to analyse the performance of existing systems or proposed designs operating under surcharge conditions.

A general outline of the four methods is given in Table 18.4, which itemises component parts of each method. In modelling the different components such as rainfall and overland flow, the methods incorporated many of the formulae derived for the *Flood Studies Report*. Rainfall depth–duration–frequency data and storm profiles, calculations of net rainfall and the use of catchment wetness indices were all applied in the detailing of the *Wallingford Procedure*.

The simulation method proved useful to practising engineers since the vexed problem of pipe surcharging is included in the analysis. As indicated in Table 18.4, the major difference from the hydrograph method lies in the modelling of the pipe flow (Bettess *et al.*, 1978). The same flow equations are used namely the Colebrook–White equation for velocity (Ackers, 1969) and the Muskingum–Cunge routing procedure (see Section 14.2.3), but instantaneous discharges are calculated throughout the sewer system at a given time increment instead of complete hydrographs being routed sequentially from one pipe to another. Thus the interactions between surcharged pipes can be modelled.

The simulation method allows for the storage of surcharged water within manholes and on the ground surface flooded to a uniform depth over an assumed area contributing to the pipe length. The extent of the temporary surface flooding related to calculated flood volumes must be assessed on the ground.

Surcharging occurs when incoming flow is greater than full-bore pipe capacity or when a raised tailwater level causes a backwater effect. It is assumed that the manhole losses are proportional to velocity head in the pipe, then head loss, Δh, in the surcharged pipe is composed of pipe friction loss plus losses at both manholes over a time increment dt.

$$\Delta h = \left(\frac{L\lambda}{d} + k_m \right) \frac{V^2}{2g} + \frac{1}{g}\frac{dV}{dt} \tag{18.8}$$

Table 18.4 Methods and models in the Wallingford Procedure. (Reproduced from National Water Council (1981) Design and Analysis of the Urban Storm Drainage: The Wallingford Procedure, by permission.)

Method	Model					
	Rainfall models	Overland flow models	Pipe flow models	Sewer ancillaries models	Construction cost model	Flood alleviation benefit model
Modified Rational Method	Intensity— duration— frequency relationship	Percentage runoff model+ time of entry	Pipe full velocity	Storm overflow		—
		Complete surface runoff model	Muskingum— Cunge	Storm overflow Storage tank Pumping station	TRRL resource cost model	—
Hydrograph Method	Rainfall profiles	Sewered sub-area model may be used for selected sub-areas				
Optimizing Method	As for Modified Rational Method	As for Modified Rational Method	Pipe full velocity			—
		Complete surface runoff model	Muskingum— Cunge and surcharged flow	As for Hydrograph Method plus Tailwater level		Middlesex Polytechnic Flood Hazard Research Project Model (not included in programs)
Simulation Method	As for Hydrograph Method	Sewered sub-area model (without surcharging) may be used for selected subareas				

where L and d are pipe length and diameter (m) and V is flow velocity ($\mathrm{m\,s^{-1}}$), g is gravity acceleration ($\mathrm{m\,s^{-1}}$), λ is the friction coefficient ($8gSR/V^2$) ($\mathrm{s^{-1}}$), (with R the hydraulic radius and S the hydraulic gradient), km is the head loss coefficient for manholes (with values of 0.15 for a straight manhole, 0.50 for 30° bend and 0.90 for 60° bend manhole).

From the storage equation $dS/dt = I - O$ where I is the total flow into the upstream manhole from the upstream pipes and direct surface runoff from its subcatchment. With h, the difference in levels in upstream and downstream manholes and the storage equation, the flow over the chosen time increment can be simulated given the

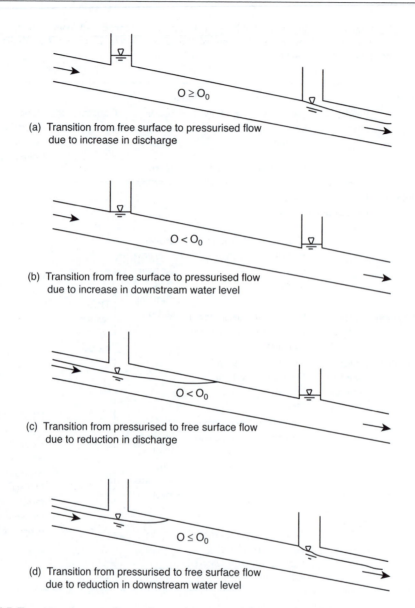

(a) Transition from free surface to pressurised flow
    due to increase in discharge

(b) Transition from free surface to pressurised flow
    due to increase in downstream water level

(c) Transition from pressurised to free surface flow
    due to reduction in discharge

(d) Transition from pressurised to free surface flow
    due to reduction in downstream water level

Fig. 18.7 Transitions between free surface and pressurised flow in a pipe. (Reproduced from National Water Council (1981) *Design and Analysis of Urban Storm Drainage: The Wallingford Procedure*, by permission.)

manhole and pipe geometries. The transition phases to and from free surface flow and surcharged (pressurized) flow are demonstrated in Fig. 18.7.

The state of the sewer system with the volumes stored at each surcharged manhole is determined by repeating the calculations for each time increment until the performance of the system has been established over the whole period of the storm event.

The Wallingford Procedure now incorporates several models for runoff generation. The original model is an empirical design formula for percentage runoff (PR) that was developed after monitoring of some 510 storms on 17 catchments during the period 1974–9. The model gave the total storm runoff depth as a product of design rainfall, catchment area and PR, hence it is essentially a variant of the rational method. The formula for PR included a parameter to describe the percentage impervious area within the catchment, along with soils information derived from *Flood Studies Report* maps and a catchment wetness index. The original PR formulation led to some difficulties in application. It is not always easy to work out what to include in total catchment area (e.g. whether to count areas connected by piped drainage systems) and hence how to define impervious area. The original model also assumed that the PR remained constant during a storm event. This is unrealistic, particularly for longer duration events. In response to these issues, a revised model was developed in the late 1980s and is now available in the mainstream software tools for application of the Wallingford Procedure.

The revised model remains known amongst urban drainage modelling practitioners as the 'new UK runoff model'. It separates permeable and impermeable areas. Permeable areas are represented using a simple soil moisture accounting model that calculates an increase in runoff during an event as the catchment wetness increases. It uses a dynamic antecedent precipitation index (API), which is updated during the event calculation. This is divided by a notional soil storage capacity (typically a fixed value of 200 mm) to obtain a fractional equivalent soil moisture deficit that is multiplied by rainfall to derive the predicted runoff. The API is based on the balance between rainfall and evaporation over the previous 30-day period, with a time decay factor applied that varies according to soil type to represent the different rates of drying out of different soil types. This model provides consistency for both event-based and longer 'continuous simulation' of runoff, where soil moisture can be depleted by evaporation when sequences of events are modelled over a period of time.

The model allows an initial rainfall depth to be subtracted to represent the wetting of surfaces and filling up depressions before runoff begins. Each surface, such as paved surfaces, roofs or well-drained roads has a characteristic depression storage depth that depends on the slope and the type of the surface, and is typically in the range 0.5–2.0 mm. If the model is run for continuous simulation, then the depression storage is dried out by evaporation after rainfall ceases.

Impermeable areas are split into two fractions. One is assumed to produce runoff with a runoff coefficient of 1.0, representing a direct connection to the drainage system. The other fraction is treated as if it were permeable, and is incorporated within the permeable area soil moisture accounting model. The fractions assumed to generate direct runoff vary according to surface type. For normal paved surfaces, the value is 0.6, which was calibrated in development of the model. Values for other surfaces are estimates. A user can calibrate these parameters when setting up a drainage model.

Further details about the Wallingford Procedure runoff models can be found in the guidance notes published and updated from time to time by the Wastewater Planning Users Group (WaPUG,[1] part of the Chartered Institute of Water and Environmental Management, CIWEM).

18.6 Urban drainage system modelling

From early computer programs written in the 1970s and 1980s, desktop software packages for urban runoff modelling have evolved over the past decades. An important line of development was the WASSP family of products, which from the early 1980s provided a computer implementation of the Wallingford Procedure. This led to the WALLRUS package and subsequently HydroWorks and InfoWorks. In the UK there is now a range of urban drainage modelling software in use, including software packages such as MIKE URBAN and SWMM. Two of the most widely used software packages are InfoWorks and WinDES, both of which have been developed in the UK and include support for UK runoff models and design standards.

Modern urban drainage software applications allow users to design, analyse and run scenario tests on large, complex systems of linked surface and subsurface drainage. They are capable of modelling the control features commonly found in urban drainage networks, including gullys, conduits, open channels, weirs of various types, orifices and penstocks. Some packages include optimisation methods to assist in sizing of pipes and setting channel capacities. There are also now capabilities to link the drainage network with two-dimensional (2D) depth-averaged flow models to simulate overland flood flow pathways.

An example of the type of complex drainage network that can be modelled is shown in Fig. 18.8, which is a pipe and gulley network for an area in the south-west of England.

It is typical of urban drainage data that despite the best efforts of the responsible operating authorities, there can still be uncertainty about the where some drains are,

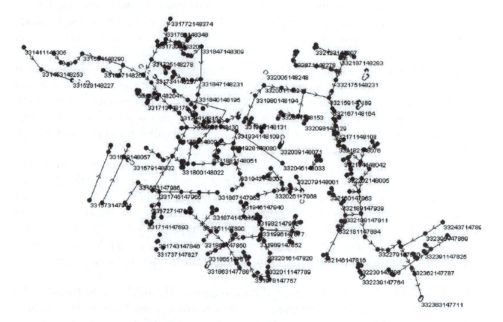

Fig. 18.8 Schematic connectivity diagram for a modelled urban drainage system (data used with permission of Somerset Drainage Boards Consortium).

their levels, material condition and even which drains are connected. In this example, investigation centred on a major culverted drain denoted 'WR' of approximately 700 m length that has been linked with reported flooding incidents. In order to assess the capacity of the whole urban drainage system, all known highway gullys were included in the model. This enabled the model to be used to distinguish surface water flooding, where water ponds and flows over the surface as it is unable to enter the underground systems, from sewer and culvert flooding where water exits from overloaded sewers and culverts.

The hydrological inputs to the system were discretised into a series of sub-catchments. For urban sub-catchments, the 'new UK' runoff model was applied because it is suited to the mixed permeable and impermeable surfaces present in the study area, and models the change in runoff production as permeable areas become saturated during the storm event. In this case, areas drained by highway drainage and permeable surfaces were assessed as one layer of sub-catchments, while impermeable roof areas were assessed separately. In line with the methods adopted by some UK water companies, antecedent wetness was represented by the net antecedent precipitation index for the 30 days preceding an event (NAPI30), which was estimated using the 95 percentile NAPI30 value for summer and winter seasons. This was calculated using gauged rainfall from a local site for 1998 to 2007. Average evaporation values of 3 mm day^{-1} in summer and 1 mm day^{-1} in winter were used.

For rural areas, *ReFH* unit hydrograph models (Chapter 13) were used to generate inflows to the drainage system based on *FEH* design rainfall profiles. One of the major sources of uncertainty in small catchment runoff estimation can be the difficulty of establishing the true catchment area. In this study, a combination of flat topography and artificial land drainage made it difficult to determine the exact drained area at the upstream inflow to the 'WR' culvert system. Hence two possible inflow areas were modelled at this location, one assuming a 40 ha sub-catchment and one assuming a 190 ha sub-catchment.

Water surface profiles for the culvert at 'WR' are shown in Fig. 18.9 for the two possible upstream drained areas. Of particular note is a constriction from 600 mm to twin 400 mm diameter pipes downstream of a manhole at 229 m, where the culvert passes under a gas main. The twin pipes form a constriction in pipe full capacity, from 0.46 m^3 s^{-1} to 0.31 m^3 s^{-1}. It can be seen from the modelling that this constriction would lead to surcharging and surface flooding for the 100-year return period (1 per cent AEP) storm event should the larger of the assumed upstream catchment areas be the true drained area.

In addition to allowing hydrological input scenarios to be tested, the drainage model in this case was used to assess the implications of differing levels of sediment build up within the WR culvert. Table 18.5 shows the predicted volumes of flood water produced during design storms of 2–100-year return periods with the culvert modelled either in its 'clean' state (with data taken from a survey following cleaning operations) or in one of two sedimentation scenarios. The impact of allowing sediment depths to build up would not necessarily be noticed for a 2-year event. However, at a 10-year return period and above, a substantial increase in flood risk was found.

With a system-wide network model in place, it is also possible to study the implications of changes to the connectivity of the system (Fig. 18.10). In this case, a number

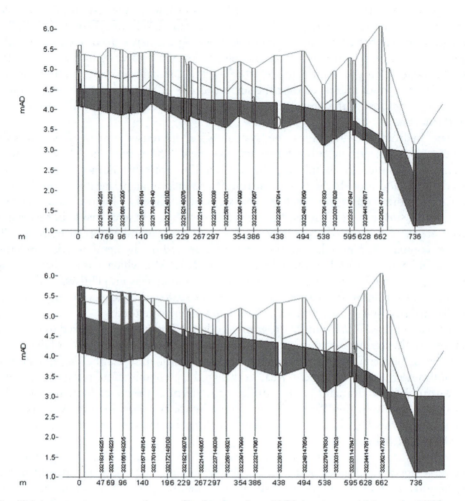

Fig. 18.9 Long section water surface profiles for the culvert 'WR' showing sensitivity under 1/100-year rain storm to assumed drained areas of 40 ha (top panel) and 190 ha (lower panel). Data used with permission of Somerset Drainage Boards Consortium.

of drainage areas were identified that had no known outfall to the main 'WR' culvert system and hence could be unable to discharge storm water effectively, leading to flood risk. These 'orphaned' drains were connected to the 'WR' culvert and the design events run for storm return periods as above (see Table 18.6). The culvert appears to have some available capacity to accept these additional flows without raising flood risk to properties during the 100-year event. The solutions tested would require major capital works and hence detailed feasibility studies, including flow survey. The drainage system model enables a set of hydrological 'design' inputs to be tested against various scenarios for management or development of the sewer network thus providing a cost-effective basis for planning and future investment.

Table 18.5 Total flood water volume over the whole modelled area for different scenarios about maintenance of the 'WR' culvert system

Return period in years	Total flood volume (m³)		
	Clean (sediment depths from post-cleaning survey)	Sediment 20% of pipe/channel height	Sediment 40% of pipe/channel height
2	463	483	512
10	1226	1237	2174
30	1765	2291	4660
100	3991	5472	9489

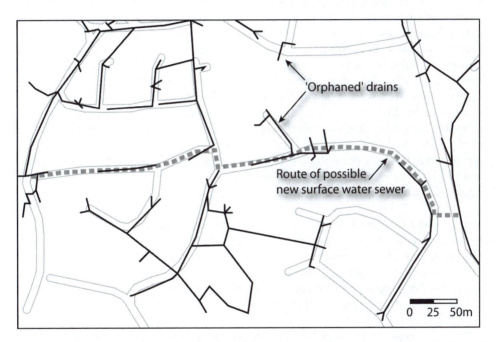

Fig. 18.10 Possible new surface water sewer (grey dotted line) to improve connectivity of 'orphaned' drains. Data used with permission of Somerset Drainage Boards Consortium.

18.7 Sustainable urban drainage systems (SUDS)

Historically, urban drainage has been based on surface drains and underground pipes, which are designed to allow water to flow away from the built-up area as quickly as possible to prevent flooding. There is then a risk that high-intensity rain storms may overwhelm the system, leading to flooding, either because the drains do not have a designed capacity to convey water at a high enough rate or because of degradation in the condition of the pipe network, possibly including sewer collapses or siltation. Of course the fact that the piped network is underground makes it all the more difficult

Table 18.6 Changes in total flood volumes over the whole modelled area for design events with two different drainage network connectivity scenarios

Return period (1 in x years)	Total flood volume (m³)	
	Network connectivity as-is	'Orphaned' areas connected
2	463	201
10	1226	464
30	1765	843
100	3991	2081

to inspect and maintain the infrastructure. Drainage systems often take foul water to waste water treatment works in a sewerage system separate from the surface runoff drains. In some cases, surface runoff is routed together with foul water through a combined main sewer. In both cases, heavy rainfall can exceed the capacity of the system causing discharge of the foul water and surface water runoff through combined sewer overflows (CSOs), leading to contaminated water entering receiving water-courses such as lakes and rivers. High sediment and solute loads can by-pass urban water-courses and be discharged directly into a downstream river, causing strong spatial and temporal variations that can affect ecological status and channel morphology (Old *et al.*, 2006) In contrast to these conventional systems, 'SUDS' are urban drainage systems that follow sustainable development principles. Sustainable drainage offers benefits brought by mitigating the adverse impacts of urban development on storm runoff, particularly through:

- reduced runoff rates, hence reduced downstream flood risk;
- encouraging groundwater recharge in areas where urban drainage might otherwise cause rainfall and runoff to be exported from the catchment too quickly;
- improving water quality by holding back overflows from CSO spills and preventing direct discharge of high concentrations of contaminated water;
- providing amenity and habitat within in the urban environment.

The basic philosophy of SUDS is to reduce the impact of development and maintain as far as possible a natural runoff regime through a 'little and often' approach. The main components of SUDS are designed to manage localised runoff throughout the developed area, rather than conveying storm water artificially and then requiring large control structures or flood storage, as might be the case in more conventional drainage systems. A comprehensive guide to SUDS, including design principles, construction, maintenance and examples is the 'SUDS Manual' published by CIRIA, formerly known as the Construction Industry Research Association (Woods-Ballard *et al.*, 2007). Typical SUDS components are:

- filter strips – wide, gently sloping vegetated areas adjacent to impermeable surfaces;
- swales – broad, shallow channels planted with grass or other vegetation to convey or store runoff and allow infiltration;

- infiltration basins – depressions similar to swales but designed for storage and infiltration rather than conveyance;
- detention basins – larger basins to provide runoff storage, usually dry though they may contain permanent ponds;
- wet ponds – basins designed to have permanent ponds of water to treat water quality, often providing wildlife or amenity value and additional runoff storage;
- constructed wetlands – ponds with shallow water and vegetation planted to enhance water quality and habitats;
- filter drains and perforation pipes – trenches filled with gravel or other permeable material typically placed along paved areas or across slopes;
- infiltration devices – temporary, localised storage of runoff to allow infiltration into the ground;
- pervious surfaces – paved surfaces that allow infiltration of rainwater into a permeable storage layer;
- green roofs – vegetated roofs planted over a drainage layer.

Fig. 18.11 shows a typical SUDS feature in an area of new housing development.

Achieving a sustainable urban drainage system requires planning, careful development and on-going management. There is a series of elements that make up the whole system and that should be considered during development planning. *Source control* refers to control of runoff at a very small spatial scale, through soakaways and filter strips or similar infiltration mechanisms, green roofs and permeable paving. At a slightly larger scale, *site control* refers to routing and attenuation of runoff from an

Fig. 18.11 Drainage ponds associated with new housing development.

area to larger soakaways or detention ponds. Larger still, runoff from a whole site or a group of developments may be routed into a wetland or balancing pond. In addition to these runoff control and attenuation strategies, reduction in runoff production can be achieved by good practice in maintenance, for example, clearing trash from car parks to prevent blockage of flow routes and rainwater harvesting.

By combining many small features in a large SUDs system, the aim is to prevent the system failing completely in the event of any one component not performing. The links in the system should include surface routing through natural swales and trenches, if possible, rather than artificial pipes. This promotes a highly distributed, attenuating drainage system that should be resilient if well maintained. A SUDS system therefore requires maintenance, such as vegetation cutting and removal of debris and sediment, to ensure long-term sustainability. The maintenance regime should be an integral part of the management plan for a development. In general, where sustainable development principles have been embraced, there is a shift from 'design' and 'periodic maintenance' of individual structures towards concepts of 'whole system' management, where the operation of the system is planned and managed over the entire life cycle of the development. This shift in thinking is enshrined in the UK in policies on planning and development appraisal. One practical example of this shift that is of relevance to urban hydrology is found in the revision of guidance on culvert design; the *Culvert Design Guide*, published in 1997 by CIRIA, has been updated and extended in the newer *Culvert Design and Operation Guide* (emphasis added) taking a somewhat wider environmental and whole life-cycle view.

In addition to their benefits for flood management, SUDS can also improve water quality through a number of processes. Many pollutants are bound to sediments, which can be trapped by SUDS features or removed by filtration through soil and using geotextile layers. Heavy metals, solid wastes, organic material, hydrocarbon and pesticides can all be removed from runoff by a combination of filtration, sediment deposition, biodegradation and plant uptake. Again, maintenance is required to ensure sustainable effectiveness, e.g. by removal of plants and replanting to stop metals being returned to the water-course and removal of sediment to prevent flushing of contaminated solids in a subsequent flood event.

18.8 Urban flood risk mapping

The sources and pathways of flooding can be particularly complex in the urban environment (Fig. 18.12). Flood risk pathways in an urban system can comprise raised defences, surface flow pathways and the below-ground drainage system. How flood flows in these systems interact in terms of conveyance and storage and the consequences of the flooding are key challenges.

A study reported by Hankin *et al.* (2008) conducted as part of *Making Space for Water* (see Chapter 16) reviewed flood risk modelling and mapping approaches for multiple sources of flooding. To produce accurate models and maps requires large amounts of data collection and flood maps generated for non-fluvial and coastal sources on a wide scale using current flow modelling and data acquisition techniques are likely to be less accurate than the existing models for fluvial and coastal flooding. Table 18.7 gives an overview of possible approaches for sources of flooding other than rivers and coasts based on the findings of the report.

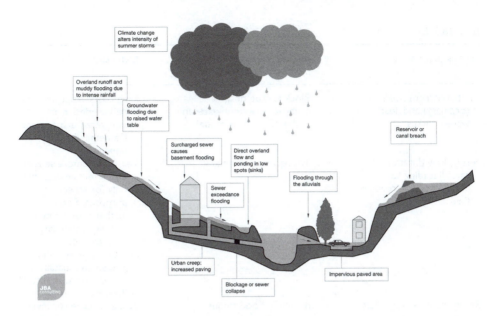

Climate change
alters intensity of
summer storms

Overland runoff and
muddy flooding due
to intense rainfall

Groundwater
flooding due
to raised water
table

Surcharged sewer
causes
basement flooding

Direct overland
flow and
ponding in low
spots (sinks)

Sewer
exceedance
flooding

Reservoir or
canal breach

Flooding through
the alluvials

Urban creep:
increased paving

Blockage or sewer
collapse

Impervious paved area

JBA
consulting

Fig. 18.12 Sources and pathways of flooding in the urban environment.

Table 18.7 Methodologies investigated by the Making Space for Water (see Chapter 16) Flooding from Other Sources Scoping Study (Hankin et al., 2008) for the feasibility of wide-scale potential flood hazard mapping

Flooding process	Methods reviewed for mapping	Comments
Direct runoff, integrated urban drainage	Blanket rainfall and 2D overland routing	Demonstrated for direct runoff with simple assumptions about drainage exceedance in terms of rainfall return period
	Different levels of integration of sewer and overland routing	Would require collation of a large amount of asset data and co-operation between data owners. Not currently feasible for a national product
Sewerage and drainage system flooding from pipe capacity exceedance	Different levels of integration of sewer and overland routing	As above
Sewerage and drainage system flooding from 'other causes' (blockage and collapse)	As above, with blockage scenarios or probability matrices	Not feasible for national probabilistic mapping in the near future, although further investigation by the DTI SAM project (http://www.dti-sam.co.uk/)
Restricted outlets from drainage systems due to high flood levels in the receiving watercourse	Control the levels in receiving watercourses from outlets of hydraulic sewer model	Would also require a process for combined probability analysis and national access to data sets as above

Table 18.7 Cont'd

Flooding process	Methods reviewed for mapping	Comments
Surcharge from small (ordinary) and 'lost' watercourses	Block escape or limit capacity of floodwater in the 2D routing model	High costs. Approach can be included in an integrated urban model if a detailed survey is undertaken
Heavy, long duration rainfall leading to ponding on e.g. roads or fields	Topographic screening	Yes, with remote sensed digital terrain data, but only for screening purposes. Requires further development of technique with larger rural dataset
	Name search	No, this approach did not give strong spatial correlation with historic flooding
Rising groundwater levels in major aquifers	Borehole flood frequency analysis	Yes, at a screening level, and subject to spatial coverage and record length of borehole level time series
	Buffer around historical records	Yes, this would give detail of the specific locality of flooding for flood warning purposes
	Geostatistical methods	Yes, but at a very coarse level. The interpolation of BFIHOST (see Chapter 13) is useful where the geology varies significantly, but for areas on the chalk, it is uniformly high
Groundwater rebound owing to rising water table and failed or ceased pumping	Theoretical modelling using the Theis equation	Screening level for known problem areas. This would require collation of abstraction records and borehole data
Groundwater flooding through alluvial aquifers	Generic groundwater modelling	Further research is needed into how the approach can be used to extend the National Flood Map in areas likely to be affected
Canal breach, over-topping or leakage	Simple risk assessment with 2D overland routing	Possible with costs of additional survey data requirements
Reservoir dam break (for reservoirs covered and not covered by the Reservoir Act 1975)	Risk ranking method for prioritisation of reservoirs in need of risk mapping; 2D overland routing where inertial effects not dominant (as for canal embankments)	Subject of current scoping study to specify a national method for reservoir inundation modelling (RIM)

2D, two-dimensional.

The combined risk from fluvial and sewer flooding can be a combination of overland flow routing and the urban drainage infrastructure. While the most detailed studies can afford to gather the data and invest the time needed to build complex urban drainage models, in many places there is also a need for more rapid assessment tools. Some suitable approaches were reviewed and tested as part of a study in 2008 (Balmforth *et al.*, 2006) in response to the Pitt Review into flooding of the summer of 2007 in the UK. The study assessed five different methods of varying complexity. This investigated how well the methods replicated known flooding for location and spatial extent. The methods tested were:

- topographic index analysis;
- 2D overland routing of a spatially uniform rainfall event;
- decoupled hydraulic sewer model and one-dimensional overland routing;
- decoupled hydraulic sewer model and 2D overland routing;
- coupled hydraulic sewer model and 2D overland routing.

A national map of surface water flood risk was produced in 2008 to obtain a rapid assessment of areas naturally susceptible to surface water flooding Design rainfall profiles for a 6-h duration, 1 per cent AEP storm were created using the rainfall depth duration frequency (DDF) models given in Vol. 2 of the *Flood Estimation Handbook* (see Chapter 9) on a 5 km × 5 km grid over the whole of the UK.

The design rainfall for each grid tile was then applied to a 2D hydraulic model to route the water over a digital terrain model (DTM). The resulting flood depths were used as an indicator of risk by applying thresholds at 0.1 m, 0.3 m and 1.0 m depth. The resulting map, although not incorporating detailed information about the operation of sewer network, has been found to be consistent with recoded flooding incidents in many places and provides a high-level screening tool for more detailed investigations; Fig. 18.13 shows an example for a city centre area where roads, railway lines and some natural topographic drainage lines at risk of flooding are visible.

18.9 Integrated urban drainage pilots

As noted in Chapter 16, there is a complex mix of responsibilities, institutional and funding arrangements that divide responsibilities between water companies, the Environment Agency, planning departments in local government, housing developers, householders and internal drainage boards. An integrated approach is seen as necessary to ensure that these different stakeholders communicate and work together to manage flood risk and water quality in the urban environment. The Government's 'Making Space for Water' programme therefore identified a need for 'integrated urban drainage management' (IUDM) as an important part of sustainable water management and development.

To develop new approaches to IUDM, Defra supported 15 pilot projects in 2007/08 which aimed to:

- understand the causes of flooding in urban areas and the best ways of managing urban drainage to reduce flooding;

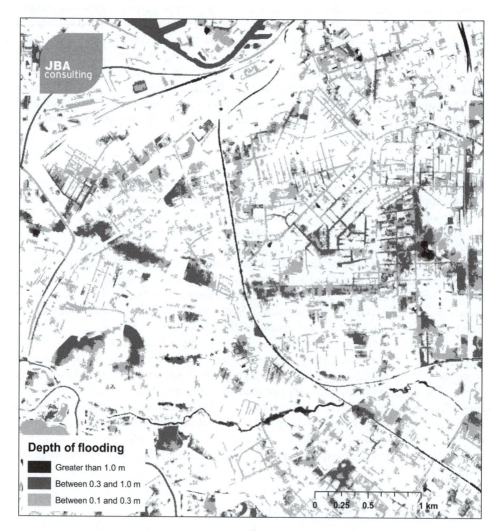

Fig. 18.13 Mapping of areas naturally vulnerable to surface water flooding.

- examine the effectiveness of partnership working between various drainage systems currently and how this partnership can be improved to find solutions to flooding problems;
- test the effectiveness of new approaches to urban flood risk management, including: the use of hydraulic models, surface water management plans, SUDS and the managed routing of drainage exceedance flows.

By 2009, experience with the IUDM pilot studies already showed benefits in terms of partnership working, applications of a range of modelling tools and development of models and risk mapping of areas vulnerable to surface water flooding of use for development control departments and emergency planning. The IUDM pilots found that measures to reduce surface water flood risk and improve water quality can be more effective and cost beneficial when developed by stakeholders working in partnership than when acting individually. For new developments, drainage strategies were produced to safeguard downstream areas, protect the development and are adaptable to climate change.

The IUDM pilot studies also identified some of the problems that hinder integrated urban drainage management. Data and models were often found to be poor, incomplete or inaccessible. A number of methods to assess flood risk were trialled but without conclusive findings about which ones are best in which situations. The current fragmented institutional arrangements and responsibilities can make it difficult to coordinate and to fund integrated improvements to urban drainage management. Also, the IUDM pilots found that urban drainage issues are often endemic, being built into past infrastructure and planning decisions, and may be very hard to improve without town centre redevelopment, which may take years to realise.

Whilst some IUDM studies focus on very detailed analysis of surface and subsurface drainage networks, others adopt a more hydrological approach based on understanding the role of the urban drainage within a whole catchment context. One of the IUDM pilot studies took this approach to consider flooding problems, which have occurred on a regular basis in the Red House Farm Estate in Gosforth within the Ouseburn catchment to the north of Newcastle. Here, a Muskingum–Cunge flow routing model (see Chapter 14) was used to represent the propagation of flood events through the Ouseburn catchment. Rather than build detailed drainage system models of the catchment, surface drains and sewers, the behaviour of the urban drainage network was represented by hydrological inputs to the catchment routing model, in the form of unit hydrograph models that were calibrated to represent discharges from detailed urban drainage network models where necessary.

Based on these model results, it was determined that levels in the Ouseburn could interact with discharges from a CSO at Red House Farm to create conditions for flooding. One option for flood management in the area is to use SUDS methods such as flood storage ponds near to areas of new housing development within the catchment. One such pond was found to attenuate design flows as shown in Fig. 18.14 for a 25-year return period design event, which in this case reduced to approximately a 10-year peak flow.

During the IUDM pilot study, the public were involved via leaflet drops, press pieces and a drop-in workshop and site tour attended by the professional partners working on the project. Public concerns about sources of runoff, channel maintenance and the functioning of the SUD ponds were factored into the study so that they could be investigated explicitly. As a result of this, the IUDM study led to proposals for more flexible use of the SUDS to provide river flood storage as well as solely urban drainage attenuation. Whilst the viability of these suggestions is subject to further detailed study, the IUDM pilot showed that the integrated approach helped in identifying such approaches that join together urban drainage, planning and river maintenance, activities generally falling under the responsibility of different bodies.

18.10 Industry guidance

There are a number of technical guides that provide up-to-date information about methods used for urban drainage modelling and also set out recommendations about

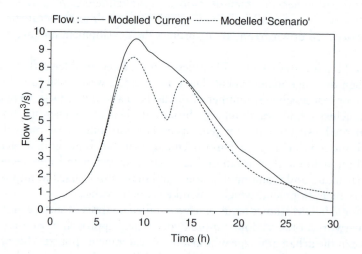

Fig. 18.14 Attenuation of flood hydrograph by sustainable urban drainage systems (SUDS) flood storage.

Table 18.8 Key guidance documents for urban drainage hydrology

The SUDS manual (CIRIA report C697), 2007	Guidance on the planning, design, construction, operation and maintenance of sustainable urban drainage systems. Includes landscaping, biodiversity, public perception and community integration, water quality treatment and sustainable flood risk management
Designing for exceedance in urban drainage: good practice (CIRIA report C635), 2006	Advice for design and management of urban sewerage and drainage systems to reduce the impacts that arise when flows occur that exceed their capacity. Includes both underground systems and overland flood conveyance. Advice on risk assessment procedures and planning for extreme events
Code of Practice for the Hydraulic Modelling of Sewers, 3rd edn (WaPUG; the Wastewater Planning Users Group, part of the Chartered Institute of Water and Environmental Management), 2002	Covers all aspects of model building for hydraulic analysis and testing, flow surveys and verification, and documentation

the choice and application of methods as recommended by relevant technical working groups within the UK water industry. Many of the guidance notes relevant to urban drainage have been produced by CIRIA. Three key documents of direct relevance for hydrologists are given in Table 18.8.

In some cases the specific guidance contained within these documents has been based on a combination of scientific, practical and historical considerations. There can therefore be variations in the advice given to practitioners in different situations. Ultimately, the hydrologist should form a clear understanding of the assumptions and data that are used in the various methods so as to be able to judge their suitability in different circumstances.

For example, some guidance notes recommend that for estimating design peak flow rates for small rural ('greenfield') catchments of less than 200 ha (2 km^2) the methods for small catchment flood estimation published in Institute of Hydrology Report No. 124 (IH124; Marshall and Bayliss, 1994) should be used, whereas for larger catchments, the *FEH* is recommended. The IH124 formulae have the advantage of drawing on a sample of gauged data specifically from smaller catchments. However, despite this empirical base, analysis of the results presented in IH124 shows that the peak flow relationships obtained for small catchments were not unequivocally superior to those of similar models available at the time based on the *Flood Studies Report* (Natural Environment Research Council, 1995; e.g., when compared with the existing FSR (Supplementary Report 16) empirical equation for time-to-peak, the IH124 results were better in about half of the 24 smallest catchments and worse in others).

The newer *FEH* methods are designed to allow a hydrologist to incorporate the most up-to-date and more representative available data for a particular catchment. The required catchment descriptors can be defined with due care down to a scale of approximately 50 ha, and, given a reasoned judgement about the representativeness of the data, an *FEH* peak flow estimate could be scaled by area to provide runoff values for smaller areas that are consistent with those for the wider catchment. Statistical analysis of the residuals about the *FEH* equations for estimation of median annual flood show little sign of bias or strongly increasing uncertainty towards the smaller catchment areas and so the empirical evidence does not make it obvious that the model is unsuitable for use in areas smaller than 200 ha. This is not to say that one method is universally or automatically better than another but that, with a good understanding of the methods and the catchment area, a hydrologist may consider a range of applicable methods and choose appropriately between them.

Note

1 http://www.ciwem.org/groups/wapug/

References

Ackers, P. (1969) *Tables for the Hydraulic Design of Pipes*, 2nd edn. Hydraulics Research Station, HMSO, London (3rd edn. 1977).

Balmforth, D., Digman, D., Butler, D. and Shaffer, P. (2006) *Integrated Urban Drainage Pilots: Scoping Study*. Department for Environment, Food and Rural Affairs, London.

Bettess, R., Pitfield, R. A. and Price, R. K. (1978) A surcharging model for storm sewer systems. In: Helliwell, P. R. (ed.) *Urban Storm Drainage*, Pentech Press, London, pp. 306–16.

Brun, S. E. and Band, L. E. (2000) Simulating runoff behavior in an urbanizing watershed. *Computers, Environment and Urban Systems* 24, 5–22.

CIRIA (2007) *The SUDS Manual* (CIRIA report C697). Classic House, London.

CIRIA (2006) *Designing for Exceedance in Urban Drainage: Good Practice* (CIRIA report C635). Classic House, London.

Endreny, T. A. (2005) Land use and land cover effects on runoff processes: urban and suburban development. In: Anderson, M. G. (ed.) *Encyclopedia of Hydrological Sciences*. John Wiley & Sons, Chichester, pp. 1775–1804.

Hankin, B., Waller, S., Astle, G. and Kellagher, R. (2008) Mapping space for water: screening for urban flash flooding, *Journal of Flood Risk Management*, 13–22. doi: 10.1111/j.1753-318X.2008.00003.x.

Institute of Hydrology (1999) *Flood Estimation Handbook*. Institute of Hydrology, Wallingford.

Marshall, D. C. W. and Bayliss, A. C. (1994) *Flood estimation for small catchments*. Institute of Hydrology, Report No. 124. Institute of Hydrology, Wallingford.

National Water Council (NWC). (1976) *Working Party on the Hydraulic Design of Storm Sewers: A Review of Progress, March 1974–June 1975*, 26 pp. NWC, London.

National Water Council (NWC). (1981) *Design and Analysis of Urban Storm Drainage: The Wallingford Procedure*. 5 vols. NWC, London.

Natural Environment Research Council (1975) *Flood Studies Report*. 5 Vols.

Old, G. H., Leeks, G. J. L., Packman, J. C., Smith, B. P. G., Lewis, S. and Hewitt, E. J. (2006) River flow and associated transport of sediments and solutes through a highly urbanised catchment: Bradford, West Yorkshire. *Science of the Total Environment*, 360, 98–108. doi:10.1016/j.scitotenv.2005.08.028.

Packman, J. C. (1980) *The Effect of Urbanisation on Flood Magnitude and Frequency*. Institute of Hydrology, Report No. 63. Institute of Hydrology, Wallingford, 117 pp.

Robinson, M., Boardman, J., Evans, R., Heppell, K., Packman, J. and Leeks, G. (2000) Land use change. In: Acreman, M. C. (ed.) *The Hydrology of the UK: Study of Change*. Routledge, London, 303 pp.

Rose, S. and Peters, N. E. (2001) Effects of urbanization on streamflow in the Atlanta area (Georgia, USA): a comparative hydrological approach. *Hydrological Processes* 15, 1441–1457.

Tabony, R. C. (1980). Urban effects on trends of annual and seasonal rainfall in the London area. *Meteorological Magazines* 109, 189–202.

Transport and Road Research Laboratory (TRRL) (1976) *A Guide for Engineers to the Design of Storm Sewer Systems*. TRRL Road Note 35, 30 pp.

Watkins, L. H. (1962) *The Design of Urban Sewer Systems*. DSIR Road Research Technical Paper No. 55. HMSO, London, 96 pp.

Woods-Ballard, B., Kellagher, R., Martin, P., Jefferies, C., Bray, R. and Shaffer, P. (2007) *The SUDS Manual*. CIRIA C697. CIRIA. Classic House, London.

Chapter 19

Hydrology, climate and catchment change

19.1 Global climate change

Climate change is arguably the most important environmental problem facing the world's population. In 1988, the Intergovernmental Panel on Climate Change (IPCC) was established through a resolution of the United Nations General Assembly

> with concern that the emerging evidence indicates that continued growth in atmospheric concentrations of 'greenhouse' gases could produce global warming with an eventual rise in sea levels, the effects of which could be disastrous for mankind if timely steps are not taken at all levels (United Nations General Assembly, 1988).

The IPCC has since helped create a collective knowledge of climate change that shapes public policy for sustainable development and raises awareness of the threats to human security posed by a changing climate and, therefore, the need for action to avoid such threats in the future. The IPCC 4th Assessment 'Scientific Basis' report (IPCC, 2007b) describes some key features of the many climate models in use around the world, along with the main uncertainties and improvements that are being made. In the UK, much of the climate change information used for impacts studies derives from models developed at the Met Office Hadley Centre, which have been summarised by Pope *et al.* (2007).

19.1.1 Assessment of global climate change

The IPCC has released a series of 'Assessment Reports' to provide an integrated view of climate change, updated to reflect developments in scientific knowledge and in policies about climate risk. The Fourth Assessment Report (AR4) was presented in 2007 (IPCC, 2007a). It included summaries of observed changes in climate indicators, projected future changes, projected future impacts, causes of climate change and discussion of options for action to manage the risk.

The AR4 considers that warming of the climate system is unequivocal, as is now evident from observations of increases in global average air and ocean temperatures, widespread melting of snow and ice and rising global average sea level (see Fig. 19.1). Furthermore, impacts of warming are being detected in most continents and oceans, although much of the observational evidence uses data collected since around 1970 and so it is not always possible to set measurements in their longer term context. Indeed, one of the difficulties in understanding evidence of climate change impacts, especially at regional (i.e. sub-continental) scales, is that any systematic change can be

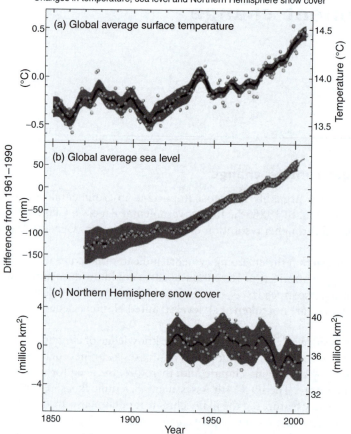

Changes in temperature, sea level and Northern Hemisphere snow cover

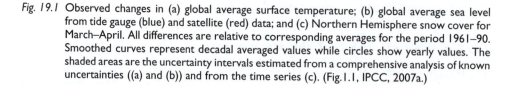

Fig. 19.1 Observed changes in (a) global average surface temperature; (b) global average sea level from tide gauge (blue) and satellite (red) data; and (c) Northern Hemisphere snow cover for March–April. All differences are relative to corresponding averages for the period 1961–90. Smoothed curves represent decadal averaged values while circles show yearly values. The shaded areas are the uncertainty intervals estimated from a comprehensive analysis of known uncertainties ((a) and (b)) and from the time series (c). (Fig.1.1, IPCC, 2007a.)

combined with natural variability and also obscured by adaptations or non-climate factors.

19.1.2 Attribution

When changes in climate are observed or suspected, there remains a question of *attribution*. Is the change a result of human activity or part of a cyclical or random natural process? In AR4 the IPCC considers that

> Most of the observed increase in global average temperatures since the mid-
> 20th century is very likely due to the observed increase in anthropogenic

greenhouse gas concentrations. It is likely that there has been significant anthropogenic warming over the past 50 years averaged over each continent (except Antarctica).

The conclusion is supported by comparison of multiple climate model simulations driven either by natural forcing factors alone (e.g. solar activity) or with the addition of anthropogenic forcing, mainly greenhouse gas (GHG) emissions. This attribution depends on the accuracy of the climate model outputs, which are recognised as being uncertain. However, the evidence for anthropogenic increases in GHGs is considered compelling.

Modelling at smaller scales than the global or continental scale remains difficult. This has implications for hydrologists because most analysis of the impacts of climate change in hydrology will be required at smaller regional and catchment scales.

19.1.3 Global projections

Changes are projected to continue into the future. It is important to note that climate change models are not in general used to make 'forecasts' of future climate, but rather they predict what may occur conditional on a set of assumptions about the future trajectory of GHG emissions and other factors. These emissions projections are known as scenarios. Many climate studies use a standard set of scenarios derived from the IPCC *Special Report on Emissions Scenarios* (IPCC, 2000) known as the 'SRES' scenarios, which make differing assumptions about future social and economic development. These scenarios are very difficult to evaluate in terms of probability, but can be regarded as a set of plausible story lines about what the future might look like under different circumstances. The IPCC scenarios project increases in atmospheric CHG concentrations of between 25 and 90 per cent from 2000 to 2030. Thereafter some scenarios see a reduction in concentrations to around year 2000 levels or below, whereas others indicate further increases. The scenarios lead to a range of projections of future warming, generally of around 0.2°C per year global mean temperature to 2020 but then diverging, with a likely range of between +1.1 and +6.4°C change for the decade 2090–99 relative to 1980–99.

Uncertainty is an issue of enormous importance and complexity in climate science, arising from inherent unpredictability (e.g. unknown future political movements, unpredictable economic cycles, chaotic behaviours of natural systems), lack of knowledge (e.g. incomplete descriptions of physical processes in models) and 'value uncertainty' (e.g. measurement errors, inappropriate spatial or temporal averaging, missing data) . One of the responses to the uncertainties is to evaluate not just a single best estimate of the system but to consider a range of possible outcomes. There are broadly two ways in which this can be done. A scenario-based approach focuses around assessment of vulnerability to a number of discrete 'stories' about the future, which tests robustness against different possible outcomes, without believing any particular one to be a forecast. A probabilistic approach is conceptually different. It attaches probability distributions to possible future changes to help inform decisions about plans and actions.

Although the two approaches are conceptually different, there may be many similarities in practice in the methodologies used. The main difference is in the mind set that each approach implies. It is worth considering the circumstances in which a probability-based approach is appropriate, given that it is not possible to characterise fully the statistical distribution of possible futures. There is a conceptual challenge in assigning likelihoods or probability weights to emissions scenarios, and also the practical issues of deciding what sources of uncertainty to include.

Table 19.1 illustrates likely impacts of these changes into the late twenty-first century based on the SRES scenarios and interpretation of model outputs. It is notable that very many of the impacts are relevant for a hydrologist, including water resources, flood risk, ecological changes, health and social impacts of floods, and drought and impacts on agriculture and sea level. A future climate may alter river flow regimes, with potential changes in low flows, water yields and flood frequency. The hydroecology of rivers and other water bodies could change because of these changes in flow regime and also increases in water temperature. There may also be changes in the patterns and scale of water demand for domestic consumption and irrigation. There is therefore a need for more localised climate change analysis.

19.2 UK climate change impacts on river flows

Advances in climate modelling technology have gradually increased the spatial and temporal resolution of climate models. Global models are based on numerical simulation of coupled atmospheric and ocean fluxes, and increasingly detailed representations of other earth system components such as aerosols, dust, sea ice, soil moisture, river flows, biological carbon cycles and atmospheric chemistry. At the time of the IPCC 4th Assessment Report, global models of this type have been run with spatial resolutions of 0.3° latitude or longitude (IPCC, 2007b), corresponding to approximately 33 km over the British Isles.

For climate impacts studies, another strategy is to combine a global atmosphere–ocean general circulation models (GCMs) with a higher resolution regional climate model (RCM). The RCM takes its boundary conditions from data generated by the GCM. In 2009 the UK Climate Impacts Programme (UKCIP) released a suite of climate projections, UKCP09,[1] based on this 'dynamic downscaling' approach with 25-km resolution grids that include greater regional detail and also aggregated climate statistics computed for administrative regions and river basins. One advantage of the approach is that an RCM may be computationally cheaper to run than a high-resolution, fully coupled GCM or 'earth system' model. This allows for a greater number of model simulations to be performed, which in turn permits the use of ensemble prediction methods to investigate the uncertainty in the simulations. A typical output from UKCP09 is shown in Fig. 9.2 in terms of projected changes in winter rainfall in 2050. The changes are shown for three different emissions scenarios with probabilities derived from the ensemble predictions. UKCIP consider that changes outside the 10–90 per cent quantiles can be considered as unlikely, while the 50 per cent quantile can be taken as a best estimate.

19.2.1 Flood flow impact scenarios

One of the first studies to assess the impacts of climate projections and uncertainty in flood flows in the UK was by Cameron *et al.* (2000). They showed that the shift in estimates of the 0.01 annual exceedance probability flood ($T = 100$, see Chapter 11) might

Table 19.1 Examples of possible impacts of climate change due to changes in extreme weather and climate events, based on projections to the mid- to late twenty-first century. These do not take into account any changes or developments in adaptive capacity. The likelihood estimates in column 2 relate to the phenomena listed in column I. (IPCC, 2007a, Table 3.2.)

Phenomenon[a] and direction of trend	Likelihood of future trends based on projections for the twenty-first century using SRES scenarios	Examples of major projected impacts by sector			
		Agriculture, forestry and ecosystems	Water resources	Human health	Industry, settlement and society
Over most land areas, warmer and fewer cold days and nights, warmer and more frequent hot days and nights	Virtually certain[b]	Increased yields in colder environments; decreased yields in warmer environments; increased insect outbreaks	Effects on water resources relying on snowmelt; effects on some water supplies	Reduced human mortality from decreased cold exposure	Reduced energy demand for heating; increased demand for cooling; declining air quality in cities; reduced disruption to transport due to snow and ice; effects on winter tourism
Warm spells/heat waves. Frequency increases over most land areas	Very likely	Reduced yields in warmer regions due to heat stress; increased danger of wildfire	Increased water demand; water-quality problems, e.g. algal blooms	Increased risk of heat-related mortality, especially for the elderly, chronically sick, very young and socially isolated	Reduction in quality of life for people in warm areas without appropriate housing; impacts on the elderly, very young and poor
Heavy precipitation events. Frequency increase over most areas	Very likely	Damage to crops; soil erosion, inability to cultivate land due to waterlogging of soils	Adverse effects on quality of surface and groundwater; contamination of water supply; water scarcity may be relieved	Increased risk of deaths, injuries and infectious, respiratory and skin diseases	Disruption of settlements, commerce, transport and societies due to flooding; pressures on urban and rural infrastructures; loss of property
Areas affected by drought increases	Likely	Land degradation; lower yields/crop damage and failure; increased livestock deaths; increased risk of wildfire	More widespread water stress	Increased risk of food and water shortage; increased risk of malnutrition; increased risk of water- and food-borne diseases	Water shortage for settlements, industry and societies; reduced hydropower generation potentials; potential for population migration
Intense tropical cyclone activity increases	Likely	Damage to crops; windthrow (uprooting) of trees; damage to coral reefs	Power outages causing disruption of public water supply	Increased risk of deaths, injuries. water- and food-borne diseases; post-traumatic stress disorders	Disruption by flood and high winds; withdrawal of risk coverage in vulnerable areas by private insurers; potential for population migrations; loss of property
Increased incidence of extreme high sea level (excludes tsunaims)[c]	Likely[d]	Salinisation of irrigation water, estuaries and freshwater systems	Decreased freshwater availability due to saltwater intrusion	Increased risk of deaths and injuries by drowning in floods; migration-related health effects	Costs of coastal protection versus costs of land-use relocation; potential for movement of populations and infrastructure; also see tropical cyclones above

SRES,

[a] See Intergovernmental Panel on Climate Change (2007b) Table 3.7 for further details regarding definitions.

[b] Warming of the most extreme days and nights each year.

[c] Extreme high sea level depends on average sea level and on regional weather systems. It is defined as the highest 1% of hourly values of observed sea level at a station for a given reference period.

[d] In all scenarios, the projected global average sea level at 2100 is higher than in the reference period. The effect of changes in regional weather systems on sea level extremes has not been assessed.

UKCP09 predicted change (%) in winter precipitation in 2050 at 10%, 50% and 90% probability levels

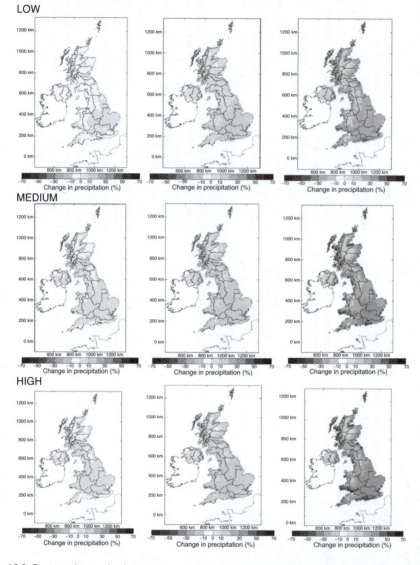

Fig. 19.2 Projected quantile changes in winter precipitation in the UK under three different emisisions scenarios (UKCIP, 2009).

be small relative to the uncertainty in estimating the correct magnitude under current conditions. The risk of over-topping a flood defence might, however, be reduced significantly. More recently, outputs from the Hadley Centre RCM (HadRM3) generated on a ~25-km cell-size grid were used by Kay *et al.* (2006) to investigate the impact in the 2080s of one climate change scenario on flood flows for 15 catchments in Great Britain (Fig.19.3). The study, like most impact studies in hydrology, used time series outputs of the relevant climate model variables (e.g. grid cell average rainfall, temperature, wind speed) to drive a calibrated rainfall–runoff model. This 'continuous simulation' approach is necessary to account for the variation in the forcing variables that occurs at different temporal scales. For example, it allows the soil moisture state in the hydrological model to vary in response to both seasonal changes in water balance and also changing conditions within a sequence of rain storms. Annual average rainfall increased over all but one of the catchments and eight showed an increase in flood frequency at most return periods. However, modelled peak river flows of a given return period were found to decrease for a number of the catchments in the south and east of England, despite an increase in winter mean and extreme rainfall because of increased summer and autumn soil moisture deficits, related to temperature change. Other catchments, further north or west, showed increased peak flows, with changes of over 50 per cent at the 50-year return period in some cases.

An important point to note from this analysis is that the impacts of climate change scenarios at a catchment level may be quite variable geographically. Care needs to be taken when interpreting these results, as they are based on a single RCM experiment (using driving data from one GCM under a single-emissions scenario). The results therefore represent just one 'snapshot' or realisation of many plausible future scenarios. Other emissions scenarios and GCM or RCM simulations may give quite different results, and ensemble runs would ideally be required to understand the sampling uncertainty.

19.2.2 Ensemble simulation and probabilistic climate projections

Ensemble simulations involve producing a group of parallel model simulations to characterise uncertainty (see discussion papers in Collins and Knight, 2007). Ensembles can be made with the same model but different initial conditions to characterise the uncertainty associated with climate variability, or by several models to represent the impact of uncertainty about choice of model formulation. More detailed ensemble simulations may also vary model parameters systematically to explore the modelling uncertainty. Run time is a major constraint for these methods because, in principle, an impractically large number of simulations may be needed to provide a stable estimate of the uncertainty. One way around this problem is to make a limited number of simulations with complex climate models and then use the results to 'train' a simpler statistical model emulator (Rougier et al., 2007).

This was the approach taken in producing the UKCP09 projections introduced earlier. The UKCP09 data derives from ensemble simulations, which provide not only a single trajectory for each climate model output but also a probability distribution. The probabilistic projections reflect some of the uncertainty associated with the climate model used to produce most of the data by varying its parameter systematically over several hundred simulations. The simulations are conducted for a period including both historical and future dates, and the results weighted to reflect how well each ensemble member agrees with real historical observations. In addition, a limited number of simulation results from other climate models are incorporated to recognise, to some degree at least, the uncertainty apparent in the differences between climate models.

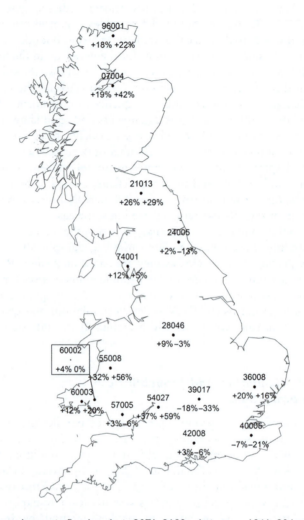

Fig. 19.3 Percentage changes in flood peaks in 2071–2100 relative to a 1961–90 baseline simulated by Kay *et al.* (2006) using regional climate model data to drive a rainfall runoff model. Five digit numbers are UK National Water Archive gauging station catalogue numbers, percentages give change for a 10 year (left) and 50 year (right) return period. Reprinted from Journal of Hydrology, Volume 318, Kay, A. L., Jones, R. G. and Reynard, N. S., RCM rainfall for UK flood frequency estimation. II. Climate change results, pages 163–172, Copyright (2006), with permission from Elsevier.

Within UKCP09, a *weather generator* has been provided to simulate daily or hourly time series of weather variables corresponding to the future scenarios at 5-km grids across the country (Kilsby *et al.*, 2007). This includes a stochastic rainfall model of the type outlined in Section 9.8 with parameters based on historical data that are then modified according to the UKCP09 predictions of percentage changes in monthly precipitation.

The probabilistic projections are made for specific GHG emissions scenarios, but the likelihood of these scenarios is not known and so the probability distributions of the model results are entirely conditional on the choice of scenario. It is therefore useful for hydrologists to understand that any analysis of impacts based on the probabilistic data will not give a predictive probability, but rather an indication of some (not all) of the uncertainty about the modelled impacts for a given scenario.

19.2.3 UK guidance on climate change and flood estimation

As noted above, modelling climate change impacts on river flows, and particularly flood flows, is facilitated by the use of a continuous water balance accounting model and long series of baseline and 'future climate' weather data. Although continuous simulation approaches have been applied in practice for some flood risk studies in the UK (Faulkner and Wass, 2005), it remains the case that for most studies the design event or statistical methods such as *FEH* and *ReFH* (see Chapter 13) are used. Hence, for most flood studies in practice, a simpler approach is needed.

The method currently adopted is to apply a simple change factor to design flow estimates as a sensitivity test within the scope of the Defra Project Appraisal Guidance (PAG3 – Economic Guidance). The rationale and choice of sensitivity test is set out in the document '*Supplementary Note to Operating Authorities – Climate Change Impacts, October 2006*' (Defra, 2006), which covers both coastal and fluvial flood risk. The guidance note gives 'indicative sensitivity ranges' for key relevant variables, based on research on impacts models and on reference scenarios developed by the UK Climate Impact Programme[2] (Hassell *et al.*, 2002; Table 19.2). The guidance was set as a precautionary upper envelope for changes in river flows over the next 50 years, and was applicable to all regions of Britain. This has since been extended to 2115, reflecting a lack of scientific evidence to suggest any alternative figure.

The approach greatly simplifies the patterns of change found in more detailed regional studies. Research will continue to assess the suitability of this guidance and it may be that, in future, a more sophisticated, regionally based analysis emerges. This is likely to include testing earlier guidance against a range of projections of future flooding that reflects uncertainties associated with the climate modelling.

19.2.4 Water resources

Regulators and water companies need to be able to undertake assessments of the potential impacts of climate change on water resources over coming decades, e.g. in the assessment of yields in a Water Resources Management Plan (see Chapter 17).

Table 19.2 Indicative sensitivity ranges for climate change (Defra, 2006)

Parameter	1990–2025	2025–55	2055–85	2085–2115
Peak rainfall intensity (preferably for small catchments)	+5%	+10%	+20%	+30%
Peak river flow (preferably for larger catchments)	+10%	+20%		

The most common method for climate change impact assessment is to use regionally averaged change factors for the key variables such as rainfall, potential evapotranspiration, runoff and recharge. This approach was used for assessment of climate change impacts on water resources using the UKCIP02 scenarios mentioned earlier. These scenarios are applied using a simple perturbation method whereby the observed series of climatic inputs to a hydrological model are changed proportionally according to the UKCIP02 monthly factors. The approach is deterministic, with uncertainty only reflected in the availability of several different scenarios of future climate.

Arnell (2004) examined the impacts of the UKCIP02 scenarios on river flows by perturbing a baseline daily series of weather data for 1961–90 using change factors derived from the climate scenarios for the 2020s, 2050s and 2080s. The modified data were rainfall and weather variables used to model evaporation. The baseline and perturbed data were used to drive a daily water balance accounting model to examine changes in river flows. Substantial changes were found for mean annual flow and also the Q95 low flow (flow exceeded 95 per cent of the time; see Chapter 11). The analysis was based on one set of emissions scenarios used to drive one family of climate models and one type of hydrological model.

The UKCIP02 analysis expressed change factors as 'block' changes for each month, but had no information about variability at other time scales, e.g. daily or inter-annual. The block change factors were sufficient to investigate changes in mean flows. However, as would be expected in a simulation methodology, variability in the time sequences would tend to increase the variance in the simulated quantities. For example, Arnell (2003) had found that including relative variability within the weather data, on annual or decadal scales, made projected decreases in low flows more extreme. The lack of information about short-term variability in the change factor approach is particularly important for simulation of impacts on flood flows, where both the seasonal climate and short-term rainfall intensity variations affect the distribution of extreme high flows. This is why the high-flow impact simulations discussed above were based directly on time-series outputs from a climate model rather than change factors.

The Environment Agency concluded that the UKCIP02 medium high emissions scenario would lead to a significant impact on average river flows across England and Wales by the 2050s. Winter increases of 10–15 per cent were indicated, but lower flows in most rivers from April to December, with late summer and early autumn seeing decreases of 50–80 per cent in some catchments. Overall, this could mean a drop in annual river flows of up to 15 per cent, the '... biggest challenge we need to overcome to ensure that there is enough water for people and the environment' (Environment Agency, 2008). Geographical patterns in the changes are summarised in Fig. 19.4 for the months of January and July.

This analysis has not yet, at the time of writing, been repeated using the UKCP09 probabilistic regional climate projections. However, Wilby and Harris (2006) have shown how this type of information could be used in a study involving an ensemble of four GCMs, two emissions scenarios, two statistical methods for downscaling GCM outputs, two alternative hydrological models and two distributions of hydrological model parameters to compute probability distribution functions for low flows in the River Thames for the 2080s (Fig. 19.5).

The results underline limitations of climate change impact assessments based on a single climate model or impact model. For example, applying either the HadCM3

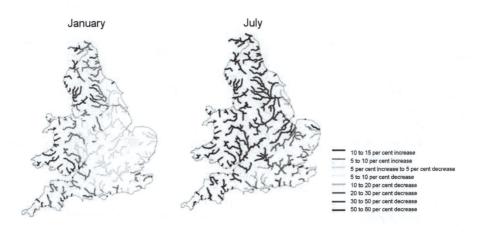

January   July

— 10 to 15 per cent increase
— 5 to 10 per cent increase
5 per cent increase to 5 per cent decrease
5 to 10 per cent decrease
— 10 to 20 per cent decrease
— 20 to 30 per cent decrease
— 30 to 50 per cent decrease
— 50 to 80 per cent decrease

Fig. 19.4 Percentage change in mean monthly flow between now and the 2050s using the medium high UKCIP02 scenario (a) January, (b) July. Adapted from Environment Agency (2008).

or CGCM2 climate model simulations in isolation would yield contrasting river flow scenarios for the Thames basin. Hence there is a need for the water industry and governments to assess multiple uncertainties in the context of water resource planning and flood risk management.

19.3 Catchment change

In many parts of the world the landscape is far from a natural environment, but is the product of hundreds or thousands of years of human management. The way in which the land is managed has important effects on hydrological processes, influencing how water moves through the hydrological cycle and the balance between storage, evaporation, runoff, river flows and groundwater. Throughout the evolution of the hydrological sciences, hydrologists have therefore been interested in understanding the way in which land management affects catchments and river flows. This interest has been driven not just by curiosity about physical processes but by the practical need to make predications about hydrological responses under different forms of management, often to assess the risk of flooding or the implications of change for water resources.

19.3.1 Urbanisation

Perhaps the most important hydrological influences of land management are related to whether the primary land use is rural or urban. Urban hydrology is considered separately in Chapter 18.

19.3.2 Agriculture

Agricultural land occupies 74 per cent of the area of the UK (based on 2005 data), as shown in Table 19.3. Throughout history there have been expansions in agricultural land use, the latest being the Victorian and mid-twentieth-century periods. In 1872,

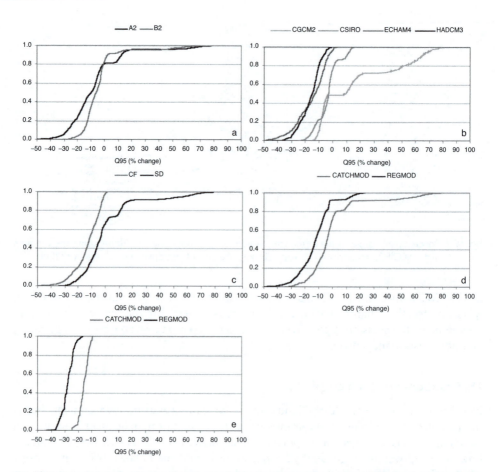

Fig. 19.5 Cumulative distribution functions of changes in low flows (the Q95 flow) by the 2080s reflecting uncertainty in (a) the emission scenario only, (b) the GCM only, (c) the downscaling technique only, (d) the hydrological model structure only, and (e) the hydrological model parameters (under the A2, HadCM3, CF scenario). Reprinted from Wilby, R. L., and I. Harris (2006), A framework for assessing uncertainties in climate change impacts: Low-flow scenarios for the River Thames, UK, Water Resources Research, 42, W02419. Copyright (2006) American Geophysical Union. Reproduced by permission of American Geophysical Union.

the ratio between the area under crops and that of improved grassland was 0.42. This dropped to 0.27 in the 1930s during the depression years, but rose again to 0.53 in 1988, driven by post-war policies to ensure security of food production. In the mid-twentieth century, agriculture in the UK became more intensive. The pre-war landscape of small fields, hedgerows and natural meandering rivers changed in many places to be replaced a more 'industrial' agricultural economy, with larger fields, compacted soils due to use of heavy machinery, land drains and modified, maintained river channels.

Since then, there has been a shift towards less intensive and more environmentally sustainable methods of food production, linked to European policy, which has

Table 19.3 Land cover by percentage and total land area of the UK (Environment Statistics Service, Department for Environment, Food and Rural Affairs)

Land cover type	Percentage cover	Notes
Crops and fallow	18.90	
Grasses and rough grazing	51.52	
Other agriculture	3.50	Set aside and other land on agricultural holdings (e.g. farm roads, buildings, gardens)
Forest and woodland	11.65	Private and state-owned land
Urban and unspecified	14.43	Includes urban and other purposes (such as transport and recreation, semi-natural environments, e.g. sand dunes, grouse moors and non-agricultural grasslands, and inland waters)
Total cover (1000s ha)		
Total land	24 251	
Inland water	325	

seen a decline in crop area from 5.3 million hectares in 1987 to the 2005 estimate of 4.6 million hectares. In the mid-1980s British agricultural policy changed following modification of the European Common Agricultural Policy (CAP) to allow the creation of agri-environment schemes. These provide incentives to farmers to adopt practices that protect and enhance the farmland environment and wildlife through the creation of environmentally sensitive areas (ESAs) and the Countryside Stewardship Scheme (CSS).

Changes in agricultural land management in the UK from the 1960s to 2005 have been reviewed by Posthumus *et al.* (2008). Whilst the total land agricultural land area has declined, there have also been important shifts in the make-up of the total land use. For instance, Fig. 19.6 shows increased areas under wheat and oilseed rape.

These changes have hydrological implications because of the associated changes in farming practices, in particular a move from spring to winter sown cereals. This causes increased working on the soil in the autumn and, with fewer fields containing stubble over the winter, a greater cover of bare soil in early winter. A significant development over the past 40 years is the increasing size, power and work rate of farm machinery. Heavy machinery has the potential to cause soil compaction, with consequent risk of increased runoff rates and soil erosion resulting in 'muddy floods' (Verstraeten and Poesen, 1999; Boardman and Poesen, 2006). However, it does not follow automatically that increased mechanisation causes greater runoff rates or sediment loads. Posthumus *et al.* (2008) note that modernisation of farming practices has included reductions in field travel by combining multiple operations in one pass and reduction of soil damage through use of wide, low-ground pressure tyres. Farmers have gained awareness of the benefits of avoiding soil compaction in terms of increased yields. The depth of compaction caused by machinery on the soil surface depends in part on loading as determined by axle weight, tyre pressures and the degree of wheel slippage (Söhne, 1958; Davies *et al.*, 1973; Voorhees *et al.*, 1986). However, soil moisture content is also important. The structural strength of soil varies with soil moisture content. Soils lose structural stability when wet. As soil moisture content increases, so does

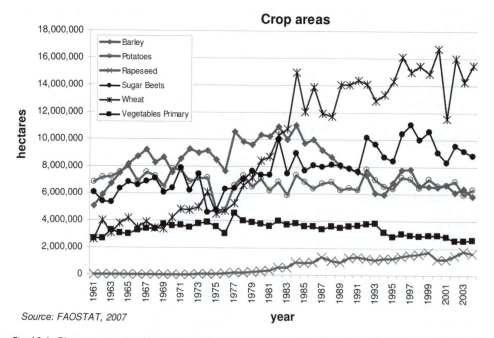

Source: FAOSTAT, 2007

Fig. 19.6 Changes in crop areas in the UK, 1961–2005 (Posthumus *et al.*, 2008). Crown copyright, 2008.

the pore water pressure, reducing friction between particles, which can then slide over each other more easily.

One of the most important aspects of agriculture in the British Isles has been the widespread use of field drainage systems to bring improvements to crop yields, grazing and the ability to work the soil with heavy machinery. Some field drainage systems are very old. For example, the 'Backcut', a 3.4-km long subsurface drain at Long Preston on the River Ribble, dates back to a Land Enclosure Act of 1799, but was later covered over in the mid-nineteenth century. There are many such old drains in formerly wet areas now used for agriculture, of varying degrees of continued effectiveness.

Drainage activity was intense in the nineteenth century. In the twentieth century the rate of field drainage continued to be high, supported by government funding, with peak rates of around 100 000 ha per year in the 1970s (Robinson *et al.*, 2000). Modern field drainage tends to comprise systems of subsurface clay or perforated pipes linked into larger arterial drains. Fig. 19.7 shows the concentration of this drainage work in the east of England and the Midlands, where water level management is often undertaken by internal drainage boards (IDBs; see Chapter 16). In the uplands, many areas have been drained by cutting ditches into peat moorland (moorland 'gripping') for sheep grazing or in the management of grouse moors.

The effect of subsurface drainage at a field scale is to lower the water table in the soil, reduce soil moisture content in the unsaturated zone and allow water to discharge more rapidly to nearby water-courses. It may reduce groundwater recharge

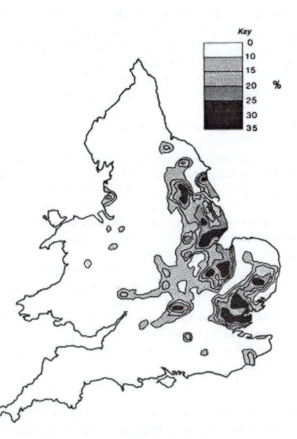

Fig. 19.7 Percentage of agricultural land in England and Wales with pipe drainage installed during the period 1971–1980. Reprinted from The hydrology of the UK: a study of change, Chapter 2, Robinson, M., Boardman, J., Evans, R., Heppell, K., Packman, J. and Leeks, G., Land Use Change, Pages 30–54, Copyright (2000) with permission from Routledge.

but increase channel flow rates in flood conditions, at least for areas with moderately permeable soils and well-connected field boundary drains (Robinson *et al.*, 2000). The importance of this and other forms of 'hydrological connectivity' is one of the themes of research into the aggregated, catchment-scale impact of changes in land management practices. In the uplands, the impacts of changes in drainage practice on runoff and downstream river flows have been found to be variable (Newson and Robinson, 1983). Lane *et al.* (2003) discuss how the connectivity of the catchment drainage system helps to understand how land management can affect the hydrological response at the catchment scale.

19.3.3 Forestry

There is a long history of investigation of the hydrological behaviour of forests, which has only intensified over the last century. Two contrasting views on the hydrological role of forests have been summarised by Jewitt (2005). At one time a widely accepted idea was that forests act like 'sponges', soaking up water during rainy spells and releasing it evenly during dry periods. In this school of thought, forests are good for water resources by promoting groundwater recharge, maintaining dry period river flows, improving water quality and moderating floods. An alternative view sees forest roots as 'pumps', where trees act more to remove water via transpiration during dry periods and reduce overall runoff yields, whilst having little influence on the occurrence of large flood events.

A number of authors and studies have produced evidence that effectively challenges the 'sponge' model. Hibbert (1967) examined data from 39 catchment monitoring experiments in different parts of the world and concluded that, in general, reductions in forest cover led to increased water yield (and vice versa), but that the response of the hydrological regime to afforestation or deforestation is very uncertain from place to place. Bosch and Hewlett (1982) updated Hibbert's review with new information and data from an additional 55 experiments. They confirmed the original findings and added greater confidence to the generalisation, noting:

> The direction of change in water yield following forest operations can be predicted with fair accuracy since no experiments, with the exception of perhaps one, have resulted in reductions in water yield with reductions in cover, or increases in yield, with increases in cover. The approximate magnitude of changes can also be estimated. Pine and eucalypt forest types cause on average 40 mm change in water yield per 10% change in cover and deciduous hardwood and scrub ~25 and 10 mm, respectively.
>
> (Bosch and Hewlett, 1982)

In the UK, there has been an increase in forest and woodland area over the past century, with a particular surge in coniferous plantations from the 1950s onwards (Fig. 19.8.)

Since the early work of Law (1956) in the Forest of Bowland in Lancashire, there have been a number of long-term UK studies of the hydrological impacts of forest management, including Plynlimon in mid-Wales (Kirby et al., 1991; Neal, 1997), Llanbrynmair in Wales (Hudson et al., 1997), the Coalburn and Irthing catchments in northern England (Robinson, 1986; Archer and Newson, 2002; Archer, 2003), Balquhidder in Scotland (Johnson and Whitehead, 1993; Calder, 1993) and Leadburn, Scotland (David and Ledger, 1988).

At the Plynlimon experiment, monitoring began to provide data in 1969, with the aim of determining whether coniferous plantations increased evaporation and decreased stream flows relative to grassland. Here, two nearby upland catchments were intensively monitored in a paired catchment experiment designed to separate the effects of land management changes from climatic variability. Hudson et al. (1999) describe how 27 years of data from this paired catchment experiment provided detailed information about the water balance in the British uplands and the implications of forestry operations for water resources.

The Plynlimon study included both catchment-scale and small-plot-scale data and covered a complete rotation cycle, from planting through clear felling to regeneration. At the beginning of the study, the water balance difference between precipitation and discharges from the areas covered with new plantation exceeded the grassland amount

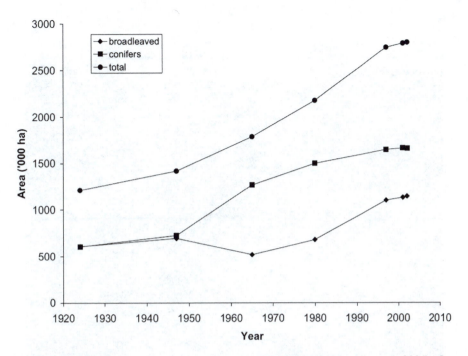

Fig. 19.8 Areas of woodland and forest in the UK, 1924–2002 (O'Connell *et al.*, 2004). Crown copyright, 2005.

by approximately 61 per cent but this difference dropped to about 18 per cent by 1985, prior to the start of clear-felling. O'Connell *et al.* (2004) interpreted interception losses from the forest canopy to be the main factor explaining the differences. During the period following clear-felling losses from the fallow forest areas appear to have been less than those of the neighbouring grassland catchment area. It is clear that the impacts of forestry operations on catchment water balance change over time along with the forest life cycle. Even so, the study 'justified the concern of the water industry in Britain that afforestation and water supply may be incompatible' (Hudson *et al.*, 1997) in respect of the impacts of upland plantations on water resources.

19.4 Studies of land management change and flood flows in the UK

19.4.1 Can land management be an effective runoff management option?

The advent of a 'whole catchment' approach to flood risk management has concentrated attention on the impact of rural land use management on catchment flood response. A comprehensive review by O'Connell *et al.* (2004, 2007a) looked at the results from over 80 studies based on field monitoring at various scales, numerical modelling and empirical data analysis. One of the key conclusions of the study was

Fig. 19.9 Field corner storage pond at Nafferton Farm, managed by Newcastle University. Runoff from the field above is collected in the pond, which discharges into surface water drains to the right of the photograph.

that there is substantial evidence that changes in land use and management practices affect runoff generation at the local scale, but only limited evidence was found that local changes in runoff manifest in the surface water drainage network as changes in stream flow regime.

There are measures that can be taken to mitigate local flooding by delaying runoff. Examples include the maintenance of grass buffers, temporary storage ponds (Fig. 19.9), ditches and 'leaky weirs' that may be built structures (Fig. 19.10) or use willow or other natural vegetation.

An integrated approach is needed in applying these measures to obtain maximum overall benefit for flood mitigation and also in reducing erosion and promoting water quality. Nafferton Farm in Northumberland, UK, is the site of a full-scale integrated runoff management experiment trialling a combination of measures, such as those illustrated above, that can potentially be implemented cost effectively on farm land. Fig. 19.11 shows an overview of the experimental features and monitoring at the site.

These experiments have shown evidence of reducing total phosphorus loads in storm runoff by up to 77 per cent, largely by trapping suspended sediments (Jonczyk *et al.*, 2008). Runoff reduction studies are continuing to determine whether these measures

Fig. 19.10 'Leaky weirs' installed in surface water drainage ditches at Nafferton Farm to test runoff attenuation measures.

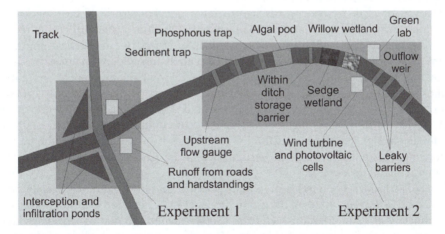

Fig. 19.11 Overview of runoff management structures installed at Nafferton Farm. (Jonczyk *et al.*, 2008. Reproduced with permission of the British Hydrological Society.)

could have an appreciable benefit for downstream flood risk. It is clear that monitoring of hydrological processes is needed at multiple scales to provide new information about hydrological responses for different magnitudes of storm event, different environments and different land management practices.

More information is expected as a result of detailed experimental studies such as the Pontbren catchment in mid-Wales, where there has been monitoring of rainfall, soil moisture content, flow in field drains and overland flow. Early results from Pontbren (Jackson *et al.*, 2008) suggest that field-scale interventions such as planting strips of trees might offer some benefits for runoff management visible at the small (c. 12 km^2) catchment scale.

19.4.2 Catchment scale analysis of land management and flood risk

The evidence of observations of soil compaction and the impacts of features such as tree belts and storage ponds suggests land management can reduce peak runoff rates at a small scale. The question is then whether small-scale interventions can be useful as part of a larger catchment-scale flood mitigation strategy. This implies a need to understand how peaks in runoff from upland catchments are routed into the downstream flow and how the effects of land management measures change with changing magnitude of storm event.

It is expected that relative effects of land use management interventions will decrease with increasing magnitude of event (Jackson *et al.*, 2008). This view would be consistent with studies of process controls on the probability distribution of flood flows suggesting that, in the most extreme events, the rainfall distribution controls the magnitude–frequency relationship for flows but that the processes associated with catchment soil moisture storage modify this distribution in the lower magnitude events (Beven, 1986; Sivapalan *et al.*, 1990).

Analysis of peak runoff records has so far produced very little firm evidence of catchment-scale impacts of land use management. O'Connell *et al.* (2007a) noted that lack of evidence does not necessarily imply lack of effect, but may reflect limitations of the data and analytical techniques. There remains no widely accepted model for predicting impacts of land management change on flood risk and there are limitations in the methods available for estimating the uncertainty in predictions. Analysis of temporal sequences of annual maximum flood peak data have failed to show convincing, systematic and statistically significant changes that could be attributed to catchment or climate change (e.g. Robson, 2002). The analysis is made difficult by the fact that different processes such as climate variability and changes in runoff production could be interacting at different scales.

There is considerable uncertainty about whether, and how, land management measures could be implemented on the ground to meet strategic flood risk mitigation aims. O'Connell *et al.* (2007a) and Beven *et al.* (2008) have found that conclusive evidence of a large-scale influence remains elusive. If further studies do show more significant evidence of an impact on flood risk, it still remains to be seen what the cost and practicality would be of implementing the changes required, or what the costs and benefits would be relative to more conventional approaches such as flood storage or raised embankments.

There is a continuing need for further monitoring of hydrological responses in the field at a range of scales and for different land management practices. Several long-term research programmes are under way to provide new information that will improve understanding of processes and may help to develop better predictive models. Examples include the CHASM (Catchment Hydrology And Sustainable Management) programme (O'Connell *et al.*, 2007b), the Pontbren programme mentioned earlier and the Sustainable Catchment Management Programme (SCaMP) managed by United Utilities.[3]

19.4.3 Practical approaches to modelling change

Some attempts have been made to provide interim approaches to model land management impacts on flood risk using readily available modelling methods. One such approach is to modify the parameters of conceptual rainfall–runoff models to reflect changes in land management practice, which has been the basis for a study at the 120 km² catchment draining through Ripon in North Yorkshire (Fig. 19.12).

Ten sub-catchment rainfall-runoff models were applied, each representing an area with similar physical characteristics, such as topography, rainfall, soils and land use. The probability distributed moisture (PDM) model was used, with each sub-catchment linked via a flood routing model (see Chapter 14) in order to simulate flows at the

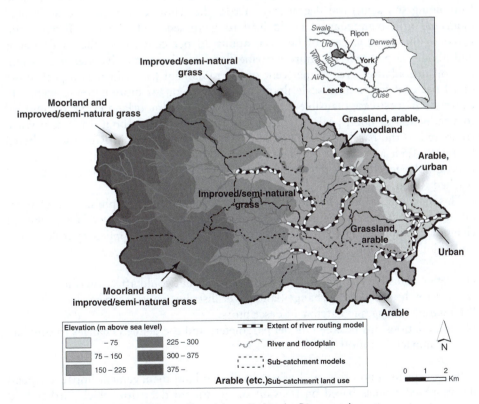

Fig. 19.12 Sub-catchment rainfall runoff models applied in the Ripon catchment.

outlet to the study catchment at Alma Weir on the River Skell in Ripon. The PDM and routing models were calibrated assuming spatially uniform parameters to the current 'baseline' conditions, which were assumed to be degraded to some extent given the general agricultural intensification and moorland improvements that have taken place over the last 50 years or so. Design events with 10-year, 50-year and 100-year return periods were simulated using a separate stochastic rainfall model fitted to rain-gauge datasets from the catchment.

Table 19.4 identifies four scenarios representing plausible changes to the land management or land use in the catchment. The scenarios reflect practices that would lead to the degradation of the soils within the catchment (compaction or crusting of vulnerable soils, conversion of improved grassland to maize production) or a change to the management of the moorland in the headwaters (maintenance of moorland grip drainage or moorland grip blocking). Impacts of the scenarios were modelled by alterations to parameters of the PDM models based on changes in hydrological descriptors suggested by Packman *et al.* (2004) and inferred links to the rainfall–runoff model parameter values given by Calver *et al.* (2005). The impact of moorland grip drainage blocking in controlling the generation and rate of runoff was also investigated via a sensitivity test in which the mean response time of fast-flow routing parameters in each moorland sub-catchment was varied by ±1 h. For each scenario, Table 19.4 shows modelled changes in peak flow rate and timing.

The modelling results indicated that the worst-case plausible degradation scenario (combining soil structural degradation across the whole catchment and additional moorland grip maintenance) could lead to increased peak flows in Ripon compared to the baseline case of between about 20 per cent for smaller scale floods and about 10 per cent for more extreme floods. A less extensive scenario assuming soil degradation over 30 per cent of the catchment led to increased peak flows of 10 per cent for smaller scale floods and 3 per cent for more extreme events. In contrast, the best-case plausible improvement scenario (moorland grip blocking) led to a reduction of flood peak magnitudes in Ripon by up to about 8 per cent when compared to the baseline case. The timing of the flood peak in Ripon was altered by up to 1.5 h as a result of the scenarios, though changes to the timing of the hydrographs generated in the moorland areas were attenuated by the time they had reached Ripon.

This modelling approach is not as sophisticated as the detailed, physics-based methods used in some research studies, but provides an approach to modelling catchment change accessible using tools that are in widespread use in applied hydrology and underpinned by empirical relationships. The key steps are:

(1) define plausible changes to land management, that could affect runoff;
(2) specify the scale of the change (land cover distribution, soil types);
(3) quantify changes in catchment descriptors;
(4) use relationships between model parameters and the defined changes to simulate the impact on runoff.

Improvements upon this, and other, studies of land management impacts depend on improved models based on relevant observational data, for which further field monitoring is vitally important.

Table 19.4 Land management scenarios for Ripon catchment

Scenario	Impact on soils and catchment	Associated PDM model parameters		Peak flow increase			Advance of peak (h)		
		Maximum soil moisture storage capacity	Runoff response time	10 years	50 years	100 years	10 years	50 years	100 years
1	Increased percentage runoff	Reduced	Reduced	10%	3%	3%	−0.25	0.25	0.25
2	Increased percentage runoff	Reduced	Reduced	21%	16%	10%	−0.25	1	1
3	Grip maintenance, loss of peat	No change	reduced	12%	9%	5%	0	0.5	0.5
4	Grip blocking	No change	increased	−1%	−7%	−5%	0	−0.5	−0.25

Scenario key:
1. Soil structural degradation applied to 50% of all degradable soils, about 35% of the catchment area.
2. Combined degradation (soil structural degradation and moorland gripping). A 'worst-case scenario'.
3. Moorland degradation (grip maintenance), 40% of the catchment area.
4. Moorland improvement (grip blocking), 40% of the catchment area.

Notes

1 http://ukclimateprojections.defra.gov.uk/
2 http://www.ukcip.org.uk/
3 http://www.unitedutilities.com/scamp.htm

References

Archer, D. (2003) Scale effects on the hydrological impact of upland afforestation and drainage using indices of flow variability: the River Irthing, England. *Hydrology and Earth System Sciences* 7, 325–338.

Archer, D. and Newson, M. D. (2002) The use of indices of flow variability in assessing the hydrological and instream habitat impacts of upland afforestation and drainage. *Journal of Hydrology* 268, 244–258.

Arnell, N. W. (2003) Relative effects of multi-decadal climatic variability and changes in the mean and variability of climate due to global warming: future streamflows in Britain. *Journal of Hydrology* 270, 195–213. doi:10.1016/S0022-1694(02)00288-3.

Arnell, N. W. (2004) Climate change impacts on river flows in Britain: the UKCIP02 scenarios. *Journal of the Chartered Institute of Water and Environmental Management* 18, 112–117.

Beven, K. (1986) Runoff production and flood frequency in catchments of order n: an alternative approach. In: Gupta, V. K. (ed.) *Scale Problems in Hydrology*. D. Reidel, Dortdrecht, pp. 107–131.

Beven, K., Young, P., Romanowicz, R., O'Connell, E., Ewen, J., O'Donnell, G., Homan, I., Posthumus, H., Morris, J., Hollis, J., Rose, S., Lamb, R. and Archer, D. (2008) *Analysis of Historical Data Sets to Look for Impacts of Land Use and Management Change on Flood Generation*. R&D Technical Report FD2120/TR, Department for Environment, Food and Rural Affairs, London.

Boardman, J. and Poesen, J. (2006) *Soil Erosion in Europe*. John Wiley & Sons, Chichester. http://doi.wiley.com/10.1002/0470859202.

Bosch, J. M. and Hewlett, J. D. (1982) A review of catchment experiments to determine the effects of vegetation changes on water yield and evapotranspiration. *Journal of Hydrology* 55, 3–23.

Calder, I. R. (1993) The Balquhidder catchment water balance and process experiment results in context – what do they reveal? *Journal of Hydrology* 145, 467–477.

Calver, A., Crooks, S., Jones, D., Kay, A., Kjeldsen, T. and Reynard, N. (2005) *National River Catchment Flood Frequency Method Using Continuous Simulation*. Defra/Environment Agency R&D Technical Report FD2106/TR, Department for Environment, Food and Rural Affairs, London.

Cameron, D., Beven, K. and Naden, P. (2000) Flood frequency estimation under climate change (with uncertainty). *Hydrology and Earth System Sciences* 4, 393–405.

Collins, M. and Knight, S. (eds) (2007) Ensembles and probabilities: a new era in the prediction of climate change. *Philosophical Transactions of the Royal Society A.,* August 15, 2007, no. 365.

David, J. S. and Ledger, D. C. (1988) Runoff generation in a plough drained peat bog in S. Scotland. *Journal of Hydrology* 99, 187–199.

Davies, D. B., Finney, J. B. and Richardson, S. J. (1973) Relative effects of tractor weight and wheel-slip in causing soil compaction. *European Journal of Soil Science* 24, 399–409. doi: 10.1111/j.1365-2389.1973.tb00775.x., published online 28 July 2006.

Defra (2006) *Flood and Coastal Defence Appraisal Guidance: FCDPAG3 Economic Appraisal Supplementary Note to Operating Authorities – Climate Change Impacts*. Updated May 2007, http://www.defra.gov.uk/.

Environment Agency (2008) *Water Resources in England and Wales – Current State and Future Pressures*. Report GEHO1208BPAS-E-E. Environment Agency, Bristol.

Faulkner, D. and Wass, P. (2005) Flood Estimation by Continuous Simulation in the Don Catchment, South Yorkshire, UK. *Water and Environment Journal* 19, 78–84. doi: 10.1111/j.1747-6593.2005.tb00554.x.

Hassell, D., Murphy, J. M., Lowe, J., Jones, R., Mitchell, T., Turnpenny, J., Lu, X., Jenkins, G. and Hulme, M. (2002) *Climate Change Scenarios for the United Kingdom: The UKCIP02 Scientific Report*. Tyndall Centre for Climate Change Research, Norwich, 120 pp.

Hibbert, A. R. (1967) Forest treatment effects on water yield. In: Sopper, W. E. and Lull, H. W. (eds) *International Symposium for Hydrology*. Pergamon, Oxford, 813 pp.

Hudson, J. A., Crane, S. B. and Robinson, M. (1997) The impact of growth of new plantation forest on evapotranspiration and streamflow in the Llanbrynmaier catchments, mid-Wales. *Hydrology and Earth System Sciences* 1, 463–475.

Hudson, J. A., Crane, S. B. and Blackie, J. R. (1999) The Plynlimon water balance 1969–1995: the impact of forest and moorland vegetation on evaporation and streamflow in upland catchments. *Hydrology and Earth System Sciences* 1, 409–427.

Intergovernmental Panel on Climate Change (IPCC) (2000) *Emissions Scenarios, A Special Report of IPCC Working Group III*. IPCC, Geneva.

Intergovernmental Panel on Climate Change (IPCC) (2007a) *Climate Change 2007: Synthesis Report. Contribution of Working Groups I, II and III to the Fourth Assessment Report of the Intergovernmental Panel on Climate Change* (Core Writing Team: Pachauri, R. K. and Reisinger, A. (eds)). IPCC, Geneva, 104 pp.

Intergovernmental Panel on Climate Change (IPCC) (2007b) *Climate Change 2007: The Physical Science Basis. Contribution of Working Group I to the Fourth Assessment Report of the Intergovernmental Panel on Climate Change* (Solomon, S., Qin, D., Manning, M., Chen, Z., Marquis, M., Averyt, K. B., Tignor, M. and Miller, H. L. (eds)). Cambridge University Press, Cambridge, 996 pp.

Jackson, B. M., Wheater, H. S., Mcintyre, N. R., Chell, J., Francis, O. J., Frogbrook, Z., Marshall, M., Reynolds, B. and Solloway, I. (2008) The impact of upland land management on flooding: insights from a multiscale experimental and modelling programme. *Journal of Flood Risk Management* 1, 71–80.

Jewitt, G. (2005) Water and forests. In: Anderson, M. G. and McDonnell, J. J. *Encyclopedia of Hydrological Sciences*. Wiley, Chichester. Chapter 186, pp 2895–2910.

Johnson, R. C. and Whitehead, P. G. (1993) An introduction to the research in the Balquidder experimental catchments. *Journal of Hydrology* 145, 231–238.

Jonczyk, J., Quinn, P. F., Rimmer, D. L., Burke, S. and Wilkinson, M. (2008) Farm Integrated Runoff Management (FIRM) plans: a tool to reduce diffuse pollution. *British Hydrological Society, Proceedings of the 10th National Hydrology Symposium*, Exeter. http://www.hydrology.org.uk/water_quality.htm

Kay, A. L., Jones, R. G. and Reynard, N. S. (2006) RCM rainfall for UK flood frequency estimation. ll. Climate change results. *Journal of Hydrology* 318, 163–172. doi:10.1016/j.jhydrol.2005.06.013.

Kilsby, C. G., Jones, P. D., Burton, A., Ford, A. C., Fowler, H. J., Harpham, C., James, P., Smith, A. and Wilby, R. L. (2007) A daily Weather Generator for use in climate change studies. *Environmental Modelling and Software* 22, 1705–1719.

Kirby, C., Newson, M. D. and Gilman, K. (1991) *Plynlimon Research: the First Two Decades*. Report 109, Institute of Hydrology, Wallingford.

Lane, S., Brookes, C. J., Hardy, R. J., Holden, J., James, T. D., Kirkby, M. J., McDonald, A. T., Tayefi, V. and Yu, D. (2003) Land management, flooding and environmental risk: new approaches to a very old question. *Proceedings of the Chartered Institution of Water and Environmental Management*, National Conference, September 2003.

Law, F. (1956) The effect of afforestation upon the yield of water catchment areas. *Journal of the British Waterworks Association* 38, 484–494.

Neal, C. (ed.) (1997) Special issue: Water quality of the Plynlimon catchments. *Hydrological and Earth System Sciences*, 1(3).

Newson, M. D. and Robinson, M. (1983) Effects of agricultural drainage on upland streamflow: case-studies in mid-Wales. *Journal of Environmental Management* 17, 333–48.

O'Connell, P. E., Beven, K. J., Carney, J. N., Clements, R. O., Ewen, J., Fowler, H., Harris, G. L., Hollis, J., Morris, J., O'Donnell, G. M., Packman, J. C., Parkin, A., Quinn, P. F., Rose, S. C., Shepherd, M. and Tellier, S. (2004) *Review of Impacts of Rural Land Use and Management on Flood Generation: Impact Study Report*. R&D Technical Report FD2114/TR, Department for Environment, Food and Rural Affairs, London.

O'Connell, P. E., Ewen, J., O'Donnell, G. and Quinn, P. (2007a) Is there a link between agricultural land-use management and flooding? *Hydrology and Earth System Sciences* 11, 96–107.

O'Connell, P. E., Quinn, P. F., Bathurst, J. C., Parkin, G., Kilsby, C., Beven, K. J., Burt, T. P., Kirkby, M. J., Pickering, A. and Robinson, M. (2007b) Catchment Hydrology and Sustainable Management (CHASM): an integrating methodological framework for prediction, Predictions in Ungauged Basins: PUB Kick-off (Proceedings of the PUB Kick-off meeting held in Brasilia, 20–22 November 2002). *IAHS Publication* 309, 53–62.

Packman, J. C., Quinn, P. F., Hollis, J. and O'Connell, P. E. (2004) *Review of Impacts of Rural Land Use and Management on Flood Generation. Short-term Improvement to the FEH Rainfall–Runoff Model: Technical Background*. Defra R&D Project Record FD2114/PR3.

Pope, V., Brown, S., Clark, R., Collins, M., Collins, W., Dearden, C., Gunson, J., Harris, G., Jones, C., Keen, A., Lowe, J., Ringer, M., Senior, C., Sitch, S., Webb, M. and Woodward, S. (2007) The Met Office Hadley Centre climate modelling capability: the competing requirements for improved resolution, complexity and dealing with uncertainty. *Philosophical Transactions of the Royal Society A* 365, 2635–2657. doi: 10.1098/rsta.2007.2087.

Posthumus, H., Morris, J. and Holman, I. (2008) *FD2120: Analysis of Historical Data Sets to Look for Impacts of Land Use Management Change on Flood Generation, A1. Technical Appendix on Drivers of Change in Rural Catchments*. R&D Project Record FD2120/PR. Department for Environment, Food and Rural Affairs, London.

Robinson, M. (1986) Changes in catchment runoff following drainage and afforestation. *Journal of Hydrology* 86, 71–84.

Robinson, M., Boardman, J., Evans, R., Heppell, K., Packman, J. and Leeks, G. (2000). Land use change. In: Acreman, M. C. (ed.) *The Hydrology of the UK: Study of Change*. Routledge, London, 303 pp.

Robson, A. J. (2002) Evidence from trends in UK flooding. *Philosophical Transactions of the Royal Society A* 360, 1327–1343.

Rougier, J. C. and Sexton, D. M. H. (2007) Inference in ensemble experiments. *Philosphical Transactions of the Royal Society A* 365, 2133–2144.

Sivapalan, M., Wood, E. F. and Beven, K. J. (1990) On hydrologic similarity 3: a dimensionless flood frequency model using a generalized geomorphological unit hydrograph and partial area runoff generation. *Water Resources Research* 26, 43–58.

Söhne, W. (1958) Fundamentals of pressure distribution and soil compaction under tractor tyres. *Agricultural Engineering* 39, 272–281, 290.

United Nations General Assembly (1988) *Protection of Global Climate for Present and Future Generations of Mankind*. Resolution A/RES/43/53, 6 December 1988.

Verstraeten, G. and Poesen, J. (1999) The nature of small-scale flooding, muddy floods and retention pond sedimentation in central Belgium. *Geomorphology* 29, 275–292. doi: 10.1016/S0169-555X(99)00020-3.

Voorhees, W. B., Nelson, W. W. and Randall, G. W. (1986) Extent and persistence of sub-soil compaction caused by heavy axle loads. *Soil Science Society of America Journal* 50, 428–433.

Wilby, R. L. and Harris, I. (2006) A framework for assessing uncertainties in climate change impacts: low-flow scenarios for the River Thames, UK. *Water Resources Research*, 42, W02419, doi:10.1029/2005WR004065.

Chapter 20

The future of hydrology in practice

20.1 The context of hydrology in practice

20.1.1 The legal context

Hydrological analysis and prediction is usually carried out for a purpose, to inform decisions about future investments and requirements in the management of water resources. These decisions are made within the context of legal requirements that stem from high-level social, economic and environmental policies adopted by governments. In the European Union (EU), there are a variety of legal directives that influence the management of water. Of these, the most important recent directives are the Water Framework Directive[1] and the Floods Directive,[2] discussed in Chapters 16 and 17 and elsewhere in this book. Both of these directives set out a time table for improving the management of water quality and floods, although the details of implementation are decided individually by each of the nations in the European Union.

In the case of the Water Framework Directive, it is required that every designated water body (rivers, lakes, groundwater, coastal and estuarine bodies) in the EU be brought to 'good ecological and chemical status' for 'sustainable use' by 2015, unless a water body is considered to be heavily modified by man in which case implementation can be deferred to 2027. The directive does, in itself, not define good ecological and chemical status, but requires that each national body with responsibility for implementation provides a definition. The implementation of the Water Framework Directive is an immense undertaking. In England and Wales alone there are 4885 designated river water bodies, 736 lakes, reservoirs and standing freshwater bodies, 300 groundwater bodies, and 233 coastal and estuarine waters.

In the case of the Floods Directive, it is required that areas at significant risk of flooding be identified by 2011, with flood risk maps being prepared by 2013 and flood risk management plans showing measures to decrease the likelihood or impact of flooding by 2015. It is also intended that the flood risk management plans should be updated every 6 years thereafter.

In England and Wales, the responsible body for implementation of the Water Framework and Floods Directive is the Environment Agency. Since privatisation of the water industry in the UK by the Water Resources Act (1995), however, there is also another legal framework that more directly affects the water utility companies. This is a 5-year review process in which the Office of Water Services (Ofwat) reviews the performance and plans of each of the water utilities through their asset management plans (AMPs). This review process has a huge commercial significance since it involves Ofwat approving the level of service charges to be paid by consumers to each of the utility companies as a consequence of past performance and future investment plans. These investment plans might include plans to reduce leakages in the pipe system, to improve effluent treatment at specific sites or mitigate the effects of waste water treatment overflows during floods, to reduce sewer flooding, and many other different possible investment strategies. In terms of both water quality and flood risk in meeting the requirements of the Water Framework and Floods Directives, these investment plans will clearly have an impact. At the time of writing, the plans for AMP5 (2010–15) had been approved in late 2009, and the process of developing the Ofwater guidelines and utility plans for AMP6 (2015–2020) is underway.

In prioritising investments to improve water quality and reduce the impacts of floods, it is necessary to have predictions about what the effects and benefits of that investment might be. Clearly hydrological analysis and models will be needed to underpin such predictions into the future, but will need to incorporate some assumptions about the expected future changes to the catchments being considered.

20.1.2 The stakeholder context

One interesting feature of the Water Framework Directive legislation was its requirement to carry out consultations with all the relevant stakeholders and professional partners for each designated water body. Consultation is also part of the Catchment Flood Management Plan process described in Chapter 16. It is increasingly recognised that improvement plans are most likely to be successful if the relevant stakeholders are involved early in the planning stage with an input to the choice of alternatives. This can, of course, be a difficult process, since different stakeholders may well have quite different and conflicting expectations of the outcomes. A classic example is in the treatment of diffuse and point source pollution in effecting improvements to water quality by reducing the quantity of nutrients reaching a river water body. Most of the diffuse sources are from agricultural inputs (and some household septic tanks); most of the point sources are from water utility and industrial effluents. In many catchments, attaining good ecological and chemical status will require improvements to both diffuse and point sources, but both will have a cost, either in loss of agricultural production or, more directly, in installing improved treatment plant for point sources. The latter are easy to identify but, once a certain level of treatment has been attained, further investment may not be cost effective relative to reducing the nutrient inputs from diffuse sources. This is a more difficult problem because there are many more individual stakeholders involved and because, even if a policy were implemented that reduced applications of nutrients to absolute minimum levels, storage in the soil and groundwater might mean that no improvements to river water quality would be seen for many years or even decades. Exploring some of these issues can also be informed by hydrological analysis and prediction. In the catchment draining to the Kaggebo Bay in southern Sweden, for example, predictive models of the hydrology and water quality have been developed in collaboration with local people and been used to explore 'what-if' scenarios of future management strategies to reduce the eutrophication of the Bay (Andersson et al., 2008).

Another example is provided by the recent work on defining a flood model and flood mitigation strategies for the Ryedale catchment in Yorkshire (Ryedale Flood Research Group, 2008). This catchment has seen floods in 1927, 1930, 1931, 1932, 1960, and in more recent times, during March 1999, summer and autumn 2000, August 2002, June 2005 and June 2007. The events of autumn 2000 and June 2007 (see Section 9.1.4) were particularly damaging. Working with local people right from the initiation of the project, a model was developed that was informed by the knowledge of local stakeholders. This was then used with groups of stakeholders to explore a variety of mitigation strategies, such as the positioning of bunds in different locations. The outcomes from the study were presented in a public exhibition in Pickering, one of the towns most affected by flooding.

In fact, stakeholder interest in many catchments in the UK is now represented by charitable bodies called River Trusts,[3] whose aim is to act in the public interest in effecting improvements to the environmental quality of the river. Many of the River Trusts have their origins in particular incidents that required many different organisations to work together to provide and implement solutions. Their continuing commitment to a particular catchment includes funding local improvements to improve habitats, such as river bank fencing projects, and educational projects. Some, such as the Eden Rivers Trust in Cumbria, have also funded research projects.

On a much larger scale, the Mersey Basin Campaign,[4] which includes both the Mersey and Ribble catchments, was founded in 1985 with a 25-year programme to improve river quality across the Mersey Basin so that all rivers and streams would be clean enough to support fish by 2010; to stimulate attractive waterside developments; and to encourage people living and working in the Mersey Basin to value and cherish their water-courses and waterfront environments. The Campaign has helped to organize funding for both urban waterfront developments and rural environmental improvements in a catchment that had a legacy of industrial decay and rivers with some of the worst water quality in the UK. Some of the successes have been spectacular, included the waterside developments at Salford Quays, which required a programme of managing water quality in the highly polluted former docks. The Campaign was always intended as a partnership of public, local authorities, industry and other interested stakeholders such as the Environment Agency.

These types of initiatives will be increasingly important in the management of the water environment in the future. Improvements to water quality and habitats are still required and require the understanding and predictive capabilities of the hydrologist to feed into the management process. Successful management of such improvements, however, requires that the stakeholders involved are on board despite the fact that they might have conflicting interests. It is therefore important that decision-making is consultative and adaptive as projects progress. Good practice in this context requires that such decisions are informed by good hydrology, in so far as that is possible in predicting future conditions often with limited data.

20.1.3 The context of future change

A clear example of this is the need to take account of future climate change in making planning and investment decisions. In planning future flood mitigation strategies, for example, it is clearly important if the risk of flooding is changing over the projected lifetime of a project. Future climate projections for the UK, based on the UKCP09 regional climate model runs,[5] suggest that there will be wetter winters in the future and generally drier summers, but perhaps with more intense local events. However, such changes will be superimposed on the natural climate variability that already makes the assessment of the frequency of flooding rather uncertain, given the relatively short records that are generally available. There is also uncertainty about whether, under a changing climate forcing, the nature of hydrological processes might change and whether man's influence in terms of changing land use and management might be more important than a changing climate. As noted earlier, it has proven rather difficult to identify significant changes in flood frequency, whether due to climate or changing land management, in the UK during the period of observations given the high variability from year to year (Robson, 2001). There do appear to be changes in the flow regimes of different catchments in the last two decades (e.g. Climent-Soler *et al.*, 2009), but it remains difficult to separate out the reasons for such changes.

Thus makes the collection of hydrological data, both water quantity and water quality, all the more important in the management of any water resource, both in developed and developing countries. If we suspect that hydrological responses are changing, then that change needs to be tracked, predicted if at all possible, and adaptive management strategies adopted. Unfortunately, in many countries, hydrological data collection has not been a high priority and there has been a decline in hydrological networks and in the quality control of the data collected. A different problem exists in some developed countries where data are collected but are not necessarily freely available to stakeholders and potential users. If data has to be paid for, even if it has been collected using public resources, then there will be fewer analyses and checking of that data. In fact, technological advances in the robustness of continuous monitoring devices and networking mean that it would be quite possible to make hydrological data available in real time on the internet. The USA sets a good example in this respect: data that are collected using state and federal resources are required to be made freely available. Common standards for the storage and exchange of data are being developed under the CUAHSI[6] programme in the USA in collaboration with the OpenMI[7] initiative in Europe.

20.2 The complexities of hydrology

20.2.1 Complexities of process

Way back in Section 1.2 of this book we briefly outlined current understanding of the processes by which hillslopes and catchments respond to rainfalls. The mix of surface and subsurface flow processes is often complex and will change over storm and seasonal timescales with the wetting and drying of the soils. Complicating features, such as the occurrence of perched water tables or preferential flows through soil macropores and percolines were also noted. Section 1.2 did not say much about the additional complexities added by human interventions, such as the maintenance of channels, introduction of agricultural field drains and the dramatic changes to hydrological processes induced by urbanisation. Even in urban areas, however, the system can be very complex. The generation of surface runoff on impermeable surfaces such as roofs, pavements and roads may be relatively simple, but some of those surfaces will drain to soakaways, some to pipe systems linked directly to rivers, some to the sewer network and some may simply discharge on to the soil.

We understand a considerable amount about these complexities in a qualitative way. The difficulty is coming up with descriptions of the processes that can be used to make quantitative predictions. As you have seen in Chapters 12 and 13, most quantitative models of runoff processes are relatively simple. They do not attempt to describe the flow processes in any detail, but concentrate on predicting the volume and timing of runoff hydrographs. Given good input data, they can often be quite successful in doing so. However, as noted earlier in this chapter, hydrologists are increasingly asked to contribute to studies of other water quality and ecological characteristics in a catchment. In these types of studies, how the hydrograph is generated might also be important. Flow pathways can influence the nutrient content, pH and sediment concentrations in the river, which can then influence the ecological status of a water body. Thus in future it is likely that the hydrologist will need to be much more concerned with flow pathways than in the past, while recognising that it can be difficult to capture the complexity of flow pathways using the measurement techniques currently available. Thus, new ways of assessing how hillslopes and catchments work might need to be developed.

20.2.2 Complexities of scale

One of the issues involved in doing so is that of scale. Many measurement techniques in hydrology are at the 'point' scale. Rain gauges, weather variables, soil moisture and water table levels are all measurements made at the point scale. Measurements of parameter values needed in hydrological models are also often made at the point or local scale. Examples are small sample measurements of hydraulic conductivity, and cross-section measurements of roughness coefficients in rivers. However, for most practical applications of hydrology, we are much more interested in what is happening at the larger catchment scale, with all the variability and heterogeneity and complexity of points within that larger scale. As yet, there is no hydrological theory that allows us to go from point scale measurements to the larger scale, only rather simple (generally linear) interpolation techniques and approximate models that assume that small scale theory can be used at larger scales. It is going to be very difficult to develop such a theory in the future; so much of the local variability and heterogeneity and complexity is simply unknowable with current measurement techniques (see Section 20.4 below).

20.2.3 Model complexity

The process and scale issues are directly related to issues of model complexity. Since we can perceive small scale complexity of flow pathways in the real hydrological system, there is a desire to reflect that complexity on hydrological models properly. The increasing availability of computer power, with faster processors and more memory, have fuelled that desire to add complexity, including processes that depend on the hydrology such as nutrient and sediment mobilisation and transport. The problem is that, as more process representations and variables are added, then more and more parameter values need to be specified in order to make a run of the model. Where those parameters are not easily estimated prior to running the model (which is the case for most parameters in hydrological models) then that means that there are more and more parameters to be estimated by calibrating the model against any available observations. There is then a danger that, unless new data sources become available, a model will be 'over-parameterised', which means that there is not enough information content in the observations to support the identification of values of all the parameters. If only a discharge record is available to calibrate a model, for example, then experience suggests that the information content will only support the identification of five or six parameters, and even then the potential for error in the input data and interactions between the values of different parameters may mean that some of these values may be rather uncertain. This therefore suggests that, in future, there may be an increasing requirement to assess the uncertainty associated with model predictions (see e.g. Pappenberger and Beven, 2006).

20.3 The uncertainties in hydrology

20.3.1 Uncertainties in data

There are many sources of uncertainty in hydrological systems, even if we think only about the water and its flow pathways. The most important, however, are uncertainties in hydrological data, since these are basic inputs to any hydrological analysis or modelling application.

Basic to any hydrological analysis is the water balance equation:

Inputs = Discharge + actual evapotranspiration + change in storage

+ other potential fluxes

Considering each of the terms of this equation in turn, it is easy to see that each of them is more or less uncertain, even when they are measured using the most up-to-date techniques outlined in the earlier chapters of this book. The rainfall, evapotranspiration and change in storage terms generally need interpolating from point measurements to the catchment scale, while discharge is not normally measured directly but is converted from water level measurements by a rating curve. As discussed in Section 10.8, snow accumulation and melt can be particularly difficult to estimate and interpolate as part of the inputs to a catchment. The other potential fluxes term, such as losses to deeper groundwater systems, are often neglected because they cannot be measured at all. Thus, if we did try to measure each of the terms of the water balance equation with a view to assessing the overall balance, then it is often the case that the equation can only be closed by allowing for uncertainty in each of the terms. Uncertainties in data are also important if we want to interpret changes over time or in space. For example, it would be very useful to have a measurement technique that would allow an assessment of the incremental discharges over successive (short) stream reaches. Where this has been done in small catchments using tracer dilution methods, it has revealed some very interesting results (see Section 11.3). Tracer dilution is not really viable for continuous measurements at larger scales and even expensive ultrasonic discharge installations are only accurate to about 5 per cent, which makes the derivation of increments over short reaches highly uncertain (even if it was feasible financially). Thus data limitations will remain important in hydrology for the foreseeable future.

20.3.2 Uncertainties in models

Uncertainties in data will feed into the predictions of any hydrological or hydraulic model, both in the calibration of model parameters and in the way uncertainties in input data affect the model prediction. In the past it has been common to optimise model parameters in calibration and use only the 'best' model found in prediction. This is still common practice today, although increasingly some attempt is made to evaluate the prediction uncertainties around the outputs from the 'best' model. We should, in fact, expect the predictions to be uncertain because of the difficulty of representing the complexity of hydrological processes and the input uncertainties. Now many Agencies, including in the UK, are moving towards risk-based management strategies and decision-making that seek to assess the probabilities and consequences of what can sometimes be a wide range of possible alternative scenarios generated using hydrological (and other) models. In this context it may be increasingly required that data and prediction uncertainties are assessed as part of a hydrological project. While this is beyond the scope of the current text, a full discussion of uncertainty estimation techniques in environmental modelling is provided by Beven (2009).[8]

20.3.3 Uncertainties in decision-making

If uncertainties in data and model predictions are significant, then they may affect decision-making within a risk-based management context. There are a number of different ways of making decisions or ranking alternative strategies within a risk-based framework where the uncertainties can be quantified (again, see Beven, 2009), but it may be the case that in some applications the uncertainties may be so large that another type of decision assessment might be appropriate, e.g. by applying the precautionary principle. Flood damages in all countries in the developed world, for example, would be must less if land planners had been more precautionary about allowing development in flood plains. In the UK alone it is estimated that property worth some £800 000 000 000, including 10 per cent of the housing stock, is at risk of flooding. The UK summer floods of 2007 led to insurance claims of some £4 000 000 000 (Environment Agency, 2010).

An important principle in making decisions in the face of uncertainty is the concept of a robust decision. This is a decision that allows a reassessment of strategy over time but which does not preclude a change to another option as a result of that reassessment. Robustness in this sense is an important part of adaptive management, leaving possible options open in case the characteristics of the system change in unexpected ways.

20.4 The future data requirements of hydrology

20.4.1 Coping with data limitations

Crucial to making real progress in hydrological analysis and prediction will be new sources of data in terms of new measurement techniques, more accurate measurement techniques or more widespread use of existing measurement techniques. Measurements are, however, expensive to both install and maintain, so that there have been increasing pressures in many countries to *reduce* the hydrological monitoring network rather than expanding into new areas of measurement. In many developing countries, the number of hydrological measurement stations, often installed by past colonial administrations or by aid programmes, is in decline. This is unfortunate when it is expected that climate and land management change will lead to hydrological change. It is necessary to monitor and analyse such changes as part of adaptation strategies. Unfortunately, there are also many countries where it is very difficult to obtain data for analysis, either because it is not in digital form or because there is a policy not to release it. As noted in Chapter 2, The Global Runoff Data Centre (GRDC) holds an extensive database of daily or monthly river flow data from 156 countries but depends on data being supplied by the different national hydrological agencies.

This tendency for sites to be lost has been counteracted to some extent by the fact that some forms of sensor have become cheaper, more robust, and more easily networked, opening up possibilities for networks of 'pervasive' and mobile sensors including, e.g. water depth and quality variables. Some internationally supported programmes, such as the global network of FLUXNET stations,[9] provide a useful data resource but at a limited number of sites. In addition, the water environment has been receiving more attention from the remote sensing community and there are a number of satellites recently launched or planned that will make more hydrologically relevant remote sensing images of different types available (see later).

However, data limitations will remain and, as noted above, must inevitably lead to uncertainty in both hydrological analysis and prediction. Depending on the type of analysis or prediction, this uncertainty might be more or less important. A quick assessment of a project will generally give an indication of how important uncertainty might be to a final decision (such as the implementation of the NUSAP assessment methodology in the Netherlands National Institute for Public Health and the Environment (RIVM); see van der Sluijs et al., 2005[10]).

Where uncertainty might be important, then it might also be the case that some resources can be made available to collect additional data to try to constrain the uncertainty in informing any decision to be made. This then presupposes some understanding how different types of data might have value in constraining uncertainties in this way. However, searching the hydrological literature suggests that little or no such guidance is available (there is some related work on rain-gauge network design to reduce uncertainty in areal estimates of rainfall inputs and the siting of additional ground water wells to improve the calibration of groundwater models). This will be an important topic for hydrological research in the future.

20.4.2 Remote sensing

One new source of data that has the potential to improve hydrological understanding dramatically, particularly in remote parts of the world where traditional data collection is sparse, is airborne and satellite remote sensing. Airborne synthetic aperture radar (SAR) and light detection and ranging (LIDAR) sensors have, for example, dramatically improved data on surface topography, and consequently led to improvements in the representation of flood plains in hydraulic models of flood inundation (although neither can penetrate the water surface so that knowledge of channel-cross section geometry is still a constraint in such applications).

Active and passive microwave sensing can also be used to provide patterns of soil moisture. The microwave emissivity of a surface is related to the dielectric constant, which changes as a result of changes of water content. This relationship, however, does not show much sensitivity when the soil is either very wet or very dry, and will also be affected by water in and on any vegetation cover for the surface. Thus, the technique will be most useful in areas where vegetation is sparse and where the water content of the soil varies significantly. Even then, since microwaves in the ranges used will penetrate only a few centimetres into the soil, only a surface soil moisture content can be derived, not changes in the profile of soil moisture. There have been attempts to model the profile under the surface in a way that uses the remote sensing data to update the model states (e.g. DeLannoy *et al.*, 2007), but these will be subject to the uncertainties of hydrological models already noted.

A further technique based on data from the two GRACE (gravity anomaly and climate experiment) satellites, launched in 2002, has been used to estimate changes in total subsurface storage of water by means of changes in gravity anomaly (after taking account of changes in atmospheric mass storage based on the ECMWF atmospheric model). This allows global maps to be produced of changes in water storage on land (and ocean elevations).[11] At a more local level, however, the utility of the images is limited by a pixel size of the order of 60 km by 60 km (the size of a large UK catchment) and uncertainty in the estimates.

At the catchment scale, as well as topographic data, remote sensing has been used to define land use categories based on the different characteristics of the surface in different wavelength bands. The spectral sensors on various Landsat satellites (since 1972) and the MODIS sensors on the NASA Terra (since 2000) and Aqua (since 2002) satellites have been used for this purpose. The derived maps give only vegetation type, however, with some uncertainty in the classification. Any hydrological value of such maps is then dependent on the specification of parameter values for each vegetation class. Thermal infrared wavelengths can also be used to detect vegetation stress and in energy balance closure models to estimate spatial patterns of evapotranspiration (see Section 10.3.2). Again, the availability of such data would have enormous value in hydrology, but energy balance closure is subject to similar uncertainties to the water balance closure noted above, such that there may be significant uncertainty associated with these estimates.

The European Space Agency (ESA) soil moisture and ocean salinity (SMOS)[12] satellite launched in November 2009 uses a passive microwave sensor to provide global coverage of surface soil moisture every 3 days, but with a pixel resolution of the order of 35 km by 35 km, so again of limited utility at local catchment scales to the hydrologist. Perhaps of greater interest will be the NASA/ESA surface water and ocean topography (SWOT)[13] satellite (planned for launch sometime in the period 2013–16), which will provide information on water surface heights with a pixel resolution of less than 100 m by 100 m every 8 days, albeit with a estimated uncertainty of ± 1–3 cm (after local averaging). For larger rivers, with long duration flood peaks, this will allow water level and spatial gradient data to be assimilated into flood prediction models. It is also suggested that spatial patterns of stream discharge can be derived from the water levels. This has already been attempted for the River Amazon using coarser resolution data from the TOPEX/POSEIDON satellites, which were originally designed for ocean observation (see Zakharova *et al.*, 2006).

There will also continue to be improvements in ground-based remote sensing, most importantly from a hydrological perspective in the estimation of rainfalls. A new generation of Doppler radar systems is allowing better treatment of anomalies in the radar signal and more accurate calibrations of the relationship between reflectivity and rainfall rate. The first Doppler system in the UK is the Thurnham radar in Kent (see Rico-Ramirez *et al.*, 2009). With a single radar, however, there is still a possibility that, when rainfalls are heavy, the signals will be attenuated sufficiently for near-radar raincells to mask those further away. One way around this is to install several cheap X-band radar systems with overlapping coverage of a catchment area from different directions (e.g. Anagnostou *et al.*, 2004). Another interesting innovation is the use of information about attenuation of existing microwave communication signals to estimate rainfall rates (Leijnse *et al.*, 2008; Zinevich *et al.*, 2008). Given the coverage for mobile communications of such networks, this has the potential to be a type of pervasive sensor system for rainfall estimation.

To summarise the application of remote sensing in hydrology, there is no doubt that the most valuable contribution has been from rainfall radar and SAR/LIDAR determined detailed topographic data. Most of the satellite-based imaging is, as yet, likely to be most useful at continental or global scales. For larger rivers, the SWOT satellite, if launched successfully, might be a very useful source of data but, because of the 100 m planned resolution, in the UK for example, it will be of greater use in estuaries and coastal waters than in most rivers.

20.4.3 Pervasive sensor systems

The SWOT satellite will, however, cost about $450 m. While this cost will be shared among a number of countries and the coverage of the satellite will be global, for local hydrological assessments there may be much more cost-effective ways of improving data coverage using networked pervasive sensors. The cost and reliability of sensors and wireless networking have been improving over time. There are now experimental installations of multiple depth and water quality sensors, providing data in real time, that are providing much more detail on water levels and quality. These networks will be useful for applications including model calibration, flood forecasting and pollution alert systems. It remains to be seen how far these types of systems might be able to constrain the uncertainties in hydrological analysis and prediction. For example, although networked measurements of water levels in a river system can be useful, to obtain discharges (which would be still more valuable), they will still need to be converted using a rating curve. This necessarily adds some uncertainty to the discharge estimates. In addition, robust, cheap and reliable sensors for some variables (such as total dissolved phosphorus, or the isotopes of water) are not yet available. Network technology is now cheap enough, however, that the additional cost of linking new sensors to a network, as they become available, is relatively small.

20.4.4 The subsurface: geophysical characterisation

One of the major problems in studying the flow processes within a catchment is obtaining information about what might be happening below the surface. Some of the geophysical techniques available for investigating the subsurface have been outlined in Chapter 6. To date, the hydrological use of these techniques has been limited to rather local sampling, although elsewhere (such as in petrochemical exploration) some techniques have been used on much larger scales. The difficulty in using such techniques (as previously noted for remote sensing) is in the inversion of the geophysical signals to derive variables of hydrological interest, such as water table depths or unsaturated water contents of a volume of soil. The inversions depend on an uncertain model of the geophysical properties and result in systems of equations that are difficult to solve uniquely. However, we shall undoubtedly learn more about the use of such geophysical methods and they are likely to receive much more attention in future. An extensive review of the application of geophysical methods in hydrological applications is given by Robinson *et al.* (2008) while more detail on specific techniques can be found in Vereecken *et al.* (2006).

20.5 The future modelling requirements of hydrology

20.5.1 Improving process representations for flow

We have already noted how current hydrological models tend to be rather simplified representations of our qualitative understanding of the complex hydrological processes in a catchment. We have also noted, however, that adding to model complexity will also generally add more model parameters that need values to be identified before a model run can be made. This can lead to models that are over-parameterised where the available observations cannot support the identification of the additional parameters. This then raises two questions that need to be addressed in the future. The first is how complex a model needs to be in practice in order to make useful predictions for decision-making. The second is how to make more observational data available to support more realistic, but more complex, process representations.

The first question depends very heavily on the type of application. For flood forecasting, for example, we are often interested only in predicting water levels at sites at risk with an adequate lead time for warnings to be issued. In larger basins in which the time delay between rainfall and peak discharge is longer than the required lead time, this can be done by rather simple models. Romanowicz *et al.* (2008), for example, use simple non-linear transfer function models that predict level directly without imposing any water balance constraints. In small basins, however, successful flash flood warnings will depend on having forecasts of future rainfall and modelling the rainfall–runoff processes for all the small catchments that might be subject to flooding by such events. This is not yet possible with any accuracy, although meteorological agencies will generally issue alerts for heavy rain that might cause local flooding on the basis of their weather forecasting systems. In this type of application, improving local predictions of rainfall over the next few hours might be more important than improving models of the runoff processes. However, representations of flow processes and pathways will become much more important as other water quality and ecological variables are considered.

Following the previous discussion of future data needs in hydrology, it is difficult to see much changing in the near future in terms of the second question. Much more needs to be learned about the value of different types of data to provide information about the nature of hydrological processes at the catchment scale. The most useful information might be observations that allow the contributions of water from different sources to the hydrograph to be determined (see the discussion of mixing models in Section 11.3). This requires, however, that it be possible to characterise those different sources (including the precipitation inputs) in some way, usually in terms of isotope or geochemical characteristics. It can then be expensive to collect and analyse sufficient samples, although recent developments have resulted in much cheaper equipment for stable isotope analysis (see, e.g. Lyon *et al.*, 2009).

20.5.2 Improving process representations for water quality and hydroecology

Similar considerations of complexity and the cost of obtaining additional observations apply to improving representations of water quality and hydroecological models. Some relatively cheap and robust sensors are available for water quality variables (see Section 8.4); others such as the system used to obtain continuous records of total phosphorus in the Oona catchment in Ireland remain expensive with significant running and maintenance costs (see Jordan *et al.*, 2007). Collecting even basic quantitative ecological data to support model development and testing can be particularly expensive but this is certainly an interesting area for future development.

In the subsurface, except for small laboratory samples, the resolution of geophysical imaging can be relatively coarse. Tracer experiments suggest that, at least in some circumstances, the preferential flow of water and pollutants can be important in the subsurface with implications for subsurface/surface water interactions, water quality and stream ecology. More observational information and better theory and models will be needed to describe such processes in the future.

20.5.3 Representing everywhere

Hydrological modelling will, however, change in the future. The statistical analysis of river flow extremes has previously concentrated on one location in space at a time, but techniques are now emerging to generalise the analysis of hydrological extremes over a very wide range of geographical scales (Heffernan and Tawn, 2004; Keef *et al.*, 2009). There is now sufficient computing power available at low enough cost to be able to invest in implementing hydrological and water quality 'models of everywhere'. Global *Earth System Models* and the Earth Simulator in Japan[14] already require hydrological models of everywhere as a boundary condition for their atmospheric and ocean models, but these are at rather coarse resolution (10 km by 10 km for the Earth Simulator) and difficult to test in detail. The International Flood Analysis System[15] initiative in the Global Center of Excellence for Water Hazard and Risk Management (ICHARM) in Japan is also attempting to provide flood alerts globally using remote sensing estimates of rainfalls and model parameters. Perhaps of greater interest are the national and large catchment hydrological models that are being developed at much finer resolutions. Such models are already being used at national scales in Denmark[16] and Sweden and are gradually being refined as new information becomes available.

Once such a model of everywhere is available, the approach to modelling becomes much more a learning process about places than about developing generic model structures (see Beven, 2007). This is because, when such a model is first applied, then most places will not have observational data available (the situation in Denmark was rather interesting in this respect: because water resources in Denmark are dominated by groundwater sources, there were a very large number of well sites across the country with water table and pumping rate data available). Thus, initial estimates of catchment or sub-catchment responses might be rather uncertain, e.g. if models of ungauged sub-catchments are based on the type of regionalisation methods similar to those used in the *Flood Estimation Handbook* as described in Chapter 13. The predictions of these models can then be visualised, critically examined and presented to local stakeholders. As noted earlier, such a system can also be used to play 'what-if' games with stakeholders to guide decisions about future catchment management. If, because of some particular event occurring, it appears that the predictions are in error for a particular place or catchment, then this will provide a driver to re-examine the model of that place, or the data used to drive the model predictions. By such a process of continuing evaluation, learning and updating, the representation of the hydrology should gradually be improved everywhere. What starts off as a rather virtual hydrological reality based on the initial characterisation of places should, in time, become a more realistic representation of the relevant processes (at the time of writing the Danish national model is already in its fourth implementation after only a few years). This type of approach also allows that quite different representations might be required for different places, and that the system might be changing over time.

This is not the way in which most hydrological models are set up to be used now, and there are some interesting issues about how to implement a flexible system to allow this type of learning process when the predictions required might vary for different types of application. In particular, the range of applications required might require predictions across a range of different scales, informed by information at larger and smaller scales, and taking account of the relevant uncertainties at each scale and the way in which the uncertainty can be constrained by any available observations. What is clear is that, as such systems start to be implemented, the future of hydrology in practice will be very interesting.

Notes

1 http://www.defra.gov.uk/ENVIRONMENT/water/wfd/
2 http://www.defra.gov.uk/environ/fcd/eufldir/default.htm
3 See http://www.associationofriverstrusts.org.uk/about/art_history_objectives.htm
4 See http://www.merseybasin.org.uk/
5 See http://ukcp09.defra.gov.uk/
6 See http://www.cuahsi.org/
7 See http://www.openmi.org/reloaded/
8 A decision tree for uncertainty estimation methods can also be found at www.floodrisknet.org.uk/methods/ and a guide to software for different methods at www.uncertain-future.org.uk.
9 See http://www.fluxnet.ornl.gov/fluxnet/index.cfm
10 see also http://www.nusap.net
11 See http://gracetellus.jpl.nasa.gov/month_mass.html
12 See http://www.esa.int/esaLP/SEM4V1NKPZD_LPsmos_0.html
13 See http://bprc.osu.edu/water/
14 See http://www.jamstec.go.jp/esc/index.en.html
15 See http://www.icharm.pwri.go.jp/research/ifas/index.html
16 See, for example, http://www.vandmodel.dk/fs_national_water_model.pdf

References

Anagnostou, E. N., Anagnostou, M. N., Kruger, A., Krajewski, W. F. and Miriovsky, B. (2004) High-resolution rainfall estimation from X-band polarimetric radar measurements. *Journal of Hydrometeorology* 43, 106–118.

Andersson, L., Olsson, J. A., Arheimer, B. and Jonsson, A. (2008) Use of participatory scenario modelling as platforms in stakeholder dialogues. *Water SA* 34, 439–447.

Beven, K. J. (2007) Working towards integrated environmental models of everywhere: uncertainty, data, and modelling as a learning process. *Hydrology and Earth System Science* 11, 460–467.

Beven, K. J. (2009) *Environmental Modelling: An Uncertain Future?* Routledge: London.

Climent-Soler, D., Holman, I. R. and Archer, D. R. (2009) Application of flow variability analysis to identify impacts of agricultural land-use change on the River Axe, southwest England. *Hydrological Research* 40, 380–393.

De Lannoy, G. J. M., Houser, P. R., Pauwels, V. R. N. and Verhoest, N. E. C. (2007) State and bias estimation for soil moisture profiles by an ensemble Kalman filter: effect of assimilation depth and frequency. *Water Resources Research* 43, W06401, doi:10.1029/2006WR005100.

Environment Agency (2010) The costs of the summer 2007 floods in England, Flood and Coastal Erosion Risk Management Research and Development Programme Project Report SC070039/R1, Environment Agency: Bristol.

Genereux, D. P., Hemond, H. F. and Mulholland, P. J. (1993) Spatial and temporal variability in streamflow generation on the wets fork of Walker Branch watershed. *Journal of Hydrology* 142, 137–166.

Heffernan, J. E. and Tawn, J. A. (2004) A conditional approach to modelling multivariate extreme values (with discussion). *Journal of the Royal Statistical Society B*, 66, 497–547.

Jordan, P., Arnscheidt, A., McGrogan, H. and McCormick, S. (2007) Characterising phosphorus transfers in rural catchments using a continuous bank-side analyser, *Hydrology Earth System Science* 11, 372–381.

Keef, C., Tawn, J. and Svensson, C. (2009) Spatial risk assessment for extreme river flows. *Journal of the Royal Statistical Society C (Applied Statistics)* 58. 0035–9254/09/58000.

Leijnse, H., Uijlenhoet, R. and Stricker, J. N. M. (2008) Microwave link rainfall estimation: effects of link length and frequency, temporal sampling, power resolution, and wet antenna attenuation. *Advances in Water Resources* 31, 1481–1493.

Lyon, S. W., Desilets, S. L. E. and Troch, P. A. (2009) A tale of two isotopes: differences in hydrograph separation for a runoff event when using dD versus $d^{18}O$. *Hydrological Processes* 23, 2095–2101. doi:10.1002/hyp.7326.

Pappenberger, F. and Beven, K. J. (2006) Ignorance is bliss: 7 reasons not to use uncertainty analysis. *Water Resources Research* 42, W05302. doi:10.1029/2005WR004820.

Rico-Ramirez, M. A., Gonzalez-Ramirez, E., Cluckie, I. and Han, D. (2009) Real-time monitoring of weather radar antenna pointing using digital terrain elevation and a Bayes clutter classifier. *Meteorological Applications* 16, 227–236.

Robinson, D. A., Binley, A., Crook, N., Day-Lewis, F., Ferré, P. T., Grauch, V. J. S., Knight, R., Knoll, M., Lakshmi, V., Miller, R., Nyquist, J., Pellerin, L., Singha, K. and Slater, L. (2008) Advancing process-based watershed hydrological research using near-surface geophysics: a vision for, and review of, electrical and magnetic geophysical methods. *Hydrological Processes* 22, 3604–3635.

Robson, A. J. (2002) Evidence for trends in UK flooding, *Phil. Trans. R. Soc. Lond. A 2002* 360, 1327–1343, doi: 10.1098/rsta.2002.1003

Romanowicz, R. J., Young, P. C., Beven, K. J. and Pappenberger, F. (2008) A data based mechanistic approach to nonlinear flood routing and adaptive flood level forecasting. *Advances in Water Resources* 31, 1048–1056.

Ryedale Flood Research Group (2008) *Making Space for People*. Available at http://knowledge-controversies.ouce.ox.ac.uk/news/Making_Space_for_People.pdf

Sivapalan, M., Wagener, T., Uhlenbrook, S., Zehe, E., Lakshmi, V., Liang, X., Tachikawa, Y. and Kumar, P. (eds) (2006) *Predictions in Ungauged Basins: Promises and Progress*. IAHS Publication No. 303. IAHS Press, Wallingford.

van der Sluijs, J., Janssen, P. H. M., Petersen, A. C., Kloprogge, P., Risbey, J. S., Tuinstra, W., van Asselt, M. B. A. and Ravetz, J. R. (2005) *RIVM/MNP Guidance for Uncertainty Assessment and Communication: Tool Catalogue for Uncertainty Assessment*. Utrecht University and RIVM.

Vereecken, H., Binley, A., Cassiani, G., Revil, A. and Titov, K. (2006) *Applied Hydrogeophysics*. Springer, New York, 383pp.

Zakharova, E. A., Kouraev, A. V., Cazenave, A. and Seyler, F. (2006) Amazon River discharge estimated from TOPEX/Poseidon altimetry. *Comptes Rendus Geoscience* 338, 188–196.

Zinevich, A., Alpert, P. and Messer, H. (2008) Estimation of rainfall fields using commercial microwave communication networks of variable density. *Advances in Water Resources* 31, 1470–1480.

Appendix
Useful tables

Table A1 Mean daily short-wave radiation by latitude and month (W m^{-2})

Latitude	Jan	Feb	Mar	Apr	May	Jun	Jul	Aug	Sep	Oct	Nov	Dec
60°N	40.0	102.9	200.1	317.3	417.4	468.8	446.0	360.2	243.0	134.4	57.2	25.7
50	105.8	171.5	263.0	363.1	443.1	474.6	460.3	391.7	297.3	203.0	125.8	88.6
40	177.2	240.1	317.3	394.5	454.5	477.4	466.0	420.2	345.9	265.9	194.4	160.1
30	231.6	300.2	365.9	420.2	460.3	471.7	463.1	434.5	385.9	320.2	260.2	225.8
20	308.8	354.5	400.2	434.5	448.8	451.7	451.7	440.3	411.7	368.8	323.0	297.3
10	365.9	397.4	423.1	434.5	428.8	423.1	426.0	428.8	423.1	405.9	374.5	357.3
0	417.4	428.8	434.5	420.2	397.4	383.1	388.8	408.8	426.0	428.8	417.4	408.8
10	454.5	448.8	431.7	397.4	357.3	334.5	343.1	374.5	411.7	440.3	448.8	451.7
20	480.3	457.4	414.5	357.3	305.9	277.3	288.7	331.6	388.8	437.4	468.8	483.1
30	491.7	451.7	385.9	311.6	245.9	214.4	225.8	277.3	351.6	423.1	477.4	500.3
40	494.6	431.7	348.8	254.4	183.0	148.7	160.1	217.3	305.9	394.5	471.7	508.9
50	483.1	403.1	297.3	191.5	117.2	82.9	97.2	154.4	248.7	357.3	457.4	503.1
60°S	471.7	360.2	237.3	122.9	51.5	25.7	37.2	88.6	185.8	308.8	431.7	500.3

Table A2 Standard length of day, N (h), by month for different latitudes in the northern hemisphere

Month	50	52	54	56	58	60
Jan	8.55	8.25	7.95	7.60	7.20	6.70
Feb	10.00	9.85	9.65	9.50	9.25	9.00
Mar	11.80	11.80	11.75	11.75	11.70	11.70
Apr	13.70	13.80	13.95	14.10	14.30	14.50
May	15.35	15.60	15.90	16.20	16.60	17.05
Jun	16.25	16.55	16.95	17.40	17.95	18.60
Jul	15.85	16.10	16.45	16.85	17.35	17.90
Aug	14.40	14.60	14.80	15.00	15.25	15.55
Sep	12.60	12.65	12.70	12.75	12.80	12.85
Oct	10.70	10.60	10.50	10.40	10.25	10.10
Nov	9.00	8.80	8.55	8.25	7.90	7.50
Dec	8.05	7.80	7.40	7.00	6.50	5.95

Table A3 Properties of water (at standard pressure of 100 kPa).
(After Lide, D.R. (ed.) (1991) *Handbook of Chemistry and Physics*. CRC Press. Boca Raton.)

T (°C)	T (K)	Density, ρ $(kg\,m^{-3})$	Dynamic viscosity, μ $(10^6\,Pa\,s)$	Kinematic viscosity $\upsilon = \mu/\rho$ $(10^{-6}\,m^2\,s^{-1})$	Surface tension, σ $(Pa\,m^2)$
0	273.2	999.84	1793.0	1.793287	75.64
3.98	277.2	1000.00			
10	283.2	999.70	1307.0	1.307392	74.23
20	293.2	998.21	1002	1.003797	72.75
30	303.2	995.65	797.7	0.801185	71.20
40	313.2	992.22	653.2	0.658322	69.60
50	323.2	988.03	547.0	0.553627	67.94
60	333.2	983.20	466.5	0.474471	66.24
70	343.2	977.78	404.0	0.413181	64.47
80	353.2	971.82	354.4	0.364677	62.67
90	363.2	965.35	314.5	0.325789	60.82
100	373.2	958.40	281.8	0.294032	58.91

Table A4 Properties of air: variation of saturated vapour pressure (e_s), slope of the temperature–saturated vapour pressure curve (Δ), the ratio (γ/Δ), and the latent heat of water with temperature (λ)

T_a (°C)	T_a (K)	Density of dry air $(kg\,m^{-3})$	Density of air saturated with vapour at temperature T_a $(kg\,m^{-3})$	e_s (hPa)	Δ $(hPa\,K^{-1})$	γ/Δ	λ $(10^6\,J\,kg^{-1})$	Psychrometric constant γ at atmospheric pressure of 100 kPa $(kPa\,°C^{-1})$
−10	263.2			2.8627	0.2262	2.829	2.525	
−5	268.2	1.316	1.314				2.513	0.0643
0	273.2	1.292	1.289	6.1078	0.4438	1.456	2.501	0.0646
5	278.2	1.269	1.265	8.7192	0.6082	1.067	2.489	0.0649
10	283.2	1.246	1.240	12.272	0.8222	0.7934	2.477	0.0652
15	288.2	1.225	1.217	17.044	1.098	0.5967	2.466	0.0655
20	293.2	1.204	1.194	23.373	1.448	0.4549	2.453	0.0658
25	298.2	1.183	1.169	31.671	1.888	0.3505	2.442	0.0662
30	303.2	1.164	1.145	42.430	2.435	0.2731	2.430	0.0665
40	313.2	1.128	1.096	73.777	3.933	0.1707	2.406	0.0671

Table A5 Tables of conversion factors to standard SI units

Length (dimension L)

			SI unit m (metre)
1	mile	=	1609
1	yard (yd)	=	0.914
1	foot (ft)	=	0.3048
1	inch (in)	=	0.0245

Area (dimension L^2)

		SI unit m^2 (square metre)
1 hectare (ha)	=	10 000
1 square foot (ft^2)	=	0.09290
1 acre	=	404.7

Volume (dimension L^3)

		SI unit m^3 (cubic metre)
1 litre (L)	=	0.001
1 cubic foot (ft^3)	=	0.02832
1 acre-foot	=	123.35
1 imperial gallon	=	0.004546
1 US gallon	=	0.003785

Velocity (dimensions $L\,T^{-1}$)

		SI unit $m\,s^{-1}$ (metre per second)
1 $ft\,s^{-1}$	=	0.3048
1 $km\,h^{-1}$	=	0.2778
1 $km\,day^{-1}$	=	0.01157
1 $mile\,day^{-1}$	=	0.01862

Discharge (dimensions $L^3\,T^{-1}$)

	SI unit $m^3\,s^{-1}$ (cumec)
1 ft^3s^{-1} (cfs)	= 0.0283
1 $ML\,day^{-1}$ (megalitre per day)	= 0.001157
1 acre-foot per day	= 0.001428

Pressure (dimensions $M\,L^{-1}\,T^{-1}$)

		SI unit Pa ($Pascal = Nm^{-2} = m^{-1}\,kg\,s^{-2}$)
1 cm water	=	98.07 ($\approx$1 hPa)
1 bar	=	100 000 (100 kPa)
1 millibar (mbar)	=	100 (= 1 hPa)
1 mm mercury (mm Hg)	=	133.322 (= 1.33322 hPa)

Energy flux or power (dimensions $M L^2 T^{-3}$)

	SI unit $J s^{-1}$ (Joules per second $= N m s^{-1} = m^2 kg s^{-3}$)
1 watt per square metre ($W m^{-2}$) =	1

Temperature

	SI unit K (Kelvin)
T degrees Celsius (°C) =	$T + 273.15 K$
T degrees Farenheit (°F) =	$5(T - 32)/9 + 273.15 K$

Note: Dimensions are mass (M), length (L) and time (T)

Index